Marine Biology

Marine Biology

Edited by **Suzy Bullock**

SYRAWOOD
PUBLISHING HOUSE
New York

Published by Syrawood Publishing House,
750 Third Avenue, 9th Floor,
New York, NY 10017, USA
www.syrawoodpublishinghouse.com

Marine Biology
Edited by Suzy Bullock

International Standard Book Number: 978-1-68286-285-8 (Hardback)

Printed in China.

Contents

Preface

Marine biology is a significant field for understanding lives of organisms inhabiting deep seas and oceans. This book is compiled in such a manner, that it will provide in-depth knowledge about the theory and practice of marine biology. The objective of this book is to give a general view of the different areas of fisheries science, marine flora and fauna, marine conservation and technological advancements in marine biology research. It contains contributions of some of the eminent environmentalists, academicians and researchers. The extensive content of this book provides the readers with a thorough understanding of the subject.

After months of intensive research and writing, this book is the end result of all who devoted their time and efforts in the initiation and progress of this book. It will surely be a source of reference in enhancing the required knowledge of the new developments in the area. During the course of developing this book, certain measures such as accuracy, authenticity and research focused analytical studies were given preference in order to produce a comprehensive book in the area of study.

This book would not have been possible without the efforts of the authors and the publisher. I extend my sincere thanks to them. Secondly, I express my gratitude to my family and well-wishers. And most importantly, I thank my students for constantly expressing their willingness and curiosity in enhancing their knowledge in the field, which encourages me to take up further research projects for the advancement of the area.

Editor

Aquatic Environmental Contamination: The fate of Asejire Lake in South-Western Nigeria

Titilayo O. Aladesanmi, Oluwatosin G. Oladipo and Godwin A. Ali

National Centre for Technology Management (NACETEM), Federal Ministry of Science and Technology. Obafemi Awolowo University Campus, Ile-Ife, Nigeria.

In Nigeria major cities face serious water pollution crises, in which lack of environmental control of water-dependent activities play an important part. This generates unpleasant implications for health and economic development since most urban and rural areas depend upon lakes and rivers for their water supplies. A study of catfish from Asejire Lake (located at the outskirt of Ibadan, a major city in Oyo State of South-West Nigeria) was carried out to assess the level of contamination due to effluents from various industries in Ibadan, Oyo State particularly the Nigerian Bottling Company, Plc (NBC). The industrial site is located close to the lake and is known to be passing effluent to it. Heavy metals analysis was carried out by the use of an atomic absorption spectrophotometer (AAS) using an air-acetylene flame in the Alpha-4 ChemTech spectrophotometer. The result shows that Fe had the highest concentration (11.9 ppm), this was evidenced by an absorption peak at 406 nm and the high level of Fe after its AAS metal determination experiment. The other metals gave the following concentration: Ni (0.9 ppm), Mg (0.47 ppm), Mn (0.45 ppm), Zn (0.35 ppm), Ca (0.06 ppm), Cu and Co (0.00 ppm). These metals had a threshold in which the body can take before they will be lethal, as a result, accumulation and continuous consumption of these aquatics and drinking of the water can lead to coma or death. Fish stocks are at the upper end of the food chains and are vital food supplies to local populations and thus they present a major source of contaminants to local communities. Therefore this project highlights the need for environmental regulation and policy intervention in other to prevent the risk attached to accumulation of these contaminants via the water and the aquatic life. This paper focused on the instruments needed by the policy maker to be better informed and understand how to create a cleaner environment for people so as to increase the economy, health and well being of the people.

Key words: Asejire lake, pollution, contamination, catfish.

INTRODUCTION

Over the last years, in Nigeria a considerable population growth has taken place, accompanied by a steep increase in urbanization, industrial and agricultural land use. This has entailed a tremendous increase in discharge of a wide diversity of pollutants to receiving water bodies and has caused undesirable effects on the different components of the aquatic environment and on fisheries (Tolba, 1982). Organic pollution of inland waters in Nigeria, in contrast to the situation in developed countries of the world, is often the result of extreme poverty and economic and social under development.

Unfortunately, there are very few water quality studies for most Nigerian inland waters. In general, the available data come from scattered investigations which were carried out by individuals and by very few scientific projects concerned with Nigerian waters. Few reviews exist on their state of pollution (Dejoux et al., 1981; Dejoux, 1988; Burgis and Symoens, 1987; Davies and Gasse, 1988). Throughout the world, human use of water and bad planning have led to drier and polluted rivers, lakes, and groundwater resources with dramatic effects on the natural ecosystems. Nigeria's vast freshwater resources are among those most affected by environmental stress imposed by human population growth, urbanization, and industrialization. Disposal and management of wastes in Nigeria present serious environmental problems. The usual methods of waste disposal in the country are: land filling, dumpsites, land spreads, water disposal, and incineration. Each of these methods has serious environmental implications because of their potential to pollute and contaminate underground and surface water bodies in the country (Olanike, 2003).

Major cities in Nigeria face serious water pollution crises, in which lack of environmental control of water-dependent activities (including domestic, agricultural, and industrial) play an important part. Fish and marine resources in the country face total collapse or extinction, due to over-fishing and destruction of marine life and natural habitats by pollution of water bodies. Unregulated and excessive use of pesticides for fishing and the deliberate disposal and dumping of toxic and hazardous wastes into water bodies are significant causes of massive fish kills and loss of aquatic life and habitats in the country.

The protection of water quality and aquatic ecosystem as a vulnerable resource, essential to sustain life, development and environment is of utmost importance to prevent further pollution and degradation of Nigeria's freshwater resources. Integrating operational measures for safeguarding adequate levels of protection of endemic habitats remains a major challenge (Olanike, 2003).

Due to population and industrial growth, inland waters (rivers, lakes, etc.) become often the recipient of organic matter in amounts exceeding their natural purification capacity; while in the past natural purification and dilution were usually sufficient. Sewage and other effluents rich in decomposable organic material, cause primary organic pollution. Secondary organic pollution is defined as the surplus of organic matter, which is the sum of undecomposed organic material introduced into the water body with primary pollution and of the material resulting from an extremely increased bioproductivity within the polluted ecosystem itself (Stirn, 1973).

As stated by Dejoux et al. (1981), organic wastes mineralize in the receiving water bodies and the resulting nutritive elements stimulate plant production, leading to eutrophication. In this situation, the biomass increases considerably and goes beyond the assimilation limit by herbivores. This secondary organic pollution is considerably greater than the primary organic load. The excessive production of organic matter leads to the build up of "sludge" and the mineralization process consumes all dissolved oxygen from the water column, which causes fish kills. Consequently, organic pollutants are called oxygen-demanding wastes. The relatively high temperatures in tropical countries accelerate this process.

Rain water transports soil to streams, rivers and lakes by erosion processes, including dissolved and particulate organic matter.

Decomposition of this organic matter continues during transport and in the sediments, giving new soluble organic and inorganic matter. The quantities of organic matter transported, its characteristics and composition vary from one region to another. A man-made transport mode of organic material to natural receiving waters is sewage pipes. Man himself is unable to use all the energy stored in food and his wastes are often discharged into the water without treatment. It is well known that untreated sewage creates a public health danger, being a potential for epidemics of water-borne diseases, such as typhoid fever, and also causes a serious loss of the recreational value of the inland waters (Stirn, 1973; Shuval, 1986).

In addition to the ever-increasing urbanization, industry and development of agriculture and forestry contribute considerably to the organic loads, which pose a hazard for inland waters and fisheries. Accordingly, domestic sewage and organic industrial wastes, as well as wastes from agricultural and forestry products are considered as main sources of organic pollution of African waters. Alabaster (1981) pointed out that agriculture is being further developed in some African countries, leading to an extension of existing industries involved in the processing of plant and animal products and to an increase in the highly oxidizable discharges.

Lake Asejire is located in Southwestern Nigeria in Egbeda L.G.A. of Oyo State (Figure 1). It is a manmade lake constructed on River Osun in 1972. River Osun is one of the series of West African rivers which do not drain into Niger system but discharge into coastal lagoons and creeks bordering the Atlantic Ocean (Ayoade et al., 2006). Lake Asejire has elevated levels of contaminants due to effluents from various industries and breakdown of ecological balance caused by widespread destruction of flora and fauna diversities (Lameed and Obadara, 2006). One of such industries is the Nigerian Bottling Company, Plc (NBC) situated in the outskirt of Ibadan, Oyo state.

A study was carried out by Lameed and Obadara (2006) to assess the activities of Nigerian Bottling Company, Plc (NBC), Coca Cola Industry (both positive and negative) on the biodiversity of the area of Asejire. Parameters analyzed were Dissolved Oxygen Concentration, pH, temperature, turbidity and some heavy metals. Results showed that family Palmae (flora) and Archachatinidae (fauna) have the largest population abundance with 24.32 and 18.64%, respectively. The water sample collected showed a pH of 5.7, 6.3 and 6.5 at different points of the Lake. This is an indication of acidic water and might be toxic for human consumption except treated to comply with the World Health Organization recommended pH 7, for safe and drinkable water (Lameed and Obadara, 2006).

Figure 1. Asejire Lake where the fish samples were collected.

Figure 2. The African Cat fish used for this study.

Asejire Lake, which is the focus of this study, has elevated levels of contaminants due to effluents of various industries in Ibadan, Oyo State. Previous researchers have established the presence of pollutant in the water. In the local traditional settings around the river Osun, from which Asejire lake was built, it is a common thing for cassava tubers to be steeped in the flowing river for processing into other food products. This is a major source of cyanide pollution in this water. Despite this high level of pollution, the lake still support aquatic lives and serves the populace. Cyanide, a rapidly acting poison is one of the likely chemical pollutants present in the water. There is a need to assay various parameters to obtained information on the level of pollution of the water and the health implication.

MATERIALS AND METHODS

Asejire Lake; the study area is located in Egbeda Local Government area of Oyo state in Nigeria (Figure 1). The study was carried out by analysing fish samples from the lake to determine the level of pollutant and its possible effect on man. The part of the fish used for the analysis was the liver because it is solely responsible for detoxification of any poison or pollutant in the fish (Figure 2). The parameters that were assessed include heavy metals, temperature, and pH.

Heavy metals

The sample was examined for the presence of various divalent metal ions: Iron (Fe), magnesium (Mg), manganese (Mn), calcium (Ca), cobalt (Co), zinc (Zn), nickel (Ni) and copper (Cu). These metals were estimated by atomic absorption spectrophotometer

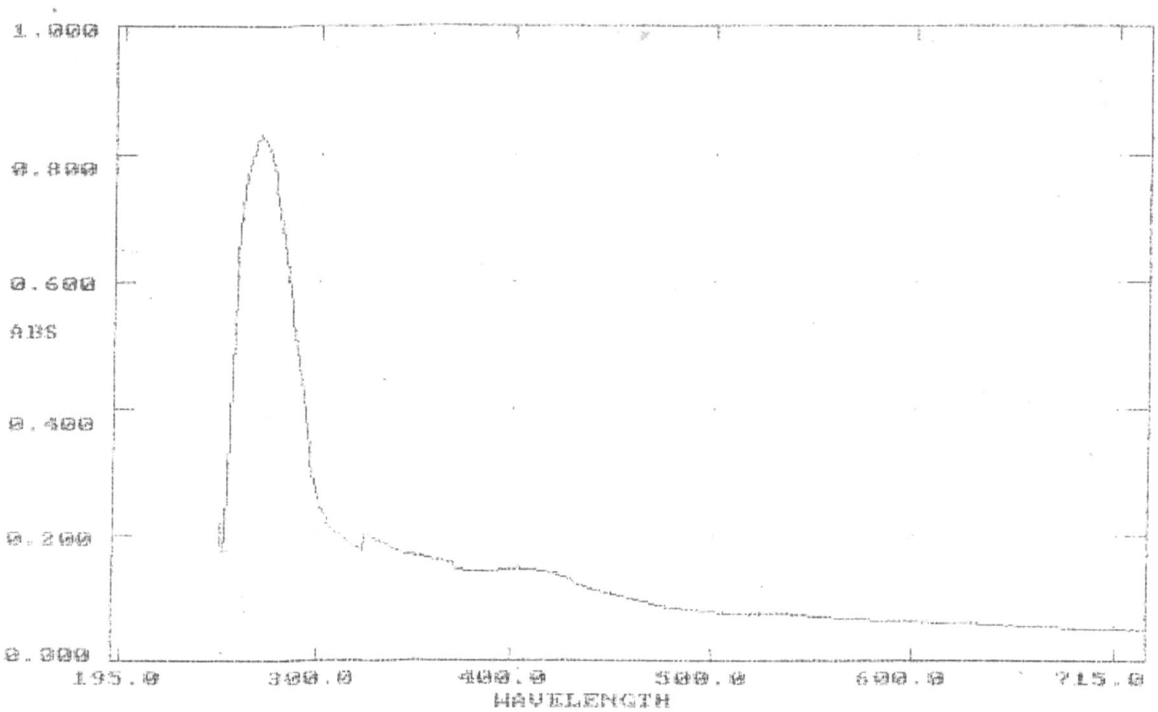

Figure 3. Spectrophotometric scanning.

using an air-acetylene flame in the Alpha-4 ChemTech spectrophotometer (Figure 3). The liver sample was digested in the presence of mixture of concentrated HNO_3 and $HClO_4$ (1:1). Prior to analysis, the sample was extensively dialysed against double change of distilled water to remove any ions contributed by the buffer according to the methods of Kaur et al. (2006).

Temperature

The optimum temperature and the heat stability of the sample was determined as follows:

Optimum temperature

The sample was assayed at temperatures between 0 and 70°C to investigate the effect of temperature on the activity of the liver and to determine the optimum temperature at which the detoxification of pollutants can occur. The sample was first incubated at the indicated temperature for 10 min before initiating reaction and assaying for the residual activity using the spectrophotometer.

Determination of heat stability

The heat stability of the sample was also determined by incubating 0.6 ml of the sample for 1 h at 30, 40, 50, 60 and 70°C respectively. 0.1ml was withdrawn at 10 min interval and assayed for residual activity. The activity at 30,40,50,60 and 70°C was expressed as a percentage of activity of the sample incubated at 30°C which was the control.

pH

The effect of the hydrogen ion concentration (pH) on the African catfish liver sample was performed according to the methods of Agboola and Okonji (2004). The sample was assayed using the following buffers of different pH values: 50 mM citrate buffer (pH 4.0-6.5), 10 mM phosphate buffer (pH 7.0-8.5), and 50 mM borate buffer (pH 9.0-11) with the aid of a pH meter.

RESULTS

Heavy Metal

The divalent metal content determined by atomic absorption spectroscopy showed that Fe had the highest concentration (11.9 ppm), followed by Ni (0.9 ppm), Mg (0.47 ppm), Mn (0.45 ppm), Zn (0.35 ppm), Ca (0.06 ppm), Cu and Co (0.00 ppm). The enzyme was also scanned across a wide range of wavelength (200-700 nm) in order to determine the presence of any peak. A sharp peak was obtained at about 406 nm which indicates the presence of divalent ions.

Effect of temperature on catfish liver

The rhodanese enzyme was assayed at temperature between 0 and 70°C. The assay mixture was first incubated at the indicated temperature for 10 min before initiating the reaction and assayed with a spectrophotometer:

A. The effect of temperature on the activity of catfish liver (Figure 4): Activity-temperature profile showing optimum temperature.
B. The Arrhenius plot of logarithm of activity against the reciprocal of incubation temperature (1/Kelvin).

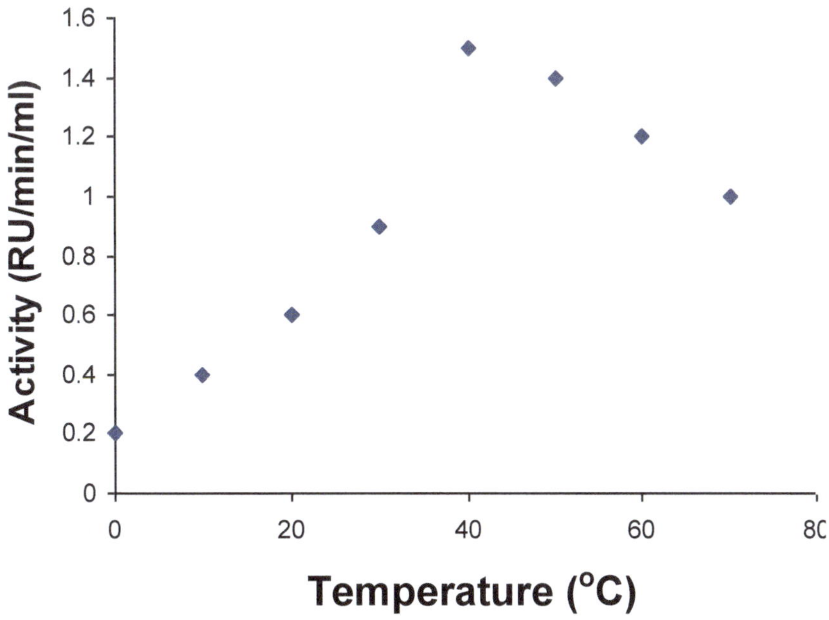

Figure 4. Activity-temperature profile showing optimum temperature.

The Arrhenius plot of catalytic activity of rhodanese consisted of two linear segments of different slopes with the larger slope at high temperatures. The temperature at which transformation (sudden break in the plot (Figure 5) occurred was 40°C. The two slopes of the plot yield apparent activation energies of 7.3 and 72.9 kcal/mol respectively.

Figure 6 shows the effect of temperature on heat stability of the liver of catfish. The activity at 30,40,50,60, and 70°C was expressed as a percentage of activity of the sample incubated at 30°C which was the control. The activity at 30°C was 100%. At 40 and 50°C, the activities were about 70 and 60% respectively and this gradually decreased with time. The activity was about 35% at 60°C and 20% at 70°C, this also decreased with time.

Effect of pH on catfish liver

The influence of pH on the rate of enzyme activity is shown in Figure 7. The enzyme activity when assayed at pH 5.0 showed no detectable increase in activity. Considerable increase in activity was obtained between pH 5.0 and 6.5, which later decreased gradually. The optimum pH was obtained at pH 6.5.

The pH optimum was determined using 50 mM citrate buffer (pH 4.0-6.5), 10 mM phosphate buffer (pH 7.0-8.5) and 50 mM borate buffer (pH 9.0-11) as shown in Figure 7. The curve represents the experimental data.

DISCUSSION

The African catfish inhabit calm waters from lakes, streams, rivers, swamps to floodplains, many of which are subject to contaminations from various sources such as effluents from manufacturing companies (Bruton, 1979; Clay, 1979). The local traditional settings around the river Osun, from which Asejire lake was built comprises mainly of fishermen and farmers. It is a common thing for cassava tubers to be steeped in the flowing river for processing into other food products. This is a major source of cyanide in this water. There are reports that fish were the most sensitive aquatic organisms to cyanide. Adverse effects on swimming and reproduction were observed between 5 and 7.2 µg free cyanide per litre; lethal effects usually occurred between 20 and 76 µg/L. Cyanide hazards to fish, wildlife, livestock and man are well documented. Massive kills of fresh water fish by accidental discharges of cyanide wastes are fairly common (Holden and Marsden, 1964; Leduc 1978; Towill et al., 1978; EPA 1980). Free cyanide is the primary toxic agent in the aquatic environment. Moreover, studies on the food composition of African catfish have shown that they are either omnivores or predators (Bruton, 1979; Clay, 1979). Micha (1973) found that African catfish feed mainly on aquatic insects, fish and higher plant debris. Some of these food items contain cyanogenic glycosides which upon hydrolysis release cyanide or hydrocyanic acid (HCN) (Wokes and Willimott, 1951; Montgomery, 1965).

This study established the presence of contaminants in Asejire Lake which in turn resulted in the contamination of the aquatics. An investigation into the presence of heavy metals in the preparations by spectrophotometric scanning and atomic absorption spectrophotometry showed a considerable concentration with Fe as the highest. This was evidenced by an absorption peak at 406 nm and the high level of Fe after its AAS metal determination experiment. These metals have a threshold in which the body can take before they will be lethal, as a

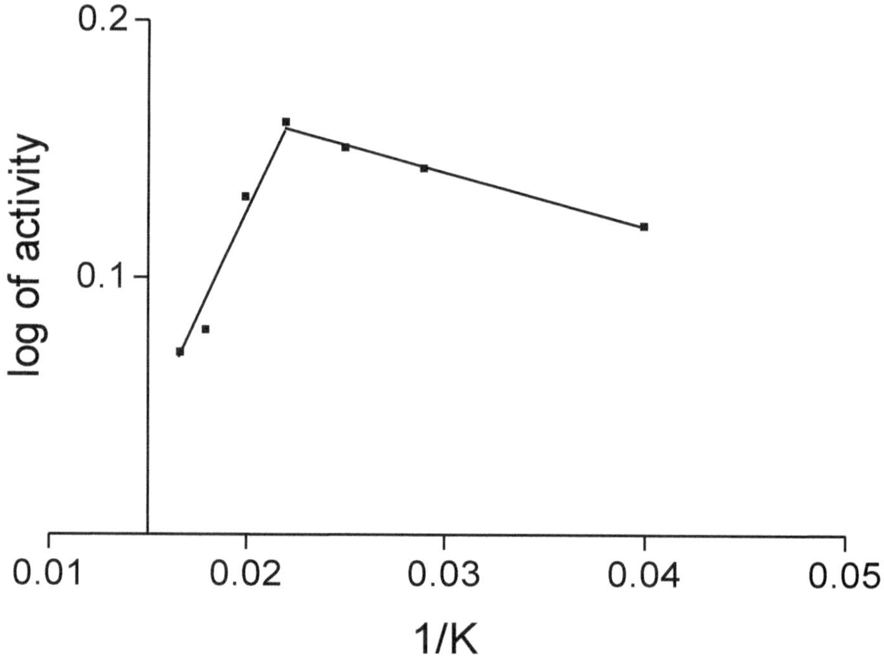

Figure 5. Arrhenius plot of logarithm of activity against the reciprocal of incubation temperature (1/Kelvin).

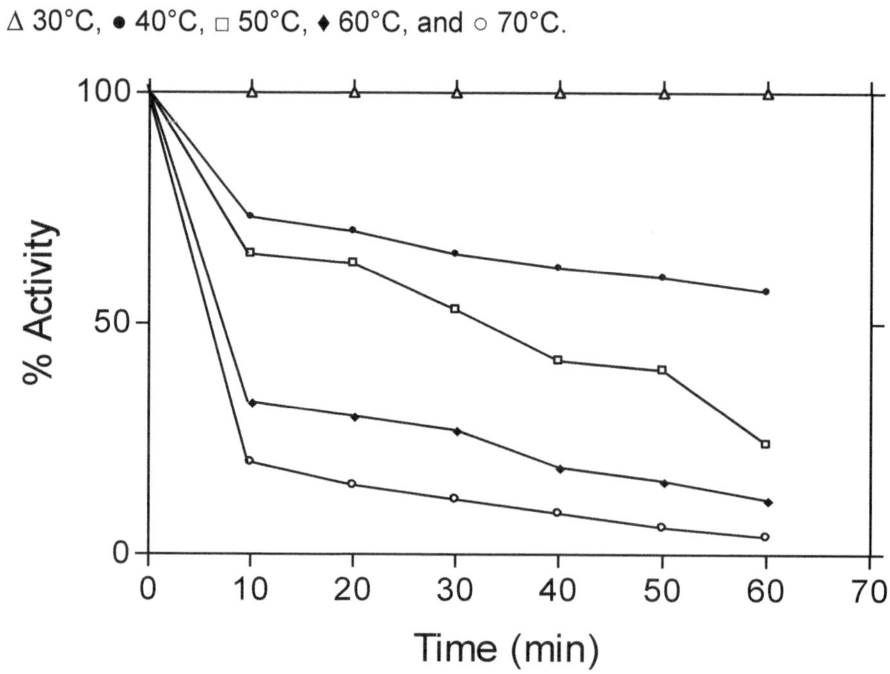

Figure 6. Effect of temperature on the stability of catfish liver.

result, accumulation and continuous consumption of these aquatics and drinking of the water can lead to coma or death.

The optimum temperature obtained was 40°C for rhodanese from the liver of the catfish from Asejire Lake. With the level of pollution in this water, there will be various metabolic activities going on in it which results in the release of heat. This ambient temperature might have conditioned the enzyme to function at higher temperatures. Arrhenius plot of effects of temperature on reaction rate

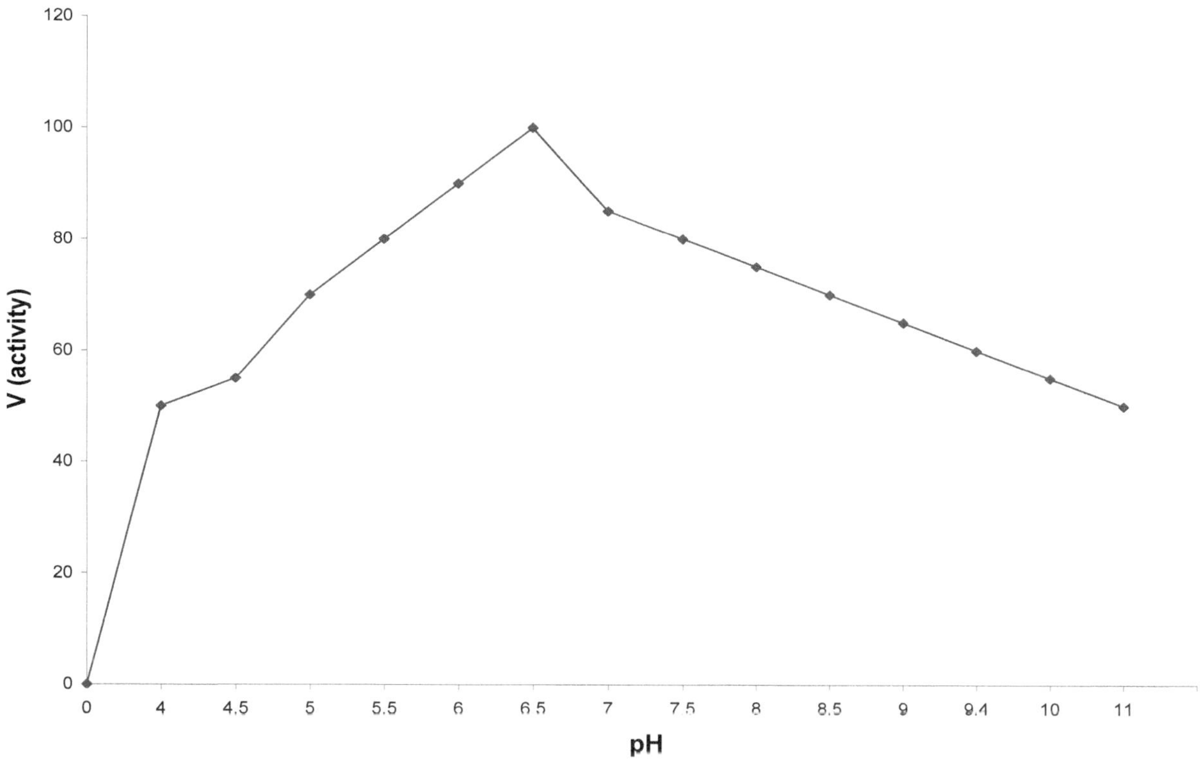

Figure 7. Effect of pH on catfish liver.

consists of two linear segments with a break occurring at 45°C. The apparent activation energy values from these slopes are 7.3 and 72.9 kcal/mol respectively.

Pollutants in water include wide spectrum of chemicals and pathogens which sometimes alters the acidity, conductivity and temperature of the water. The water sample collected from Asejire Lake showed a pH of 5.7, 6.3 and 6.5 at different points of the Lake (Lameed and Obadara, 2006). They suggested this is an indication of acidic water which might be toxic for human consumption except treated to comply with the World Health Organization recommended pH 7, for safe and drinkable water. An optimum pH of 6.5 was obtained for the liver of the catfish from Asejire Lake. This is as a result of the acidity of the water as stated above.

Conclusion and recommendation

Despite all these contamination, Asejire Lake supports various aquatics and serves the populace, the people in the environ feed on these contaminated aquatics and drink the water from the lake, the accumulation of which can lead to coma or death. The adverse effect of river pollution is the spread of water borne diseases. The uncontrolled disposal system renders surface waters and underground water systems unsafe for humans and poses a threat to human life and is therefore against the principles of sustainable development. The major goals of environmental regulation from the inception has been to reduce pollution, there have been no clearly established, coordinated policy framework and standards for attaining such goal especially through resource pricing, incentives and taxes. Rather, heavy reliance has been placed on qualitative legal rules.

Proper and adequate regulating policies will go a long way in ameliorating the hazard caused by water pollution. The advantages of clean environment would be available only if the generators of pollutants are encouraged to invest in pollution prevention and abatement technologies with the help of a judicious mix of regulatory policies, economic incentives and fiscal instruments. There is need to change the behaviour of industries/firms by selecting suitable levels of effluent charges and pollution taxes, and by supporting the industries with soft loans for investments in effluent treatment for common treatment facilities. The need to enhance proper waste management through measures that encourage minimisation, recycle and reuse of processed waste by both the industries and people involved can not be compromised. The industries /firms are to co-operate with federal, state and local governments and other relevant stakeholders in sharing emerging environmental issues, finding solutions to environmental problems and developing cost-effective, scientifically-based environmental standards. The absence of suitable national guidelines and standards on environmental pollution contributed to the high level of Asejire Lake pollution, so government should create an environmental monitoring network in other to regulate the activities of the lake pollutant. The Government should also set objectives, targets and performance standards in order to promote continual improvements in environmental management performance and the prevention of pollution.

REFERENCES

Alabaster JS (1981). Review of the state of aquatic pollution in Eastern African inland waters. CIFA Occas. Pap. 9; 36.

Ayoade AA, Fagade SO, Adebisi AA (2006). Dynamics of limnological features of two man-made lakes in relation to fish production. Afri. J. Biotech. 5(10):1013-1021.

Burgis MJ, Symoens JJ (1987). African wetlands and shallow water bodies/Zones humides etlacspeu profonds d'Afrique. Directory/Repertoire. Trav. Doc. Inst. Fr. Rech. Sci. Dév. Coop. 211, 650.

Davies B, Gasse F (1988). African wetlands and shallow water bodies/Zones humides et lacs peu profonds d'Afrique. Trav. Doc. Inst. Fr. Rech. Sci. Dév. Coop. 211, 502.

Dejoux C, Deelstra H, Wilkinson R (1981). Pollution. In: The ecology and utilization of African inland waters, UNEP Rep.Proc.Ser., 1: 149–61.

Dejoux C (1988). La pollution des eaux continentales africaines. Expérience acquise, situation actuelle et. perspective. Trav. Doc. Inst. Fr. Rech. Sci. Dév. Coop. 213, 513.

Kaur M, Singh K, Rup PJ, Kamboj SS, Saxena AK, Sharma M, Bhagat M, Sood SK, Singh J (2006). A tuber lectin from Arisaema jacquemontii Blume with anti-insect and antiproliferative properties. J Biochem. Mol. Biol. 39(4):432–440

Lameed GA, Obadara PG (2006). Eco-Development Impact of Coca-Cola Industry on Biodiversity Resources at Asejire Area, Ibadan; Nigeria. J. Fish. Int. 1(2-4), 55-62.

Olanike KA (2003). Consequences of Pollution and Degradation of Nigerian Aquatic Environment on Fisheries Resources. The Environmental. 23(4), 297-306.

Shuval HI (1986). Thalassogenic diseases. UNEP Reg.Seas Rep.Stud. 79, 44.

Stirn J (1973). Ecological consequences of marine pollution. In Proceedings of the Fifth Intérnational Colloquium on Medical Oceanography, p. 25–56.

Tolba MK (1982). Development without destruction. Evolving environmental perceptions. Dublin, Tycooly. Nat. Resour. Environ. Ser. 12, 197.

2

Comparison of East African and Iran natural feeding condition based on the chemical and biochemical properties of lake algae

Ezzati, Ramin

Department of Plant Biology, Faculty of Biological Sciences, Kharazmi University, Tehran, Islamic Republic of Iran.

In the present world where pollutants like chemical fertilizers affect environmental conditions, a kind of organic fertilizer with enough micro and macro elements is very important. African soil is reported by many researchers to be important and having large capacity due to its nutrition materials. The possible effect of variety soils transported through atmosphere besides sunlight and cloud to the chemical and biochemical compositions of *Cladophora Glomerata* alga is determined for the first time in the scientific world in this study. *C. Glomerata* is one of the important algae that grow in Iran's condition; it has a good content of carbohydrates and lipids. In this scientific work, we used soils from Zanzibar (Tansania region east African), Guilan (north of Iran), Tehran (center of Iran) and two type of soils from Zabol and Saravan regions (southeastern of Iran) in a condition simulated like atmospheric ones. The amounts of reduced Fe, Mn, Cu, Zn, pH, Ec, carbohydrate and lipid were examined. Talaghan Dam water and soil were used as our control. We studied the chemical and biochemical composition of *C. Glomerata* in 5 different feeding conditions, using comparison method. This alga best adapts to Zanzibar, Saravan and Zabol soil feeding conditions as fertilizer compared to other soils.

Key words: *Cladophora glomerata*, biochemical properties, Iran soils, African soils, feeding condition.

INTRODUCTION

Health and diet are two factors which have acquired particular significance along with scientific and technological advances in medicine. This opens the doors for further discussion on the benefits of nutritional and dietary supplements. The *Cladophora glomerata* is a great example of a dietary supplement (Guiry et al., 2008); studies on effect of macro elements on nutritional values have been reported in Chaetomorpha linum alga (Menendez et al., 2004). Algae are significant source of human food, especially in Asia and are cultivated for nutrition and pigments for supplemental use as human food and animal feed (Promya et al., 2008). Factors such as cell size and shape, rate of ingestion, digestibility and biochemical composition determine the nutritional quality of algae and their utility as food (Whyte, 1987; Doroudi et al., 2003).

Cladophora is a genus of reticulated filamentous Ulvophyceae (green algae). The genus, Cladophora contains many species that are very hard to differentiate and classify mainly because of the great variation in their appearance which is affected by habitat age and environmental conditions. Unlike Spirogyra, the filaments of Cladophora branch and do not undergo conjugation. There are two multicellular stages in its life cycle: a

Figure 1. *Chladophora Glomerata.*

haploid gametophyte and a diploid sporophyte which are highly similar. The only way to differentiate the two stages is to either count their chromosomes or examine their offspring. The haploid gametophyte produces haploid gametes by mitosis and the diploid sporophyte produces haploid spores by meiosis. The only visible difference between the gametes and spores of *Cladophora* is that the gametes have two flagella and the spores have four. The *Cladophora* species can be a major nuisance, causing major alteration to benthic conditions linked particularly to increased phosphorus loading (Taweesak, 2007).

Scientific classification

Algae grow on underwater rocks and thrive in clear spot water in river basin (Figure 1) (Traichaiyaporn et al., 2010).

A large number of algae are used for human food in their larval stages as crustaceans and some fish species and for zooplankton used in mariculture food chains as live feeds (Brown et al., 1997). Carbohydrates are one of the most important components in many foods. Carbohydrates may be presented as isolated molecules or they may be physically associated or chemically bound to other molecules. Individual molecules can be classified according to the number of monomers that they contain like monosaccharides, oligosaccharides or polysaccharides. Molecules in which the carbohydrates are covalently attached to proteins are known as glycoproteins, whereas those in which the carbohydrates are covalently attached to lipids are known as glycolipids. Some carbohydrates are digestible by humans and therefore provide an important source of energy, whereas others are indigestible and therefore do not provide energy. Indigestible carbohydrates form part of a group of substances known as dietary fiber, which also includes lignin. Consumption of significant quantities of dietary fiber has been shown to be beneficial to human nutrition, reduces the risk of certain types of cancer, coronary heart disease, diabetes and constipation. As well as being an important source of energy and dietary fiber, carbohydrates also contribute to the sweetness, appearance and textural characteristics of many foods. It is important to determine the type and concentration of carbohydrates in foods for a number of reasons (Bellis et al., 2007). The amount of chrbohydrate in algae structure can increase when atmospheric nutrient full of dust is up taken by algae (Ezzati and Iranmanesh, 2011).

Lipids are one of the major constituents of foods and are important in our diet for a number of reasons. They are a major source of energy and provide essential lipid nutrients. Nevertheless, over consumption of certain lipid components can be detrimental to our health like cholesterol and saturated fats. In many foods, the lipid component plays a major role in determining the overall physical characteristics, such as flavor, texture, mouthfeel and appearance. For this reason, it is difficult to develop low fat alternatives of many foods because once the fat is removed some of the most important physical characteristics are lost. Finally, many fats are prone to lipid oxidation which leads to the formation of off flavors and potentially harmful products.

MATERIALS AND METHODS

This report analyzes the breeding methodology and production rate of *C. glomerata* utilizing common feeding conditions and various soil growth media. The proper growth media applied in this research partly constitute soils collected from East Africa and different regions of Iran; the sources of soil were from Zanzibar (Tansania), Saravan (Iran), Tehran (Iran), Zabol (Iran) and Guilan (Iran), as well as Taleghan Dam (Iran) which is a common and suitable location to grow algae.

The *Chladophora* samples used in this research were provided by the team of Kharazmi University Biological Science Faculty in June 2012. Algae samples were grown in different feeding conditions within the dimensions of 45 by 55 cm.

The soil samples collected from the Tansania around Zanzibar city (6° 9′ 57″ S, 39° 11′ 57″ E), Zabol, Saravan, Tehran and Guilan regions are individually added to bi-distillate water, centrifuged under 2500 rpms for a period of 15 mins, passed through a 0.50 µm membrane filter; and then placed in feeding chambers.

Some biochemical and chemical parameters for example carbohydrate and lipid were determined by standard AOAC (2000) methods and reduced iron, magnasium, zinc, pH and EC were determined by in-house method based on Tee et al. (1991). Laboratory of Environmental Engineering in Kharazmi University Faculty of Biological Sciences was used.

For all analyses, the mean and standard deviation were calculated. Analysis of variance (ANOVA) and Duncan's Multiple Range Test (DMRT) were adopted for statistical analyses and comparative studies of proximate compositions and micro elements under different natural feeding conditions.

RESULTS

In this study, the value of the physical texture of soil and electrical conductivity were determined (Table 1). In the result, Guilan soils have a large part of sandy soil; Saravan soils texture were determined as clay type and

Table 1. Texture properties and electrical conductivity of soils sample

Soil sample location	Soil texture	Electrical conductivity (μS/cm)
Guilan (Iran)	Clay%27-Sand%42-Silt%25	452
Tehran (Iran)	Clay%46-Sand%31-Silt%23	894
Saravan (Iran)	Clay%48-Sand%28-Silt%24	11569
Zabol (Iran)	Clay%44- Sand%30-Silt%26	15780
Zanzibar (Tanzania)	%28Clay-Sand%39-Silt%23	12027

Table 2. The level of pH at different sample.

Parameter \ Soil sample	Guilan (Iran)	Tehran (Iran)	Saravan (Iran)	Tansania (Tanzania)	Zabol (Iran)
pH	6.7	7.8	8.4	8.6	8.5

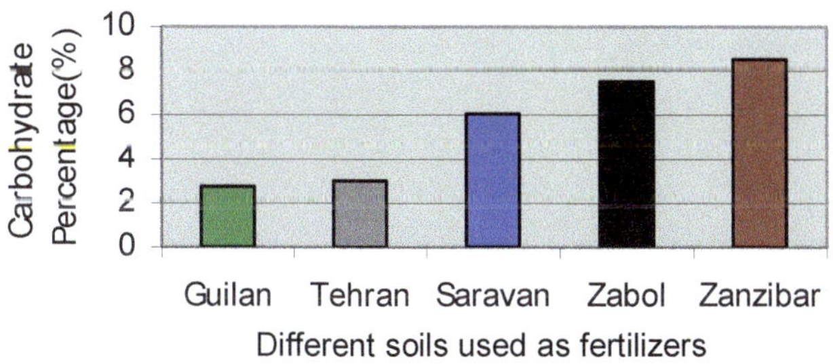

Figure 2. Displays algae carbohydrate rate during day 25.

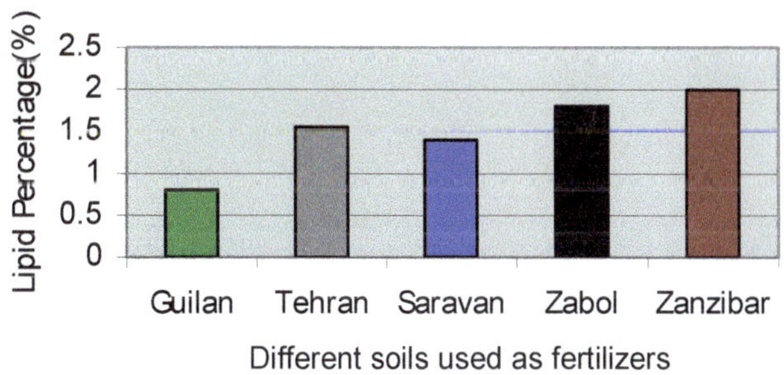

Figure 3. Displays algae lipid rate during day 25.

Zabol soils contain 26% silt.

The final step of this research uses 5 feeding conditions, all placed within laboratory condition. Table 2 displays Chladophora's carbohydrate rate during day 25 in various feeding conditions. Carbohydrate rate in Chladophora fed by Tansanian soil is about 8.77%, although Guilan soil carbohydrate level was measured to be about 2.88% (Figure 2).

The lipid in Chladophora grown in different conditions is also measured during day 25 (Figure 3). As shown in Table 2, Tansanian soil has a big potential to accumulate about 34.69% lipid in the Cladophora, but Guilan soils' capacity is about 27.14%.

In this kind of soil, the amount of reduced Iron absorbed by Chladophora showed different ranges between 27.14 to 34.17%. The Tansanian soil has 34.17% lipid (Figure 4).

Same results are seen in Figures 5 and 6. In these figures, the amount of Mn absorbed by algae that were fed by Tansania soil solution is more than the ones in other soils. Guilan soil has not shown a good performance to up take Mn by algae. The value of Zn was same as Mn. All of soil pH properties were weak alkaline but Guilan soil pH properties were in neutral condition.

DISCUSSION

This scientific performance like previous researches has shown that some kind of soil in the atmosphere with the appropriate sunlight and cloud can be used as a natural fertilizer for plant and algae. This study has shown again the ability of some soil to have plant nutrition. Zanzibar, Zabol and Saravan soils by having a large capacity for reducing Fe^{3+} to Fe^{2+} are better than Guilan and Tehran soils; and also absorption of Mn and Zn is same as Iron. Absorption of Fe, Mn and Zn by culturing of an edible freshwater *C. glomerata* depends on using different natural feeding condition of algae in laboratory. This situation is the same for the amount of carbohydrate and lipid. As we have shown Tansania, Zabol and Saravan soils with a large capacity of absorb different elements can act as organic fertilizer. The results show that by

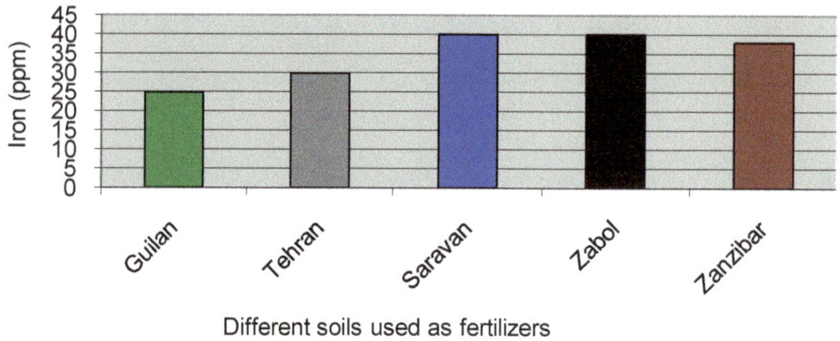

Figure 4. The level of algae Iron during day 25.

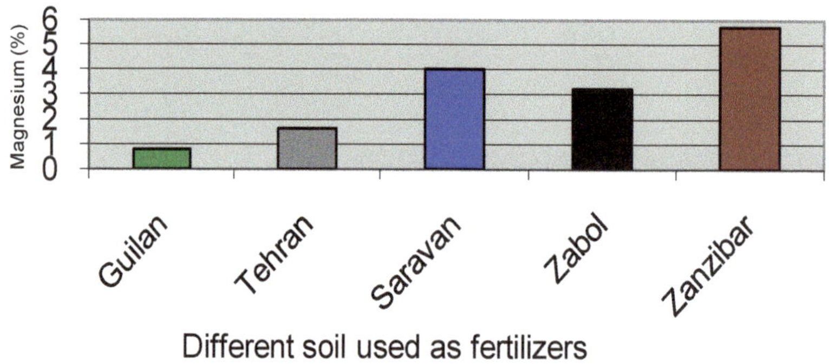

Figure 5. The level of Magnesium algae during day 25.

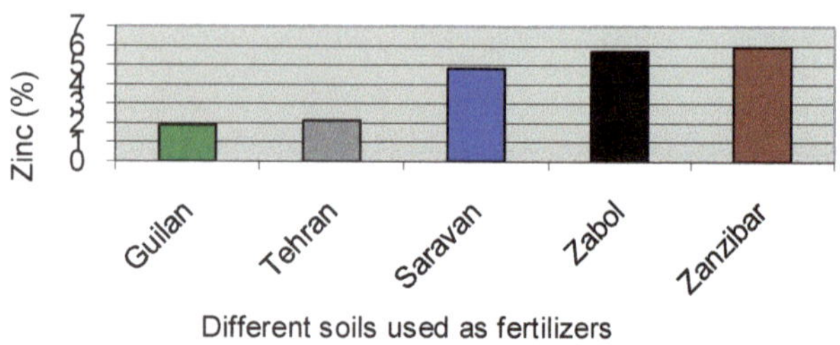

Figure 6. The level of algae zinc during day 25.

using Tansania, Zabol and Saravan natural feeding conditions instead of Tehran and Guilan soils, carbohydrate, lipid, Fe, Mn and Zn contents increased in *C. glomerata* structure.

By transferring this technology and through the exchange of Iranian and African scientist, we can gain a huge potential of natural macro and micro elements as nutrient materials from Zanzibar, Zabol and Saravan soils, and also by using this soils as complex we can gain a good potential of nutrient material for aquaculture plant like seaweed and algae which have suitable potential value of nutrient for human and we can protect our world from pollutant.

ACNOWLEDGEMENT

This research project would not have been possible without the support of many people. The author wishes to express his gratitude to Prof. Dr. Shahrbanoo Oryan, Dean of Biosciences Faculty, Kharazmi University, Tehran / IRAN who was abundantly helpful and offered invaluable assistance, support and guidance.

REFERENCES

Bellis VJ, McLarty DA (2007) Ecology of *Cladophora Glomerata* KÜTZ in Southern Ontario± DOI: 10.1111/j.1529-8817.

Brown MR, Garland CD, Jeffrey SW, Jameson ID and Leroi JM (1993). The gross and amino acid compositions of batch and semi-continuous cultures of Isochrysis sp. (clone T.ISO), Pavlova lutheri and Nannochloropsis oculata. J. Appl. Phycol. 5:285-296.

Doroudi M, Southgate and Lucas J (2003) Variation in clearance and ingestion rate by larvae of the black-lip pearl oyster (Pinctada margaritifera, L.) feeding on various microalgae. Aquac. Nutr. 9:11-16.

Ezzati R (1993). Effect of different ecological condition on chemical and physical parameter of some apple varity. Master Thesis. Ankara-Turkey.

Ezzati R, Iranmanesh M (2011). What kind of iran desert dust affect on microalgae growth and developmen. International Conference on Algal Biomass, Biofuels and Bioproduct. Westin St. Louise. USA. pp. 17-20.

Guiry FA (2008). Effect of nitrogen growth, chlorophyll content and tissue composition of the macroalga Chaetomorpha. Sci. Mar. 66:233-242.
Menendez M, Herrera J, Comin FA (2004). Effect of phosphorus supply on growth and development of the macroalgae (O.F. Miill.) Kütz in a Mediterranean coastal lagoon. Sci. Nov. 59:355-364.
Promya JS, Traichaiyaporn, Deming RL (2008). The optimum N:P ratio of kitchen wastewater and oil-extracted fermented soybean water for cultivation of Spirulina platensis: pigment content and biomass production. Int. J. Agric. Biol. 10:437-441.
Taweesak (2007). Culture of a Green Alga Genus Cladophora (Kai) as Feed for the Mae-Kong Giant Catfish Pangasianodon gigas, Chevey. The Thailand Research Fund, Bangkok, Thailand.
Tee ES, Lim CL (1991). Carotenoid composition and content of Malaysian vegetables and fruits by AOAC and HPLC methods. Food Chem. 41:303-339.
Traichaiyaporn SB, Waraegsiri, Promya J (2010). Culture of a Green Alga Genus Cladophora (Kai) as Feed for the Mae-Kong Giant Catfish Pangasianodon gigas, Chevey. The Thailand Research Fund, Bangkok, Thailand
Whyte JNC (1987) Biochemical composition and energy content in six species of phytoplankton used in mariculture of bivalves. Aquaculture 60:231-241.

3

The effect of pesticides on fish fauna of Bhopal lower lake (M. P.)

Mukesh Kumar Napit

Department of Zoology Swami Vivekanand Government College Berasia, Bhopal (M.P.) India.

Studies on the toxicity of common pesticides and some industrial pollutants on fishes have been reviewed. Pollutants such as pesticides cause diseases, behavioral abnormalities, cancer and gene mutations in fishes. Physiological malformation, histological, hematological and biochemical changes have also been observed in fishes. Pesticides significantly affect the early life stages of fishes. Toxic effects of pesticides vary in different organs of the fish. Liver, gill, kidney are tissues that can accumulate high level of pollutants as well as other factors, such as salinity temperature, hardness, etc.

Key words: Bhopal lower lake, fishes, pesticides, toxicity.

INTRODUCTION

Industrial waste has long been recognized as serious pollutant of the aquatic environment. Pesticides have toxic effect on exposed organism. The presence of pesticides beyond permissible limits in water has been reported worldwide. Our study focuses on the toxic effect of pesticides on fish.

In lake, steps have been taken for the monitoring of pesticides and other pollutants level in water sample. Monitoring of pollutants has been done in the lake. Water samples collected from various sources in different sites of lake were analyzed and pesticides were found within the permissible level. Pesticides concentration in submerged plants and fish at various sites of Lower lake were observed only downstream sites and fish collected at Lake Weir, Lake dam, Lake hills, regions, which was possible due to discharge of pesticides containing influent from various industries along sites of the lake.

Most of the sites of lake are seriously polluted by industrial effluents. Effluents are waste products in liquid form resulting from industrial processing. They are released by different industries such as petrochemical complex, fertilizer factories, oil refineries, pulp paper, textile mills, sugar mills, steel mills and tanneries, etc. All the chemicals of industrial waste are toxic to animal and many cases of death or sub-lethal pathology of liver, kidney, gonads, nervous tissues of fishes have been reported.

By discharge from the effluent inflows, amount of pesticides in water show an increase. They are present in water in dissolved condition form only at low levels, since pesticide compounds have low solubility. Mineral suspension and precipitation substances are able to store pesticides ions on their outer surface. Pesticides can also be found in water organisms. They can be taken up by higher organism through the food chain and sink to bottom as sediment.

Pesticides are pollutants which affect the aquatic fish. Presence of pesticides show alteration of behavior, bio-accumulation of pesticides in the body of fish histopathological and biochemical alterations in fish. Pesticides also affect early life stages of fish (Table 1).

MATERIALS AND METHODS

Water samples were collected seasonally during June 2011 to July

Table 1. Toxicity of pesticides on fishes.

S/N	Fish fauna	Effect of pesticides	Affected organism
1	*Catla catla*	Endosulfan carbonyl	Significant histological alteration in gill.
2	*Labeo rohita*	Chlorophyrifos	Biochemical changes of total protein and glycogen observed.
3	*Puntius punctatus*	Endosulfan and Diazinon	Affect of arginine and tryptophane showing the interaction of pesticides with cellular proteins depletion.
4	*Puntius punctatus*	Endosulfan and Diazinon	Alteration of calcium content in the stomach after pesticides treatment.
5	*Puntius punctatus*	Rogon (dimithoate)	Abnormal behavior pattern in fish
6	*Cyprinus carpio*	Carbofuran 16ppm	Decrease in total erythrocyte count, total leucocytes count and hemoglobin count.
7	*Cyprinus carpio* and *Puntius ticto*	Aldrin, Dieldrin BHC and DDT	Bioaccumulation of chlorinated pesticides in fish tissue gill, liver muscle and kidney observed.
8	*Mystus vittatus*	Dimecron and Thiodon	Rate of food intake, absorption and metabolism decrease from the control value
9	*Heteropneustes fossilis*	Endosulfan 0.00075 ppm 0.00050 ppm 0.000375 ppm	Increased concentration, of toxicant showed the decrease in liver glycogen. Hepatic cells are damaged due to depletion of glycogen.
10	*Heteropneustes fossilis*	Dimecron	Significant decrease in Hb%, RBC numbe and O2 carrying capacity of blood.
11	*Clarias batrachus*	Carbaryl and Phorate	Cholesterol level in serum decreased during exposure period
12	*Clarias batrachus*	Phorate 0.27ppm	Physiological and histological disorder in testis and ovary of insecticide exposed fish.
13	*Gambusia affinis*	Dimecron 0.0068ppm	Histopathological changes such as hepatic lesion with necrosis pyconic nuclei vasculation damaged blood vessel in alimentary canal, liver, kidney and gill.
14	*Oreochromus mossambica*	Endosulfan	Endosulfan induced dysfuncation of osmoregulation processes, resulting in alteration of ionic composition of blood.
15	*Nandus nandus*	Endosulfan and Diazinon	Significant histological alteration in gill and fin region.

2012 from polluted and non polluted selected sites using rottener water sampler and were estimated by standard methods as given by APHA (1985). Fishes were collected, arranged, preserved, identify and classify based on the work of Jhingran (1982), with slight modification by Day (1080) and Jayaram (1999).

RESULTS AND DISCUSSION

All pest destruction agents are collectively known as pesticides. The most frequent used pesticides are insecticides, herbicides and fungicides. Pesticides are of two type, naturally occurring and synthetic pesticide. Synthetic pesticides include chlorinate organophosphate and carbonate. Organophosphate is most toxic to vertebrate. Organophosphates inhibit cholinesterase an enzyme essential for transmission of nerve impulse across the synapse. Dichlorodiphenyltrichloroethane (DDT) and other organic chlorine chemical may have an influence on endocrinal system. DDT is highly stable chlorinated hydrocation or extremely low degradability. DDT is one of the most known pesticides previously used worldwide. The fat solubility and extreme persistence of DDT have led it to be stored in the body of almost all organisms. The use of DDT has been declared illegal by almost all countries. Pesticides also caused toxic effect on fish. Behavioral abnormalities include cancer, hematological and biochemical change which have been observed in pesticides exposed fish.

Abnormal behavior of *Puntius punctatus* due to exposure of Rogor insecticide have been previously reported. Bioaccumulation of Aldrin, Dieldrin, benzene hexachloride (BHC), and DDT in gill, liver, muscle and kidney of *Cyprinus carpio* and *Puntius ticto* was observed with alter the enzyme activity. Effect of pesticide was observed in the acetyl cholinesterase activity in the brain of *C. carpio* (finger lings).

Histology of gill was altered due to the effect of pesticide (endosulfan carbarly). Pesticides exposed fish also shows hematological changes. Decrease of erythrocyte and leucocytes count was observed in pesticide (carbofuran) exposed *C. carpio.* Dimecron exposed *Heteropneustes fossilis* showed significant decrease in Hb%, red blood cell (RBC) number and O_2 carrying capacity of blood.

In *Catla catla, Nandus nandus* and *Labeo rohita* significant histological and biochemical alteration in gill and fin region were observed due to the effect of endosulfan, diazinon and chlorophyrifos. Carbaryle and phorate, decreased cholesterol level in serum of *Clarias batrachus.* Endosulfan exposed *Oreochromus mossambica* showed dysfunction of osmoreguration processes, resulting in alteration of ionic composition of blood (Table 1).

Conclusion

Pesticides have been recognized as serious pollutants of aquatic environment. They affect fish directly by accumulation in their body. They cause serious impairment in metabolic, physiological and structural changes in different organs (Table 1). It may affect fish indirectly by transfer to the next tropic level of food chain. The accumulation of pesticides in the tissues of fishes can result in chronic illness and cause potential damage of population. Fish are able to accumulate and retain pesticides and other pollutants from their environment. Accumulation of pesticides in the tissue of fish is dependent upon exposure concentration as well as other factors such as salinity, temperature, hardness and metabolism of fish.

Pesticides affect specific vital organs such as liver, gill and kidney. Liver contains the highest pesticides concentration because it is an organ of storage and detoxification of pesticides. Liver plays an important role in detoxification and also act as an active site of pathological effects, induced by contaminants. Different degrees of pesticides accumulate in various tissues of gill and kidney. Fish may accumulate pesticides by absorption through gills. It has been observed that the concentration of pesticides in gill reflect the concentration of pesticides in water in which fish species live.

REFERENCES

APHA (1985). Standard methods of examination of water and wastewater American Public Health association Washington D.C. New York.

Day F (1989). Fish fauna of British India. Vol. 1& 2.

Jayaram KC (1999). Fresh water fishes of Indian region, India,. Narendra publishing House, New Delhi. pp. 365-380.

Jhingran VG (1982). Fish and fisheries of India. Hindustan Publishing Corporation Delh. pp. 345-355.

4

Effect of Cypermethrin on some hematological parameters and prediction of their recovery in a freshwater Teleost, *Cyprinus carpio*

Sahar Masud[1] and I. J. Singh[2]

[1]Division of Livestock Production and Management, F.V. Sc & AH, SKUAST-Jammu, India.
[2]Department of Fishery Biology, C.F.Sc., G.B.P.U.A. & T. Pantnagar, Uttarakhand, India.

The present investigation was conducted to evaluate the lethality of cypermethrin "pyrethroid" (25% EC with 100% purity) on common carp, *Cyprinus carpio* (fingerlings). The effect was assessed on the basis of impact of short term exposures of its below safe concentrations (1/2 (0.05 µl/L) and 1/10 (0.01 µl/L) parts of safe concentration) on some hematological parameters. The sub-lethal exposure studies were done on the same day and after 7 days of exposure. Hematological recovery was also studied by maintaining the pesticide-exposed fish in a fresh water system for an additional 7 days. The hematological analysis showed significant reduction in red blood cells (RBCs) count and hemoglobin (Hb) value while total white blood cells (WBCs) count were significantly increased in the fish *C. carpio*. Complete recovery was obtained in all the parameters after a recovery period of 7 days.

Key words: Cypermethrin, *Cyprinus carpio,* hematology, recovery.

INTRODUCTION

Cypermethrin, a synthetic pyretheroid has become one of the most important insecticides in wide-scale use. In 1988, pyrethroids amounted to 40% of the sales for insecticides for cotton treatment in the world (cypermethrin 8%). Pyrethroids are synthetic analogs of pyrethrins belonging to non-systemic chemical group of insecticides. This group can be classified into two categories- Type I and II, depending on their structure, properties and mechanism of toxicity (Burr and Ray, 2004). Pyrethroids generally affect central and peripheral nervous system. The primary site of action of these pesticides is the sodium channel in the nerve membrane. They cause sudden and prolonged progressive increase in Na^+ permeability of the nerve membrane resulting in long lasting chain of impulses in the sense organs and frequency dependent depression of nerve impulses in nerve fibers (Roberts and Hudson, 1998). The pyrethroid insecticides are extremely toxic to fish with 96 h LC_{50} values generally below 10 µg/l. Corresponding LD_{50} values in mammals and birds are higher (hundred to several thousand mg/kg). Sensitivity of fish to the pyrethroids is dependent on their relatively slow metabolism and delayed elimination (Bradbury and Coats, 1989).

Cypermethrin [CAS: 52315-07-8, Chemical Name: (R,S)-alpha-Cyano-3- phenoxybenzyl - 2,2-dimethyl (1R,1S) - cis, trans - 3 - (2,2 - dichlorovinyl)cyclopropane carboxylate] is a class II - moderately toxic, highly active and broad spectrum, non accumulative pyrethroid insecticide, which is effective in public health and animal husbandry, and targets a wide range of pests in agriculture. The products containing cypermethrin are classified as Restricted Use Pesticides (RUP) by the US Environmental Protection Agency (US EPA, 1989) because of their very high toxicity to fish. The LC50values for small fish and other aquatic organisms typically lie below 1 µg/l. This is mainly because it is metabolized and eliminated significantly more slowly by fish than mammals or birds. Their mean life in water is two weeks but they get rapidly absorbed by aquatic organisms with a significant toxicity (Rand and Petrocelli, 1985; Phillip and Rajasree, 1996).

The assessment of the ecotoxicological risks caused by pesticides to ecosystems is based on toxicity data and the effects of pesticide preparations on non-target organisms. Fish are among the group of non-target aquatic organisms. Blood parameters are considered pathological indicators of the whole body and therefore are important in diagnosing the structural and functional status of fish exposed to toxicants (Adhikari et al., 2004). It was reported that the blood parameters of diagnostic importance are erythrocyte and leucocytes counts, haemoglobin, haematocrit and leucocyte differential counts would readily respond to incidental factor such as physical stress and environment stress due to water contaminants (Bhatnagar and Bana, 1992; Ralio and Nikinmaa, 1985). Some authors (Reddy and Bashamohideen, 1989; Chauhan et al., 1994; Agarwal and Chaturvedi, 1995) have reported a decrease in hematocrit, hemoglobin and red blood cells values of some fish after their exposure to insecticides. The information suggests that hematological parameters could be used as potential biomarkers of pyrethroid insecticides. Hematology of *Cyprinus carpio* has not been much documented, so this paper would provide an important contribution to the knowledge of the specimen constitution variation. The aim of the present investigation was to assess and contribute to knowledge on the hematological changes in *C. carpio* following the short term exposure of 7 days to 1/2 (0.05 µl/L) and 1/10 (0.01 µl/L) part of safe concentrations of cypermethrin and recovery response after withdrawal of exposure.

MATERIALS AND METHODS

Collection and maintenance of test organism

Live specimens (fingerlings) of *C. carpio*, average length 10.78 (range 9.0 to 12.5 cm) and average weight 14.73 g (range from 11.5 to 19.5g), were obtained from the Instructional Fish Farm, College of Fisheries, Pantnagar and were acclimatized in 1500 L capacity cemented tank for one month before they were transferred to the test aquaria. During acclimatization, the fish were fed with conventional fish feed (rice bran and soya cake in1:1 ratio) at the rate of 10% body weight. Water quality characteristics in the experimental units were recorded according to APHA (1980).

Test pesticide specification used

Agriculture grade cypermethrin, trade name Challenger 25 (25% EC with 100% purity) was used for the experiment. Micropipette was used to measure the doses of cypermethrin for test exposure.

Experimental design for static bioassay

Forty five fingerlings of *C. carpio*, divided into three groups of 15 fish in each, were used for the experiment in circular fiber tanks of 500 L capacity.

Exposure of fish to cypermethrin

Two groups were exposed to 1/2 (0.05 µl/L) and 1/10 (0.01 µl/L) part of safe concentrations of cypermethrin for seven days, while third group served as control. During the experimentation, feeding and adding of fresh dose of cypermethrin were done fresh after changing water on alternate day. After exposure for seven days, the treated fish were kept in fresh water (without cypermethrin) for another seven days with daily renewal of water and feeding. Fish were observed twice daily for mortality and other behavioral changes.

Hematological study

Five fishes from each group were sampled initially at the beginning (on the same day) of the experiment while five fishes from each group were sacrificed after 7 days of exposure. Remaining five fishes in each group were kept in continuously flowing clean water for recovery response for another one week and were sacrificed at the end of the experiment. During each sampling blood samples were obtained by severance of caudal peduncle and collected in Appendorf tubes containing EDTA anticoagulant (Mgbenka et al., 2003) for determination of RBC, WBC and hemoglobin content. Physico-chemical characteristics like dissolved oxygen (DO), pH and temperature were recorded twice weekly (APHA, 1980).

Total count of RBC

Total red blood cells (tRBCs) were counted using Neubaur haemocytometer. Blood was diluted 1:200 with Hayem's fluid (Mishra et al., 1077). Erythrocytes were counted in the loaded haemocytometer chamber and total numbers were reported as 10^6 mm^{-3} (Wintrobe, 1967).

Total count of WBC

Total white blood cells (WBC) were counted using Neubaur haemocytometer (Mgbenka et al., 2003). Blood was diluted 1: 20 with WBC diluting fluid and placed in haemocytometer. 4 large (1sq mm) corner squares of the haemocytometer were counted under the microscope. The total number of WBC was calculated in mm^3 $x10^3$ (wintrobe, 1967).

Estimation of hemoglobin (Hb)

It was done by using Sahli's (1962) method. Graduated hemoglobin tube was filled with N/10 HCl upto mark 10 then blood was sucked in hemoglobin pipette upto mark 20 cu.mm. The blood was poured into hemoglobin tube already containing N/10 HCl. After leaving for 10 min, the mixture was diluted with N/10 HCl with stirring continuously with a glass rod. The addition of N/10 HCl was continued drop by drop into the hemoglobin tube till the color matches with that of the standard brown glass rod. The reading was taken on hemoglobin tube showing percentage of hemoglobin.

Statistical analysis

Hematological changes were tested by using one way ANOVA (analysis of variance). Post hoc test were carried out using Duncan's multiple comparison procedure. Significance was tested at 5% level. All the statistical analysis were performed via SPSS 14.0 for windows.

Table 1. RBC, WBC and hemoglobin in blood of *C. carpio* during exposure to 1/2 (0.05 µg/L) and 1/10 (0.01 µg/L) part of safe concentrations of cypermethrin and recovery pattern after one week.

Concentrations (µg/L)	Exposure response		Recovery response
	0 days	One week	One week
Total erythrocyte count (x 10^6 /mm^3)			
Control	2.95 ± 0.02	2.94 ± 0.019	2.95 ± 0 .011
0.01	2.63 ± 0.05	2.46 ± 0.019	2.86 ± 0.017
0.05	2.44 ± 0.029	2.32 ± 0.023	2.55 ± 0 .023
Total Leukocyte count (x 10^3 mm^3)			
Control	18.44±0.029*	18.48±0.021*	18.48±0.019*
0.01	18.73±0.024*	22.08±0.063*	18.69±0.059*
0.05	21.24±0.038*	23.29±0.027*	19.71±0.056*
Hemoglobin (g %)			
Control	8.38±.040*	8.46±.040*	8.43±.015*
0.01	7.73±.034*	7.35±.035*	8.15±.029*
0.05	7.42±.072*	6.85±.058*	8.11±.085*

N = 5 fish in each group, * Significance level tested at 5% level.

RESULTS AND DISCUSSION

Exposure of common carp to cypermethrin results in disturbances in hematology. Below safe concentrations of cypermethrin significantly decrease the RBC content. The mean values of RBC's were 2.63 (0.01 µl/L), 2.44 (0.05 µl/L), 10^6 mm^{-3} and 2.49 (0.01 µl/L), 2.32 (0.05 µl/L), 10^6 mm^{-3} within few hours and after 7 days of exposure, respectively. However, fishes showed significant recovery within 7 days after withdrawal of cypermethrin treatment with the mean values of RBC's being 2.86 (0.01 µl/L) and 2.55 (0.05 µl/L), 10^6 mm^{-3}, respectively (Table 1 and Figure 1). Treatment with cypermethrin was found to cause a drastic reduction in the total count of RBC's. The reduction was dosage depended; as concentration of cypermethrin increased the RBC level declined. The values mentioned above showed a significant decrease as compared to the control ($P<0.05$). Also, a significant reduction was recorded in hemoglobin of cypermethrin treated group. The control fishes showed mean value of 8.38 g % for hemoglobin. Fishes exposed to below safe concentrations of cypermethrin showed the hemoglobin mean value 7.73 (0.01 µl/L), 7.42 (0.05 µl/L), 7.35 (0.01 µl/L), 6.85 (0.05 µl/L), g % on the same day and after 7 days of exposure, respectively. However, significant increase in hemoglobin was recorded when treated fish were kept in cypermethrin free fresh water for 7 days and the mean values were 8.43 (0.01 µl/L) and 8.11 (0.05 µl/L), respectively (Table 1 and Figure 2). The values for treatment showed a significant decrease when compared with the control ($P<0.05$).

Changes in hematological parameters might have been brought about by cypermethrin as an anemic condition due to decreased synthesis of Hb and RBC number in hemopoietic organs. The reduction of RBC is mainly due to development of hypoxic condition during the treatment which intern leads to increase in destruction of RBC or decrease in rate of formation of RBC due to non availability of Hb content in cellular medium (Chen et al., 2004). The damaging of toxicant on erythrocyte may be secondary, as resulting from a primary action of toxicant on erythropoietic tissues on which there exist a failure in red cell production and or due to increase in the erythrocyte destruction (Verma et al., 1982; Agarwal and Srivastava, 1980). Saxena and Seth (2002), Paul (2004), Bunn et al. (1996) and Kumari and Banergi (1986) also reported that the RBC count and Hb concentration decrease may depend on age of animal, stress condition, sex and availability of food in a particular medium

Total WBC count was significantly increased in cypermethrin exposed group. The results of the total count of white blood cells revealed that the blood of the control fish showed a mean value of 18.44, mm^3 x 10^3. The fishes exposed to safe concentrations of cypermethrin showed the mean values of WBC to be 18.73 (0.01 µl/L), 21.24 (0.05 µl/L) and 22.08 (0.01 µl/L), 23.29 (0.05 µl/L), mm^3 x 10^3 on the same day and after 7 days of exposure of cypermethrin, respectively. However, treated fish recovered from ill-fated effect of cypermethrin within 7 days after withdrawal of exposure with the mean values of 18.48 (0.01 µl/L) and 19.71 (0.05 µl/L) respectively (Table 1 and Figure 3). The values mentioned above showed a significant change ascompared to the control ($p<0.05$).

Increase in WBCs count occurred as a pathological response since these WBCs play a great role during infestation by stimulating the haemopoietic tissues and the immune system by producing antibodies and chemical substances working as defense against infection

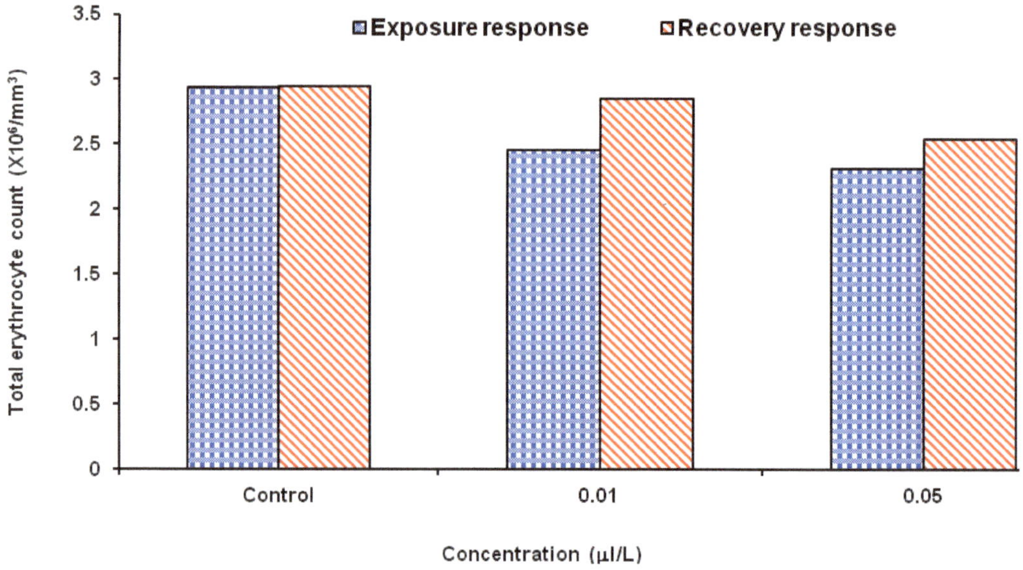

Figure 1. Total RBC during exposure to 1/2 (0.05 µg/L) and 1/10 (0.01 µg/L) part of safe concentrations of cypermethrin and recovery pattern after one week.

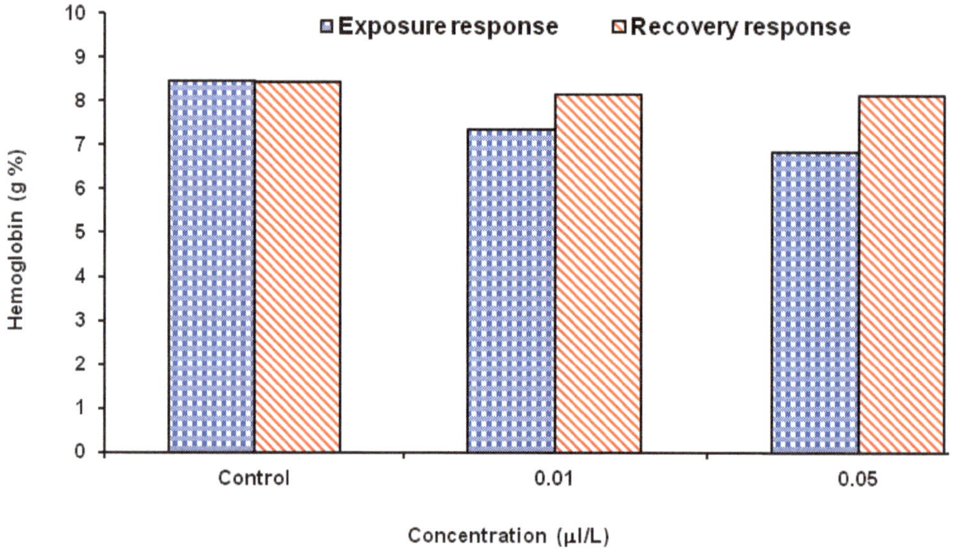

Figure 2. Hemoglobin during exposure to 1/2 (0.05 µg/L) and 1/10 (0.01 µg/L) part of safe concentrations of cypermethrin and recovery pattern after one week.

(Wedmeyer and Wood, 1974; Lebelo et al., 2001, Hassen, 2002). WBC is important cells in the immune system, because of their main defensive function. The WBCs respond immediately to the change in medium due to xenobiotic transformation. During toxic exposure period of cypermethrin, the WBC counts were enhanced. It indicates that fish can develop a defensive mechanism to overcome the toxic stress.

The present study suggested that the perturbations in these blood indices attributed to a defense reaction against toxicity of cypermethrin through the stimulation of erythropoesis or may be due to the disturbances that occurred in both metabolic and haemopoetic activities of fish exposed to below safe concentrations of cypermethrin. The toxicant caused hematological disturbance which could lead to impairment of the fish ability to combat diseases, reduce its chances for survival and potential for growth and reproduction.

The improvement in blood parameters of the test fish in response to transfer to cypermethrin-free freshwater for 7 days after acute exposure suggested that cypermethrin entering the system did not accumulate in the body and was slowly eliminated, resulting in recovery from the pesticidal toxicity. Similar observations were recorded in

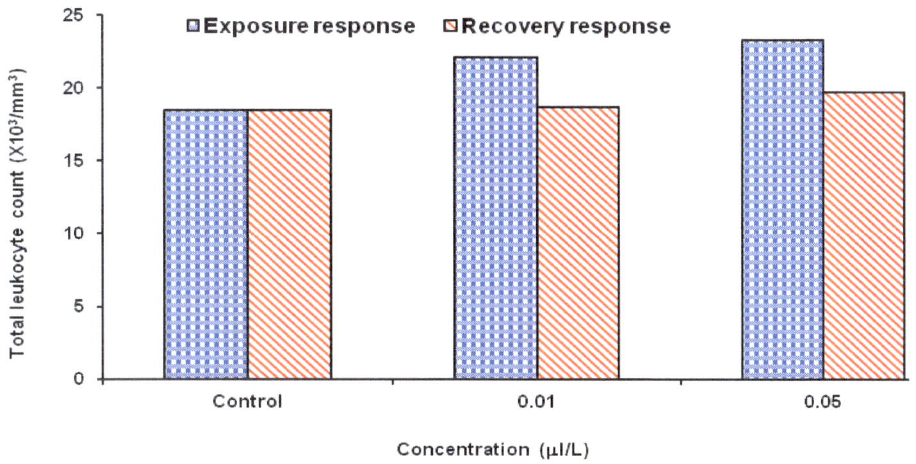

Figure 3. Total WBC during exposure to 1/2 (0.05 µg/L) and 1/10 (0.01 µg/L) part of safe concentrations of cypermethrin and recovery pattern after one week.

the studies carried out earlier by Adhikari et al. (2004) on the toxicity of cypermethrin and carbofuran on hematological parameters and recovery in *Labeo rohita*.

Conclusion

The above observations clearly demonstrate that the short term exposure of *C. carpio* to cypermethrin at even very low concentrations was sufficiently effective in disrupting physiological processes of fish and it may be recovered from the ill effect of pesticide by providing healthy environment.

REFERENCES

Adhikari S, Sarkar B, Chatterjee A, Mahapatra CT, Ayyappan S (2004). Effect of cypermethrin and carbofuran on certain hematological parameters and prediction of recovery in a freshwater teleost, Labeo rohita (Hamilton). Ecotoxicol. Environ. Saf. 58:220-226.

Agarwal K, Chaturvedi LD (1995). Anomalies in blood corpuscles of Heteropneustes fossilis induced by alachlor and rogor. Adv. Bios. 14: 73-80

Agarwal SJ, Srivastava AK (1980). Haematological responses in a freshwater fish to experimental manganese poisoning. Toxicology 17: 97-100

APHA (1980). Standard Methods for the Examination of Water and Wastewater, 15th edition, Washington, D.C. American Public Health Association. pp. 1134.

Bhatnagar MC, Bana AL (1992). Respiratory distress to Clarias batrachus (Linn.) exposed to endosulfan a histological approach. J. Environ. Biol. 13: 227-231.

Bradbury SP, Coats JR (1989). Comparative toxicology of the pyrethroid insecticides. Rev. Environ. Contam. Toxicol. 108:133-177.

Bunn KE, Thompson HM, Tarrant KA (1996). Effect of agrochemical on the immune system. J. Bull. Environ. Contam. Toxicol. 57(4): 632-639.

Burr SA, Ray DE (2004). Structure activity and interaction affects of 14 different pyretheroids on voltage- gated chloride ion channels. Toxicol. Sci. 77: 341-346.

Chauhan RRS, Saxena KK, Kumar S (1994). Rogor induced haematological alterations in Cyprinus carpio. Adv. Bios.13:57-62.

Chen X., Yin D, Hu S, Hou Y (2004). Immunotoxicity of penta chlorophenol on macrophage immunity and IgM secretion of the crucian carp (Carassius auratus). J. Bull. Environ. Contam. Toxicol. 73: 153-160.

Hassen FEZM (2002). Studies on diseases of fish caused by Henneguya infestation, Ph.D Thesis, Faculty of Veterinary Medicine, Suez Canal University, Egypt.

Kumari K, Banergi V (1986). Effect of sublethal toxicity to Zinc, Hg and Cd on peripheral haemogram in A. testudineus. Uttar Predesh. J. Zool. 6(2): 241-250.

Lebelo SL, Saunders DK, Crawford TG (2001). Observations on blood viscosity in striped bass, Morone saxtilis (Walbaum) Associated with fish Hatchery conditions. Kansa Acad. Sci. 104: 183-194.

Mgbenka BO, Oluah NS, Umeike I (2003). Effect of Gammalin 20 (Linndane) on differential white blood cells count of the African catfish Clarias albopunctatus. Bull. Environ. Contam. Toxicol. 71:248-254.

Mishra N, Pandey PK, Datta Munshi JS, Singh BR (1977). Haematological parameter of an air breathing mud eel, *Amphipnous cuchia* (Ham). J. Fish Biol. 10(6): 567-573.

Paul EA (2004). Toxicity of a synergized formulation of sumithrin to Daphnia magna. J. Bull. Environ. Contaim. Toxicol. 72(6): 1285-1289.

Phillip GH, Rajasree VH (1996). Action of cypermethrin on tissue transamination during nitrogen metabolism in Cyprinus carpio. Ecotoxicol. Environ. Saf. 34, 174-179.

Ralio E, Nikinmaa M (1985). Effect of sampling on blood parameters in the rainbow, Salmo gairdueri. J. Fish. Biol. 26(6): 725-732.

Rand GM, Petrocelli SR (1985). Fundamentals of aquatic toxicology. Hemisphere Publisphere Publishing Corporation. pp. 666.

Reddy MP, Bashamohideen M (1989). Fenvalerate and cypermethrin induced changes in the haematological parameters of Cyprinus carpio. Acta Hydrochim. Hydrobiol. 17:101-107.

Roberts T, Hudson D (1998). Metabolic Pathway of Agrochemicals, Part 2: Insecticides and Fungicides. The Royal Society of Chemistry: Cambridge.

Sahli (1962). Determination of haemoglobin by acid haematin method. In: Practical haematology (Ed: Dacie J.V. and Lewis S.M.) 5th Edn. Churuchill, London.

Saxena KK, Seth N (2002). Toxic effects of cypermethrin on certain hematological aspects of fresh water fish Channa punctatus. Bull. Environ. Contam. Toxicol. 69:364-369.

United States Environmental Protection Agency (USEPA) (1989). Pasticide Fact Sheet 199: Cypermethrin. Office of Pesticides and Toxic substances, Washington, DC.

Verma SR, Sarita R, Dalela RC (1982). Indicators of stress induced by pesticides in Mystus vittatus Haematological parameters. Indian J. Environ. Health 24: 58-64.

Wedmeyer GA, Wood J (1974). Stress a predisposing factor in fish disease, P: 399. U.S. Fish/ Wildlife Service.

Wintrobe MM (1967). (6th Eds.) clinical hematology. Lea and Febiger, Philadelphia, Library of Congress, Print USA.

5

Fish distribution in relation to environmental characteristics in the Aby-Tendo-Ehy lagoon system (Southeastern Côte d'Ivoire)

Barthélemy K. Koffi, Benié R. D. Aboua, Tidiani Koné and Mamadou Bamba

Laboratoire d'Hydrobiologie, UFR-Biosciences, Université Félix Houphouët-Boigny, 22 BP 582 Abidjan 22, Côte d'Ivoire.

Although West African lagoons are known to provide significant fishery resources to populations, few studies have been done on processes responsible for their fish composition, abundance and spatial distribution. The aim of this study was to assess fish species distribution along the Aby-Tendo-Ehy lagoon system in Côte d'Ivoire. Sampling sites were distributed in the lagoon and grouped into three zones from a freshwater zone with high contributions of freshwater (zone 1) to a mixo-eurihaline water zone (zone 3), with a transition site (zone 2) located between the first two zones. Environmental characteristics showed gradual decrease of mean pH from zone 1 to 3, whereas mean conductivity, total dissolved solids and transparency measurements increased from zone 1 to 3. Salinity data were 0 all along the sampling period in zones 1 and 2 and increased up to 3 ppt in zone 3. A total of 67 species belonging to 29 families were caught. Overall fish diversity decreased from zone 1 (n=58) to 3 (n=40), with 45 species caught in zone 2. The number of species in different fish trophic categories (invertivorous, herbivorous, piscivorous and omnivorous species) decreased from the zone 1 to 3. Fish compositions in zones 1 and 2 were dominated by freshwater species, while in the zone 3 it was dominated by estuarine dependent species. The canonical correspondence analysis performed showed a clear-cut influence of some environmental variables on fish ecological categories distribution: the zone under continental freshwater discharges had higher pH and temperature measurements and housed more freshwater, estuarine resident, marine migrant and estuarine dependent freshwater species, while the site with marine influences had higher conductivity, salinity, transparency and dissolved oxygen data and was mainly associated with estuarine dependent marine species.

Key words: Ichthyofauna, species distribution, environmental variables, Aby lagoon, Côte d'Ivoire.

INTRODUCTION

Estuaries are transition zones between freshwater and marine environments with an extreme spatial and temporal variability of physical, chemical and biological characteristics. In these environments, the diversity of fish community is an outcome of ecosystem processes reflecting changes in productivity, energy pathways and material flow, disturbance regimes, abiotic stress and biological interaction (Lowe-McConnell, 1987). Research on fish assemblages in estuaries has shown that salinity plays a major role in shaping assemblage structure (Marshall and Elliott, 1998; Barletta et al., 2005; Selleslagh and Amara, 2008), although only a few studies encompass the full salinity gradient, that is, from ocean to tidal freshwater. Its influence on fish is often due to the tolerance and preference of species for this variable (Elliott et al., 1990).

Other environmental variables such as temperature, depth, transparency, dissolved oxygen, pH, conductivity and total dissolved solids can also play important roles in determining fish assemblages (Pombo and Rebelo, 2002; Aboua et al., 2010). Fish are known to play a fundamental role in the functioning of shallow waters through a series of physico-chemical (e.g. nutrient release, sediment turbation) and biological mechanisms (for example, competition, predation) (Gelós et al., 2010). Fish may affect the spatial distribution of their prey through the behavioural cascades processes (Schmitz et al., 2004; Súarez et al., 2004; Gelós et al., 2010).

Many human uses of estuaries (transportation, wastewater, pollution, overfishing, etc.) are potentially in conflict with the aquatic resources (Vinagre et al., 2004; Baeta et al., 2005). They not only have an impact on fish communities, but in the same way also rebound in human communities associated with the exploitation of these resources (Costa and Elliott, 1991). Thus, fish community diversity is a basic ecological indicator of health, knowledge of which is necessary for the correct exploitation, regulation and management of fishing resources since it can provide a first approach to the health level of the estuarine system and allows for the identification of response patterns to possible environment impacts.

Although, the importance of fish in these ecosystems is well recognized, few studies have attempted to characterize the fish communities and to elucidate the processes responsible for their composition and abundance in West Africa. The objectives of this paper were to establish fish species composition and abundance in different zones of the Aby-Tendo-Ehy lagoon system and to assess the impact of environmental factors on fish assemblages.

MATERIALS AND METHODS

Sample collection

The Aby-Tendo-Ehy lagoon system (5°05'-5°22'N; 2°51'-3°21'W), located in southeast of Côte d'Ivoire encompasses an area of 424 km^2 and is 24 km long north-south direction and extends 54 km wide east-west direction (Figure 1). The Aby lagoon is the most extended on the studied lagoon system and covers 305 km^2. It has a total shoreline of 24.5 km long, and is 15.5 km wide. The Tendo lagoon, which is the median part of the studied system, is more stretched, with a length of 22 km and a width varying between 1.5 and 3.5 km. Its surface is 74 km^2. The Ehy lagoon is located in the eastern side of the lagoon system and has a surface area of 45 km^2 (Chantraine, 1980). The overall Aby-Tendo-Ehy lagoon complex is lined with mangroves, and communicates with the sea by the Assini channel. It is supplied with freshwater inputs by the rivers Bia in the northwest and Tanoe in the east.

A total of 6 sampling sites were selected, distributed along the lagoon in three areas (two sites per zone) according to different environmental characteristics related to horizontal salinity range (Figure 1). Seu-Anoï et al. (2011) describes the salinity gradients found in the lagoon. Sampling zone 1, in the most inland part of the system (Ehy lagoon) had no marine water influences, but was under a considerable contribution of freshwater from the Tanoe Ehy Swamp Forest and the Tanoe River. Zone 2 (Tendo lagoon) sampling sites were in transition positions within the study area, while zone 3 (in Aby lagoon) were greatly influenced by marine waters and corresponded to the closest zone to the ocean. Monthly fish samplings were carried out between March 2012 and February 2013 using two sets of 12 weighted monofilalement gill nets (bar mesh sizes 6, 8, 10, 12, 15, 20, 25, 30, 35, 40, 45 and 50 mm), each measuring 30 m long by 1.5 m deep. Nets were set overnight (17-7 h). Fish collected were identified to species level following Paugy et al. (2003a, b) and Decru et al. (2012), counted and measured to the nearest 1 mm body length. The depth (cm) was recorded at each sampling station. Water temperature (°C), dissolved oxygen (mg/l), pH, conductivity (µS/cm), and total dissolved solids (mg/l) were recorded between 7 and 9 h using a Sper Scientific multiparameter while salinity (ppt) and transparency (cm) were recorded with a Hydrobios refractometer and a Secchi disc, respectively. The study area is characterized by two rainy seasons (April-June and October-December) and two dry seasons (July-September and January-March) (Konan et al., 2014).

Data analysis

To categorize the fish composition within the coastal lagoon, fish species were grouped into ecological categories following Albaret (1994, 1999) and Chabanne (2007), and classified as estuarine resident fish (ER) (those that inhabit estuarine waters throughout their life cycle), estuarine dependent marine fish (EDM) (marine species which are predominantly found in lagoons at some stages of their life cycle), estuarine dependant freshwater fish (EDF) (freshwater species which are predominantly found in lagoons at some stages of their life cycle), estuarine nondependent marine fish (EN) (species commonly found in both estuarine and coastal inshore areas and do not depend to estuarine environment to complete their life cycles), accessory marine visitor fish (AV) (marine species which are regularly caught in lagoon but not abundant in the catch), occasional marine visitor fish (OV), and freshwater fish (FW)

Fish classification into herbivorous, invertivorous, omnivorous, piscivorous, and planktivorous categories were made following Trewavas (1983), King (1994), Diomandé et al. (2009), Yao et al. (2010), Kouamélan et al. (2012), and FishBase (2013). Species richness (which is the number of species sampled at each sampling zone) was used to determine the structure and ecological dynamics of the community. Jaccard's coefficient (J) was used to measure similarity between the sampling zones. Mann-Whitney U tests, based on ecological categories occurrence, were used for univariate comparisons.

Environmental influences on the ecological categories of each sampling zone were assessed with a canonical correspondence analysis (CCA) performed on log-transformed [log10(x+1)] data (ter Braak, 1986).

The CCA was done using CANOCO software for windows (version 4.5). This analysis was based on fish ecological categories occurrence data and physico-chemical data as well.

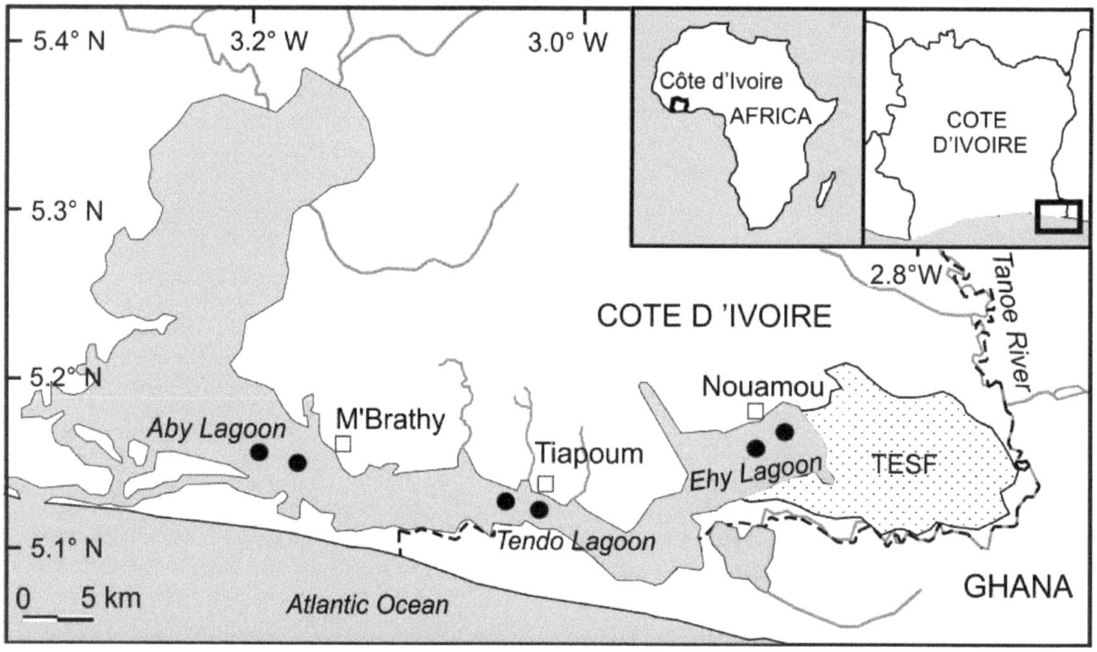

Figure 1. Sampling sites (•) in the Aby-Tendo-Ehy lagoon system (Côte d'Ivoire). TESF = Tanoe Ehy Swamp Forest.

Table 1. Environmental parameters [minimum (min), maximum (max), mean and standard deviation (SD)] in the different sampling zones of the Aby-Tendo-Ehy lagoon system (Côte d'Ivoire) between March 2012 and February 2013. T°C = temperature, DO = dissolved oxygen, Salt = salinity, Cond = conductivity, TDS = total dissolved solids, and Transp = transparency.

Sampling zone		Environmental characteristic							
		T°C	DO (mg/l)	pH	Salt (ppt)	Cond (µS/cm)	TDS (mg/l)	Transp (cm)	Depth (cm)
	Min	25.5	1.18	5.3	0	48.2	25.1	20	64
Zone 1	Max	35.8	7.35	9.15	0	123.1	65.5	120	298
(Ehy Lagoon)	Mean	30.13	5.17	7.27	0	65.47	34.98	43.86	147.87
	SD	3.41	1.51	0.84	0	16.05	8.88	16.37	64.15
	Min	25.3	2.03	5.31	0	60	30	24	68
Zone 2	Max	31.8	6.91	8.55	0	497	249	82	510
(Tendo lagoon)	Mean	28.22	4.74	7.02	0	156.76	84.91	42.69	289.75
	SD	1.87	1.07	0.8	0	119.82	64.74	14.41	153.37
	Min	25.6	3.18	4.29	0.5	70.7	36.8	37	46
Zone 3	Max	30.9	7.72	8.78	3	1546	804	101	264
Aby Lagoon	Mean	28.5	5.45	6.65	1.73	594.27	349.08	63.43	179.6
	SD	1.36	0.98	0.92	0.87	416.22	227.75	17.39	63.1

RESULTS

Environmental characteristics

Overall environmental parameters showed spatial trends in Aby-Tendo-Ehy lagoon system (Table 1). Mean pH decreased slightly from zone 1 (pH=7.27) to 3 (pH=6.65). Zones 1 and 2 had freshwater all along the sampling period (salinity=0 ppt) versus brackish water in zone 3 (with a salinity varying between 0 and 3 ppt). Furthermore, means of conductivity (65.47, 209.69 and 594.27, respectively for zones 1, 2 and 3) and total

Table 2. Ecological categories, trophic groups, occurrence percentage (%F), mean abundance (mean ab.±SD) of species collected in different sampling zones of the Aby-Tendo-Ehy lagoon system (Côte d'Ivoire) between March 2012 and February 2013.

Taxa	Zone 1: Ehy lagoon			Zone 2: Tendo lagoon			Zone 3: Aby lagoon		
	F (%)	Mean ab.	SD	F (%)	Mean ab.	SD	F (%)	Mean ab.	SD
Estuarine resident (ER)									
Monodactylus sebae (I)	18.18	0.25	0.71	40.00	2.08	5.85	72.73	5.75	0.55
Sarotherodon melanotheron (Pt)	100.00	15.42	5.29	100.00	9.58	8.13	100.00	5.92	5.18
Tilapia guineensis (I)	72.73	9.25	2.76	90.00	4.17	5.32	72.73	5.67	8.65
Tylochromis jentinki (I)	100.00	14.67	11.63	80.00	2.08	2.90	100.00	5.33	5.60
Awaous lateristriga (U)	18.18	0.25	0.71	0.00	0.00	0.00	63.64	0.92	0.98
Porogobius schlegelii (I)	45.45	1.58	0.96	80.00	4.83	1.81	63.64	1.58	2.36
Eleotris vittata (U)	36.36	0.33	0.49	50.00	0.58	0.89	81.82	3.25	3.61
Eleotris senegalensis (U)	0.00	0.00	0.00	0.00	0.00	0.00	9.09	-	-
Estuarine dependent freshwater fish (EDF)									
Pellonula leonensis (I)	81.82	3.17	4.84	90.00	20.75	1.33	100.00	30.17	11.71
Chrysichthys maurus (I)	72.73	1.42	2.10	60.00	4.25	7.20	36.36	1.08	1.71
Chrysichthys nigrodigitatus (I)	100.00	32.17	3.70	100.00	6.67	5.64	90.91	8.58	8.14
Chrysichthys teugelsi (I)	9.09	-	-	0.00	0.00	0.00	0.00	0.00	0.00
Hemichromis fasciatus (P)	100.00	6.50	6.12	100.00	3.08	1.88	100.00	5.33	1.67
Tilapia mariae (Pt)	72.73	1.42	1.81	80.00	2.00	2.20	9.09	-	-
Estuarine dependent marine fish (EDM)									
Ethmalosa fimbriata (Pt)	9.09	-	-	0.00	-	-	63.64	1.33	2.36
Hemiramphus balao (Pt)	0.00	0.00	0.00	0.00	0.00	0.00	9.09	-	-
Trachinotus teraia (I)	27.27	0.25	0.00	0.00	0.00	0.00	18.18	0.25	0.71
Plectorhincus macrolepis (U)	9.09	-	-	0.00	0.00	0.00	9.09	-	-
Pomadasys jubelini (I)	18.18	0.17	0.00	70.00	1.50	0.41	45.45	1.08	1.34
Liza falcipinnis (Pt)	90.91	12.25	1.62	60.00	1.67	2.25	72.73	3.25	3.36
Citharichthys stampflii (I)	18.18	0.33	1.41	10.00	-	-	100.00	11.50	9.26
Estuarine nondependent marine fish (EN)									
Elops lacerta (P)	63.64	0.75	0.76	50.00	2.00	5.22	54.55	1.75	2.43
Caranx hippos (I)	63.64	0.67	0.38	20.00	1.25	7.78	54.55	3.00	4.47
Eucinostomus melanopterus (I)	27.27	0.50	1.73	30.00	0.33	0.58	90.91	5.25	6.04
Mugil cephalus (Pt)	9.09	-	-	0.00	0.00	0.00	9.09	-	-
Occasional marine visitor fish (OV)									
Hemicaranx bicolor (U)	9.09	-	-	0.00	0.00	0.00	9.09	-	-
Accessory marine visitor fish (AV)									
Hyporamphus picarti (Pt)	0.00	0.00	0.00	0.00	0.00	0.00	9.09	-	
Lutjanus goreensis (I)	0.00	0.00	0.00	0.00	0.00	0.00	9.09		
Synaptura lusitanica (U)	0.00	0.00	0.00	10.00	-	-	0.00	0.00	0.00

* = introduced species, ** = hybrids, H = herbivorous, I = invertivorous, O = omnivorous, P = piscivorous, Pt = planktivorous, U = unknown.

dissolved solids (34.98, 150.92 and 349.08, respectively for zones 1, 2 and 3) increased importantly from zone 1 to 3. Mean transparency varied relatively slightly from zones 1 and 2 (43.86 and 42.69 cm, respectively) to zone 3 (63.43 cm).

Fish composition and ecological categories

Overall 5175 individuals belonging to 67 species and 29 families were collected during the study period. Regarding the ecological categories, 8 permanent residents, 6 estuarine dependents freshwater, 7 estuarine dependents marine, 4 estuarine non-dependents marine, 1 occasional, 3 accessory visitors and 38 freshwater species occurred in the studied lagoon (Table 2). Among these, two introduced species in Côte d'Ivoire (*Heterotis niloticus* and *Oreochromis niloticus*) and one hybrid (*Tilapia zillii* × *Tilapia guineensis*) were recorded. The

Table 2. Contd.

Species	Zone 1: Ehy lagoon			Zone 2: Tendo lagoon			Zone 3: Aby lagoon		
	F (%)	Mean ab.	SD	F (%)	Mean ab.	SD	F (%)	Mean ab.	SD
Freshwater fish (FW)									
*Heterotis niloticus** (I)	81.82	0.92	0.44	10.00	-	-	9.09	-	-
Papyrocranus afer (I)	90.91	2.00	2.88	20.00	0.25	0.71	0.00	0.00	0.00
Mormyrus rume (I)	27.27	0.33	0.58	0.00	0.00	0.00	0.00	0.00	0.00
Marcusenius furcidens (I)	9.09	-	-	0.00	0.00	0.00	0.00	0.00	0.00
Marcusenius senegalensis (I)	63.64	3.33	0.00	30.00	0.58	2.31	0.00	0.00	0.00
Marcusenius ussheri (I)	27.27	0.50	1.73	10.00	-	-	0.00	0.00	0.00
Mormyrops anguilloides (P)	18.18	0.33	1.41	10.00	-	-	0.00	0.00	0.00
Petrocephalus bovei (I)	0.00	0.00	0.00	10.00	-	-	0.00	0.00	0.00
Hepsetus akawo (P)	90.91	2.67	3.08	60.00	2.33	3.44	27.27	0.83	2.31
Brycinus longipinnis (I)	63.64	1.42	2.30	30.00	0.42	1.15	9.09	-	-
Brycinus nurse (I)	36.36	0.17	0.00	10.00	-	-	0.00	0.00	0.00
Brycinus imberi (I)	18.18	0.75	1.50	10.00	-	-	0.00	0.00	0.00
Brycinus macrolepidotus (H)	18.18	0.33	1.41	0.00	0.00	0.00	0.00	0.00	0.00
Distichodus rostratus (H)	0.00	0.00	0.00	40.00	0.50	0.58	0.00	0.00	0.00
Barbus wurtzi (U)	0.00	0.00	0.00	10.00	-	-	9.09	-	-
Barbus snoeksi (U)	9.09	-	-	30.00	0.25	0.00	0.00	0.00	0.00
Labeo coubie (Pt)	90.91	4.25	1.64	70.00	6.75	4.27	18.18	0.25	0.71
Labeo parvus (Pt)	9.09	-	-	40.00	0.33	0.00	9.09	-	-
Parailia pellucida (I)	9.09	-	-	60.00	26.92	2.36	27.27	5.50	6.36
Schilbe intermedius (P)	9.09	-	-	30.00	0.92	3.06	0.00	0.00	0.00
Schilbe mandibularis (I)	100.00	10.50	10.76	100.00	14.33	14.69	81.82	5.08	4.60
Heterobranchus isopterus (O)	45.45	0.42	0.50	0.00	0.00	0.00	0.00	0.00	0.00
Heterobranchus longifilis (O)	18.18	-	-	0.00	0.00	0.00	0.00	0.00	0.00
Clarias anguillaris (O)	81.82	0.83	0.33	0.00	0.00	0.00	0.00	0.00	0.00
Clarias buettikoferi (O)	63.64	1.75	0.84	10.00	-	-	0.00	0.00	0.00
Malapterurus electricus (P)	18.18	0.25	0.71	0.00	0.00	0.00	0.00	0.00	0.00
Synodontis bastiani (O)	0.50	-	-	30.00	0.42	0.58	0.00	0.00	0.00
Synodontis koensis (H)	9.09	-	-	0.00	0.00	0.00	0.00	0.00	0.00
Synodontis punctifer (H)	0.00	0.00	0.00	10.00	-	-	0.00	0.00	0.00
Synodontys schall (O)	0.00	-	-	10.00	-	-	0.00	0.00	0.00
Parachanna obscura (P)	0.70	1.50	1.23	20.00	0.25	0.71	9.09	-	-
Thysochromis ansorgii (U)	9.09	-	-	0.00	0.00	0.00	0.00	0.00	0.00
Chromidotilapia guntheri (I)	2.92	7.50	0.45	70.00	2.17	2.93	36.36	5.83	17.99
*Oreochromis niloticus** (Pt)	27.27	0.42	0.58	0.00	0.00	0.00	18.18	0.33	1.41
Tilapia zillii (H)	90.91	6.42	2.14	100.00	12.17	4.22	100.00	13.58	13.46
*Tilapia hybride*** (U)	9.09	-	-	0.00	0.00	0.00	0.00	0.00	0.00
Ctenopoma petherici (I)	72.73	1.00	0.93	10.00	-	-	0.00	0.00	0.00
Mastacembelus nigromarginatus (U)	36.36	0.42	0.50	0.00	0.00	0.00	9.09	-	-

highest monthly average of total fish abundance was obtained in zone 1 with 172.54 ± 137.35 individuals while the lowest was recorded in zone 3 with 147.54 ± 86.54 individuals.

Fifty-eight of the 67 species collected occurred in zone 1 (Table 1). *Chrysichthys nigrodigitatus*, *Hemichromis fasciatus*, *Sarotherodon melanotheron*, *Schilbe mandibularis* and *Tylochromis jentinki* were the most frequent (100%), followed by *Clarias anguillaris*, *Heterotis niloticus*, *Pellonula leonensis*, (81.81%), *Chromidotilapia guntheri*, *Chrysichthys maurus*, *Ctenopoma petherici*, *Tilapia guineensis* and *Tilapia mariae* (72.72%). Species with the highest monthly mean abundance data were *C. nigrodigitatus* (32.17±3.70), *S. melanotheron* (15.42±5.29), *T. jentinki* (14.67±11.63), *Liza falcipinnis* (12.25±1.62), *S. mandibularis* (10.50±10.76), and *T. guineensis*(9.25±2.76).

Forty-five of the 67 species collected occurred in zone 2. *C. nigrodigitatus*, *H. fasciatus*, *S. melanotheron,* and *S. mandibularis* were recorded in all the samples (100%). They were followed by *Tilapia zillii* (83.33%), *P. leonensis, T. guineensis* (75%), *Porogobius schlegelii*, *T. mariae* and *T. jentinki* (66.67%). The most abundant species were *Parailia pellucida* (26.92±2.36), *P. leonensis* (20.75±1.33), *S. mandibularis* (14.33±14.69), *T. zillii* (12.17±4.22), and *S. melanotheron* (9.58±8.13).

The lowest number of species was recorded in zone 3 (40 out of 67). The most frequent species in this area were *Citharichthys stampflii*, *C. nigrodigitatus*, *Hemichromis fasciatus*, *Pellonula leonensis, Sarotherodon melanotheron*, *Schilbe mandibularis, T. zillii,* and *T. jentinki* (100%). The species that were most abundant were *P. leonensis* (30.17±11.71), *T. zillii* (13.58±13.46), *C. stampflii* (11.50±9.26) and *C. nigrodigitatus* (8.58±8.14).

The Jaccard's coefficient of similarity (J) indicated that zones 1 and 2 were more similar in their species composition [$J_1(z_1,z_2) = 0.63$] than the comparisons between zones 1 and 3 [$J_2(z_1,z_3) = 0.53$], and between zones 2 and 3 [$J_3(z_2,z_3) = 0.51$].

Based on feeding habit data recorded in the literature the numbers of species in different trophic categories were higher (except for planktivorous species) in zone 1 than zone 2 and zone 3 with 26 invertivorous (versus 23 and 18 respectively), 7 piscivorous (versus 6 and 4), 5 omnivorous (versus 3 and 0), and 3 herbivorous (versus 3 and 1) (Table 2).

In terms of ecological categories, fish composition was dominated by freshwater species in zones 1 and 2 (respectively 34 and 26 out of 67 species) and by estuarine dependent fishes (including ER, EDF, and EDM) (20 out of 67) in zone 3. These estuarine dependent species were almost equally present in the three zones with 19 and 15 species, respectively in zones 1 and 2. The non-dependent marine fish species diversities were almost comparable, with 4 species in both zones 1, 3 and 3 species in zone 2 out of 4 species collected. For occasional marine visitor fishes, out of 4 species collected, 1 was recorded in zones 1 and 2, against 3 in zone 3.

Furthermore, univariate Mann-Whitney U tests on the occurrences of the ecological categories was performed to compare dry and wet season's samples between sampling zones. The results showed significant difference ($p<0.05$) only for estuarine nondependent fishes (EN) between zones 1 and 3 in dry season ($p=0.006$) and estuarine dependent marine fish (EDM) between zones 2 and 3 in wet season ($p=0.02$). Indeed, EN was represented in zone 1 by 66 specimens, versus 10 in zone 3. In this group, *Caranx hyppos* and *Eucinostomus melanopterus* exhibited a higher occurrence data differences with, respectively, 25 specimens in zone 1, against 5 in zone 3, and 35 specimens in zone 1, versus 1 in zone 3. The EDM was represented in zone 2 by 10 specimens, against 128 in zone 3. Specifically for this group, there were: 0 *C. stampflii* in zone 2, versus 87 in zone 3; 5 *L. falcipinnis* in zone 2, versus 16 in zone 3; and 0 *Ethmalosa fimbriata* in zone 2, versus 9 in zone 3.

Influence of environmental variables

Results of CCA showed that first and second axes accounted respectively for 86.1 and 13.9% of the total variance for the environmental-species ecological categories relationship. Based on first axis which expressed maximum percentage variance, there is a clear-cut separation of species ecological categories and environmental characteristics between Ehy (zone 1) and Aby (zone 3) lagoons (Figure 2): The first zone, with higher pH and water temperature measurements is associated with freshwater, estuarine resident (ER), marine migrant (MM) and estuarine dependent freshwater (EDF) species. Aby lagoon had higher values of conductivity, salinity, transparency, and dissolved oxygen and was associated with EDM species. The Tendo lagoon, which had an intermediate position, could be distinguished from other sampling sites mainly by its higher water depth based on axis 2 of the CCA.

DISCUSSION

Physical and chemical data showed spatial trends in Aby-Tendo-Ehy lagoon system, based on sampling sites chosen in this study. The fact that there was no variation in the salinity of zones 1 and 2 further showed that these parts of the lagoon are freshwater environments, as compared to the zone 3 where the salinity varied between 0 and 3 ppt. The spatial variation in the salinity level corresponded to results of N'Goran (1998) and Seu-Anoï et al. (2011) for the Aby-Tendo-Ehy lagoon system. According to Chantraine (1980) salinity level in coastal Coastal lagoon depends on freshwater supply: lagoon salinity is relatively lower when freshwater inputs are higher. Hauhouot (2004) indicated that the Tanoe River, which is connected to Ehy lagoon (zone 1), contributes to 63% of the Aby-Tendo-Ehy lagoon system. This data could explain why the overall salinity measurements along the Aby-Tendo-Ehy complex increased from sampling zone 1 to 3.

Species richness recorded for the Aby-Tendo-Ehy lagoon system is 67 in this study. It was lower than 83 species reported by Charles-Dominique (1993), but higher than 60 species noted by Daget and Iltis (1965), and 33 species mentioned by N'Goran (1998). Species recorded by Daget and Iltis (1965) and Charles-Dominique (1993) but not caught in the present study in the Aby-Tendo-Ehy lagoon included *Arius latiscutatus, Epinephelus aeneus, Sphyrena afra, Polydactylus*

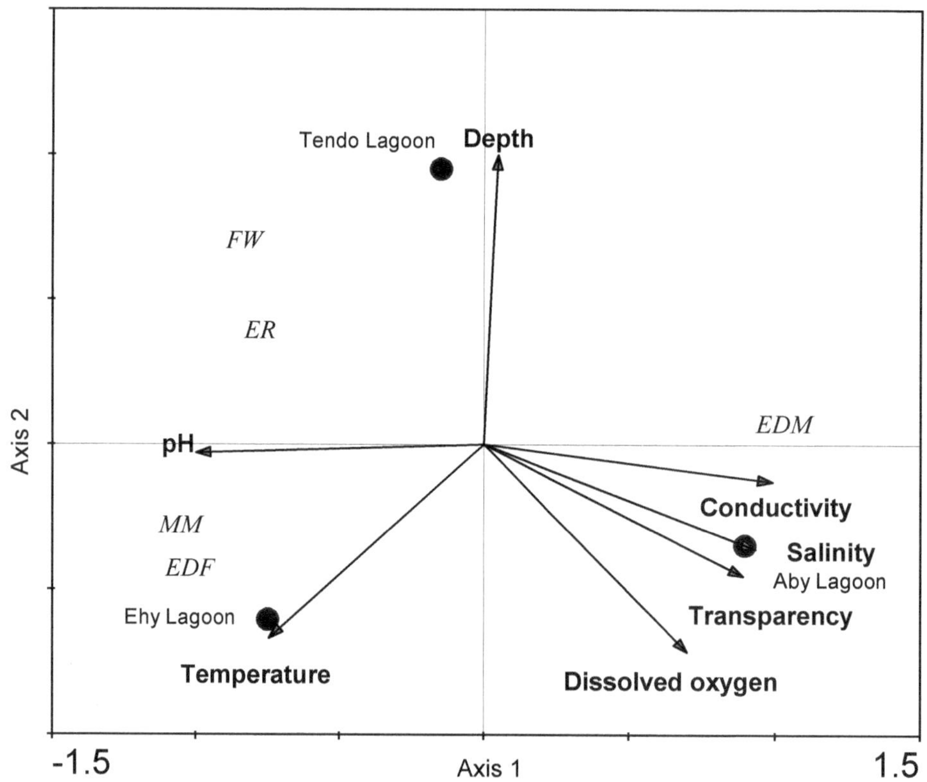

Figure 2. Canonical correspondence analysis applied to the environmental variables and to the fish ecological categories in the Aby-Tendo-Ehy lagoon system (Côte d'Ivoire). EDM = estuarine dependent marine fish, EDF= estuarine dependent freshwater fish, MM = marine migrant, ER = estuarine resident and FW = freshwater fish species.

quadrifilis, Pseudotholitus elongatus, Eleotris senegalensis, Scomberomorus tritor, Liza grandisquamis, Drepane africana, Strongylura senegalensis, Galeoides decadactylus, Cynoglossus senegalensis, Lutjanus dentatus, Caranx rhonchus, Dasyatis margarita, Trachinotus ovatus, Sardinella maderensis, Bathygobius soporator, Caranx senegallus, Ilisha africana, Lichia amia, Mugil curma, Pegusa triophtalma, Sardinella aurita, Megalops atlantica, Gymnothorax afer, Dalophis cepholopeltis, Myrophis plumbeus, Neolebias unifasciatus, Aplocheilicthys splilauchen, Epiplatys chaperi, Nimbapanchax petersii, Enneacampus kaupi, Gerres nigri, Pomadasys perotaei, Oblada melanura, Scarus hoefleri, Gobionelus occidentalis, Periophtalmus barbarus, Acanthurus monroviae and *Sphoeroides spengleri*. Many reasons such as the differences in sampling gear [experimental gillnets for the present study, against commercial fishing captures considered by Charles-Dominique (1993) and N'Goran (1998)] (Koné et al., 2003), the variation in fishing pressure due to human population increase (Albaret and Laë, 2003), and the environmental characteristic modifications (Charles-Dominique, 1993) could explain these species diversity differences.

In zone 3 which was much more investigated by previous studies, N'Goran (1998) and Charles-Dominique (1993), based on artisanal fisheries, indicated that the most abundant species was clupeid *E. fimbriata* (60 to 80% of total catch), versus *P. leonensis* for the present study. N'Goran (1998) and Charles-Dominique (1993) noticed a gradual decline in the capture of *E. fimbriata* in the Aby lagoon and attributed this to the increasing fishing intensity. In addition to these reasons, fishing gears characteristics in the present study could also have caused some differences: *P. leonensis* is a small sized species that was caught only by gillnet with small mesh size (06 and 08 mm), which is prohibited for commercial fisheries in the Aby-Tendo-Ehy lagoon. This result raised the question of some small-sized species that can be abundant but not exploited due to regulations on fishing gear.

Several studies suggested that tropical fish communities do not exhibit a clear organizational pattern (Goulding et al., 1988; Saint-Paul et al., 2000; Súarez et al., 2004). However, this study does not support the concept of unpredictability in the richness of fish community along Ehy-Tendo-Aby lagoon system: the number of species was found to decrease from zone 1 under freshwater influence to zone 3 with marine water influence. Furthermore canonical correlation analysis showed that sampling zone characteristics explain the variation in species richness: lowest values of salinity, transparency, dissolved oxygen and conductivity were associated with zones 1, whereas the highest values occurred in zones 3. This study also showed a variation in the number of different trophic categories. Piscivorous, herbivorous and omnivorous species number was higher in zone 1 with a reduced conductivity, salinity and transparency and a higher pH and temperature data. Planktivorous species were almost equally (species number) found in zones 1 and 3. Fish may affect diversity and water transparency by indirect trophic cascading effects: by consuming zooplankton and plant-attached macroinvertebrate grazers, fish may indirectly enhance phytoplankton and periphyton biomass (Jones and Sayer, 2003; Gelós et al., 2010), thus increasing water turbidity. But in the case of Ehy lagoon (zone 1) low transparency could also be the consequence of the turbulence phenomenon that occurred in this part of the lagoon due to high nutrient loads from both continental drainage and flooding tides (Neves et al., 2011; Konan et al., 2014).

Indeed, it is recognized that during the rainy season, runoff brings to aquatic environments a significant amount of organic material (which may include plant materials and/or insects) and nutrients from land that enrich these environments (Castillo-Rivera, 2013). The Tanoe River which is directly connected to zone 1 and the surrounding swampy forest of this zone act as important nutrient input sources. Insects and other terrestrial organisms, fruit and plant debris would also likely fall in the water under the effect of wind and rain (Konan et al., 2014). This possible importance and heterogeneity of food resources in this area could explain, partly, the high diversity both in species and trophic category diversities observed in zone 1.

In conclusion, this study recorded 67 species in the Aby-Tendo-Ehy lagoon system. Overall, fish diversity was higher in the most continental part of the studied lagoon than in the zone that was closer to the ocean. Species distribution along the Aby-Tendo-Ehy lagoon system was influenced by some environmental characteristics: the zone under the influence of freshwater discharges had higher pH and water temperature data and was mainly associated with freshwater, estuarine resident, marine migrant, and estuarine dependent freshwater species; the zone under marine water influences presented higher conductivity, salinity, transparency and dissolved oxygen data and had higher proportion of estuarine dependent marine species.

Conflict of Interests

The author(s) have not declared any conflict of interests.

ACKNOWLEDGEMENTS

This work is part of the research project entitled "Evaluation of refuge and nursery roles of the Tanoe Ehy swamp forest for fishes of adjacent lagoon and marine ecosystems" and was funded by the Program d'Appui Stratégique à la Recherche Scientifique, Côte d'Ivoire (PASRES) and Centre Suisse de Recherches Scientifiques en Côte d'Ivoire (CSRS) through its RASAPCI program (Recherche et Action pour la Sauvegarde des Primates en Côte d'Ivoire, CSRS, Côte d'Ivoire). The authors are grateful to Berté Issouf, Djiriéoulou Kémomadjèhi Claver, Money Attoubé Ida and Simmou Yomi Junior for their support during field sampling.

REFERENCES

Aboua BRD, N'zi KG, Kouamélan EP, Berté S, Bamba M (2010). Organisation spatiale du peuplement de poissons dans le Bandama. Int. J. Biol. Chem. Sci. 4(5):1480-1493.

Albaret JJ (1994). Les poissons : biologie et peuplements. In: Durand J.R., Dufour P., Guiral D., Zabi S.G.F. (eds) Les milieux estuariens. ORSTOM, Paris. pp. 239-279.

Albaret JJ (1999). Les peuplements des estuaires et des lagunes. In: Lévêque C., Paugy D. (eds) Les poissons des eaux continentales africaines : diversité, biologie, écologie, utilisation par l'homme. IRD, Paris. pp. 325-249.

Albaret JJ, Laë R (2003). Impact of fishing on fish assemblages in tropical lagoons: the example of the Ebrie lagoon, West Africa. Aquat. Living Resour. 16:1-9.

Baeta A, Cabral HN, Neto JM, Marques JC, Pardal MA (2005). Biology, population dynamics and secondary production of the green crab, *Carcinus maenas* (L.) in a temperate estuary. Est. Coast. Shelf Sci. 65:43-52.

Barletta M, Barletta-Bergan A, Saint-Paul U, Hubold G (2005). The role of salinity in structuring the fish assemblages in a tropical estuary. J. Fish Biol. 66:45-72.

Blaber SJM (2000). Tropical estuarine fishes: exploitation and conservation. Oxford, Blackwell Science. p. 372.

Castillo-Rivera M (2013). Influence of rainfall pattern in the seasonal variation of fish abundance in a tropical estuary with restricted marine communication. J. Water Resour. Prot. 5(3A):311-319.

Chabanne D (2007). Les catégories bioécologiques des espèces de poissons des estuaires et lagunes de l'Afrique de l'Ouest. Master en Sciences et Technologie; Université des sciences et techniques du Languedoc de Montpellier 2. p. 18.

Chantraine JM (1980). La lagune Aby (Côte d'Ivoire). Morphologie, hydrologie, paramètres physicochimiques. Doc. Sci. Cent. Rech. Océanogr. Abidjan 11(2):39-77.

Charles-Dominique E (1993). L'exploitation de la lagune Aby (Côte d'Ivoire) par la pêche artisanale. Dynamique des ressources, de l'exploitation et des pêcheries. Thèse de Doctorat. Université de Montpellier 2. p. 407.

Costa MJ, Elliott M (1991). Fish usage and feeding in two industrialised estuaries-the Tagus, Portugal, and the Forth, Scotland. In: Elliott M, Ducrotoy J-P (eds) Estuaries and coasts. Spatial and temporal intercomparisons. Olsen and Olsen, Fredensborg, Denmark. pp. 289-297.

Daget J, Iltis A (1965). Poissons de Côte d'Ivoire (eaux douces et eaux saumâtres). Bull. Inst. fond. Afr. Noire 74. 1-385.

Decru E, Vreven E, Snoeks J (2012). A revision of the West African *Hepsetus* (Characiformes: Hepsetidae) with a description of *Hepsetus akawo* sp. nov. and a redescription of *Hepsetus odoe* (Bloch, 1794). J. Nat. Hist. 46 (1-2):1-23.

Diomandé D, Doumbia L, Gourène G (2009). Stratégies alimentaires de deux espèces de poissons-chats dans l'hydrosystème fluvio-lacustre de la Bia: *Synodontis bastiani* et *S. schall* (Bloch & Schneider, 1801). Eur. J. Sci. Res. 27 (1):66-76.

Elliott M, O'Reilly MG, Taylor CJL (1990). The forth estuary: a nursery and overwintering area for North Sea fishes. Hydrobiologia 195:89-103.

FishBase (2013). A global biodiversity information system on finfishes. DVD, Royal Museum for Central Africa, Tervuren, Belgium. Published in June 2013.

Gelós M, Teixeira-de Mello F, Goyenola G, Iglesias C, Fosalba C, García-Rodríguez F, Pacheco JP, García S, Meerhoff M (2010). Seasonal and diel changes in fish activity and potential cascading effects in subtropical shallow lakes with different water transparency. Hydrobiologia. DOI 10.1007/s10750-010-0170-6.

Goulding M, Carvalho ML, Ferreira EG (1988). Rio Negro. Rich life in poor water: Amazonian diversity and food chain ecology as seen through fish communities. The Hague: SPB Academic Publishing. p. 200.

Hauhouot C (2004). Les pressions anthropiques sur les milieux naturels du sud-est ivoirien Human pressure on natural environment in South-East Ivory-Coast. Geo. Eco Trop. 28 (1-2):69-82.

Jones JI, Sayer CD (2003). Does the fish-invertebrate periphyton cascade precipitate plant loss in shallow lakes? Ecology 84:2155-2167.

King RP (1994). Seasonal dynamics in the trophic status of *Papyrocranus afer* (Günther, 1868) (Notopteridae) in a Nigerian rainforest stream. Rev. Hydrobiol. trop. 27 (2):143-155.

Konan YA, Bamba M, Koné T (2014). Aspects qualitatifs et quantitatifs de l'alimentation de *Clarias buettikoferi* (Siluriformes ; Clariidae) dans la forêt des marais Tanoé-Ehy (Côte d'Ivoire). Cybium 38 (1):61-68.

Koné T, Teugels GG, N'Douba V, Kouamélan EP, Gooré Bi G (2003). Fish assemblages in relation to environmental gradients along a small West African coastal basin, the San Pedro River, Ivory Coast. Afr. J. Aquat. Sci. 28:163-168.

Kouamélan EP, Berté S, Aboua BRD (2012). Feeding habits of the mochokid *Synodontis punctifer* Daget, 1964, in the Bandama River, Côte d'Ivoire. Afr. J. Ecol. 51:199-205.

Lowe-McConnell R (1987). Ecological studies in tropical fish communities. Cambridge University press, Cambridge. p. 382.

Marshall S, Elliott M (1998). Environmental influences on the fish assemblage of the Humber estuary. Est. Coast. Shelf Sci. 46:175-184.

N'Goran YN (1998). Statistiques de pêche en lagune Aby (Côte d'Ivoire): évolution de l'effort et des captures de 1979 à 1990. J. Ivoir. Océanol. Limnol. Abdjan 3 (1):25-37.

Neves LM, Teixeira TP, Araùjo FG (2011). Structure and dynamics of distinct fish assemblages in three reaches (upper, middle and lower) of an open tropical estuary in Brazil. Mar. Ecol. 32:115-131.

Paugy D, Lévêque C, Teugels GG (2003a). Faune des poissons d'eaux douces et saumâtres de l'Afrique de l'Ouest. Tome 1. MRAC, Tervuren. p. 457.

Paugy D, Lévêque C, Teugels GG (2003b). Faune des poissons d'eaux douces et saumâtres de l'Afrique de l'Ouest. Tome 2. MRAC, Tervuren. p. 815.

Pombo L, Rebelo JE (2002). Spatial and temporal organization of a coastal lagoon fish community, Ria de Aveiro, Portugal. Cybium 26(3):185-196.

Saint-Paul U, Zuanon J, Correa MAV, Garcia M, Fabre NN, Berger U, Junk WJ (2000). Fish communities in central Amazonian white- and blackwater floodplains. Environ. Biol. Fishes 57:235-250.

Schmitz OJ, Krivan V, Ovadia O (2004). Trophic cascades: the primacy of trait-mediated indirect interactions. Ecol. Lett. 7:153-163.

Selleslagh J, Amara R (2008). Environmental factors structuring fish composition and assemblages in a small macrotidal estuary (eastern English channel). Est. Coast. Shelf Sci. 79:507-517.

Seu-Anoï NM, Ouattara A, Koné YJ-M, Gourène G (2011). Seasonal distribution of phytoplankton in the Aby lagoon system, Ivory Coast, West Africa. Afr. J. Aquat. Sci. 36(3):321-330.

Súarez YR, Júnior MP, Catella AC (2004). Factors regulating diversity and abundance of fish communities in Pantanal lagoons, Brazil. Fisheries Manag. Ecol. 11:45-50.

ter Braak CJF (1986). Canonical correspondence analysis: a new eigenvector technique for multivariate direct analysis. Ecology 67:1167-1179.

Trewavas E (1983). Tilapiine fishes of the genera *Sarotherodon*, *Orechromis* and *Danakilia*. British Museum (Natural History) Cromwell Road, London SW7 5BD. p. 583.

Vinagre C, Franc AS, Costa MJ, Cabral HN (2004). Accumulation of heavy metals by flounder *Platichthys flesus* (Linnaeus 1758), in a heterogeneously contaminated nursery area. Mar. Poll. Bull. 49:1109-1126.

Yao SS, Kouamé KA, Ouattara NI, Gooré Bi G, Kouamélan EP (2010). Preliminary data on the feeding habits of the endemic species *Synodontis koensis* Pellegrin, 1933 (Siluriformes, Mochokidae) in a West African River (Sassandra River Basin, Côte d'Ivoire). Knowl. Manag. Aquat. Ec. 396, 04. DOI: 10.1051/kmae/2010014.

6

Benthic macroinvertebrate community of a fourth order stream in Kashmir Himalaya, India

Shazia Habib[1] and A.R. Yousuf[2]

[1]Department of Environmental Science, University of Kashmir, Srinagar, India.
[2]National Green Tribunal, Government of India, India.

The study was conducted in order to determine the abundance of benthic macroinvertebrates in Doodhganga stream passing through Brenwar forest (Yousmarg) Kashmir. Altogether 6 orders under 2 phyla namely: Arthropoda and Annelida were recorded from the stream. Inspite of the torrential flow of the stream, the hard substrata like boulders and cobbles provided a stable habitat for diverse number of macroinvertebrates. The diversity of benthic community helped to assess the health of the stream. On the basis of Hilsenhoff biotic index the study area was found pristine with negligible organic pollution. The seasonal dynamics showed greater diversity and density in summers than in winters.

Key words: Benthic, diversity, pollution, stream.

INTRODUCTION

Benthic macroinvertebrates are by far the most common groups used in the assessment of water quality (Williams and Feltmate, 1992; Rosenberg and Resh, 1993; Lin and Yo, 2008; Martins et al., 2008). These animals of size 200-500 μm (Rosenberg and Resh, 1993) inhabit the bottom substrate of fresh water, estuarine and marine ecosystem (A.P.H.A, 1998). The distribution of benthic macro invertebrate species and communities is controlled by a variety of environmental factors such as habitat characteristics (Peeters and Gardeniers, 1998), water quality (Hellawell, 1986), sediment quality (Chapman, 2001), sediment grain size (Tolkamp, 1980), contaminants (Phipps et al., 1995) and by biological factors such as competition and predation (Macneil et al., 1999).

The benthic community shows a range of taxa-specific responses to environmental stressors, these may be with respect to alteration in the food webs (Goedkoop and Johnson, 1996; Lodge et al., 1998; Stockley et al., 1998) or due to floods or drought (Covich, 1993; Power, 1995; Johnson et al., 1998), that alter the species composition of the benthic macrofauna.

This paper provides useful information pertaining to the type of invertebrates inhabiting the bottom substrate, thereby reflecting the pollution level in the stream.

MATERIALS AND METHODS

Study area

Doodganga stream is the principal left bank tributary of the River

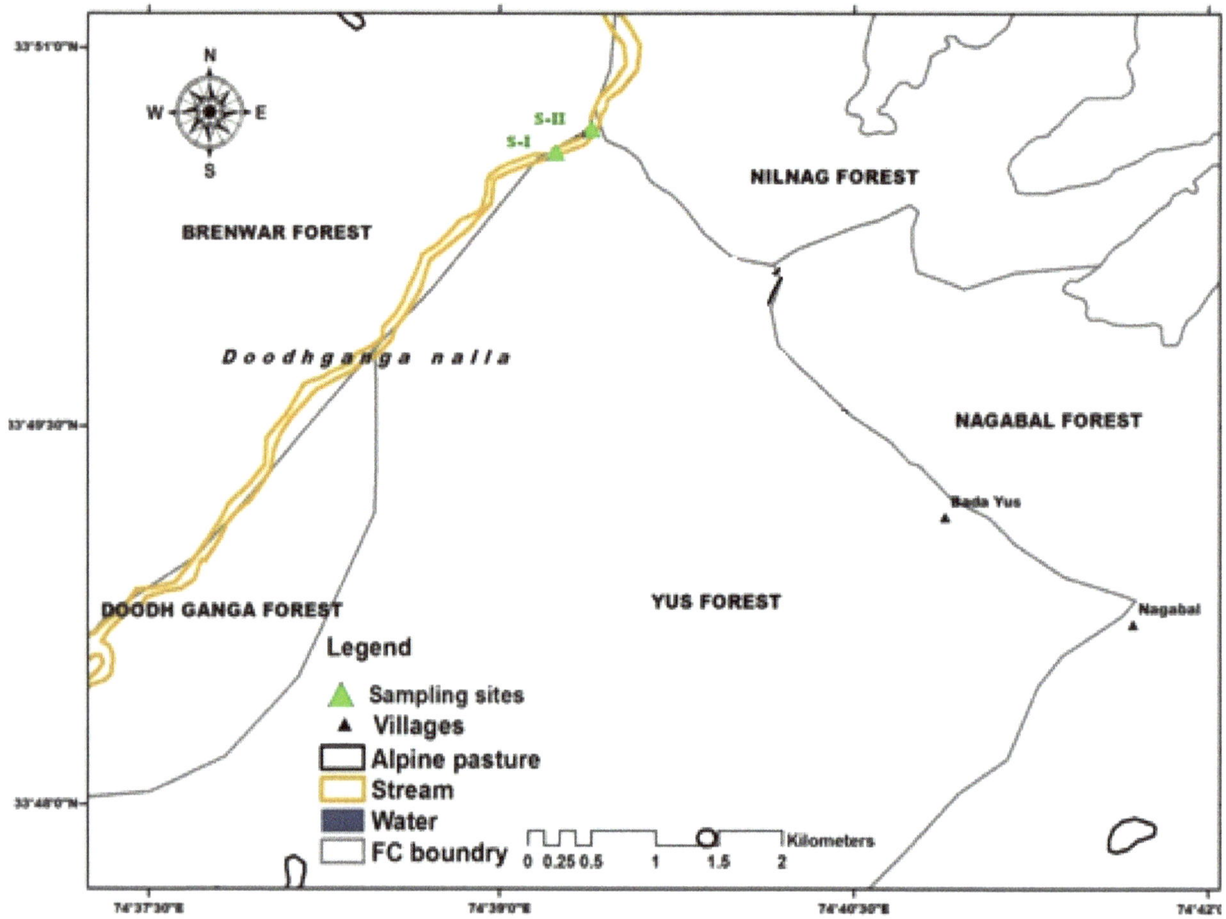

Figure 1. Doodhganga stream through Brenwar forest, Yousmarg and identification of two sampling sites.

Jehlum. The sampling sites (Figure 1) were located in the stream passing through the Brenwar forest of Yousmarg at an altitude of 2,304 m a.m.s.l. The coordinates for the two sites (I & II) were 33° 50′ 34.4″N and 74° 39′ 12.4″E and 33°50'41.2" N and 74°39'24.7" E, respectively. The study area belonged to stream order four (Hussain and Pandit, 2011) according to Rosgen classification. Sampling was carried in two different sites and the results obtained were averaged to provide a holistic information about the stream.

Methodology

Quantitative sampling was done during May, June, November and December (2010) via following the protocol for hard-bottomed streams (Hoffsten and Malmqvist, 2000; Stark et al., 2001; Ilmonen and Paasivirta, 2005). Fifteen replicates were taken from each site per month.

Samples collected were preserved in 4% formalin (for specimen with exoskeleton) and soft bodied organisms were preserved in 70% ethanol (Borror et al., 1976). The identification was done upto genus level with the help of standard works of McCafferty and Provonsha (1998), Wetzel and Likens (2000), and Ward (1992).

The population density was determined by calculating the number of individuals recorded per meter square. Biotic indices like Shannon diversity index (Shannon and Weaver, 1976), Simpson index (Simpson, 1949), evenness index (Pielou, 1966) and Hilsenhoff Biotic Index (Hilsenhoff, 1977, 1982, 1987) were also determined.

RESULTS AND DISCUSSION

The general characteristics of the stream are given in Table 1. During the period of investigation, 22 species of macrozoobenthos were recorded (Table 2). Arthropoda was found to be the most dominant group, comprising of 21 species followed by Annelida with 1 species. The former was represented by class Insecta (5 order) and Crustacea (1 order).

Highest relative density (94.5%) was recorded for phylum Arthropoda. The hard bottomed stream provided a stable habitat (Williams and Feltmate, 1992; Emere and Nasiru, 2007; Arimoro et al., 2007) to the macroinvertebrates against the increased water flow rate. The other phyla Annelida was represented by only *Erpobdella octoculata*. Here their contribution to the total invertebrate

Table 1. General characteristics of the study area.

Parameter	Average values
Depth (m)	1.23
Dissolved oxygen (mg/l)	6.52
Water flow rate (m^3/s)	2.06
Substrate type	Hard-bottomed stream (leaf litter from adjoining forest)

Table 2. Community structure of macrozoobenthos species in Doodhganga stream

Phylum	Order	Family	Taxa/Species
Arthropoda	Diptera	Athericidae	*Atherix* sp.
		Blephariceridae	*Bibiocephala* sp.
		Chironomidae	*Diamessinae* sp.
			Glyptotendipes sp.
		Empididae	*Clinocera* sp.
		Tabanidae	*Tabanus* sp.
			Chrysops sp.
		Tipulidae	*Hexatoma* sp.
			Tipula sp.
		Simuliidae	*Simulium* sp.
	Trichoptera	Glossosomatidae	*Glossosoma* sp.
		Hydropsychidae	*Hydropsyche* sp
		Limnephilidae	*Limnephilus* sp.
		Rhyacophilidae	*Rhyacophila* sp.
		Brachycentridae	*Brachycentrus* sp.
	Plecoptera	Capniidae	*Allocapnia* sp
		Chloroperlidae	*Xanthoperla sp.*
	Ephemeroptera	Baetidae	*Alainites* sp.
		Heptageniidae	*Epeorus* sp.
	Coleoptera	Chrysomelidae	Unidentified
	Amphipoda	Gammaridae	*Gammarus pulex*
Annelida	Pharyngobdellida	Erpobdellidae	*Erpobdella octoculata*

community was little, as the annelids have been observed to thrive better in soft depositing substrates which were not the present case.

The total number of individuals varied 632 ind.m^2 (May) to 185 ind.m^2 (December). During the individual sampling months (Figure 2) more or less constant trend of trichoptera and diptera dominating over other orders was reported. The macroinvertebrates belonging to these two orders are known to have a wide range of feeding behavior ranging from shredders, collectors to predators (Hutchinson, 1993) thereby occupy every possible trophic level (Williams and Feltmate, 1994; Mackie, 2001).

The benthic fauna showed a seasonal variation in density and diversity as is seen in the Figure 2. The

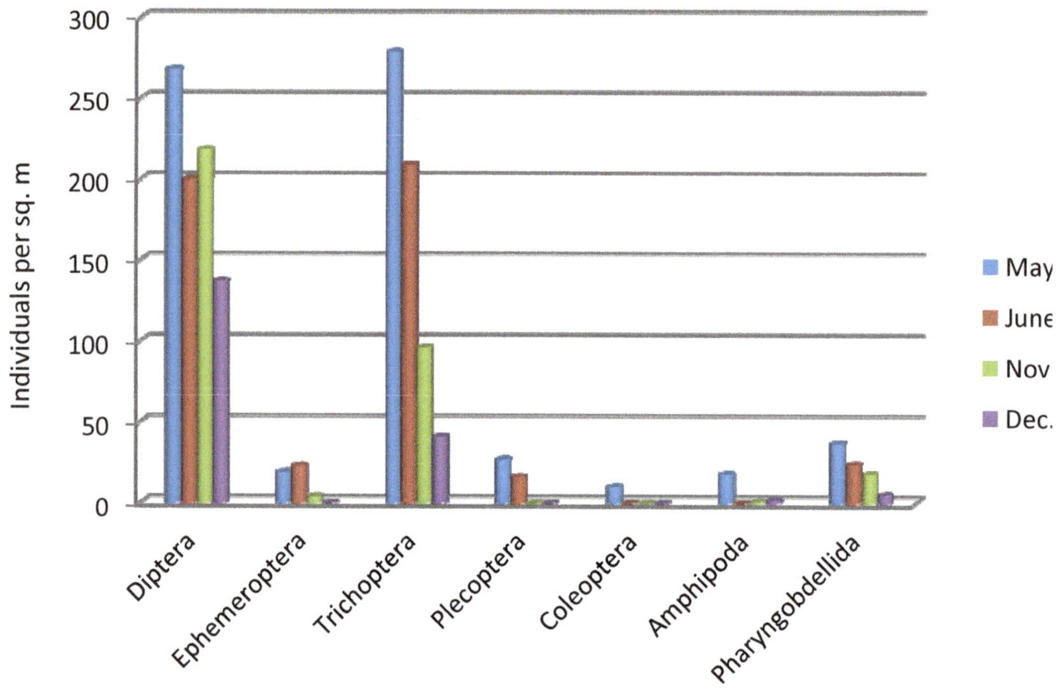

Figure 2. Population densities (number of individuals per sq. m.).

abundance was found to be highest during summer as compared to winters (Bruce et al., 2003; Lamp and Haube, 2004). The freezing temperature in winter in this high altitude temperate stream limits the density and species richness however warmer conditions in summers favors establishment of diverse fauna (Gupta and Michael, 1983; Allan, 1995; Cowell et al., 1997; Stark and Armitage, , 2004). The mean values of Shannon-, Simpson- and equitability indices were found to be 2.18, 0.16 and 0.85 respectively. The Hilsenhoff Biotic Index (3.6) indicated no apparent organic pollution in Doodhganga stream has apparently no organic pollution.

Conclusions

The study shows that the substrate type has an effect on the distribution of benthic organisms. A seasonal variation in the abundance of individuals indicates that temperature and its factors have a pronounced influence on the life cycle of invertebrates. The diversity indices and biotic index shows high abundance, richness and diversity with no organic pollution in the study area hence indicating good stream health.

Conflict of Interests

The author(s) have not declared any conflict of interests.

ACKNOWLEDGEMENT

The author is highly thankful to Prof. A.R. Yousuf for his thoughtful insight and mentoring during the project work. Thanks is also due to Prof. Azra N. Kamili for providing the necessary laboratory facilities in University of Kashmir. The author also wishes to thank Gowhar Mehraj, Department of Earth Sciences for providing the study area map.

REFERENCES

Allan JD (1995). Stream Ecology. Chapman & Hall, New York.

Arimoro FO, RB, Ikomi, Efemuna E (2007). Macro invertebrate community patterns and diversity in relation to water quality status of river Ase, Niger Delta, Nigeria. J. Fish. Aquatic Sci. 2:337-344.

Borror DJ, Delong DM, Triplehorn CA (1976). An Introduction to the Study of Insects. 4thEdn. Holt Reinhert and Winston, USA., p. 852.

Bruce CC, Amy HR, Diane ML (2003). Seasonal changes in the distribution and abundance of benthic invertebrates in six headwater streams in central Florida. Hydrobiologia 522:99-115.

Chapman PM (2001). Utility and relevance of aquatic Oligochaetes in Ecological Risk Assessment. Hydrobiologia, 463: 149-169.

Covich AP (1993). Water and ecosystems. in Water in Crisis. Oxford: Oxford University Press. pp. 40-55

Cowell BC, DM, Lynch, Remley AH (1997). Meiofauna and macrofauna in six headwater streams of the Alafia River, Florida.Publ. No. 03-101-130. Florida Institute of Phosphate Research Bartow (Fla.), 115 p.

Emere MC, Nasiru EC (2007). Macroinvertebratess as indicators of the water quality of an urbanized stream kaduna Nigeria. J. Fisheries Int., 2:152-157.

Goedkoop W, Johnson RK (1996). Pelagic-benthic coupling: Profundal benthic community response to spring diatom deposition in mesotrophic Lake Erken. Limnology and Oceanography 41:636-647.
Gupta A, Michael RG (1983). Population ecology and feeding properties of two co-existing taxa of Baetidae(Insecta : Ephemeroptera). Proc. Symp. Ecology of Animal population. Calcutta.
Hellawell JM (1986). Biological indicators of fresh water pollution and environmental management. Elsevier, London, England, 546 p.
Hilsenhoff WL (1977). Use of arthropods to evaluate water quality of streams. Tech. Bull. WI. Dept. Nat. Resour. No 100 15p.
Hilsenhoff WL (1982). Using a Biotic Index to Evaluate the Water Quality in Streams. Techinical Bulletin No. 132. Wisconsin Department of Natural Resoursces, Madison, WI.
Hilsenhoff WL (1987). An improved biotic index of organic stream pollution. Great Lakes Entomol. 20:31-39.
Hoffsten PO, Malmqvist B (2000). The macro invertebrate fauna and hydrogeology of springs in central Sweden. Hydrobiologia 436:91-104.
Hussain A, Pandit AK (2011). Hydrology, geomorphology and Rosgen classification of Doodhganga stream in Kashmir Himalaya, India. Int. J. Water Resour. Environ. Eng. 3(3):57-65.
Hutchinson G, Evelyn (1993). A Treatise on Limnology. Vol. 4. The Zoobenthos. New York: John Wiley & Sons.
Ilmonen J, Passivirta L (2005). Benthic macrocrustacean and insect assemblages in relation to spring habitat characteristic: Patterns in abundance and diversity. Hydorbiologia 533:99-113.
Johnson SL, Covich AP, Crowl TA, Estrada A, Bithorn J, Wurtsbaugh WA (1998). Do seasonally and disturbance influence reproduction in freshwater atyid shrimp in headwater streams, Puerto Rico? Proceedings of the International Association of Theoretical and Applied Limnology 26:2076-2081.
Lamp W, Haube M (2004). Aquatic Insects of Harpers Ferry National Historical Park: Assessing Environmental Associations and Ecological Vulnerability. pp. 25-28.
Lin KJ, Yo SP (2008). The effect of organic pollution on the abundance and distribution of aquatic oligochaetes in an urban water basin, Taiwan. Hydrobiologia 596(1):213-223.
Mackie GL (2001). Applied Aquatic Ecosystem Concepts. Kendall/Hunt Publishing Company. 24:744.
Macneil C, Dick JTA, Elwood RW (1999).The dynamics of predation on *Gammarus* spp. (Crustacea: Amphipoda). Biol. Rev. 74:375-395.
Martins RT, Stephan NNC, Alves RG (2008). Tubificidae (Annelida: Oligochaeta) as an indicator of water quality in an urban stream in southeast Brazil. Acta Limnol. Bras. 20(3):221-226.
McCafferty WP, Provonsha AV (1998). *Aquatic Entomology*. Jones and Bartlett Publishers Toronto Canada.
Peeters ET, Gardeniers JJP (1998). Logistic regression as a tool for defining habitat requirements of two common gammarids. Freshwater Biol. 39:605-615.
Phipps GL, Mattson VR, Ankley GT (1995). Relative sensitivity of three freshwater benthic macroinvertebrates to ten contaminants. Arch. Environ. Contam. Toxicol. 28:281 286.
Power ME (1995). Floods, food chains, and ecosystem processes in rivers.p.52-60 in Jones CG, Lawton JH eds. Linking Species and Ecosystems. New York:Chapman and Hall.
Rosenberg DM, Resh VH (1993). Introduction to Freshwater Biomonitoring and Benthic Macroinvertebrates. In: Freshwater Biomonitoring and Benthic Macroinvertebrates, Rosenberg DM and VH Resh (Eds.). Chapman and Hall, New York pp. 1-9.
Shannon CE, Weaver W (1976). Mathematical theory of communication. Univ. of Illinois, Press Urbana, p.117.
Simpson EH (1949). Measurement of diversity. Nature 163:688.
Stark BP, Brian JA (2004). Stoneflies (Plecoptera) of Eastern North America, Volume II. Chloroperlidae, Perlidae, and Perlodidae (Perlodinae). Bulletin of the Ohio Biological Survey New Series 14, Number 4. Ohio Biological Survey, Columbus, Ohio.
Stark JD, Boothroyd IKG, Hardiing JS, Maxted JR, Scarsbrook MR (2001). Protocols for sampling macroinvertebrates in wadeable streams. New Zealand Macroinvertebrate Working Group Report No 1. Prepared for the Ministry for the Environment Sustainable Management Fund Project No. 5103.
Stockley RA, Oxford GS, Ormond RFC (1998). Do invertebrates matter? Detrital processing in the River Swale-Ouse. Science of the Total Environment. 210:427-435.
Tolkamp HH (1980). organic-substrate relationship in low land streams. Thesis agricultural university- Wageningen. The Netherland. p. 221.
Ward JV (1992). Aquatic Insect Ecology, Biology and Habitat. John Wiley & Sons Inc. New York.
Wetzel RG, Likens G (2000). Limnological analyses. Springer (India) Publisher Private Limited, New Delhi.
Williams D, Feltmate B (1992). Aquatic Insects. CAB International Wallingford, U.K, p. 358.
Williams DD, Feltmate BW (1994). Aquatic Insects. Wallingford: CAB international. p. 300

Histopathological changes induced by copepoda parasites infections on the gills of economically important fish mugilidae (*Liza falcipinnis* and *Mugil cephalus*) from Ganvie area of Lac Nokoue, Republic of Benin

Nike F. Aladetohun[1,2] , Nestor G. Sakiti[2] and Emmanuel E. Babatunde[3]

[1]Federal College of Fisheries and Marine Technology, Ahmadu Bello Way, Victoria Island, Lagos, Lagos State, Nigeria.
[2]Laboratory of Parasitology and Ecology of Parasites, Department of Zoology and Genetics, Faculty of Sciences and Technology, University of Abomey Calavi, Republic of Benin, West Africa.
[3]Department of Marine Sciences, University of Lagos, Akoka, Lagos, Lagos State Nigeria.

Histopathological changes induced by copepods parasitic infection on the gills of economically important fish, *Mugil cephalus* and *Liza falcipinnis* from Ganvie area of Lac Nokoue were examined from December 2011 to July 2012. Histopathological changes shows that the nature of damage observed in the gill of both *M. cephalus* and *L. falcipinnis* remained the same. Histopathological observation reveals serious damage of lamellae and gill filaments due to attachment and feeding of copepods. The resultant hypertrophy of the underlying epithelial reducing the surface area for effective respiration, could lead to suffocation, particularly at high temperature. The histopathological changes enacted by the copepods parasites will eventually lead to reduced growth, low productivity and mortality resulting in economic loss.

Key words: Copepoda, parasite, histopathology, Mugilidae, Ganvie.

INTRODUCTION

Parasites have recently been highlighted as serious pathogenic problems in mullet fish in marine and brackish water. Among the parasites, copepod family is commonly found on fishes cultured in brackish water (Noor et al., 2012). With the developments of brackish and marine aquaculture, the importance of parasitic copepods as disease causing agents has become more obvious. Copepods parasites attached to gill filaments produce these small foci of eroded host tissue. Apparently, feeding involves external digestion (Halisch, 1940; Kabata, 1970); parasites produce digestive secretions which partially dissolve tissue, allowing easier ingestion. As the number of attached parasites increases, the destruction of respiratory epithelium progresses. Erosion can extend beyond the epithelial lining, resulting in obstructed branchial blood vessels. Irritation often results in responsive hyperplasia of epithelium, which as infestation intensifies, extends over considerable areas (Kabata, 1970; Paperna Overstreet, 1981). Both processes reduce the respiratory function of the gills.

Parasitic surveys of copepods (Figure 1) have been described from different hosts and locations in different

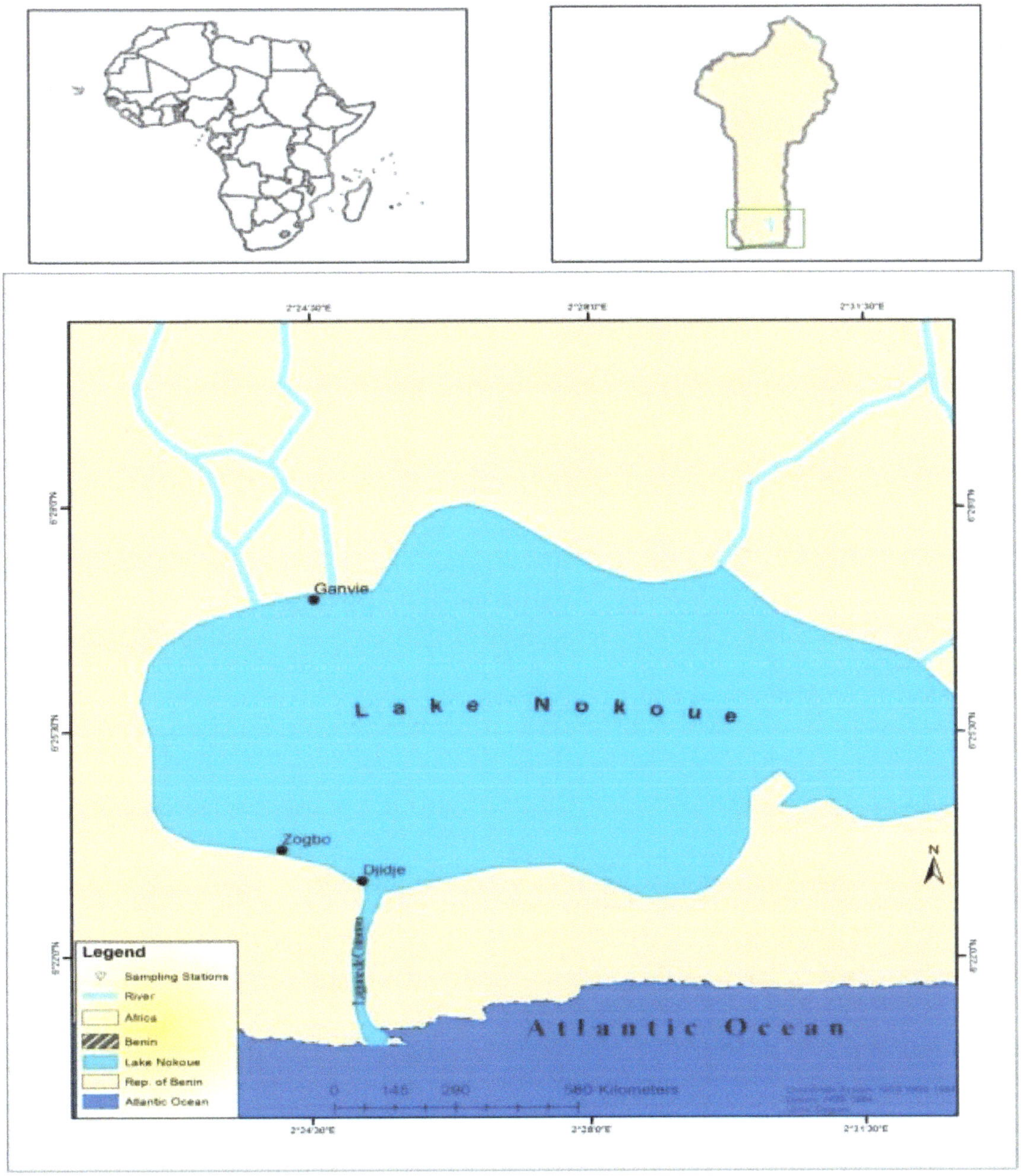

Figure 1. Location of Ganvie in lake Nokoue lagoon.

parts of the world: In Israel, Parpena and Lahav (1971), in Turkey, Koyun et al (2007), in Mexico, Suárez-Morales and Santana-Piñeros (2008), in Canada (Hogans,1989), in Sri Lanka, Vinobaba (2007), in Nigeria, Aladetohun et al. (2013) and in Republic of Benin, Aladetohun et al. (2013). However, histopathological impact of the parasites on their hosts remains to be determined. The aim of this study was to determine the host-parasite interaction, with specific reference to the pathology induced by copepod parasites on economically important fish *Mugil cephalus* and *Liza falcipinnis* in the Ganvie area of Lac Nokoue lagoon in Republic of Benin.

MATERIALS AND METHODS

The study area

The study site, Ganvie area (Figure 1) is located at the northern part of the lake Nokoue lagoon, near the floating village of Ganvie where the water is characterized by a high level of organic pollution (Laleye et al., 2003).

Lake Nokoue (Figure 1) is the largest lagoon (Moreau, 2004), it is a shallow, sub-tropical coastal lagoon (6°25N, 2°36E) with surface of 150 km and stretches 20 km in its east-west direction by 11 km n the north-south direction (Laleye et al., 2003). Lake Nokoue opens directly into the Atlantic Ocean through channel at cotonou which is about 24.5 km long.

Figure 2. Attachment of copepod in the gill of host fish.

Table 1. Prevalence and mean intensity of *Ergasilus latus* infections in *M. cephalus* and *L. falcipinni* from Ganvie area of Lac Nokoue.

Species of fish	No of fish examined	No of fish infested	No of copapod parasite	Percentage prevalence (%) for MC and LF	Mean intensity for MC and LF
M. cephalus	115	90	18		
L. falcipinnis	132	110	15	80.97	0.165
Total	247	200	33		

Histopathological analysis

Damaged fish tissues (gills) (Figure 2) were taken from the parasite attachment area of infested fishes and were cut out in fresh condition fixed in 10% buffered neutral formalin (Buoins fluid). The tissues were then processed using a 24 h automatic tissue processor for a time ranging from 17 to 19 h to process. The tissue processor contains 12 beakers, 10 glass beakers and 3 thermostatically controlled electric metal beakers containing paraffin wax.

The tissue (gills) were transferred to Beaker 1, containing 10% formal saline for complete fixation and then transferred to beaker 2-8 containing different ascending grades of dehydrating fluids (alcohol ranging from 70% alcohol to absolute alcohol (isopropyl alcohol, which helps in removing water from the tissue samples and then into beakers 9 and 10 containing clearing agents[xylene I and II] which completely clears the dehydrating agent off the tissue sample.

The tissue samples were later transferred into beakers 11 and 12 containing embedding agent, that is, molten paraffin wax which provides solid support upon embedding.

The tissues after being processed are embedded using an automatic embedding centre. Embedding is a process of submerging a tissue in a metal plastic disposable embedding mould containing molten paraffin wax, which became solidified when it was cold. This formed a support medium for the tissue during sectioning.

Sections of the tissue were cut using a microtome and were placed in a clean grease free slide which was then placed on a hot plate for 30 min in order for section to adhere to the slides. The staining method used was the H&E staining method. This method was used in order to demonstrate the general structure of the tissues. These were then dewaxed in xylene. The processed section were later taken to water by using descending grades of alcohol, that is, from absolute alcohol >95% alcohol, >70% alcohol, water. It was stained in haematoxylin for 10 min, then rinsed in water.

These were differentiated in 1% acid alcohol (a dip), rinsed in water and then blued in tap water for 5 min. This was counter stained in 1% Eosin for 2-5 min and rinsed in tap water, then dehydrated using ascending grades of alcohol (70% alcohol9, 5% alcohol absolute). These were cleared in xylen, mounted using D.P.X (a mountant) and viewed under the microscope.

RESULTS

M. cephalus and *L. falcipinnis* were found to be infested ergasilids (*Ergasilus* Latus, *Ergasilus Lizae* and *Nipergasilus bora*), attached to the gill filament (Figure 2). The list of hosts sites species with corresponding hosts parasites species are given in the Tables 1, 2, 3 and 4. The prevalence and intensity of infection varied with the host fish (Tables 1, 2, 3 and 4).

Adult females ergasilids are attached to primary lamella (gill fillaments) of the host fish with their claw-like secondary antennae close to the gill arch near the base of the fillament (Figure 2). The lamellae cells were pyknotic and degenerating as a result of second antannae insertions by the ergasili parasites (Figures 3 and 4).The infested gill rakers and gill lamellae were almost totally lost (Figures 3 and 4). It was also observed that the proliferated epithelium contained hypertrophic cells with intercellular spaces in which wandering cells such as blood granulocites, lymphocites, vacouole cells and coarse eosinophilic granular cells were evident.

Table 2. Prevalence and mean intensity of *Ergasilus lizae* infections in *M. cephalus* and *L. falcipinnis* from Ganvie area of Lac Nokoue.

Species of fish	No of fish examined	No of fish infested	No of copapod	Percentage prevalence (%) for MC and LF	Mean intensity for MC and LF
M. cephalus	115	90	20		
L. falcipinnis	132	110	7	80.97	0.135
Total	247	200	27		

Table 3. Prevalence and mean intensity of *Nipergasilus bora* infections in *M. cephalus* and *L. falcipinnis* from Ganvie area of Lac Nokoue.

Species of fish	No of fish examined	No of fish infested	No of copepod	Percentage prevalence (%) for MC and LF	Mean intensity for MC and LF
M. cephalus	115	90	56	80.97	0.77
L. falcipinnis	132	110	98		
Total	247	200	154		

Table 4. Summary of the percentage prevalence and mean intensity of *E. latus*, E. *lizae* and *N. bora*parasite in *M. cephalus* and *L. falcipinnis* from Ganvie area of Lac Nokoue.

Parasite	No of fish examined		Total	No of infested fish		Total	Percentage prevalence (%)	No of copapod parasite		Total	Mean intensity
	M. cephalus	*L. falcipinnis*		M C	L F			M C	L F		
E. latus	115	132	247	90	110	200	80.97	18	15	33	0.165
E. lizae	115	132	247	90	110	200	80.97	20	7	27	0.135
N. bora	115	132	247	90	110	200	80.97	56	98	154	0.77

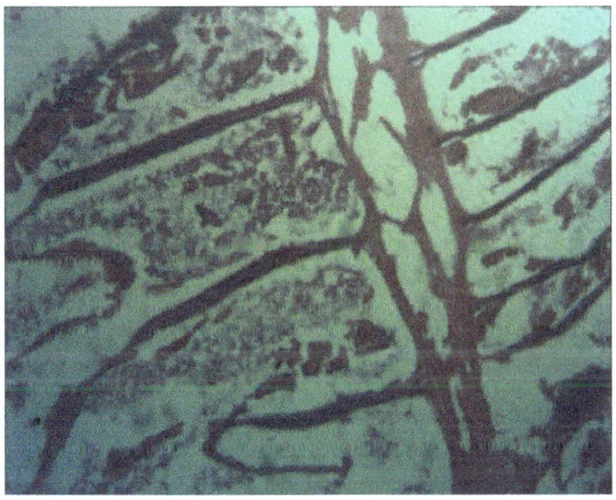

Figure 3. High destruction of gill filaments, gill lamella, gill rakers, gill arches and detachment of epithelial cells in lamella of *M cephalus*.

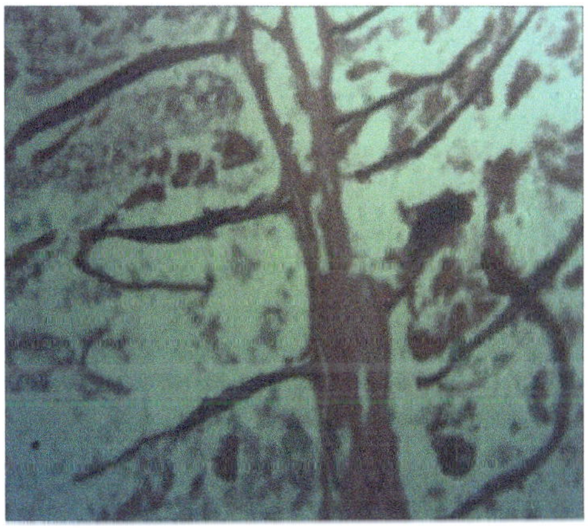

Figure 4. High destruction of gill filaments, gill lamella, gill rakers, gill arches and detachment of epithelial cells in Lamella of *L. falcipinnis*.

DISCUSSION

Copepods feed by browsing on the fish gill epithelium or by ingesting blood from ruptured blood vessels.

Apparently, feeding involves external digestion (Halisch, 1940; Kabata, 1970); parasites produce digestive secretions which partially dissolve tissue, allowing easy ingestion.

The research carried out shows the gill beneath the parasites, it seems like a cavity which is probably related to the enzymatic reaction of the gill epithelium beneath the parasite. This report coincides with that of Vinobaba (2007) in his work on histopathological changes induced by ergasilid copepod infections on the gills of food fish from Batticaloa lagoon, Sri Lanka.

In this work, the copepod parasites found caused changes on the gill fillaments through feeding. Both the gill lamella and the gill arch and gills rakers were badly damaged. This was also reported by Rameshkumar and Samuthirapandian (2013) on histopathological changes in the skins and gills of some marine fishes, Overstreet (1978), Ben-Hassin (1983), Morella and Garippa (2001) in his work on parasites of grey mullets from Mistras lagoon, Western Meditar-ranean, Ben-Hassin (1983) in his work on copepod parasites of mugilidae.

This research work reveals lost of gill fillament and totally almost missing. Parperna and Lahav (1971), reported on copepod parasite (*Ergasilus lizae*) in grey mullets in Israel and Vinobaba (2007) in his work on histopathological changes induced by ergasilid copepod infections on the gills of food fish from Batticaloa lagoon also gave the same report.

Copepod parasites attaches to the host using various appendages modified for gasping and this activity can led to secondary infection by pathogenic organism (e.g bacteria, fungi, and virus) and cause mass mortality.

High intensity of infection of these copepods may lead to serious damage of the gills and therefore show pronounced impact on the histology and lead to mortality (Vinobaba, 2007). Environmental change, especially habitat degradation by anthropogenic pollutants and oceanographic alterations induced by climatic changes and temperature dependent can influence parasitic-host interaction (Abalaka et al., 2010).

In conclusion, the resultant hypertrophy of the underlying epithelial reducing the surface area for effective respiration could lead to suffocation, particularly at high temperature. The histopathological changes enacted by the copepods parasites will eventually lead to reduced growth, low productivity and mortality resulting in economic loss.

ACKNOWLEDGEMENT

The authors wish to thank the staff of Molecular and Pathology Department, College of Medicine, University of Lagos, for the usage of their facilities and encouragement.

REFERENCES

Abalaka S, Yakasai FM, Doguwar GBN, Makonjuola KH (2010). Histopathological changes in the gills and skin of adult Clariasgariepinus exposed to ethanolic extract of Parkia biglobosa pods. Basic Appl. Pathol. 3:109-114.

Aladetohun NF, Sakiti NG, Emmanuel BE (2013). Copepod parasite in gills economically important fish mugiliidae (Mugil Cephalus and Liza Falcipinis from Lagos Lagoon, west Africa Nigeria. J. Am. Sci. 9(6):1.

Aladetohun NF, Sakiti NG, Emmanuel BE (2013). Copepod parasite in gills economically important fish mugiliidae (Mugil Cephalus and Liza Falcipinis from Lac Nokue Lagoon in republic of Benin, West African. Afr. J. Environ. Sci. Technol. 7(8):2.

Suárez-Morales E, Santana-Piñeros AM (2008). A new species of *Ergasilus* (Copepoda: Cyclopoida: Ergasilidae) coastal fishes of the Mexican Pacific. Folia Parasitol. 55:224–230.

Halisch W (1940). Anatomie und Biologie von Ergasilus minor Halisch. Z. Parasitenkd. 11:284-330

Hogans WE (1989). Mortality of cultured Atlantic salmon, *Salmo salar* parr caused by an infection of *Ergasilus labracis* (Copepoda:Poecilostomatoida) in lower St. Johns River, New Brunswick. Canada. J. Fish. Dis. 12:529-531.330.

Kabata Z (1970). Crustacea as enemies of fish. Book 1, Snieszko, S., Axelrod H.R. (ed.), Diseases of Fish, T.F.H. Publications, Jersey City, N.J.

Koyun M, Altunel FN, Öktener A (2007). *Paraergasilus longidigitus* Yin, 1954 (Copepoda: Poecilostomatoida) Infestations in the Bleak, *Alburnus alburnus* Lin.,1758 from Enne Dam Lake *Türkiye Parazitoloji Dergisi.* 31(2):158-161.

Overstreet RM (1978). Marine maladies? Worms, germs and othersymbionts from the northern Gulf of Mexico. USA: Mississippi,Alabama Sea Grant Consortium; p. 140.

Paperna I, Lahav M (1971). New records and further data on fish parasites in Israel. Bull. Fish Cult. Israel (Bamidgeh). 16: 77-86.

Paperna I, Overstreet RM (1981). Parasites and diseases of Mullets (Mugilidae). In: Oren, O.H. (ed.) Aquaculture of Grey Mullets. IBP 26, Cambridge University Press, U.K. pp. 1-19.

Rameshkumar G, Samuthirapandian R (2013) Histopathological changes in the skins and gills of some marine fishes due to parasitic isopod infestation. J. Coast. Sci. Med. 2013(1)74-80.

Vinobaba P (2007). Histopathological changes induced by ergasilid copepod infections on the gills of food fish from Batticaloa lagoon, Sri Lanka, Sri Lanka. J. Aquat. Sci. 12:77-87.

Heavy metals concentration in various tissues of two freshwater fishes, *Labeo rohita* and *Channa striatus*

S. A. Mastan

Post-Graduate Department of Biotechnology, Post-Graduate Courses and Research Centre, DNR College, Bhimavaram - 534 202 A. P., India.

Heavy metals like cadmium, zinc, copper, chromium, lead and mercury were measured in the various tissues of *Labeo rohita* and *Channa striatus* and in the water samples collected from the Kolleru Lake, Andhra Pradesh during 2009-2010. The concentrations of heavy metals in the different organs of fishes varied. In *L. rohita*, the concentrations of heavy metals were in the order of liver > kidney > gills > muscles and in *C. striatus*, it was liver > kidney > gills > muscle. The values of heavy metals concentration in the present study are within the maximum permissible levels for drinking water and fish.

Key words: Heavy metals, tissues, fish, Kolleru Lake.

INTRODUCTION

The contamination of aquatic systems with a wide range of pollutants has become a matter of concern since the last few decades (Canli et al., 1998; Dirilgen, 2001; Vutukuru, 2005; Amaraneni, 2006; Rao and Rao, 2007; Vinodhini and Narayanan, 2008; Gupta et al., 2009). The natural water bodies may extensively be contaminated with various heavy metals released from domestic, industrial effluents, idol immersion, draining of sewage, dumping of hospital, other wastes and anthropogenic activities, etc. (Conacher et al., 1993; Velez and Montora, 1998; Chandra Sekhar et al., 2004; Vinodhini and Narayanan, 2008; Malik et al., 2010; Laxmi Priya et al., 2011). Pollution of heavy metals in aquatic ecosystem is growing at an alarming rate and has become an important problem of many countries of the world. Heavy metal contamination may cause devastating effects on the ecological balance of the recipient environment and its diversity of aquatic organisms (Farombi et al., 2007; Vosyliene and Jankaite, 2006; Ashraj, 2005; Vinodhini and Narayanan, 2008). Fish occupies higher level in the food chain and is an important source of protein food for human beings. The heavy metals in aquatic ecosystem are transferred through food web into human beings. Some of heavy metals can cause health problems to fish consumers (Uysal et al., 2008; Taweel et al., 2011). Therefore, in the present study attempts have been made to assess the heavy metal concentration in water and fishes of Kolleru Lake, Andhra Pradesh.

MATERIALS AND METHODS

Study area

The study area, lake Kolleru (16° 32 and 16° 47' N and 81° 5 and 81° 21 E) is a natural wetland located between two major deltas, Godavari on the east and Krishna on the west of costal Andhra Pradesh, India. The freshwater enters the lake through a number of rivers, stream and agricultural drains. The lake has rich aquatic life, the lake has always been exploited by local population (Figure 1).

A total of one hundred and forty four fishes (ranging between 90-100 g in weight and 16-81 cm in length and three months in age) were collected with the help of a fisher man using the gill net. The water samples were collected in sterilized bottles from three sites, with one meter in depth to analyze the various physico-chemical parameters such as water temperature, pH, free carbon dioxide,

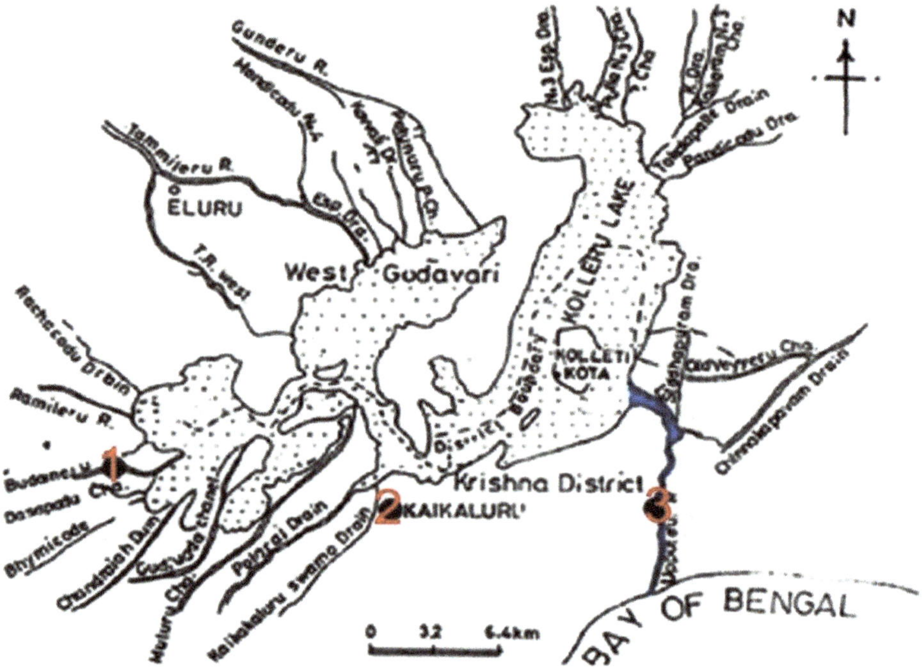

Figure 1. Geographical distribution of sampling sites in Kolleru Lake, Andhra Pradesh.

total hardness and dissolved oxygen. Dissolved oxygen (DO) was fixed at sampling site itself and estimated in the laboratory. Both fish and water samples were collected for one year from July 2009 to June 2010.

100 ml of water sample was collected in sterilized bottles from each site and was acidified with 10% HNO_3 and brought to the laboratory. The samples were filtered through what's man filter paper (No. 42) and kept in refrigerator until further analysis. The samples were subjected to analysis directly.

In each month of all the three seasons, 6 fishes of *Labeo rohita* and *Channa striatus* were caught from the lake. They were brought to the laboratory and dissected, and the organs viz. liver kidney, gills and muscles were removed with clean sterilized instruments. Tissues such as liver, kidney, gills and muscles were washed with double distilled water and put in sterilized Petri dishes to dry at 120°C in oven until they reached a constant weight. One gram of each dried tissue of liver, kidney, gills and muscles were then digested with diacid ($HN0_3$ and $HClO_4$ in 2:1 ratio) on a hot plate set at 130°C until all materials were dissolved. Digested samples were diluted with double distilled water appropriately in the range of the standards, which were prepared from the stock standard solutions of the metals (Merck). The metal concentration in the samples was measured using a Perkin-Elmer Analyst 300 Absorption Atomic Spectrophotometer (AAS). The results were expressed as µg g^{-1}dry weight and mg $L^{-1}$for fish and water, respectively.

Physico-chemical parameters of the water were analyzed as per the procedures described in APHA (2012). The obtained data were subjected to statistical analysis using a computer program (SPSS, 2001) for mean and standard deviation.

RESULTS AND DISCUSSION

The physico-chemical parameters are paramount, important and influenced by natural and manmade activities. They are also depends upon the depth of water body and ecological conditions of ecosystem. In the present study, water temperature ranged between 22 and 34.8°C from summer to winter. The pH of the water was alkaline throughout the study ranging from 7.2 to 9.0. Similar findings were recorded by Mohan et al. (2007) and Malik et al. (2010). Free carbon dioxide showed an irregular pattern and ranged from 0 to 3.8 mg L^{-l}. Dissolved oxygen (DO) is an important water parameter indicating the quality of water and organic production in lake (Wetzel and Likens, 2006). Survival of aquatic flora and fauna especially fish fauna, depends on the dissolved oxygen in water. In the present study, DO levels ranged between 6.0 and 8.6 mg L^{-l}. Low DO levels were recorded during summer season, indicating that DO level decreases with the increase in temperature. The hardness of water varied from 58-192 mg L^{-l} (Table 1) which was within the permitted levels for drinking water proposed by of Bureau of Indian standards.

In the water of Kolleru Lake, average concentration of heavy metals namely, Pb, Cd, Zn, Cu, Cr and Hg, were 0.034, 0.012, 0.304, 0.013, 0.436 and 0.001 mg L^{-l}, respectively (Table 2).

Zinc content was highest and Hg was the lowest in water. The concentration of heavy metals was in the range of Zn, Cr, Pb, Cu, Cd and Hg. Gupta et al. (2009) have reported that the concentration of Zn in the water of river Ganges at Allahabad was highest and followed by Pb, Cu, Cr and Cd, The high content of Zn and the lowest content of Hg were also reported by Jain and Sharma (2001), Sarkar et al. (2007) and Malik et al. (2010). The concentrations of heavy metals were highest in summer

Table 1. The range of water quality parameters of Kolleru Lake, A. P. in different seasons during the study period.

Physico-chemical parameter	Spring	Summer	Rainy	Winter	Permissible limits for drinking water (BIS, 1991)
Temperature (ºC)	26 - 32.1	28 - 34.8	26 - 29	22 - 25.6	-
P^H	7.6 - 7.8	7.6 - 8.0	8.2 - 8.6	8.2 - 9.6	6.5 - 8.5
Free CO_2 (mg L^{-1})	0.20 - 2.8	0.20 - 3	0 - 2	0 - 3.8	-
Total hardness (mg L^{-1})	120 - 189	120 - 192	68 - 172	56 - 86	300
D.O (mg L^{-1})	6.0 - 6.5	6.0 - 6.8	6.8 - 7.9	6.8 - 8.6	6

Table 2. Heavy metal concentration (mg L^{-1}) in water of Kolleru Lake, Andhra Pradesh, in different seasons during the study period.

Heavy metal (mg L^{-1})	Spring	Summer	Rainy	Winter	BIS, 1991 limit
Pb	0.019 ± 0.014	0.034 ± 0.002	0.027 ± 0.001	0.038 ± 0.001	0.1
Cd	0.011 ± 0.001	0.012 ± 0.001	0.001 ± 0.000	0.010 ± 0.000	0.01
Zn	0.301 ± 0.001	0.304 ± 0.002	0.279 ± 0.017	0.299 ± 0.006	15
Cu	0.012 ± 0.001	0.013 ± 0.001	0.113 ± 0.001	0.012 ± 0.001	1.5
Cr	0.014 ± 0.001	0.043 ± 0.002	0.042 ± 0.005	0.039 ± 0.001	0.05
Hg	0.001 ± 0.000	0.001 ± 0.000	0.001 ± 0.001	0.001 ± 0.001	0.001

Values expressed in mean ±SD of three replicates.

Table 3. Heavy metal concentration (µg/ g/d.w) in various tissues of *L. rohita*.

Heavy metal (µg g^{-1})	Liver	Kidney	Gill	Muscle	FAO*/WHO** guideline
Pb	1.483 ± 0.206	0.553 ± 0.032	0. 300 ± 0.010	0.223 ± 0.152	4
Cd	0.616 ± 0.249	0.250 ± 0.046	0.383 ± 0.153	0.316 ± 0.006	-
Zn	1.036 ± 0.004	0.943 ± 0.031	0.19 ± 0.0173	O.380 ± 0.020	50
Cu	1.056 ± 0.124	0.233 ± 0.042	0.090 ± 0.000	0.090 ± 0.000	10
Cr	0.696 ± 0.349	0.716 ± O.035	0.196 ± 0.015	0.023 ± 0.012	2
Hg	0.270 ± 0.552	0.070 ± 0.001	0.047 ± 0.001	0.036 ± 0.015	-

Values expressed in mean ± S.D of three replicates, unit- µg /g/d.w,*FAO (1983), **WHO (1985).

and the lowest in rainy months due to the dilution effect of water. Similar results have also been reported by Jain and Sharma (2001) and Malik et al. (2010). In the present study, it has been observed that, all the metals were below the permitted levels of BIS.

The heavy metals were accumulated at varying levels in different tissues of *L. rohita* and *C. striatus*. The concentration of heavy metals in different organs of fishes followed the decreasing order Pb > Cd > Zn > Cu > Cr > Hg. The concentration of Pb varied between 1.483 ± 0.206 and 0.223 ± 0.152 µg g^{-1}. While that of Cd, 0.616 ± 0.249 and 0.250 ± 0. 045, Zn, 1.036 ± 0.004 and 0.19 ± 0.017, Cu, 1.056 ± 0.124 and 0.0.090 ± 0.001, Cr, 0.716 ± 0.357 and 0.023 ± 0.012, Hg, 0.270 ± 0.552 and 0.036 ± 0.015 µg g^{-1} in *L. rohita* whereas the respective values for *C. striatus* were recorded as Pb 1.353 ± 0.070 and 1.30 ± 0.010, Cd 0.480 ± 0.314 and 0.160 ± 0.040, Zn 1.140 ± 0.295 and 1.013± 0.050, Cu 1.033± 0.646 and 0.028 ± 0.002, Cr 0.853 ± 0.535 and 0.326 ± 0.050, Hg 0.256 ± 0.060 and 0.123 ± 0.086 µg/g/d.w. The concentrations of metals were higher in the livers than gills and muscles (Tables 3 and 4).

Heavy metals in aquatic environment and aquatic biota pose a risk to fish consumers and other wild life. Heavy metals may enter aquatic ecosystem from different natural and anthropogenic sources including industrial or domestic sewage, storm runoff, leaching from landfills-/dump sites and atmospheric deposit (Forstner and Wittmann, 1983; Bhupander et al., 2011; Laxmi et al., 2011).

Table 4. Heavy metal concentration (µg/g/d.w.) in various tissues of *C. striatus*.

Heavy metal	Liver	Kidney	Gill	Muscle	FAO*/**WHO guideline
Pb (µg g^{-1})	1.353 ± 0.070	1.333 ± 0.704	1.30 ± 0.529	1.30 ± 0.010	4
Cd (µg g^{-1})	0.480 ± 0.314	0.160 ± 0.040	0.316 ± 0.125	0.456 ± 0.109	-
Zn (µg g^{-1})	1.013 ± 0.050	1.213 ± 0.208	1.140 ± 0. 295	0456 ± 0.109	50
Cu (µg g^{-1})	0.993 ± 0.600	1.033 ± 0.646	0.028 ± 0.002	0.530 ± 0.305	10
Cr (µg g^{-1})	0.326 ± 0.050	0.470 ± 0.160	0.596 ± 0.205	0.853 ± 0.535	2
Hg (µg g^{-1})	0.203 ± 0.125	0.256 ± 0.060	0.123 ± 0.086	0.140 ± 0.135	-

Values were expressed as mean ± S.D of three replicates, unit- µg g^{-1}/d.w. *FAO (1983), **WHO (1985).

In the present study, Pb concentration in all the tissues of *L. rohita* was higher than that of *C. striatus*. In *L. rohita*, highest concentration of Pb was recorded followed by Cu, Zn, Cr, Cd and Hg.

In nature, Cd is always associated with zinc ores (ZnS) due to its similarity with Zn. The Cd concentration in lake water and fishes has been reported by various workers (Malik et al., 2010; Bhupander et al., 2011). In the present study, Cd concentration in *L. rohita* was in the order of liver, kidney, gills and muscles. The highest level of Cd concentration was observed in liver, whilst the lower concentration was observed in muscles. In the case of *C. striatus,* highest level of Cd concentration was observed in liver and lowest concentration was observed in kidney. Similar results were also reported by Chandra Sekhar et al. (2004), Gupta et al. (2009) and Abdel-Baki et al. (2011).

Zinc is an essential trace element which plays an important role in the physiological and metabolic process of many organisms. However, in higher concentrations it can prove to be toxic. Zn showed protective effect against the Cd and Pb toxicity. The amount of Zn was highest in *L. rohita* than *C. striatus*. In *L. rohita* Zn content was higher in liver and kidney and lowest in gills. While in the case of *C. striatus*, Zn content was higher in gills followed kidney, liver and muscles. The results were in agreement with that of Gupta et al. (2009), Malik et al. (2010) and Ayejuyo et al. (2009) for *Clarias lazera*, but, were in disagreement with the work of Yang et al. (2007) for *G. nanensis* and *Ptychobarbus diposon*. Chandra Sekhar et al. (2004) reported that large fraction of Zn, Cd and Cu were associated with mobile fraction of sediment and showed grater bioaccumulation in fishes of Kolleru Lake. The concentration of heavy metals in liver and gills of *L. rohita* and *C. straitus* was found to be higher than the other organs, because liver acted as an important organ for storage and detoxification and gills acted as depot tissue. There was significant accumulation of metals in these organs, as was also reported by Yilmaz (2005), Malik et al. (2010) and Taweel et al. (2011).

Cu is an essential part of various enzymes necessary for the synthesis of hemoglobin (Sivaperumal et al., 2007); but at higher concentration it causes various health problems. In the present study, it has been observed that the concentration of Cu was higher in liver and followed by kidney, gills and muscles. While in the case of *C. striatus*, the concentration of Cu was highest in kidney followed by liver, muscles and gills. The same was also reported by Chandra Sekhar et al. (2004), Gupta et al. (2009), Abdel-Baki et al. (2011) and Laxmi Priya et al. (2011).

Cr plays an important role in carbohydrate (glucose) metabolism. The total amount of Cr was higher in *C. striatus* than in *L. rohita* .The concentration Cr was highest in liver and followed by gills and muscles. In the case of *C. striatus*, the highest accumulation of Cr was observed in muscle followed by gills, kidney and liver. Hg was the least accumulated metal in both fishes. The accumulation amount was slightly higher in *L. rohita* than in *C. striatus*. The same was also reported by Mackeviciene (2002), Malik et al. (2010) and Laxmi Priya et al. (2011). In the present study, values of all the heavy metals in the water and fish were within the permissible levels as per the codes of FAO (1983). The lake is very much suitable for fish culture.

Conclusion

The present study shows that the water quality parameter (temperature, pH, free CO_2, DO and hardness) were within the BIS standards for drinking water and fish culture. The concentration of heavy metals in the water of Kolleru Lake is under prescribed limits of BIS for drinking water. Liver was the highly metal accumulated organ, while muscle is lowest accumulated tissue of *L. rohita* and *C. striatus*. This is mainly important because muscle contributes major mass of flesh that is consumed as food. Heavy metal concentration in different tissues of fishes is below the limits of FAO. The fish of Kolleru Lake is suitable for human consumption.

REFERENCES

Abdel-Baki AS, Dkhil MA, Al-Quraishy S (2011). Bioaccumulation of some heavy metals in tilapia fish relevant to their concentration in water and sediments of Wadi Hanifah, Saudi Arabia. Afr. J.

Biotechnol. 10(13):2451-2541.
Amaraneni SR (2006). Distribution of pesticides, PACs and heavy metals in prawn ponds near Kolleru Lake wetland, India. Environ. Int., 32(3):294-302.
APHA (2012). Standard methods for examination of water and waste water. 22nd, (Edn.) APHA, AWWA, WPCF, Washington DC, USA.
Ashraj W (2005). Accumulation of heavy metals in kidney and heart tissues of *Epinephelus microdon* fish from the Arabian Gulf. Environ. Monit. Assess. 101(1-3):311-316.
Ayejuyo OO, Olowu RA, Megwa KC, Denloye AAB, Owodehinde FG (2003). Trace metals in *Clarias lazera,* water and sediments from Majidum River, Ikorodu, Nigeria. Res. Communic. Fish. 1:27-31.
Bhupander K, Mukherjee DP, Sanjay K, Meenu M, Dev Prakash SK, Sharma CS (2011). Bioaccumulation of heavy metals in the muscle tissue of fishes from selected aquaculture ponds in east Kolkata wetlands. Ann.. Biol. Res. 2(5):125-134.
BIS (1991). Drinking water specification IS: 10500:1991.New Delhi, India, Bureau of Indian Standards
Canli M, Ay O, Kalay M (1998). Level of heavy metals (Cd, Pb, Cu and Ni) in tissues of *Cyprinus carpio, Barbus capito* and *Chondrostoma regium* from the Seyhan River. Tur. J. Zool. 22(3):149-157.
Chandra Sekhar K, Chary NS, Kamal CT, Suman Raj DS, Srinivas Rao A (2004). Fractional studies and bioaccumulation of sediment bound heavy metals in Kolleru Lake by edible fish. Environ. Int. 29(7):1001-1008
Conacher HB, Page BD, Ryan JJ (1993). Industrial chemical contamination of foods, (Review). Food Addit. Contam. 10(10):129 143.
Dirilgen N (2001). Accumulation of heavy metals in freshwater: Assessment of toxic interactions. Turk. J. Chem. 25(3):173-179.
FAO (1983). Compilation of legal limits for hazardous substance in fish and fishery products (Food and Agriculture Organization), FAO Fishery Circular no: 464. pp. 5-100.
Farombi EO, Adelowo OA, Ajimoko YR (2007). Bioaccumulation of oxidative stress and heavy metal levels as indicators of environmental pollution in African Cat fish (*Clarias gariepinus*) from Nigeria Ogun River. Int. J. Environ. Public Health 4(2):158-165.
Forstner U, Wittmann GTW (1983). Metal pollution in aquatic environment, Berlin, Springer-Verlag. pp. 30-36.
Gupta A, Rai DK, Pandey RS, Sharma B (2009). Analysis of some heavy metals in the riverine water, sediments and fish from Ganges at Allahabad, Environ. Monit Assess. 157(1-4):449-58.
Jain CK, Sharma MK (2009). Distribution of trace metals in the Hindon River System, India. J. Hydrol. 253:81-90.
Laxmi Priya S, Senthilkumar B, Hariharan G, Paneer Selvam A, Purvaja R, Ramesh R (2011) Bioaccumulation of heavy metals in mullet (*Mugil cephalus*) and Oyster (*Crassostrea madrasensis*) from Pulicat lake, south east cost of India, Toxicol. Ind. Health 27(2):117-126.
Mackeviciene G (2002). Bioaccumulation of heavy metals in noble crayfish, (*Astacus astacus* L.) tissue under aquaculture conditions, Ekologia (Vilnius) 2:79-82.
Malik N, Biwas AK, Qureshi TA, Borana K (2010). Bioaccumulation of heavy metals in fish tissues of a freshwater lake of Bhopal. Environ. Monit. Assess. 160:267-276.
Mohan D, Gaur A, Choudhary D (2007) Study of Limnological and microbiology of Naya Talab Jodhpur (Rajasthan). Proceedings of DAE-BRNS, National Symposium on Limnology (NSL), February 19-21, Udaipur. pp. 64-68
Rao, S.R and Rao,P.R (2007) .Heavy metal concentrations in the sediments from Kolleru Lake, India, *Indian* J Environ Health. 43(4): 148-153.
Sarkar SK, Saha M, Takada H, Bhattacharya A, Mishra P, Battacharya B (2007). Water quality management in the lower stretch of the river Ganges, east coast of India: An approach through environmental education. J. Clean Prot. 15:1559-1567.
Sivaperumal P, Sarkar TV, Nair Viswanathan PG (2007). Heavy metal concentration in fish, shellfish and fish products from international markets of India vis-à-vis international standards. Food Chem. 102:612-618.
SPSS (2001). Statistical packages for social sciences version-11 Illinois, USA: SPSS Inc (2001).
Taweel A, Shuhaimi-Othman M, Ahmad AK (2011).Heavy metal concentration in different organs of tilapia fish (*Oreochromis niloticus*) from selected areas of Bangi, Selangor, Malaysia. Afr. J. Biotechnol. 10(55):11562-11566.
Uysal K, Emre Y, Kose E (2008). The determination of heavy metal accumulation ratio in muscle, skin and gills of some migratory fish species by inductively coupled plasma-optic emission spectrometry in Beymelek Lagoon, Antalya/Turkey. Microchem. J. 90(1):67-70.
Velez D, Montora R (1998). Arsenic speciation in manufactured seafood products: a review. J. Food Prot. 61(9):1240-1245.
Vinodhini R, Narayanan M (2008). Bioaccumulation of heavy metals in organs of fresh water fish *Cyprinus carpio* (Common carp). Int. J. Environ. Sci. Technol. 5(2):179-182
Vosyliene MZ, Janakaite A (2006). Effect of heavy metal model mixture on rainbow trout biological parameters. Ekologia 4:12-14.
Vutukuru SS (2005). Acute effect of Hexavalent chromium on survival, oxygen consumption, hematological parameters and some biochemical profile of the Indian Major carp, *Labeo rohita*. Int. J. Environ. Res. Public Health 2(3):456-462.
WHO (1985). Guidelines for drinking water quality, Vol. 8. Recommendations, WHO, Geneva
Wetzel RG, Likens GE (2006). Limnological analysis (3rd edition). New York, Springer. pp. 1-391
Yang TB, Jiang G, Xin, X (2007). Accumulation features of organ chlorine pesticides and heavy metals in fish from high mountain lakes and Lhasa River in the Tibetan Plateau environment. Turk. J. Vet. Anim. Sci. 33:151-156.
Yilmaz AB (2005). Comparison of heavy metals of grey mullet (*Mugil cephalus* L.) and sea bream (*Sparus aurata* L.) caught in Iskenderum Bay (Turkey). Tur. J. Vet. Anim. Sci. 29:257-262.

Levels of heavy metals in fish obtained from two fishing sites in Akwa Ibom State, Nigeria

Edet Okon Akpanyung, Udoudo Moses Ekanemesang, Ekam Ime Akpakpan and Ngozi Obiageli Anadoze

Department of Biochemistry, Faculty of Basic Medical Science, University of Uyo. P. M.B. 1017, Uyo, Akwa Ibom State, Nigeria.

Akwa Ibom State is located in the Niger Delta region of Nigeria and a substantial amount of the fishes consumed in the area are bought at Ifiayong and Ibaka beaches. This study was carried out to analyze heavy metals (Zn, Cu, Cd, Pb, Cr and As) in kidney, heart, gills and liver of silver catfish (*Chrysichthys nigrodigitatus)* from these locations. Analysis was performed using Atomic Absorption Spectrophotometry (AAS). The results show that the levels of Zn and Cu were significantly higher than maximum recommended levels ($p<0.05$) at both locations and were significantly higher in Ifiayong than Ibaka in all the organs ($p<0.05$). Concentrations of Pb, As, Cr and Cd were significantly higher ($P <0.05$) in Ibaka than Ifiayong in all the organs of fish analyzed. The order of accumulation of metals in the organs were as follows; Ibaka: heart/gills: Zn > Cu > Cr > Pb > As > Cd; kidney: Zn > Cu > Cr > Pb > Cd > As; liver: Zn> Cu > Cr > As >Pb > Cd. Ifiayong: gills/heart: Zn> Cu > Cr > Pb > Cd > As; kidney: Zn > Cu > Cr > As > Cd = Pb and Liver: Zn > Cu > Cr > Pb > Cd = As. In general, these results show that the levels of Zn, Cu, Pb, Cr and As in the fish bought at Ibaka were above recommended levels while those bought at Ifiayong (with the exception of Zn and Cu) were significantly lower ($p<0.05$) than standards set by WHO/FAO/UNEP. This study, therefore, suggests the need for regular monitoring and assessment of fishing sites in this region for heavy metal contamination to protect the health of consumers.

Key words: *Chrysichthys nigrodigitatus*, heavy metals, Ibaka, Ifiayong, Niger Delta region, silver catfish.

INTRODUCTION

Heavy metal contamination of aquatic ecosystem constitutes a major public health problem (Ashraf et al., 2006). Consequently, researchers and regulatory agencies carry out periodic assessment of the aquatic ecosystem to ascertain the quality of fish consumed by the populace. Heavy metals such as iron, copper, chromium, and zinc are essential for metabolic activities, but become toxic at higher concentrations, whereas lead and cadmium have no beneficial roles in living organisms but are toxic (Malik and Zeb, 2009). Nutritionally, fish consumption is widely encouraged due to its high content of omega-3 polyunsaturated fatty acids: eicosapentanoic acid (EPA), docosahexanoic acid (DHA) and protein (Clarkson, 2002). However, the presence of toxic heavy metals can counteract the positive effects of omega-3 fatty acids present in fish. There is evidence that metals accumulate in tissues of fish; therefore, levels of heavy metals measured in tissues of fish can reflect previous exposure to these pollutants (Ashraf, 2005). Heavy metals cause damage by enhancing the production of free radicals in several organs (brain, liver, kidney, heart) and interfering with cellular mechanisms against oxidation, leading to oxidative stress which has been implicated in the aetiology of several diseases (Castro-Gonzalez and Mendez-Armenta, 2008). Exposure to crude oil and its derivatives can act as a mediator in free radical generation and can induce a variety of toxic symptoms in experimental animals (Achuba and Osakwe, 2003). Heavy metal toxicity, persistence and tendency to accumulate in water and sediment, render them severe poisons for all living organisms in high concentrations (Has-Schon et al., 2006).

Many authors have reported increased concentrations of toxic metals in fish caught in Nigerian waters. For example, Farombi et al. (2007) reported high concentration of metals (Zn, Cu, Cd, Pb, As) above WHO/FAO standards in Ogun river. Alinnor (2005) reported that Aba River is polluted due to the discharge of waste from six (6) industries located close to it. High concentrations of heavy metals have also been reported in Olifant River, South Africa (Avenant-Oldewage and Marx, 2000). The present study assessed the quality of silver cat fish (*Chrysichthys nigrodigitatus*) obtained from Ibaka and Ifiayong beaches, Nigeria, by determining heavy metal concentrations in various organs (liver, heart, kidney and gills).

MATERIALS AND METHODS

Study area

The sampling areas for this study are indicated in Figure 1. Ibaka beach is located on the banks of Mbo river which is one of the major rivers in Akwa Ibom State, Nigeria. Mbo River lies within latitude 4° 39' North and longitude 8° 19' East on the southeastern Nigerian coastline. It empties into the Cross River Estuary at Ibaka in the Bight of Bonny and plays a major role in the fisheries resources of Akwa Ibom State (Essien-Ibok et al., 2010). Presently, the river is receiving serious developmental attention by both the state government (the proposed establishment of an international seaport at its banks) and the Mobil oil company that has discovered new oil deposits within the rivers catchment area (Essien Ibok et al., 2010). Conversely, Ifiayong beach is located at latitude 5° 03' North and longitude 8° 01' in Ifiayong, Uruan, Akwa Ibom State. To the best of our knowledge, Ifiayong river is devoid of oil exploration activities but may be susceptible to fertilizer run-offs from fertilizers applied to soils as well as waste discharges.

Collection of fish samples

Samples of *C. nigrodigitatus* were bought directly from fishermen at Ibaka and Ifiayong beaches. They were transported to the laboratory in ice-cold containers (0 - 4°C). The samples were authenticated by Dr J. P. Udoh, Department of Fisheries and Aquaculture, University of Uyo, Uyo. The live fishes (10 from each location) weighing between 200 and 650 g were dissected and the liver, heart, gills and kidney were removed and weighed. The organs were frozen and stored at -20°C (in a deep freezer) until required for analysis.

Heavy metal analysis

All fresh organs from the fish were dried in an oven at 105°C for 24 h to constant weight and milled with a mortar and pestle. The samples were transferred into dry labeled plastic containers and stored in a desiccator until required for digestion. A procedure similar to that described by Poldoski (1980) was used to digest the samples. This involved digesting 1 g portion of the ground samples with 10 ml HNO_3/2 ml $HClO_4$ and heating on a hot plate for one hour. After complete digestion, the residue was dissolved and diluted with 0.2 % v/v HNO_3 to 20 ml. The concentrations of metals (Zn, Cu, Cd, Pb, Cr and As) were determined in an Atomic Absorption Spectrophotometer (Bulk Scientific 205)

Statistical analysis

Data were expressed as mean ± standard deviation (SD) and analyzed with the SPSS 18.0 software. Independent T-test was used for the evaluation of data between Ibaka and Ifiayong beaches. P values <0.05 were regarded as statistically significant.

RESULTS AND DISCUSSION

Four different organs (gills, liver, kidney and heart) were analyzed for six metals (Cd, Pb, As, Cr, Zn and Cu) and the results obtained are presented in Table 1. The concentrations (ppm) of Zn and Cu were significantly higher ($P< 0.05$) in all the organs of fish obtained from Ifiayong than Ibaka with the exception of Zn concentrations in the gills which showed no significant difference at the two locations. The characteristic similarity in metal accumulation in liver, gills, kidney and heart in both locations were in the order Zn > Cu > Cr. There was slight variation in the pattern of bioaccumulation of the other metals (As, Cd and Pb) in the organs. This may be attributed to difference in the affinity of metals to fish tissue (Jezierska and Witeska, 2006), metabolic disposition of the organ (Karadede and Unlu, 2000), route of absorption (Ney and Vanhassel, 1983) and age of the fish (Khan et al., 2012).

The accumulation of Zn and Cu in the liver has been attributed to the role of this organ in the storage, metabolism and detoxification of heavy metals (Langston et al., 1998). Also metallothionine which is abundant in fish liver (Atli and Canli, 2007) has affinity for Cu and Zn and therefore may concentrate these metals (Sakulsak, 2012). Jezierska and Witeska (2001) reported that the levels of heavy metals in fish liver can be used to monitor the extent to which water is polluted by these elements. This is due to the fact that the concentration of metals in fish liver is proportional to those in the aquatic environ-

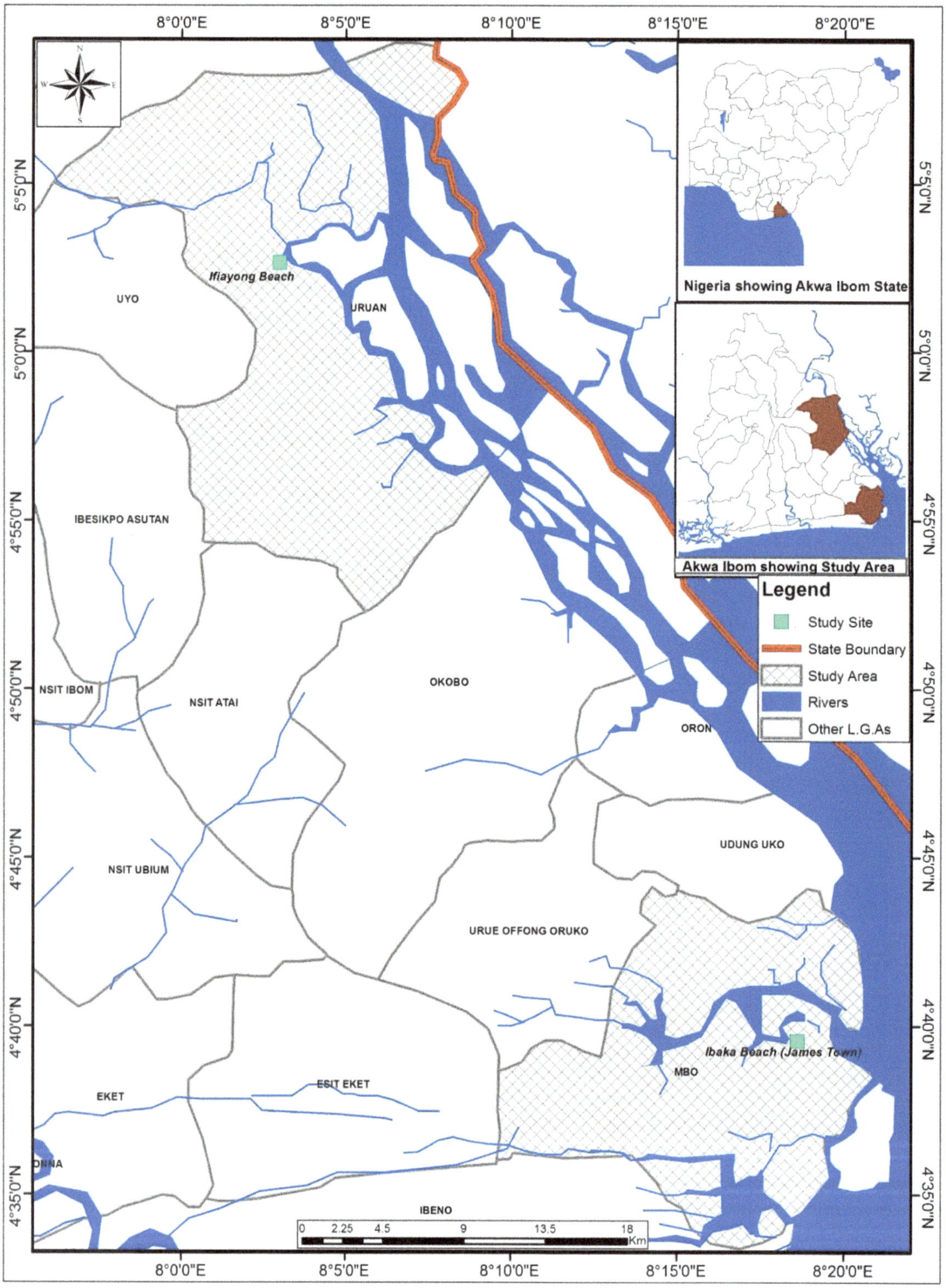

Figure 1. Study Area showing Ibaka beach and Ifiayong beach with inserted maps of Nigeria and Akwa Ibom State.

ment. The high mean concentration of Cu and Zn obtained in this study correlates with the report of other authors (Farombi et al., 2007; Etesin and Benson, 2007; Avenant-Oldewage and Marx, 2000). Zn is neurotoxic at high concentrations and causes neuronal cell death in a dose dependent manner (Chen-Jung and Su-Lan, 2003).

Table 1. Comparison of levels of heavy metals (Zn, Cu, Cd, Pb, As and Cr) (ppm) in organs of *Chrysichthys nigrodigitatus* from Ibaka and Ifiayong Rivers in Akwa Ibom State, Nigeria.

Location	Organ	Cd	Cr	Pb	As	Zn	Cu
Ibaka	Gills	0.0178±0.0055*	0.1109 ± 0.0222*	0.0193 ± 0.0017*a	0.0180± 0.0098	151.1853 ± 0.4161*	56.664± 8.4159 *b
Ifiayong		0.0129 ± 0.0016	0.0018 ± 0.0006	0.0020 ± 0.0005	0.0016 ± 0.0007	151.3052 ± 0.6325 *ab	77.5724 ± 2.7408*
Ibaka	Heart	0.0103 ± 0.0057 a	0.1360 ± 0.0143*	0.0170 a± 0.0034	0.0108± 0.0048	152.3543± 9.1324*	81.3210± 3.0650a
Ifiayong		0.0138 ±0.0005	0.0013 ± 0.0005	0.0013 ± 0.0005	0.0013 ±0.0005	170.4480 ±4.3357*a	88.8493 ±3.5911
Ibaka	Kidney	0.0229± 0.0089*	0.1334± 0.0180*	0.0311± 0.005*	0.0174 ± 0.0082	145.5368± 10.8170*	64.6692± 9.6776*
Ifiayong		0.01156 ± 0.0025	0.0021 ± 0.0019	0.0021 ± 0.0030	0.0021±0.0110*	208.6261 ± 13.8601*	80.4076 ±10.8079*
Ibaka	Liver	0.0109 ± 0.0044	0.0436± 0.0510	0.0129± 0.0032	0.0165± 0.0215	223.0013± 23.4667	81.3645 ± 5.0634
Ifiayong		0.0178 ± 0.0129	0.0010 ± 0.0000	0.0058 ± 0.0066	0.0010 ±0.0000	250.6429 ±16.7823	93.3513 ±6.7834

Values are expressed as mean ± S.D of 10 fishes. *Significantly different from liver at P< 0.05; a = significantly different from kidney at P<0.05; b = significantly different from heart at P<0.05.

The harmful toxicity of Copper is largely attributed to its cupric (Cu^{2+}) form which is commonly found in the species (Olaifa et al., 2004). The permissible limits of Zn and Cu in fish are 5.0 and 0.5 ppm respectively (UNEP, 1986).

The gills also accumulate high concentrations of Zn and Cu. Metal ions are first absorbed through the gills because they are directly in contact with the contaminated medium and also have thinnest epithelium when compared to other organs (Bebianno et al., 2004). Also, the kidney showed high concentration of Zn which according to Jaffer and Pervaiz (1989) could be based on specific metabolic process and co-enzyme catalyzed reactions involving Zn in the kidneys.

Zn and Cu concentrations were found to be higher than the WHO/UNEP/FAO permissible limits of 5.0 and 0.5 ppm respectively, (UNEP, 1986) in fish obtained from Ibaka and Ifiayong

In the present study, the level of Cr was found to be significantly higher (P<0.05) in fish obtained from Ibaka than Ifiayong in all the tissues except liver. This observation might be due to activities such as oil exploration. The health hazards associated with exposure to chromium are dependent on its oxidation state. The metal form (chromium) is of low toxicity. The hexavalent form is toxic and is known to be carcinogenic and teratogenic. Adverse effects of the hexavalent form on the skin may include ulcerations, dermatitis, and allergic skin reactions (Velma et al., 2009). Low doses of Cr produce marked inhibition of glucose reabsorption as a result of damage to convoluted proximal tubules (Stine and Brown, 2006). The mean concentration of Cr in fish from Ibaka was within the range reported by Nyirenda et al. (2011) (0.13 ± 0.01 ppm-liver and 0.11 ± 0.03 ppm-kidney) in catfish caught at Modila Dam, South Africa. The mean concentration of Cr at Ibaka was above limits specified by US EPA (0.1ppm) (US EPA, 1995) while that of Ifiayong were below the limits.

Natural as well as anthropogenic sources of Cd which includes industrial effluents as well as fertilizer and sewage sludge to farmland could increase environmental levels of Cd (Agency for Toxic Substance and Disease Registry (ATSDR), 2012). Our results show that Cd concentration was significantly higher in all the organs in fish from Ibaka than Ifiayong. This could be related to industrial activity of Mobil Oil Producing Company along Ibaka River. The relative order of bioaccumulation in the organs were as follows: Ibaka; kidney > gills > liver > heart while Ifiayong: liver > kidney > gills > heart. Generally, the mean concentration of Cd at the two locations was lower than the permissible limits of 0.05 ppm by International Program on Chemical Safety (WHO, 1992). The mean concentrations of Cd in the organs of fish at Ifiayong and Ibaka were similar to the values reported by Nyirenda et al. (2011). The Cd concentrations in Ibaka and Ifiayong were lower than that reported in the work of Farombi et al. (2007), Cogun et al. (2006) and Ashraf (2005). The results further showed that the mean concentration of As and Pb in fishes bought at Ibaka was significantly higher (p< 0.05) than that of Ifiayong in all the organs. The mean concentration of Pb and As in the gills and kidney samples of fish in both locations ranged between 0.0174-0.0311 ppm. The two organs contained the highest concentration of these metals compared to other organs. The results showed that kidney possesses the ability to concentrate heavy metals; this may be due to the presence of metal binding proteins (Kargin and Cogun, 1999). Also the high concentrations of metals in the gills may be due to its direct contact with the contaminated medium (Bebianno et al., 2004). Has-Schon et al. (2006) also reported concentrations of As to be within the range (0.001-0.019 ppm) in fish obtained from river Neretva (Croatia). In contrast to the present study, high concentrations of As had been reported by some authors (Nyirenda et al., 2011; Farombi et al., 2007; Usero et al., 2003). Our findings on mean concentration of Pb was in line with the results of Falco et al. (2006) who reported lead concentrations in various edible marine species which varied from 0.002-0.21 µ/g fw. The mean concentration of As and Pb in both locations were below the permissible limits of 0.01 and 0.29 ppm respectively (FAO/WHO, 2005; WHO, 1995). However, ATSDR (2007) states that there is no safe concentration of lead. Lead causes toxicity by replacing zinc in heme synthesis and inhibiting the function of heme-synthesizing enzymes (Goyer and Clarkson, 2001). National Academy of Science/National Research Council (1993) reported that substitution of calcium by lead resulted in toxicity of several vital enzyme systems in the central nervous system. This toxicity impaired the development and function of enzymes involved in the production and transport of neurotransmitters. Victims of Pb intoxication have been reported to manifest various forms of neurological syndrome such as lead palsy and encephalopathy, especially in children (Klaassen, 1995). Some of the clinical manifestations include muscular weakness and fatigue which are more pronounced in the fingers, wrist, toes and forearm; clumsiness, ataxia, headache, insomnia and irritability. Arsenic causes coagulation of proteins, forms complexes with coenzymes and inhibits the production of adenosine triphosphate (ATP) during respiration (Institute of Environmental Conservation and Research, 2000). It is possibly carcinogenic and high-level exposure can cause death (Ogwuegbu and Ijioma, 2003; USDOL, 2004).

Conclusion

This study shows that *C. nigrodigitatus* obtained from Ibaka and Ifiayong beaches were at various levels of contamination by heavy metals. The concentrations of these metals in the fish could endanger public health since this species of fish is most preferred by consumers in this environment. Metals with concentrations below the permissible limits may cause harm due to their synergistic effects. Therefore, regular monitoring of these rivers is desirable. Also, effective methods of waste disposal should be adopted to prevent agricultural runoff into these rivers, preserve healthy aquatic environment and ultimately protect the health of the humans who consume the fish.

Conflict of Interests

The author(s) have not declared any conflict of interests.

REFERENCES

Achuba FI, Osakwe S (2003). Petroleum-induced free radical toxicity in African cat fish (*Clarias gariepinus*). Fish. Physiol. Biochem. 29:97-103.

Agency for Toxic Substance and Disease Registry, ATSDR (2007). Toxicological profile for lead, U.S. Department of Health and Humans Services, Public Health Service, Centers for Disease Control, Atlanta, GA. pp. 403-413.

Agency for Toxic Substance and Disease Registry, ATSDR (2012). Toxicological profile for cadmium (Public Health Statement). U.S. Department of Health and Human Services, Public Health Service, Centers for Disease Control, Atlanta, GA. pp. 1-10.

Alinnor IJ (2005). Assessment of elemental contaminants in water and fish samples from Aba River. Environ. Monit. Asses. 102:15-25.

Ashraf W (2005). Accumulation of heavy metals in kidney and heart tissues of *Epinephelus microdon* fish from the Arabian Gulf. Environ. Monit. Assess. 101:311-315.

Ashraf W, Seddigi Z, Abulkibish A, Khalid M (2006). Levels of selected metals in canned fish consumed in Kingdom of Saudi Arabia. Environ. Monit. Assess.117:271-279.

Atli G, Canli M (2007). Natural occurrence of metallothionin-like proteins in liver tissues of four fish species from the Northeast Mediterranean sea. Water Environ. Res. 79(9):958-963.

Avenant-Oldewage A, Marx HM (2000). Bioaccumulation of Cr, Cu and Fe in the organs and tissues of *Clarias gariepinus* in the Olifants River, Kruger Natural Park. Water South Afr. 26:4-6.

Bebianno MJ, Geret FP, Serafim MA, Coelho MR, Gnassia-Barelli M, Romeo M (2004). Biomarkers in *Ruditapes decussatus*: A potential bioindicator species. Biomarker. 9:305-330.

Castro-González MI, Méndez-Armenta M (2008). Heavy metals: implications associated to fish consumption. Environ. Toxicol. Pharmcol. 26:263-271.

Chen-Jung C, Su-Lan L (2003). Zinc toxicity on neonatal cortical Neuron: involvement of glutathione chelate. J. Neurochem. 85:443-453.

Clarkson WT (2002). The three modern faces of mercury. Environ. Health Perspect. 110(1):11-23.

Cogun HY, Yuzereroglu TA, Firat O, Gok G, Kagrin F (2006). Metal concentration in fish species from the Northeast Mediterranean Sea. Environ. Monit. Assess. 121:431-438.

Essien-Ibok MA, Akpan AW, Udo MT, Chude LA, Umoh IA, Asuquo IE (2010). Seasonality in the physical and chemical characteristics of Mbo River, Akwa Ibom State, Nigeria. Nigerian J. Agric. Fd. Env. 6:60-72.

Etesin MU, Benson UB (2007). Cadmium, copper, lead and zinc tissue levels in Bonga Shad (*Ethmalosa fimbriata*) and Tilapia (*Tilapia guineensis*) caught from Imo River, Nigeria. Am. J. Food Technol. 2(1):48-54.

Falcó G, Llobet JM, Bocio A, Domingo JL (2006). Daily intake of arsenic, cadmium, mercury, and lead by consumption of edible marine species. J. Agric. Fd. Chem. 54:6106-6112.

FAO/WHO (2005). Expert committee on food additives, arsenic. Available at http://www.inchem.org/documents/jecfa/jeceval/jec159.htm.Accessed 27 August, 2013.

Farombi EO, Adelowo OA, Ajimoko YR (2007). Biomarkers of oxidative stress and heavy metal levels as indicators of environmental pollution in African catfish (*Clarias gariepinus*) from Nigeria Ogun River. Int. J.

Environ. Res. Pub. Hlth. 4:158-165.
Goyer RA, Clarkson TM (2001). Toxic effects of metals. In: Klaassen, C.D., ed. Casarett and Doull's Toxicology. New York: McGraw-Hill, pp. 811-868.
Has-Schön E, Bogut E, Strelec I (2006). Heavy metal profile in five fish species included in human diet, domiciled in the end flow of River Neretva (Croatia). Arch. Environ. Contam. Toxicol. 50:545-551.
Institute of Environmental Conservation and Research, INECAR (2000). Position paper against mining in Rapu-Rapu. Published by INECAR, Ateneo de Naga University, Philippines. www.adnu.edu.ph/Institutes/Inecar/pospaper1.asp. Accessed 29 August, 2013.
Jaffar J, Pervaiz S (1989). Investigation of multiorgan heavy metal content of meat of selected dairy, poultry, fowl and fish species in Pakistan. J. Sci. Ind. Res. 32:175-177.
Jezierska B, Witeska M (2001). Metal toxicity to fish, Wydawnictwo Akademii Podlaskiej, Siedlce. p.318.
Jezierska B, Witeska M (2006). The metal uptake and accumulation in fish living in polluted waters, soil and water. Pollu. Monit. Protec. Remedia. 69:107-114.
Karadede H, Unlu E (2000). Concentration of some heavy metals in water, sediment and fish species from the Ataturk dam lake (Euphrates), Turkey. Chemosphere 41:1371-1376.
Kargin F, Cogun HY (1999). Metal interactions during accumulation and elimination of Zn and Cd in tissues of the fish water fish (*Tillapia nilonitica*). Bull. Environ. Toxicol. 63:511-519.
Klaassen CD (1995): Heavy metals. In: Hardman J. and Limbird L. (ed) Goodman and Gilman's Pharmacological Basis of Therapeutics 9th ed. McGraw Hill. New York. pp. 1049-1073.
Khan B, Khan H, Muhammad S, Khan T (2012). Heavy metal concentration trends in three fish species from Shah Alam River (Khyber Pakhtunkwa Province, Pakistan). J. Nat. Env. Sci. 3:1-8.
Langston WJ, Bebianno MJ, Burt GR (1998). Metal handling strategies in molluscs. In: Langston WJ, Bebianno MJ. (Ed), Metal Metabolism in Aquatic Environments. London: Chapman and Hall. London. pp. 449-460.
Malik RN, Zeb N (2009). Assessment of environmental contamination using feathers of *Bubulcus ibis* as a biomonitor of heavy metal pollution Pakistan. Ecotoxicol. 8:522-536.
NAS/NRC-National Academy of Sciences/National Research Council (1993). Measuring lead exposure in infants, children, and other sensitive populations. Washington, DC. pp. 31-98.
Ney JJ, Vanhassel JH (1983). Sources of variability and accumulation of heavy metals by fish in a roadside stream. Arch. Envi. 12:701-706.
Nyirenda M, Itumeleng PD, Dzoma BM, Motsei LE, Ndou RV, Bakunzi FR (2011). Heavy metal levels in water, catfish (*Clarias Gariepnus*) and African eagle fish *(Haliaeetus Vocifer)* specimens from the municipal waste water fed Modimola Dam outside Mafikeng City, North West Province, South Africa. Life Sci. J. 8(S2):47-52.
Ogwuegbu MO, Ijioma M (2003). Effects of certain heavy metals on the population due to mineral exploitation. In: International conference on scientific and environmental issues In: The population, environment and sustainable development in Nigeria, University of Ado Ekiti, Ekiti State, Nigerian. pp. 8-10.
Olaifa FG, Olaifa AK, Onwude TE (2004). Lethal and sub lethal effects of copper to the African Cat fish (*Clarias gariepinus*). Afr. J. Biomed. Res. 7:65-70.
Poldoski JE (1980). Determination of lead and cadmium in fish and clam tissue by Atomic Absorption Spectrophotometry with a molybdenum and lanthanum treated pyrolytic graphite atomizer. Anal. Chem. 52(7):1147-1151.
Sakulsak N (2012). Metallothionin: an overview on its metal homeostatic regulation in mammals. Int. J. Morphol. 30(3):1007-1012.
Stine KE, Brown TM (2006). Principles of Toxicology. 2nd Ed., CRC press. USA. p 240.
UNEP (1986). Assessment of the present state of pollution by cadmium, copper, zinc and lead in the Mediterranean Sea. Document UNEP/WG, 144/11.
United States Department of Labor (US DOL) (2004). Occupational safety and health administration (OSHA); Safety and health topics: Heavy metals, USDOL publication, Washington, D.C. www.osha.gov/SLTC/metalsheavy/index.html. Accessed 28 June, 2012.
US EPA, US Environmental Protection Agency (1995). Integrated risk information system. Environmental Protection Agency, Washington, DC.
Usero J, Izquierdo C, Morillo J, Gracia I (2003). Heavy metals in fish (*Solea vulgaris, Anguilla Anguilla* and *Liza aurata*) from salt marshes on the Southern Atlantic coast of Spain. Environ. Int. 29: 949-956.
Velma V, Vutukuru SS, Tchounwou PB (2009). Ecotoxicology of hexavalent chromium in freshwater fish: A critical review. Rev. Environ. Health 24(2):129-145.
WHO (1992). Cadmium. Environmental health criteria, 134. International program on chemical safety. Geneva. Switzerland. www.inchem.org/documents/ehc/ehc134.htm. Accessed 12 June 2014.
WHO (1995). Inorganic lead. Environmental health criteria, 165. International program on chemical safety. Geneva. Switzerland. www.inchem.org/documents/ehc/ehc165.htm. Accessed 12 June, 2014.

10

Marine biotoxins and its detection

Shelar Piyusha Suresh[1], S. Gauri[2], S. Desai[3], M. Kavitha[1] and P. Padmavathy[1]

[1]Department of Fisheries Environment, Fisheries College and Research Institute, Tuticorin, Tamil nadu- 628 008, India.
[2]Department of Aquaculture, College of Fisheries, Ratnagiri-415612, Maharashtra, India.
[3]Department of Fisheries Engineering, College of Fisheries, Ratnagiri-415612, Maharashtra, India.

The incidences of intoxication due to the consumption of marine foods have been increasing in recent years. This is due to the presence of biotoxins in foods of marine origin. The biotoxins will be accumulated in the marine foods due to the consumption of toxic biota of marine origin. When this contaminated food is taken by the humans or animals, those toxins will be transferred to them causing intoxication and lethality. Among these intoxications, most of them are caused by the harmful algal blooms (HAB). In order to avoid the harmful effects from marine biotoxins, it is necessary to have the proper knowledge. In this manuscript, the different types of biotoxins, source of intoxication, characteristics of toxins, detection and control measures are discussed in detail.

Key words: Harmful algal blooms, harmful algal blooms (HAB), ciguatara fish poisoning (CFP), paralytic shellfish poisoning (PSP), diarrhetic shellfish poisoning (DSP) blooming, detection.

INTRODUCTION

Microscopic planktonic algae of the world's oceans are critical food for filter-feeding bivalve shellfish (oysters, mussels, scallops and clams) as well as for the larvae of commercially important crustaceans and fin fish. Over the last several decades, countries throughout the world have experienced an escalating trend in the incidence of "harmful algal blooms" (HABs) (Anderson, 1989; Hallegraeff, 1993). HAB events are characterized by the proliferation and occasional dominance of particular species of toxic or harmful algae. When toxic algae are filtered from the water as food by shellfish, their toxins accumulate in those shellfish to levels that can be lethal to humans or other consumers. Another type of HAB impact occurs when marine fauna are killed by algal species that release toxins and other compounds into the water. HABs include species of microscopic, usually single celled eukaryotic plants that live in estuarine and marine waters. A "bloom" occurs when algae grow very quickly or "bloom" and accumulate into dense visible patches near the surface of the water (National Office for Marine Biotoxins and Harmful Algal Blooms, 1999)

During the past two decades, the frequency, intensity and geographic distribution of harmful algal blooms has increased, along with the number of toxic compounds found in the marine food chain. Different explanations for this trend have been given such as increased scientific awareness of toxic algal species, increased utilization of coastal waters for aquaculture, transfer of shellfish stocks from one area to another, cultural eutrophication from domestic, industrial and agricultural wastes, increased mobility of humic substances and trace metals from soil due to deforestation and/or by acid precipitation (acid rain), and unusual climatic conditions (Hallegraeff et al., 1995).

A poorly defined but potentially significant concern relates to sublethal, chronic impacts from toxic HABs that can affect the structure and function of ecosystems. Adult fish can be killed by the millions in a single outbreak, with long- and short-term ecosystem impacts (Okaichi et al., 1989; Kim et al., 1999). Likewise, larval or juvenile stages of fish or other commercially important species can experience mortalities from algal toxins (White et al., 1989). HABs also cause mortalities of wild fish, seabirds, whales, dolphins and other marine animals. Non-toxic blooms of algae can cause harm, often due to the high biomass that some blooms achieve, and the deposition and decay of that biomass, leading to anoxia. Chronic toxin exposure may have long-term consequences that are critical with respect to the sustainability or recovery of natural populations at higher trophic levels (Ramsdell et al., 2005). Only a few of the many thousands of species of algae are associated regularly with toxic or harmful algal blooms (National Office for Marine Biotoxins and Harmful Algal Blooms, 1999). Shellfish poisoning syndromes include Ciguatara fish poisoning (CFP) and paralytic (PSP), diarrhetic (DSP), ciguatara (CFP), neurotoxic (NSP) and amnesic (ASP) shellfish poisoning based on human symptoms.

TOXINS

Ciguatara fish poisoning (CFP)

Ciguatera poisoning in humans and domestic animals is caused by potent neurotoxins produced by benthic dinoflagellates including *Gambierdiscus toxicu*, *Prorocentrum concavum*, *Prorocentrum hoffmannianum*, *Prorocentrum lima*, *Ostreopsis lenticularis*, *Ostreopsis siamensis*, *Coolia monotis*, *Thecadinium* and *Amphidinium carterae*. In the tropics and subtropics toxic dinoflagellates living on coral reefs are eaten by small herbivorous fish grazing on coral which in turn are eaten by larger carnivores. The poisons move up the food chain into the organs of larger top-order predators such as coral trout, red bass, chinaman fish, mackerels and moray eels and cause ciguatera fish poisoning, CFP, in people who eat these fish (Kim, 1999; Klöpper et al., 2003; Leikin, and Paloucek, 1998).

Toxins produced: Ciguatoxin, Maitotoxin

CFP produces gastrointestinal, neurological and cardiovascular symptoms. Generally, diarrhea, vomiting and abdominal pain occur initially, followed by neurological dysfunction including reversal of temperature sensation, muscular aches, dizziness, anxiety, sweating and numbness and tingling of the mouth and digits. Paralysis and death have been documented, but symptoms are usually less severe although debilitating. Recovery time is variable, and may take weeks, months, or years. Rapid treatment (within 24 h) with mannitol reported to relieve some symptoms. There is no antidote, supportive therapy is the rule, and survivors recover. Absolute prevention of intoxication depends upon complete abstinence from eating any tropical reef fish, since there is currently no easy way to measure routinely ciguatoxin or maitotoxin in any seafood product prior to consumption, (Nielsen and Tonseth, 1991; Partensky and Sournia, 1986; Partensky et al., 1988; Partensky et al., 1991; Passow, 1991; Perez et al., 2001; Rafuse et al., 2004; Schnorf et al., 2002; Tangen, 1977; Taylor et al., 1995; Tillmann, 2004).

Chemical properties

Ciguatoxins are lipid-soluble polyether compounds consisting of 13 to 14 rings fused by ether linkages into a most rigid ladder-like structure (Figure 1). They are relatively heat-stable molecules that remain toxic after cooking and exposure to mild acidic and basic conditions. Ciguatoxins arise from biotransformation in the fish of precursor gambier toxins (Lehane and Lewis, 2000; Lehane, 2000).

In areas in the Pacific, the principal and most potent ciguatoxin is Pacific ciguatoxin-1 (P-CTX-1, mol. wt. 1112). Its likely precursor is gambiertoxin-4B (GTX-4B). The main ciguatoxins in the Pacific, P-CTX-1, P-CTX-2 and P-CTX-3, are present in fish in different relative amounts (Lehane and Lewis, 2000; Lehane, 2000). Caribbean (and Indian Ocean) ciguatoxins differ from Pacific ciguatoxins. Caribbean CTX-1 (C-CTX-1) is less polar than P-CTX-1. Structures of two Caribbean ciguatoxins (C-CTX-1 and C-CTX-2) were elucidated in 1998. The structures of more than 20 congeners of ciguatoxin were elucidated. Structural modifications were mainly seen in the both termini of the toxin molecules and mostly by oxidation (Naoki et al., 2001; Yasumoto et al., 2000). Multiple forms of ciguatoxin with minor molecular differences and pathogenicity were described. CTX-1 is the major toxin found in carnivorous fish and poses a human health risk at levels above 0.1 μg/kg fish (De Fouw et al., 1999). The energetically less favored epimers, P-CTX-2 (52-epi P-CTX-3), P-CTX-4A (52-epi P-CTX-4B) and C-CTX-2 (56-epi C-CTX-1) are indicated in parenthesis. 2, 3-Dihydroxy P-CTX-3C and 51-hydroxy P-CTX-3C have also been isolated from Pacific fish (Lewis, 2001). Various species of parrotfish have previously been reported to contain a toxin less polar than CTX-1, named scaritoxin. Judging from the reported

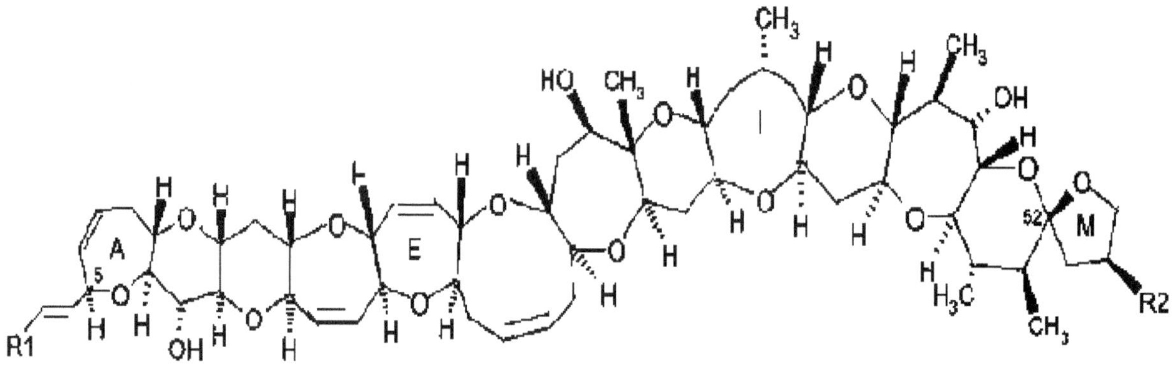

	R1	R2
P-CTX-1:	$^{1}CH_2OHCHOH$	OH
P-CTX-3 (P-CTX-2):	$^{1}CH_2OHCHOH$	H
P-CTX-4B (P-CTX-4A);	$^{1}CH_2CH$	H

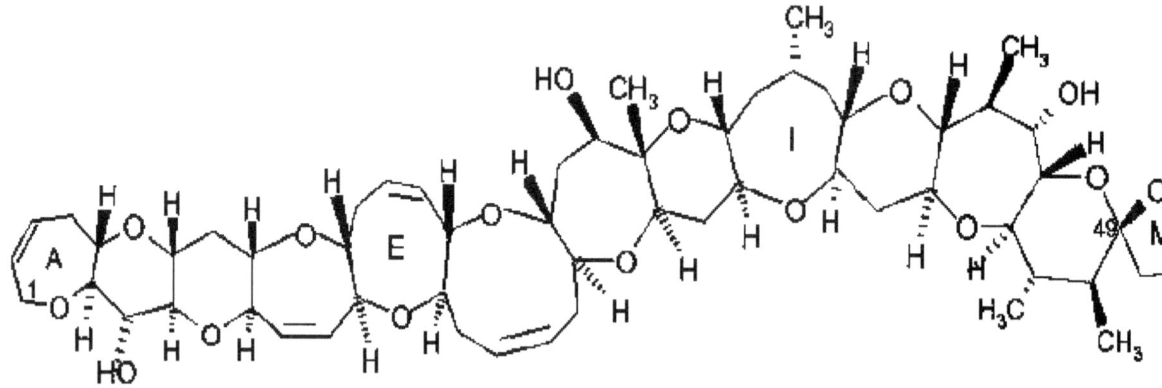

P-CTX-3C

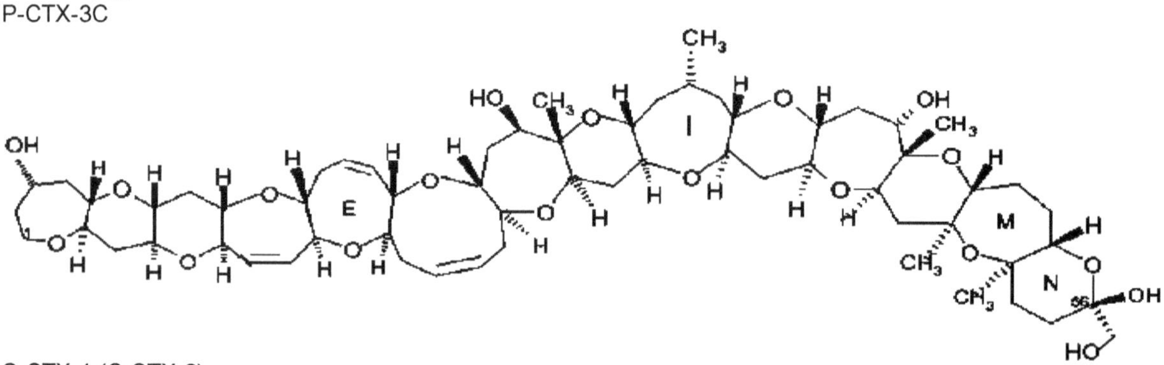

C-CTX-1 (C-CTX-2)

Figure 1. Structure of Pacific (P) and Caribbean (C) ciguatoxins (CTXs). Source: Yasumoto et al., 2000 and Lewis, 2001.

chromatographic properties, scaritoxin seems to correspond to a mixture of CTX-4A and CTX-4B (De Fouw et al., 1999).

Diarrhetic shellfish poisoning (DSP)

Diarrhetic shellfish poisoning (DSP) is produced by dinoflagellates in the genera *Dinophysis* and *Prorocentrum* like *Dinophysis*, *Prorocentrum*, *Dinophysis fortii*, *Dinophysis acuminata*, *Dinophysis norvegica*, *Dinophysis acuta* (Murakami et al., 1982; Lee et al., 1989; Jackson et al., 1993).

Toxin produced: Okadaic acid

DSP produces gastrointestinal symptoms, usually beginning within 30 min to a few hours after consumption of toxic shellfish. The illness, which is not fatal, is

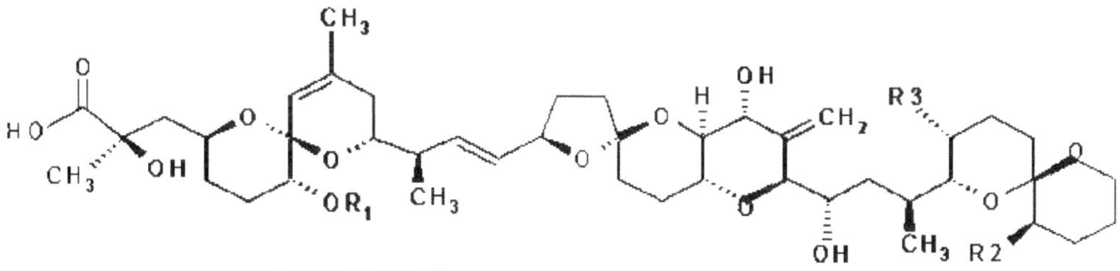

	R1	R2	R3
Okadaic acid (OA)	H	H	CH_3
Dinophysistoxin-1 (DTX1)	H	CH_3	CH_3
Dinophysistoxin-2 (DTX2)	H	CH_3	H
Dinophysistoxin-3 (DTX3)	acyl	CH_3	CH_3

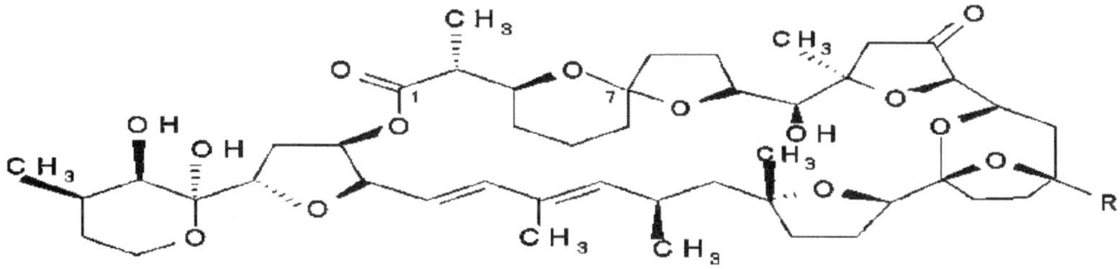

		R	C-7
Pectenotoxin-1	(PTX1)	CH_2OH	*R*
Pectenotoxin-2	(PTX2)	CH_3	*R*
Pectenotoxin-3	(PTX3)	CHO	*R*
Pectenotoxin-4	(PTX4)	CH_2OH	*S*
Pectenotoxin-6	(PTX6)	COOH	*R*
Pectenotoxin-7	(PTX7)	COOH	*S*

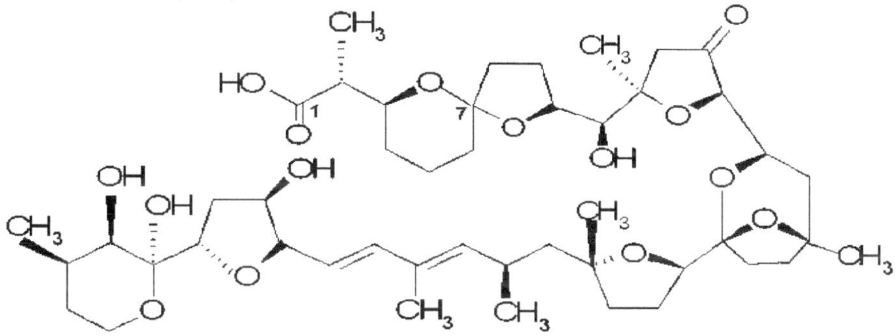

	C-7
Pectenotoxin-2 seco acid (PTX2SA)	R
7-epi-PTX2SA	S

Figure 2. Chemical structures of okadaic acid, dinophysistoxins and pectenotoxins
Source: Yasumoto et al., 2001.

characterized by incapacitating diarrhea, nausea, vomiting, abdominal cramps and chills. Recovery occurs within three days, with or without medical treatment, (Climent and Lembeye, 1993; Climent et al., 2001; Clement, 1999; Cohen, 1974; Cosper et al., 1989; Currie et al., 2000).

Chemical properties

The DSP toxins are all heat-stable polyether and lipophilic compounds isolated from various species of shellfish and dinoflagellates (Draisci et al., 1996a) (Figures 2 and 3). Although diarrhea is the most characteristic symptom of intoxication, several other effects may be of relevance and some of the toxins in the DSP complex (PTXs and YTXs) do not yield diarrhea at all (Van Egmond et al., 1993). Re-evaluation of their toxicity will probably lead to these toxins being removed from their classification as DSP toxins (Quilliam, 1998a). The different chemical types of toxins associated with the DSP syndrome comprise:

a) The first group, acidic toxins, includes okadaic acid

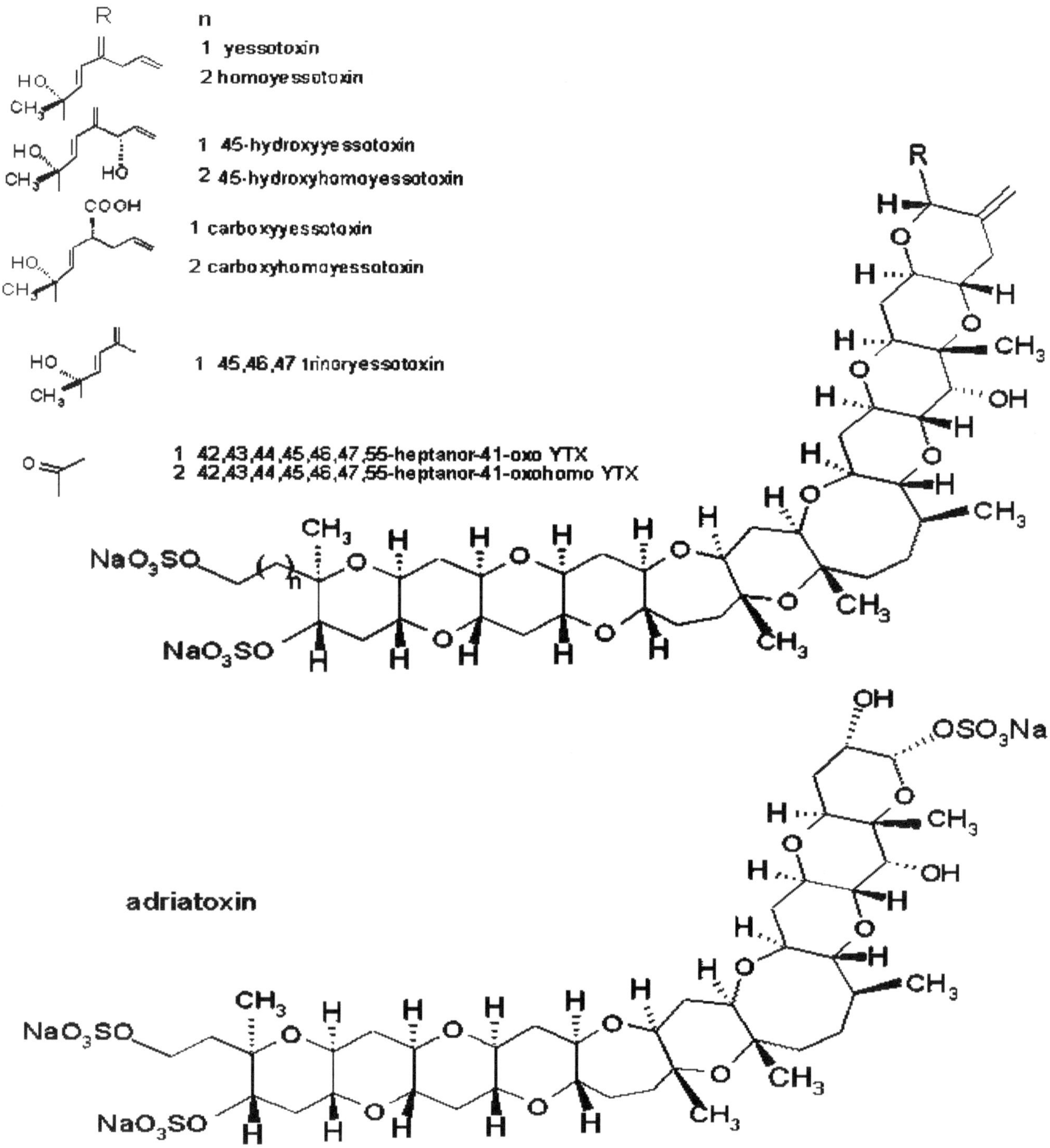

Figure 3. Chemical structures of yessotoxins and adriatoxin. Source: Ciminiello et al., 1998; 2002 and Yasumoto et al., 2001.

(OA) and its derivatives named dinophysistoxins (DTXs). Okadaic acid and its derivatives (DTX1, DTX2 and DTX3) are lipophilic and accumulate in the fatty tissue of shellfish. These compounds are potent phosphates inhibitors and this property is linked to inflammation of the intestinal tract and diarrhea in humans (Van Apeldoorn et al., 1998; Hallegraeff et al., 1995). OA and DTX1 are also tumor promoters in animal test systems (Draisci et al., 1996a; Van Egmond et al., 1993). DTX1 was first detected in *Dinophysis fortii* in Japan; DTX2 was identified in shellfish in Ireland during a DSP episode (Van Egmond et al., 1993). DTX2 was isolated also from a marine phytoplankton biomass mainly consisting of *Dinophysis acuta* (James et al., 1999). A new isomer of DTX2, named DTX2B, was isolated and identified in Irish mussel extracts (James et al., 1997). DTX3 originally described a group of DSP toxin derivatives in which saturated or unsaturated fatty acyl groups are attached to the 7-OH group of DTX1. More recently it has been shown that any of the parent toxins, OA, DTX1 and DTX2, can be acylated with a range of saturated and unsaturated fatty acids from C_{14} to C_{18} (Hallegraeff et al., 1995; Wright, 1995). In a report of an EU meeting, it was stated that chain length of the fatty acid can vary from $C_{14}$to C_{22} and that the number of unsaturation varying from 0 to 6. The most predominantly saturated fatty acid in DTX3 was palmitoyl acid (EU/SANCO, 2001). These acylated compounds also possess toxic activity. Since these compounds have only been detected in the digestive gland of contaminated shellfish, it has been suggested that they are probably metabolic products and not *de novo* products of toxin producing micro algae (Wright, 1995). Suzuki et al. (1998) demonstrated the transformation of DTX1 to 7-*O*-acyl-DTX1 (DTX3) in the scallop *Patinopecten yessoensis*. The ester bond in the acylated compounds can be hydrolyzed by heating in 0.5 M NaOH/90 percent methanol solution at 75°C for 40 min. The ester bond in DTX3 was also easily hydrolyzed by lipase and cholesterol esterase (EU/SANCO, 2001).

Two naturally occurring ester derivatives called diol esters were isolated from some *Prorocentrum* species. These diol esters did not inhibit phosphatase *in vitro*. However, it should be noted that these allylic diol esters may be somewhat labile and could be hydrolysed to yield the active parent DSP toxin (Hallegraeff et al., 1995). Draisci et al. (1998) reported the detection of another OA isomer and called it DTX2C. The structure of DTX2C is not yet elucidated. The compound was isolated from *Dinophysis acuta* collected in Irish waters.

b) The second group, neutral toxins, consists of polyether-lactones of the pectenotoxin group (PTXs). Ten (10) PTXs have been isolated until now and six out of these have been chemically identified; PTX1, -2, -3, -4, -6 and -7. Since PTX2 (PTX2,CH_3) is found in phytoplankton only (*Dinophysis fortii* in Japan and Europe) and never in shellfish, it is suggested that an oxidation occurs in the hepatopancreas of shellfish producing other PTXs (PTX1, CH_2OH; PTX3, CHO; PTX6, COOH) (Draisci et al., 1996a; Yasumoto et al., 2001; Van Apeldoorn et al., 1998). Sasaki et al. (1998) identified PTX4 and PTX7 as spiroketal isomers of PTX1 and PTX6, namely *epi*-PTX1 and *epi*-PTX6, respectively. Suzuki et al. (1998) demonstrated oxidation of PTX2 to PTX6 in scallops (*Patinopecten yessoensis*). Two new artifacts, PTX8 and PTX9, were also isolated but their structures are not yet elucidated. Daiguji et al. (1998) isolated two new pecteno toxins from the green shell mussel *Perna canalicus* from New Zealand and from *Dinophysis acuta* from Ireland and elucidated the structures as pectenotoxin-2-seco acid (PTX2SA) and 7-*epi*-pectenotoxin-2 seco acid (7-*epi*-PTX2SA), respectively.

c) The third group includes a sulphated compound called yessotoxin (YTX), a brevetoxin-type polyether, and its derivative 45-hydroxyyessotoxin (45-OH-YTX) (Draisci et al., 1996a; Van Egmond et al., 1993). Yessotoxin was first isolated from the digestive organs from scallops (*Patinopecten yessoensis*) in Japan (Ciminiello et al., 1999) and is believed to be produced by microalgae. The yessotoxins do not cause diarrhoea. Yessotoxin attacks the cardiac muscle in mice after intra peritoneal injection, while desulphated yessotoxin damages the liver (Van Egmond et al., 1993). In the digestive gland of Adriatic mussels (*M. galloprovincialis*) besides yessotoxin, two new analogues of yessotoxin, homoyessotoxin and 45-hydroxyhomoyessotoxin were identified by Ciminiello et al. (1997, 1999). Tubaro et al. (1998) also detected homoyessotoxin in *M. galloprovincialis* from the Adriatic Sea during a bloom of *Gonyaulax polyhedra* (*Lingulodinium polyedrum*). Satake et al. (1997) and Satake et al. (1999) isolated YTX and 45, 46, 47-trinoryessotoxin from cultured cells of the marine dinoflagellate *Protoceratium reticulatum*. The production of yessotoxins by *P. reticulatum* differed from strain to strain. Ciminiello et al. (1998) detected again a new analogue of YTX, adriatoxin (ATX), in the digestive glands of DSP infested Adriatic mussels collected in 1997 along the Italian coast (Emilia Romagna). In addition, four further analogues of yessotoxin, carboxyyessotoxin (COOH group on C_{44} of YTX instead of double bond), Carboxyhomoyessotoxin (COOH group on C_{44} of homoYTX instead of double bond) (Ciminiello et al., 2000a, b), 42,43,44,45,46,47,55-heptanor-41-oxo YTX and 42,43,44,45,46,47,55-heptanor-41-oxohomo YTX (Ciminiello et al., 2001, 2002) in Adriatic mussels (*Mytilus galloprovincialis*) were identified.

d) Unexplained human intoxication, with DSP symptoms, following the consumption of mussels from Killary, Ireland in 1995 was resolved by the isolation of a new toxin ($C_{47}H_{71}NO_{12}$), tentatively named Killary Toxin-3 or KT3 (Satake et al., 1998a).

NEUROTOXIC SHELLFISH POISONING (NSP)

Neurotoxic shellfish poisoning (NSP) is caused by toxins produced predominantly by *Gymnodinium* species like *Gymnodinium breve*, *Karenia brevis*. Several species of phytoplankton in New Zealand have been found to produce NSP toxins. These include *Gymnodinium c.f. breve, Gymnodinium c.f. mikimotoi* (which may include three separate species), *G. galatheanum* and a species of *Heterosigma* (Mackenzie et al., 1995a; Haywood, 1998). The identity of the causative agent in the 1993 NSP event in Northland is uncertain: both *Gymnodinium c.f. breve* and *Gymnodinium c.f. mikimotoi* were present in elevated numbers at the time (Chang, 1996; Mackenzie et al., 1995b).

Toxins produced: Brevetoxins

NSP produces an intoxication syndrome nearly identical to that of ciguatera. In this case, gastrointestinal and neurological symptoms predominate. In addition, formation of toxic aerosols by wave action can produce respiratory asthma-like symptoms. No death has been reported and the syndrome is less severe than ciguatera, but nevertheless debilitating. Unlike ciguatera, recovery is generally complete in a few days.

Monitoring programs (based on *Karenia brevis* cell counts) generally suffice for preventing human intoxication, except when officials are caught off-guard in previously unaffected areas (Passow, 1991; Perez et al., 2001; Rafuse et al., 2004; Schnorf et al., 2002; Tangen, 1977; Taylor et al., 1995).

Chemical properties

The NSP toxins, called brevetoxins, are tasteless, odorless, heat and acid stabile, lipid-soluble, cyclic polyether neurotoxins produced by the marine dinoflagellate such as *Gymnodinium breve* (or *Ptychodiscus brevis*). The molecular structure of the brevetoxins consists of 10 to 11 transfused rings; their molecular weights are around 900. Ten brevetoxins have been isolated and identified from field blooms and *G. breve* cultures (Benson et al., 1999) (Figure 4). These brevetoxins show specific binding to site-5 of voltage-sensitive Na^+ channels leading to channel activation at normal resting potential. This property of the brevetoxins causes the toxic effects (Cembella et al., 1995). PbTx-2 is the major toxin isolated from *G. breve*.

Four brevetoxin analogues (Figures 5 and 6) were isolated from contaminated shellfish. The brevetoxin analogues were analysed in cockles (*Austrovenus stutchburyi*) (BTX-B1) (Ishida et al., 1995) and Green shell mussels (*Perna canaliculus*) (BTX-B2, BTX-B3 and BTX-B4) (Morohashi et al., 1995, 1999; Murata et al., 1998) and differed from brevetoxins isolated from dinoflagellate cultures. Apparently BTX-B1, BTX-B2, BTX-B3 and BTX-B4 are metabolites formed by the shellfish itself as they were not found in field blooms or *G. breve* cultures. The presence of BTX-B2, BTX-B3 and BTX-B4 in *Perna canaliculus* does suggest that metabolic pathways in this species are more complicated than those in cockles (*A. stutchburyi*). However, the major toxins in shellfish were left unelucidated because of the extreme difficulty in isolation (Morohashi et al., 1999).

In addition to brevetoxins, some phosphorus containing ichthyotoxic compounds resembling anti cholinesterases, have also been isolated from *G. breve* (Figure 7). One example is an acyclic phosphorus compound with an oximino group in addition to a thiophosphate moiety, namely O,O-dipropyl(*E*)-2-(1-methyl-2-oxopropylidene) phosphorohydrazidothioate-(*E*) oxime (Van Apeldoorn et al., 2001).

Paralytic shellfish poisoning (PSP)

Paralytic shellfish poisoning (PSP) toxins are present in some genera of dinoflagellates and one species of cynobacteri. Several species of the genus *Alexandrium* (formerly named *Gonyaulax* or *Protogonyaulax*) are identified as contaminators in shellfish. These are *Alexandrium tamarensis*, *Alexandrium minutum* (syn. *Alexandrium excavata*), *Alexandrium catenella*, Alexandrium fraterculus, *Alexandrium fundyense* and Alexandrium cohorticula. Other clearly distinct dinoflagellates have also been recognised as sources of the STXs.

These are *Pyrodinium bahamense* and *Gymnodinium catenatum* (Mons et al., 1998). The toxicity of the dinoflagellates is due to a mixture of STX derivatives of which the composition differs per producing species and/or per region of occurrence.

Toxins produced: Saxitoxins

PSP, like ASP, is a life threatening syndrome. Symptoms are purely neurological and their onset is rapid. Duration of effects is a few days in non-lethal cases. Symptoms include tingling, numbness, and burning of the perioral region, ataxia, giddiness, drowsiness, fever, rash, and staggering.

The most severe cases result in respiratory arrest within 24 h of consumption of the toxic shellfish. If the patient is not breathing or if a pulse is not detected, artificial respiration and CPR may be needed as first aid. There is no antidote, supportive therapy is the rule and survivors recover fully. PSP is prevented by large-scale proactive monitoring programs (assessing toxin levels in mussels, oysters, scallops, clams) and rapid closures to harvest of suspect or demonstrated toxic areas. Paralytic shellfish poisoning (PSP) has been reported to occur after eating puffer fish, filter feeding shellfish and molluscs.

If ingested by humans, PSP produces neurologic symptoms such as tingling and burning of the mouth and tongue, numbness, drowsiness and incoherent speech. These symptoms occur within 30 min to two hours after ingestion and in severe cases cause ataxia, muscle weakness, respiratory paralysis and death. The Toxic Exposure Surveillance System (TESS) of the American Association of Poison Control Centres has identified 10 illnesses of presumed puffer fish poisoning due to exposure from PSP after eating puffer fish from the area of Titusville, Florida, (Klöpper et al., 2003; Leikin and Paloucek, 1998; Lembeye et al., 1993; Lembeye, 1981; Luckas et al., 2005; MacKenzie et al., 1996; Mahoney et al., 1990).

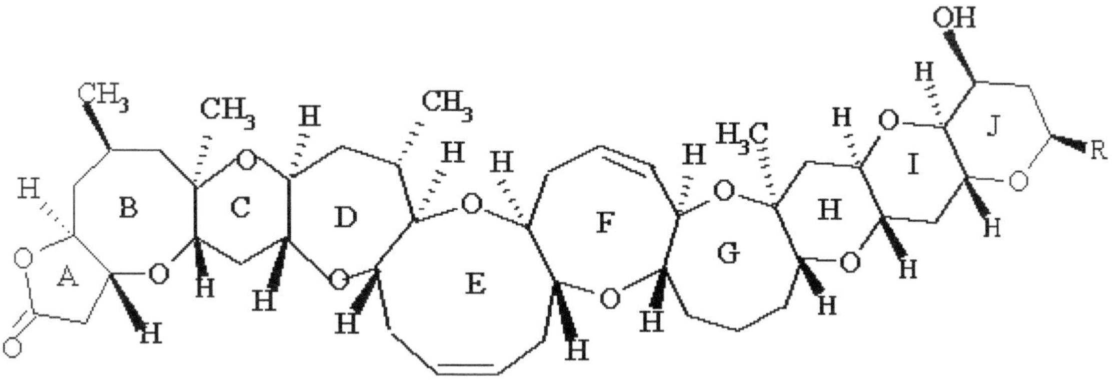

Type 1 (A) brevetoxins:

	PbTx-1,	R = $CH_2C(=CH_2)CHO$
	PbTx-7,	R=$CH_2C(=CH_2)CH_2OH$
	PbTx-10,	R=$CH_2CH(CH_3)CH_2OH$

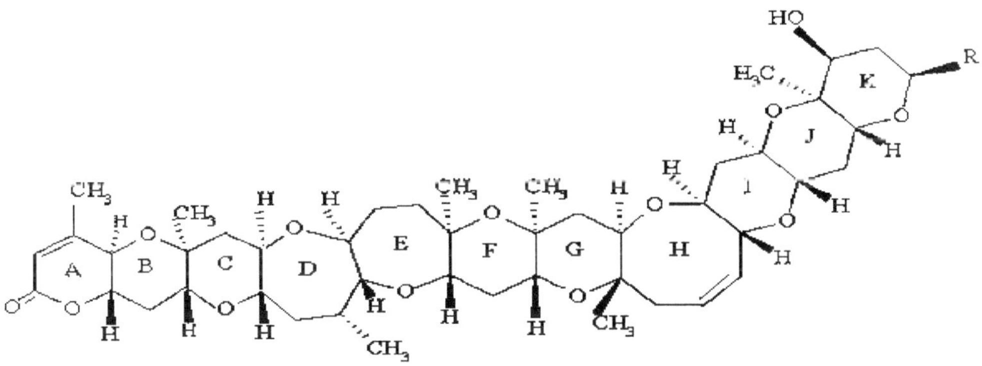

Type 2 (B) brevetoxins:

	PbTx-2	R = $CH_2C(=CH_2)CHO$
	oxidized PbTx-2	R=$CH_2C(=CH_2)COOH$
	PbTx-3	R=$CH_2C(=CH_2)$CH2OH
	PbTx-8	R=CH_2COCH_2Cl
	PbTx-9	R=$CH_2CH(CH_3)$CH2OH
	PbTx-5	the K-ring acetate of PbTx-2
	PbTx-6	the H-ring epoxide of PbTx-2

Figure 4. Chemical structures of type A and B brevetoxins (Hua et al., 1996).

Chemical properties

The PSP toxins form a group of closely related tetra hydropurine compounds that make up four subgroups: i) Carbamate (STX, neo STX and Gonyautoxins (GNTX1-4); ii) N-sulfo-carbamoyl (GNTX5-6, C1-4); iii) Decarbamoyl (dc-) (dcSTX, dcneoSTX, dcGNTX1-4); and iv) deoxydecarbamoyl (do-) (doSTX, doneoSTX and doGNTX1) components. At least 21 PSP toxins (Figure 8) mainly from marine dinoflagellates and shellfish that feed on toxic algae have been identified. Attempts to isolate PSP toxins began more than one hundred years

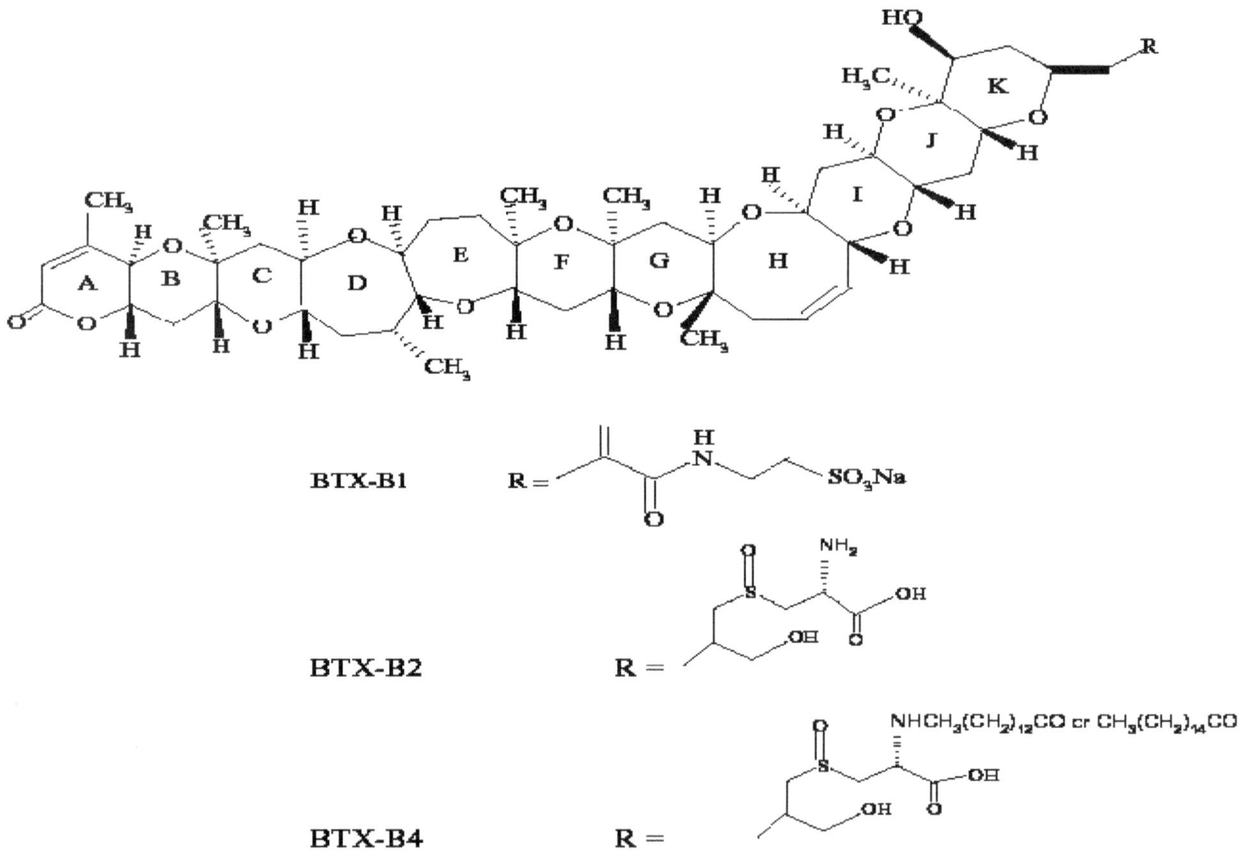

Figure 5. Chemical structures of brevetoxin analogues BTX-B1, -B2 and -B4 isolated from contaminated shellfish. Source: Yasumoto et al., 2001.

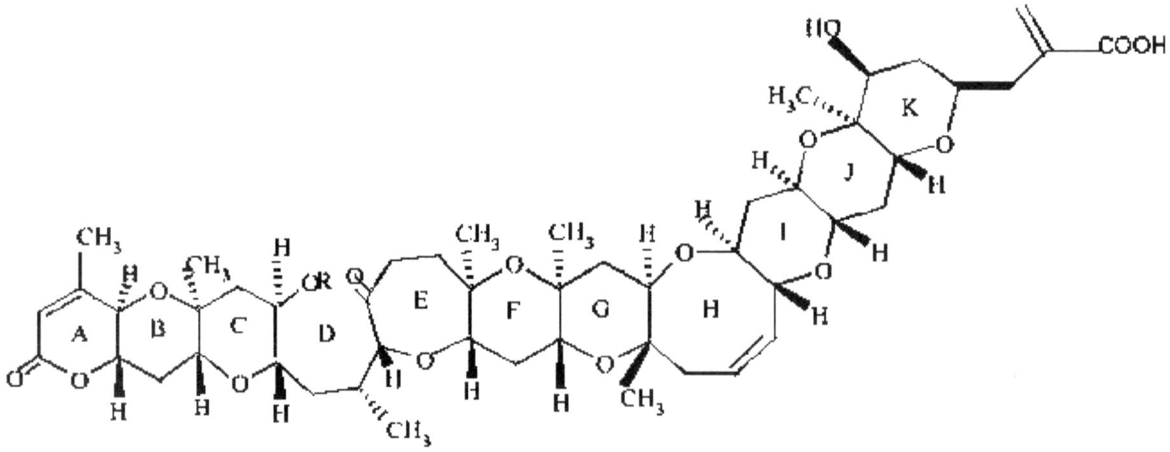

BTX-B3
R = $CH_3(CH_2)_{12}CO$ or $CH_3(CH_2)_{14}CO$

Figure 6. Chemical structure of brevetoxin analogue BTX-B3 isolated from contaminated shellfish. Source: Yasumoto et al., 2001.

ago but their occurrence as mixtures of compounds with different ionizable functionalities complicated isolation procedures and early progress was slow. The development of ion-exchange chromatography, guided by mouse bioassays, eventually led to the isolation of a water-soluble basic toxin from the Alaska butter clam (*Saxidomus giganteus*). This compound was later given the trivial name saxitoxin (STX).

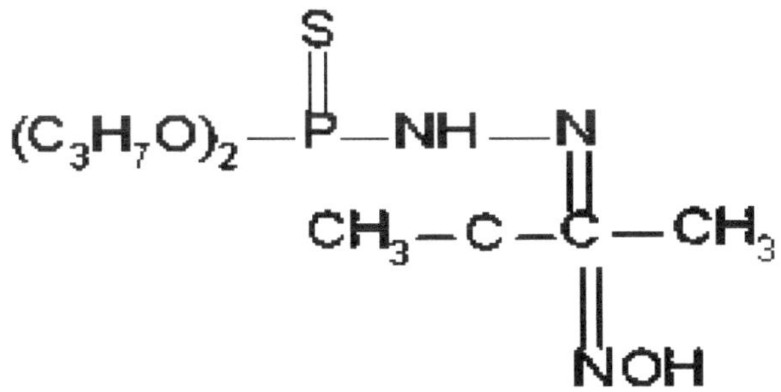

Figure 7. Phosphorus containing ichthyotoxic toxin isolated from *G. breve*. Source: Van Apeldoorn et al., 2001.

In 1975, the first crystalline derivative of STX was synthesized and the structure was studied (Bower et al., 1981). By means of X-ray crystallographic and nuclear magnetic resonance (NMR) spectroscopic studies the structure of STX was elucidated (Figure 8 for the chemical structures of STX and other PSP toxins). The dihydroxy or hydrated ketone group on the five rings is essential for its poisonous activity. Catalytic reduction of this group with hydrogen to a monohydroxy group eliminates the activity. Removal of the carbamoyl group side-chain on the six-membered ring, leaving a hydroxyl group in its place, produces a molecule with about 60% of the original toxic activity. The presence of this active hydroxyl group establishes a means for the preparation of various derivatives of STX (Mons et al., 1998). The PSP toxins are heat stable at acidic pH (with the exception of the N-sulfo-carbamoyl components) but unstable and easily oxidized under alkaline conditions (Mons et al., 1998).

Amnesic shellfish poisoning (ASP)

Amnesic shellfish poisoning (ASP) is caused by Domoic acid (Wright et al., 1989). Species within the genus *Pseudo nitzschia* produce Domoic acid. This toxin is unusual in being produced by diatoms rather than dinoflagellates and some causative organisms like *Pseudo nitzschia australis*, *Pseudonitzschia pungens*.

Toxins produced: Domoic acid

ASP can be a life-threatening syndrome. It is characterized by both gastrointestinal and neurological disorders. Gastroenteritis usually develops within 24 h of the consumption of toxic shellfish; symptoms include nausea, vomiting, abdominal cramps and diarrhea. In severe cases, neurological symptoms also appear, usually within 48 h of toxic shellfish consumption. These symptoms include dizziness, headache, seizures, disorientation, short-term memory loss, respiratory difficulty, and coma.

In 1987, four victims died after consuming toxic mussels from Prince Edward Island, Canada. Since that time, Canadian authorities have monitored both the water column for the presence of the causative diatom, and shellfish for the presence of the toxin, domoic acid. Shellfish beds are closed to harvesting when the domoic acid concentration reaches 20 µg/g shellfish meat. Fish and crab viscera can also contain domoic acid, so the risk to human consumers and animals in the marine food chain is more significant than previously believed (Rafuse et al., 2004; Schnorf et al., 2002).

ASP can be a life-threatening syndrome. It is characterized by both gastrointestinal and neurological disorders. Gastroenteritis usually develops within 24 h of the consumption of toxic shellfish and symptoms include nausea, vomiting, abdominal cramps and diarrhea. In severe cases, neurological symptoms also appear, usually within 48 h of toxic shellfish consumption. These symptoms include dizziness, headache, seizures, disorientation, short-term memory loss, respiratory difficulty and coma (Uribe, 2001, 1992; Villareal et al., 1997; Widdows et al., 1979; Windust, and Quilliam, 2004; Wommack et al., 2003).

Chemical properties

Discrimination between some species of *Pseudo-nitzschia* is virtually impossible under a light microscope because of morphological similarity between species-some species differ in details that can only be detected under an electron microscope. However, whole cell DNA probes have been developed to distinguish between species (Rhodes et al., 1998), and these are utilised by industry in risk management when deciding whether to implement voluntary closures to harvesting, pending the results of shellfish toxicity testing.

Amnesic shellfish poisoning (ASP) is caused by domoic acid (DA) (Figure 9), a naturally occurring compound belonging to the kainoid class of compounds that have been isolated from a variety of marine sources including macro- and microalgae (Wright and Quilliam, 1995). DA is a crystalline water-soluble acidic amino acid. It can be purified by a variety of chromatographic methods and contains a strong chromophore that facilitates detection by UV spectroscopy.

DA was originally discovered as a product of a red macroalgae *Chondria armata* and was later isolated from several other red macroalgae. However, these seaweeds were not the source of DA in the first reported ASP incident on Prince Edward Island in Canada in 1987. The source of DA in this outbreak of ASP was found to be the diatom Pseudo- *nitzschia* (formerly *Nitzschia*) pungens forma multi series. DA is a potent neurotoxin and the kainoid class of compounds to which DA belongs, is a class of excitatory neurotransmitters that bind to specific receptor proteins in neuronal cells causing continual

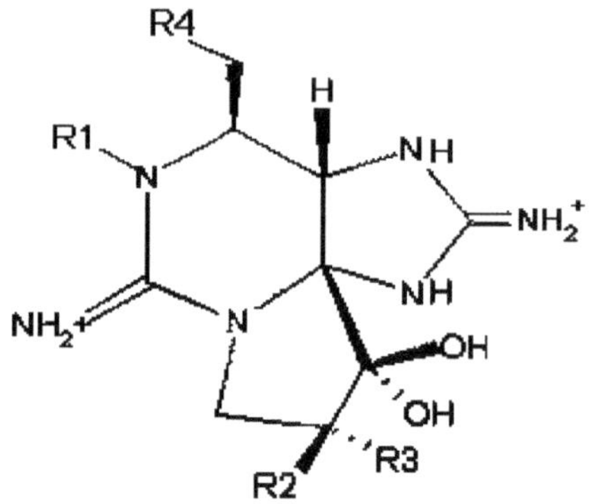

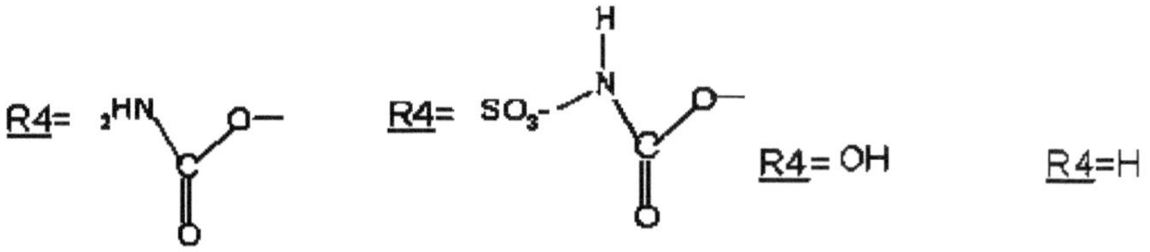

R_1	R_2	R_3	carbamate toxins	N-sulfo-carbamoyl toxins	decarbamoyl toxins	deoxy-decarbamoyl toxins
H	H	H	STX	GNTX5(B1)	dcSTX	doSTX
H	H	OSO_3^-	GNTX2	C1	dcGNTX2	
H	OSO_3^-	H	GNTX3	C2	dcGNTX3	
OH	H	H	neoSTX	GNTX6(B2)	dcneoSTX	doneoSTX
OH	H	OSO_3^-	GNTX1	C3	dcGNTX1	doGNTX1
OH	OSO_3^-	H	GNTX4	C4	dcGNTX4	

Figure 8. Chemical structures of PSP toxins. Source: Mons et al., 1998; Quilliam et al., 2001.

depolarization of the neuronal cell until cell rupture occurs (Wright, 1995).

Investigation of the kainoids present in *Chondria armata* resulted in the discovery, in minor amounts, of the geometrical isomers isodomoic acid A, B and C (Figure 9) as well as domoilactones. None of these isomers, found in seaweed, were detected in extracts of plankton or shellfish tissue. However, three other geometrical isomers (isodomoic acids D, E and F) and the C5' diastereomer (Figure 8) were isolated from both plankton cells and shellfish tissue (Wright and Quilliam, 1995; Ravn, 1995). The geometrical isomers can be prepared in the laboratory by brief exposure of dilute solutions of DA to UV light, and are therefore not considered to be *de novo* products of the plankton. Pharmacological studies indicate that these photoisomers bind less strongly to the kainate receptor proteins than DA itself suggesting that they are not as toxic as the parent amino acid. Formation of the C5' diastereomer is accelerated with warming. This C5' diastereomer shows almost the same binding efficacy to the kainate receptor as DA itself (Wright and Quilliam, 1995). Zaman et al. (1997b) reported the isolation of two new isomers of DA from the red alga *Chondria armata*, i.e. isodomoic acid G and H (Figure 9).

DETECTION OF MARINE TOXIC

In all cases, the marine HAB toxins that cause the human poisoning syndromes consist of families or groups of structurally related compounds, with individual derivatives exhibiting potencies that can significantly differ from other congeners (Van Dolah, 2001). During food web transfer, HAB toxins can also be metabolized or bio-transformed into structurally different compounds. The broad chemical and structural diversity of algal toxins and their derivatives and metabolites, coupled with differences in their potency account for many of the challenges associated with their detection in ocean observatory programs. Traditionally, biotoxin monitoring programs have relied on measurements of toxins in shellfish samples collected weekly or bi-weekly from key locations in areas affected by HABs (Shumway et al. 1988).

Toxin measurement methods can be grouped into three main types: chemical, *in vitro*, and *in vivo* assays (Hallegraeff et al., 2003). The latter (bioassays) have had a long history in HAB toxin detection, but are obviously not amenable to automation and high-throughput analysis in ocean observatories, so the only options in that context are measurements of toxin in seawater using either

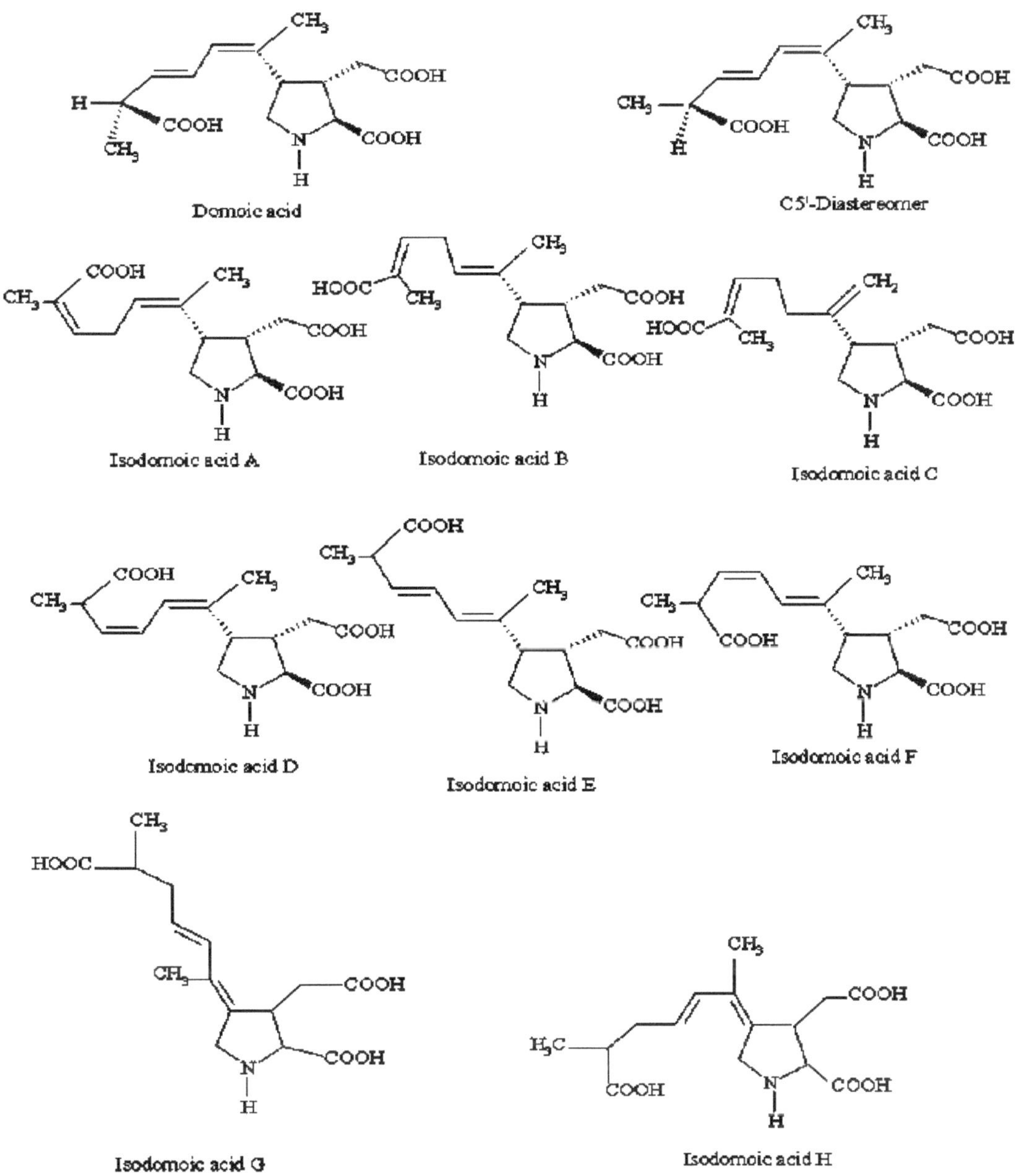

Figure 9. Chemical structures of domoic acid and its isomers. Source: Wright and Quilliam, 1995, Zaman et al. (1997b).

chemical analyses or *in vitro* assays. This immediately introduces some concerns, as considerable work will be needed to relate measurements of toxins dissolved in seawater, or in particulate form in that water, to the risk to human consumers of shellfish or fish.

Chemical methods for toxin analysis include high performance liquid chromatography (HPLC), and mass spectrometry coupled to liquid chromatographic separation (Quilliam 1996). Of these two alternatives, only mass spectrometry shows the potential for use in ocean observatories, and the challenges remain significant due to the diversity, size, and solubility of the toxins, as well as the matrices in which they occur (for example, particulate versus dissolved). Another constraint is the need to perform spectrometry in a vacuum and underwater, which poses significant engineering challenges. Progress has been good, however. For example, a small, modular mass spectrometer has been developed and mounted in an Autonomous Underwater Vehicles (AUV) (Wenner et al., 2004). That system consists of an *in situ* membrane-introduction linear-quadrupole mass spectrometer capable of detecting dissolved gases and volatile organic compounds at sub parts-per-billion concentrations. This instrument is still under development and has not been configured for HAB toxins, but future designs may permit the analysis of HAB toxins that occur dissolved in seawater (for example, brevetoxins, domoic acid, okadaic acid).

Analysis of toxins in particulate form will require a different approach, such as laser desorption mass spectrometry (LDMS), which is widely employed in analytical laboratories due to its simplicity of operation and rapid analysis times. One benefit of LDMS is that many different types of materials can be vaporized and ionized by a tightly focused laser beam (Cotter 1997). This can avoid sample purification or preparative techniques, which is critical to deployment of such technologies in a moored or mobile configuration in an Ocean Observing System (OOS), as it will greatly reduce sampling and handling requirements, and thus power drain, space needs, and reagent needs as well. LDMS has been used for the detection of bacterial spores, vegetative cells, viruses, and toxins in aerosol environments (Fenselau and Demirev, 2001), and efforts are underway to apply this method to HAB cells and dissolved toxins in seawater.

Another important and rapidly developing group of HAB toxin detection methods comprises the *in vitro* assays. One subgroup the functional assays, relies on detection of a toxin's biochemical activity while the other structural assays depends on recognition of chemical structure at the molecular level (Cembella et al., 2003; Van Dolah and Ramsdell, 2001). A variety of functional assays have been developed for the detection of HAB toxins, including cyto toxicity assays (Manger et al., 1995), enzyme inhibition assays (Della Loggia et al., 1999), and receptor binding assays (Van Dolah et al., 1994). Nevertheless, retention of the biological activity of a cell line or a receptor preparation outside the laboratory remains a significant, and thus far, insurmountable obstacle to *in situ* use of these assays (Sellner et al., 2003).

In contrast, structural assays show considerable promise for automated deployment in an observatory system. These assays rely on the structural or conformational interaction of a toxin with a recognition factor such as an antibody. Antibody-based assays have been developed for a variety of HAB toxins and many of these tests are now commercially available (Laycock et al., 2001; Cembella et al., 2003). One novel immunoassay utilizes surface plasmon resonance (SPR) in a portable system developed for rapid field quantification of toxin levels in both shellfish and seawater (Stevens et al., 2007).

The SPR assay had a limit of detection of 3 ppb domoic acid and a quantifiable range from 4 to 60 ppb. Comparison of analyses with standard HPLC protocols gave an excellent correlation. This same technology should also function for detection of domoic acid (and other algal toxins for which antibodies are available) in concentrated algal extracts or high dissolved levels in seawater. With refinement of the extraction protocols and generation of higher affinity monoclonal antibodies, detection of much lower levels of toxin should be possible, leading to eventual application of automated SPR biosensors on moorings. Another novel and potentially useful approach for *in situ* observations is a competitive immunoassay using screen printed electrodes (SPEs; Kreuzer et al., 2002; Micheli et al., 2004). Excellent sensitivity and accuracy has been achieved with HAB toxins such as okadaic acid, brevetoxin and domoic acid and for all toxins investigated, results compared favourably with other toxin analysis techniques. The advantages of speed of analysis, simplicity of design, *in situ* measurement capability, stability (storage up to four weeks prior to use), and disposability make screen printed electrodes (SPE) immune sensors good candidates for observatory instrumentation. Adaptation of this and other immunoassay technologies to robotic systems and deployment in remote locations is thus possible, but will require further development effort.

CONCLUSION

Only a small number of HAB species can be detected using optical measurements, either *in situ* or remotely from space, and therefore instruments that can detect the vast majority of HAB species need to have capabilities for sample collection, concentration, and manipulation. The chemistries and procedures for cell identification and enumeration using molecular probe assays of various types are well established.

Conflict of Interests

The author(s) have not declared any conflict of interests.

REFERENCES

Anderson DM (1989). Toxic algal blooms and red tides: a global perspective. In: Okaichi T, Anderson DM, NemotoT (eds). Red Tides: Biology, Environmental Science and Toxicology. Elsevier pp. 11-16.

Benson JM, Thischler DL, Baden DG (1999). Uptake, distribution, and excretion of brevetoxin 3 administered to rats by intratracheal instillation. J. Toxicol. Environ. Health Part A56:345-355.

Bower DJ, Hart RJ, Matthews PA, Howden MEH (1981). Nonprotein neurotoxins. Clin. Toxicol. 18:813-843.

Cembella AD, Doucette GJ, Garthwaite I (2003). *In vitro* assays for phycotoxins. *In*: Hallegraeff GM, Anderson DM, Cembella AD (eds). Manual on Harmful Marine Microalgae. IOC-UNESCO, Paris pp. 297-345.

Cembella AD, Milenkovic L, Doucette G, Fernandez M (1995). *In Vitro* biochemical and cellular assays. *In* Hallegraeff, G.M., Anderson, D.M. & A.D. Cembella, eds. Manual on harmful marine microalgae. IOC Manuals and Guides No. 33. UNESCO.

Chang F H (1996). A review of knowledge of a group of closely related economically important toxic Gymnodinium/Gyrodinium (Dinophyceae) species in New Zealand. J. Royal Soc. N.Z. 26:381-394.

Ciminiello P, Dell'Aversano C, Fattorusso E, Forino M, Magno S, Poletti R (2002). The detection and identification of 42,43,44,45,46,47,55-Heptanor-41-oxoyessotoxin, a new marine toxin from Adriatic shellfish by liquid chromatography-mass spectrometry. Chem. Res. Toxicol. 15: 979-984.

Ciminiello P, Fattorusso E, Forino M, Poletti R (2001). 42,43,44,45,46,47,55-Heptanor-41-oxohomoyessotoxin, a new biotoxin from mussels of the Northern Adriatic Sea. Chem. Res. Toxicol. 14:596-599.

Ciminiello P, Fattorusso E, Forino M, Magno S, Poletti R, Viviani R (1999). Isolation of 45-hydroxyyessotoxin from mussels of the Adriatic Sea. Toxicon 37:689-693.

Ciminiello P, Fattorusso E, Forino M, Magno S, Poletti R, Viviani R (1998). Isolation of Adriatoxin a new analogue of yessotoxin from mussels of the Adriatic Sea. Tetrahedron Lett. 39:8897-8900

Ciminiello P, Fattorusso E, Forino M, Magno S, Poletti R, Satake M, Viviani R, Yasumoto T (1997). Yessotoxin in mussels of the Northern Adriatic Sea. Toxicon 35(2):177-183.

Ciminiello P, Fattorusso E, Forino M, Poletti R, Viviani R (2000a). Structure determination of carboxyhomoyessotoxin, a new yessotoxin analogue isolated from Adriatic mussels. Chem. Res. Toxicol. 13: 770-774.

Ciminiello P, Fattorusso E, Forino M, Poletti R, Viviani R (2000b). A New Analogue of Yessotoxin, Carboxyyessotoxin, Isolated from Adriatic Sea Mussels. Eur. J. Org. Chem. 291-295.

Clément A (1999). Floraciones nocivas de *Gymnodinium* sp. Y sus efectos en la salmonicultura. Informe Técnico. Programa Monitoreo Fitoplancton INTESAL. 29 p.

Climent A, Lembeye G (1993). Phytoplankton monitoring program in the fish farming region of South Chile. In: Smayda TJ, Shimizu Y (eds), Toxic Phytoplankton Bloom in the Sea: pp. 223-228. Elsevier Science Publ. B.V.

Climent A, Seguel M, Arzul G, Guzmán L, Alarcón C (2001). Widespread outbreak of a haemolytic, ichthyotoxic *Gymnodinium* sp. in Chile. In: Hallegraeff et al. (ed), Harmful Algal Bloom.

Cohen AL (1974). Critical point drying. In: Hayat M A (ed), Principles and Techniques of Scanning Electron Microscopy 1:44-112. Van Nostrand Reinhold, Cincinatti, Ohio, Toronto, London, Melbourne.

Cosper EM, Bricelj VM, Carpenter EJ (eds). (1989). Novel Phytoplankton Blooms. Causes and Impacts of Recurrent Brown Tides and Others Unusual Blooms. Coastal and Estuarine Studies 35: 1- 799. Springer, Berlin.

Cotter RJ (1997). Time-of-flight mass spectrometry. ACS Professional Reference Books, Washington DC.

Currie BJ (2000). Clinical Toxicology: A Tropical Australian Perspective. Therapeutic Drug Monitoring. 22(1): 73-78.

Daiguji M, Satake M, James KJ, Bishop A, MacKenzie L, Naoki H, Yasumoto T (1998). Structures of new pectenotoxin analogs, pectenotoxin-2 seco acid and 7-*epi*-Pectenotoxin-2 seco acid, isolated from a dinoflagellate and greenshell mussels. Chem. Lett. pp. 653-654.

De Fouw JC, Van Egmond HP, Speijers GJA (1999). Ciguatera fish poisoning: a review. RIVM Report No.388802021 (available atwww.rivm.nl/bibliotheek/rapporten/388802021.html).

Della LR, Sosa S, Tubaro A (1999). Methodological improvement of the protein phosphatase inhibition assay for the detection of okadaic acid in mussels. Nat. Toxins 7:387-392.

Draisci R, Giannetti L, Lucentini L, Marchiafava C, James KJ, Bishop A, Healy B, Kelly SS (1998). Isolation of a new okadaic acid analogue from phytoplankton implicated in diarrhetic shellfish poisoning. J. Chromatogr. A, 798: 137-145.

Draisci R, Lucentini L, Giannetti L, Boria P, Poletti R (1996a). First report of pectenotoxin-2 (PTX-2) in algae (*Dinophysis fortii*) related to seafood poisoning in Europe. Toxicon 34(8): 923-935.

EU/SANCO (2001). Report of the meeting of the working group on toxicology of DSP and AZP 21 to 23rd May 2001, Brussels.

Fenselau C, Demirev P (2001). Characterization of intact microorganisms by MALDI mass spectrometry. Mass Spectrom. Rev. 20: 157.

Hallegraeff GM (1993). A review of harmful algal blooms and their apparent global increase. Phycologia 32:79-99.

Hallegraeff GM, Anderson DM, Cembella AD (2003). Manual on Harmful Marine Microalgae, Monographs on Oceanographic Methodology, UNESCO, Paris,11,793 S. (Monographs on oceanographic methodology ; 11).

Hallegraeff GM (1995). 1. Harmful algal blooms: a global overview. *In* Hallegraeff, G.M. et al. eds. Manual on Harmful MarineMicroalgae, pp. 1-22. IOC Manuals and Guides No. 33. UNESCO.

Haywood A (1998). Toxic Gymnodinium update. In: Proceedings of the Marine Biotoxin Science Workshop No. 10 (20 November 1998) Marine Biotoxin Technical Committee.

Hua Y, Lu W, Henry MS, Pierce RH, Cole RB (1996). On-line liquid chromatography-electrospray ionization mass spectrometry for determination of the brevetoxin profile in natural "red tide" algal blooms. J. Chromatogr. *A* 750:115-125.

Ishida H, Nozawa A, Totoribe K, Muramatsu N, Nukaya H, Tsuji K, Yamaguchi K, Yasumoto T, Kaspar H, Berkett N, Kosuge T (1995). Brevetoxin B_1, a new polyether marine toxin from the New Zealand shellfish, *Austrovenus stutchburyi*. Tetrahedron Lett. 36(5):725-728.

Jackson AE, Marr JC, McLachan JL (1993). The production of Diarrhetic shellfish toxins by an isolate of Prorocentrum lima from Nova Scotia, Canada. In: Toxic phytoplankton blooms in the sea. Smayda TJ, Shimizu Y (Eds.) Elsevier Science Publishers. pp. 513-518

James KJ, Bishop AG, Healy BM, Roden C, Sherlock IR, Twohig M, Draisci R, Giannetti L, Lucentini L (1999). Efficient isolation of the rare diarrhoeic shellfish toxin, dinophysistoxin-2, from marine phytoplankton. Toxicon 37:343-357.

James KJ, Carmody EP, Gillman M, Kelly SS, Draisci R, Lucentini L, Giannetti L (1997). Identification of a new diarrhetic toxin in shellfish using liquid chromatography with fluorimetric and mass spectrometric detection. Toxicon 35(6): 973-978.

Kim CS, Lee SG, Lee CK, Kim HK, Jung J (1999). Reactive oxygen species as causative agents in the icthyotoxicity of the red tide dinoflagellate *Cochlodinium polykrikoides*. J. Plankt. Res. 21: 2105-2115.

Klöpper S, Scharek R, Gerdts G (2003). Diarrhetic shellfish toxicity in relation to the abundance of Dinophysis spp. Ehrenberg 1839 in the German Bight near Helgoland, Marine ecology-progress series. 259:93-102.

Kreuzer MP, Pravda M, O'Sullivan CK, Guilbault GG (2002). Novel electrochemical immunosensors for seafood toxin analysis. Toxicon 40: 1267-1274.

Laycock MV, Jellett JF, Belland ER, Bishop PC, Theriault BL, Russell-Tattrie AL, Quilliam MA, Cembella AD, Richards RC (2001). Mist Alert: A rapid assay for paralytic shellfish poisoning toxins. *In*: Hallegraeff GM, Blackburn SI, Bolch CJ, Lewis RJ (eds). Harmful Algal Blooms 2000. IOC-UNESCO, Paris, pp. 254-256.

Lee JS, Igarashi T, Fraga S, Dahl E, Horgaard P, Yasumoto T (1989). Determination of Diarrhetic shellfish toxins in various dinoflagellate species. J. Appl. Phycol. 1:147-152.

Lehane L, Lewis RJ (2000). Review Ciguatera: recent advances but the risk remains. Int. J. Food Microbiol. 61: 91-125.

Lehane L (2000). Ciguatera update. Med. J. Aust. 172(4): 176-179.

Leikin J, Paloucek F (1998). Ciguatera Food Poisoning. In: Poisoning and Toxicology Compendium. Cleveland: Lexi-Comp Inc. p. 783.

Lembeye G (1981). Segunda aparición del veneno paralítico de los mariscos (VPM) asociado a la presencia de Gonyaulax catenella, en Magalanes (Chile). Anales Instituto de la Patagonia, Punta Arenas (Chile) 12: 273-276.

Lembeye G, Yasumoto T, Zhao J, Fernandez R (1993). DSP outbreaks in chilean fiords. *In*: Smayda TJ, Shimizu Y (eds), Toxic Phytoplankton Blooms in the Sea: 525 - 529. Elsevier.

Luckas B, Dahlmann J, Erler K, Gerdts G, Wasmund N, Hummert C, Hansen PD (2005). Overview of Key Phytoplankton Toxins and Their Recent Occurrence in the North and Baltic Seas, Environmental Toxicology, 20:1-17.

Mackenzie L, Rhodes L, Till D, Chang FH, Kaspar H, Haywood A, Kapa J, Walker B (1995b). A Gymnodinium sp. bloom and the contamination of shellfish with lipid soluble toxins in New Zealand. In: Harmful marine algal blooms: Proceedings of the sixth international conference on toxic marine phytoplankton, October 1993, Nantes, France. Lassus P, Arzul G, Erard E, Gentien P, Marcaillou C (Eds.) Intercept Ltd., Andover, UK. pp. 795-800.

Mackenzie L, Syhre M, Todd K, Haywood A, Truman P Garthwaite I,Towers N (1995a). DSP and NSP in New Zealand, Some recent findings. In: Proceedings of the Marine Biotoxin Science Workshop No. 4. Marine Biotoxin Technical Committee.

MacKenzie L, Haywood A, Adamson J, Truman P, Till D, Satake M, Yasumoto T (1996). Gymnodimine contamination of shellfish in New

Zealand. In: Yasumoto T, Y Oshima and Y Fukuyo (eds), Harmful and Toxic Algal Blooms: 97 - 100. Intergovernmental Oceanographic Commission of UNESCO.

Mahoney J, Olsen P, Cohn M (1990). Blooms of a dinoflagellate *Gyrodinium* cf. *aureolum* in New Jersey coastal waters and their occurrence and effects. J. Coastal Res. 6(1):121 - 135.

Manger RL, Leja LS, Lee SY, Hungerford JM, Hokama Y, Dickey RW, Granade HR, Lewis R, Yasumoto T, Wekell MM (1995). Detection of sodium channel toxins: directed cytotoxicity assays of purified ciguatoxins, brevetoxins, saxitoxins, and seafood extracts. J. Assoc. Anal. Chem. Int. 78:521-527.

Micheli L, Radoi A, Guarrina R, Massaud R (2004). Disposable immunosensor for the determination of domoic acid in shellfish. Biosensors Bioelectronics 20: 190-196.

Mons MN, Van Egmond HP, Speijers GJA (1998). Paralytic shellfish poisoning: A review. RIVM Report 388802 005. June.

Morohashi A, Satake M, Murata K, Naoki H, Kaspar HF, Oshima Y, Yasumoto T (1999). Brevetoxin B4 Isolated from Greenshell mussels *Perna canaliculus*, the major Toxin Involved in Neurotoxic Shellfish Poisoning in New Zealand. Nat. Toxins 7:45-48.

Murakami Y, Oshima Y, Yasumoto T (1982). Identification of Okadaic acid as a toxic component of a marine dinoflagellate Prorocentrum lima. Bull. Jpn. Soc. Sci. Fish. 48:69-72.

Naoki H, Fujita T, Cruchet P, Legrand A-M, Igarashi T, Yasumoto T (2001). Structural determination of new ciguatoxin congeners by tandem mass spectrometry. *In* De Koe, WJ, Samson RA, Van Egmond HP, Gilbert J, Sabino M, eds. Mycotoxins and Phycotoxins in Perspective at the Turn of the Millennium. Proceedings of the Xth International IUPAC Symposium on Mycotoxins and Phycotoxins (May 2000, Guarujá, Brazil), pp. 475-482. Wageningen, the Netherlands, Ponsen & Looyen.

National Ocean Research Leadership Council (NORLC) (1999). An Integrated Ocean Observing System: A Strategy for Implementing the First Steps of a U.S. Plan. National Oceanographic Partnership Program Office, Consortium for Oceanographic Research and Education, Washington DC.

Nielsen M, Tønseth C (1991). Temperature and salinity effect on growth and chemical composition of *Gyrodinium aureolum* Hulburt in culture. J. Plankton Res. 13:389- 398.

Okaichi T, Ochi T, Nishio S, Takano H, Matsuno K, Morimoto T, Murakami T, Shimida S (1989). The cause of fishkills associated with red tides of Chattonella antiqua (HADA) ONO. *In*: Miyachi S, Karube I, Ishida Y (eds). Current Topics in Marine Biotechnology. Japanese Society for Marine Biotechnology. pp.185-188

Partensky F, Sournia A (1986). Le dinoflagellé *Gyrodiniun cf. aureolum* dans le plancton de l'Atlantique nord: identification, écologie, toxicité. *Cryptogamie Algologie*, *7*(4): 251-275.

Partensky F, Gentien P, Sournia A (1991). *Gymnodinium* cf. *nagasakiense* = *Gyrodinium* cf. *aureolum* (Dinophycées). In: Sournia A, Belin C, Berland B, Erard-Le Denn E, Gentien P, Grzebyk D, Marcaillou-Le Baut C, Lassus P, Partensky F (eds), Le Phytoplancton Nuisible des Côtes de France (de la biologie à la prévention): 63 - 82. IFREMER, France.

Partensky F, Vaulot D, Coute A, Sournia A (1988). Morphological and nuclear analysis of the bloom-forming dinoflagellates *Gyrodinium* cf. *aureolum* and Gymnodinium nagasakiense. J. Phycol. 24: 408 - 415.

Passow U (1991). Vertical migration of *Gonyaulax catenata* and *Mesodinium rubrum*. Marine Biology 110: 455 - 463.

Perez C, Vasquez P, Perret C (2001). Treatment of Ciguatera Poison with Gabapentin. New England J. Med. 344(9): 692-693.

Quilliam MA (1996). Liquid chromatography-mass spectrometry of seafood toxins. *In*: Barcelo D (ed). Applications of LC-MS in environmental science. Elsevier, Amsterdam pp. 415-444.

Quilliam MA (1998a). General Referee reports. Committee on Natural Toxins. Phycotoxins. J. AOAC. Int. 81(1): 142-151.

Rafuse C, Cembella AD, Laycock M, Jellett J (2004). Rapid monitoring of toxic phytoplankton and zooplankton with a lateral-flow immunochromatographic assay for ASP and PSP toxins, Harmful Algae 2002. Proceedings of the Xth International Conference on Harmful Algae. Steidinger KA, Landsberg JH, Tomas CR, Vargo GA (Eds.)

Ramsdell JS, Anderson DM, Glibert PM (eds) (2005). HARRNESS, Harmful Algae Research and Response: A National Environmental Science Strategy 2005-2015. Ecological Society of America, Washington DC. 96 pp.

Ravn H (1995). HAB Publication series Volume 1. Amnesic Shellfish Poisoning (ASP). IOC Manuals and Guides 31:15.

Rhodes L, Scholin C, Garthwaite I, Haywood A, Thomas A (1998). Domoic acid producing *Pseudo-nitzschia* species educed by whole cell DNA probe-based and immunochemical assays. *In* Reguera B, Blanco J, Fernandez M, Wyatt T eds. 1997. Harmful Algae, Proceedings of the VIII International Conference on Harmful Algae (June 1999, Vigo, Spain), pp. 274-277. Xunta de Galicia and IOC of UNESCO.

Sasaki K, Wright JLC, Yasumoto T (1998). Identification and characterization of Pectenotoxin (PTX) 4 and PTX7 as spiroketal stereoisomers of two previously reported Pectenotoxins. J. Org. Chem. 63: 2475-2480.

Satake M, Ichimura T, Sekiguchi K, Yoshimatsu S, Oshima Y (1999). Confirmation of Yessotoxin and 45,46,47-Trinoryessotoxin Production by Prorocentrum reticulatum Collected in Japan. Nat. Toxins 7:147-150.

Satake M, MacKenzie L, Yasumoto T (1997). Identification of *Protoceratium reticulatum* as the Biogenetic origin of Yessotoxin. Nat. Toxins 5: 164-167.

Satake M, Ofuji K, Naoki H, Jame K, Furey A, McMahon T, Silke J, Yasumoto T (1998a). New toxic event caused by Irish mussels. *In* Reguera B, Blanco J, Fernandez M, Wyatt T, eds. 1997. Harmful Algae, Proceedings of the VIII International Conference on Harmful Algae, (June 1999, Vigo, Spain), pp. 468-469. Xunta de Galicia and IOC of UNESCO.

Schnorf H, Taurarii M, Cundy T (2002). Ciguatera fish poisoning: A doubleblind randomized trial of mannitol therapy. Neurology 58(6): 873- 880.

Sellner K, Sellner G, Doucette GJ, Kirkpatrick GJ (2003). Harmful algal blooms: causes, impacts and detection. J. Ind. Microbiol. Biotechnol. 30: 383-406.

Shumway SE, Sherman-Casswell S, Hurst JW (1988). Paralytic shellfish poisoning in Maine: Monitoring a monster. J. Shellfish Res. 7:643-652.

Stevens RC, Soelberg DD, Eberhart BT, Sylvester Spencer L, Wekell JC, Chinowsky TM (2007). Detection of the toxin domoic acid from clam extracts using a portable surface plasmon resonance biosensor. Harmful Algae; 6:166-174.

Suzuki T, Mitsuya T, Matsubara H, Yamasaki M (1998). Determination of pectenotoxin-2 after solid-phase extraction from seawater and from the dinoflagellate *Dinophysis fortii* by liquid chromatography with electrospray mass spectrometry and ultraviolet detection: Evidence of oxidation of pectenotoxin-2 to pectenotoxin-6 in scallops. J. Chromatogr. *A* 815: 155-160

Tangen K (1977). Blooms of *Gyrodinium aureolum* (Dinophyceae) in north European waters, accompanied by mortality in marine organisms. Sarsia 63:123 - 133.

Taylor F, Fukuyo Y, Larsen J (1995). Taxonomy of harmful dinoflagellates. In: Hallegraeff GM, DM Anderson & AD Cembella (eds), Manual of Harmful Marine Microalgae: pp. 283 - 309. UNESCO, France.

Tillmann U (2004). Interactions between planktonic microalgae and protozoan grazers, J. Eukaryotic Microbial. 51(2):156-168.

Tubaro A, Sidari L, Della Loggia R, Yasumoto T (1998). Occurrence of yessotoxin-like toxins in phytoplankton and mussels from northern Adriatic Sea. *In* Reguera B, Blanco J, Fernandez M, Wyatt T eds. 1997. Harmful Algae, Proceedings of the VIII International Conference on Harmful Algae, (June 1999, Vigo, Spain), pp. 470-472. Xunta de Galicia and IOC of UNESCO.

Uribe J, Garcia C, Rivas M, Lagos N (2001). First report of diarrhetic shellfish toxins in the Southern Magellanic fjords, Chile. J. Shellfish Res. 20 (1) 69-74.

Uribe J (1992). Toxic shellfish kill twelve people in Chile. Red Tide Newsletter 5(1): 3. Villareal TA, Mansfield A, Buskey EJ (1997). Growth and chemical composition of the Texas brown tide forming pelagophyte *Aureoumbr lagunensis*. In: Reguera B, Blanco J, Fernández M, Wyatt T (eds), Harmful Algae: 359 - 362. Xunta de Galicia and Intergovernmental Oceanographic Comission of UNESCO,

Spain.
Widdows J, Moore MN, Lowe DM, Salked PN (1979). Some effects of a dinoflagellate bloom (*Gyrodinium aureolum*) on the mussels, *Mytilus edulis*. J. Marine Biol. Association, U.K. 59:522 - 524.
Van Apeldoorn ME (1998). Diarrhoeic shellfish poisoning: A review. RIVM/CSR Report 05722A00. August Science Publishers B.V.
Van Apeldoorn ME (2001). Neurotoxic shellfish poisoning: A review. RIVM report 388802 023. September.
Van Dolah FM, Finley BL, Doucette GJ, Moeller PD, Ramsdell JS (1994). Development of rapid and sensitive high throughput pharmacologic assays for marine phycotoxins. Natural Toxins 2:189-196.
Van Dolah FM, Ramsdell JS (2001). Review and assessment of *in vitro* detection methods for algal toxins. J. Assoc. Anal. Chem. Int. 84:1617-1625.
Van Egmond HP, Aune T, Lassus P, Speijers G, Waldock M (1993). Paralytic and diarrhoeic shellfish poisons: occurrence in Europe, toxicity, analysis and regulation. J. Nat. Toxins 2:41-83.
Wenner PG, Bell RJ, van Amerom FHW, Toler SK, Edkins JE, Hall ML, Koehn K, Short RT, Byrne RH (2004). Environmental chemical mapping using an underwater mass spectrometer. Trends in Anal. Chem. Special issue on Deploying mass spectrometers in harsh environments; 23:288-295.
White AW, Fukuhara O, Anraku M (1989). Mortality of fish larvae from eating toxic dinoflagellates or zooplankton containing dinoflagellate toxins. *In*: Okaichi T, Anderson DM, Nemoto T (ed). Red Tides: Biol. Environ. Sci. Toxicoly. Elsevier, NY. pp. 395-398.
Windust AJ, Quilliam MA (2004). Certified reference materials for marine algal toxins, Harmful Algae 2002. Proceedings of the Xth International Conference on Harmful Algae. Steidinger KA, Landsberg JH, Tomas CR, Vargo GA (Eds.) Florida fish and Wildlife Conservation Commission and Intergovernmental Oceanographic Commission of UNESCO. 2003, submitted.
Wommack KE, Suttle CA, Wilson WH, Weinbauer M, Pédros-Alio C, Schütt C (2003). Viral diversity, CIESM 2003. Ecology of marine viruses. CIESM Workshop Monograph, n° 21:14-17.
Wright JLC, Quilliam MA (1995). 7. Methods for Domoic Acid, the Amnesic Shellfish Poisons. *In* Hallegraeff, G.M. et al. eds. *Manual on Harmful Marine Microalgae*. IOC Manuals and Guides No. 33. UNESCO. pp. 113-133.
Wright JLC (1995). Dealing with seafood toxins: present approaches and future options. Food Res. Int. 28(4):347-358.
Wright JLC, Boyd RK, De Freitas ASW, Falk M, Foxall RA, Jamieson WD, Laycock MV, McCulloch AW, McInnes AG, Odense P, Pathak VP, Quilliam MA, Ragan MA, Sim PG, Thibault P, Walter JA, Gilgan M, Richard DJ, Dewar D (1989). Identification of domoic acid, a neuroexcitatory amino acid, in toxic mussels from eastern Prince Edward Island. Can. J. Chem. 67:481-490.
Yasumoto T, Igarashi T, Satake M (2001). Chemistry of phycotoxins-Structural elucidation. *In* De Koe WJ, Samson RA, Van Egmond HP, Gilbert J, Sabino M, eds. Mycotoxins and Phycotoxins in Perspective at the Turn of the Millennium. Proceedings of the X International IUPAC Symposium on Mycotoxins and Phycotoxins (May 2000, Guarujá, Brazil), pp. 465-474. Wageningen, the Netherlands, Ponsen & Looyen.
Yasumoto T, Igarashi T, Legrand AM, Cruchet P, Chinain M, Fujita T, Naoki H (2000). Structural elucidation of ciguatoxin congeners by fast-atombombardment tandem mass spectroscopy. J. Am. Chem. Soc. 122:4988-4989.
Zaman L, Arakawa O, Shimosu A, Onoue Y, Nishio S, Shida Y, Noguchi T (1997b). Two new isomers from a red alga, *Chondria armata*. Toxicon 35(2):205-212.

11

Spatio-temporal variations in phytoplankton community structure in small water bodies within Lake Victoria basin, Kenya

Steve O. Ngodhe, Phillip O. Raburu, Boaz K. Arara, Patrick O. Orwa and Alfred A. Otieno

Department of Fisheries and Aquatic Sciences, University of Eldoret, P.O. BOX 1125, Eldoret, Kenya.

Spatio-temporal variations of phytoplankton, expressed in terms of species composition and diversity collected at various sampling sites in small water bodies (SWBs) within Lake Victoria basin, Kenya, were investigated monthly from November 2010 to June 2011, in relation to selected physical and chemical water quality parameters. Temperature, D.O, TN and TP revealed a significant difference between the dams (p<0.005) unlike pH and BOD_5. These SWBs were built during the pre-independence era and stocked with various species of fish. The dams provide water for both domestic and agricultural use. A total of 1392 phytoplankton species belonging to four families and 20 genera were identified in Kesses dam whereas in Kerita dam, a total of 376 phytoplankton species belonging to four families and 10 genera were identified. In Siaya dams, Yenga dam had three families of phytoplankton; Chlorophyceae, Euglenophyceae and Cyanophyceae with Mauna also recording four families: Chlorophyceae, Bacillariophyceae, Desmidiaceae and Cyanophyceae. All the SWBs generally registered low species diversity with majority of them recording a value of < 2. Seasonal variations in phytoplankton species composition and diversity were significant (*P* < 0.05) with low species composition and diversity occurring during the dry season, and being maximum following the end of the rainy season from November 2010 to March 2011, suggesting the possible influence of various environmental factors on the SWBs. Overall, water quality seemed to have had effect on the species diversity, dominance and richness of phytoplankton community structure.

Key words: Community structure, small water bodies (SWBs), phytoplankton, species composition and diversity.

INTRODUCTION

Phytoplankton is usually at the base of aquatic food web and is the most important factor for production of organic matter in aquatic ecosystem. Most reservoirs will require significant amount of phytoplankton to have productive and sustainable fisheries. The interplay of physical, chemical and biological properties of water most often lead to the production of phytoplankton, while their assemblage (composition and distribution) is also structured by these factors. The importance of phytoplankton in tropical reservoir ecosystems include its use in estimating potential fish yield (Descy et al., 2005), productivity (Likens, 1975), water quality (Walsh et al., 2001), energy flow (Simciv, 2005), trophic status (Reynolds, 1999) and management (Beyruth and Tanaka, 2000). These reservoirs are increasingly threatened by human activities (Cecchi, 2007; Descy and

Sarmento, 2008). In Lake Victoria, phytoplankton species composition, numerical abundance, spatial distribution and total biomass are in a direct relation with the environmental factors (Lung'ayia et al., 2000). Actually, environmental and temporal changes determine the community present in a lake (Levandowsky, 1972). Among the environmental factors are the nutrients availability and light, temperature, alkalinity and mixing depth. Talling (1966) states in the particular, the abundance of algal species changes occurs due to changes in turbulence, illumination per cell and more obviously distribution of nutrients in the water column. Phytoplankton can be divided into size classes that have different physical properties. Picoplankton have the greatest surface:volume ratio compared with nano- and microplankton (Lewis, 1976) and may, consequently, more efficiently assimilate nutrients than nano- and microplankton (Lafond et al., 1990). Smaller cells usually have greater growth potential (Bruno et al., 1983), greater biomass productivity and lower sinking rates (Stockner et al., 1987). Picoplankton is, however, the dominant size class in several environments all over the world, both in marine and oligotrophic freshwater environments (Adame et al., 2008).

Lung'ayia et al. (2001) stated that phytoplankton structure is determined by underwater light availability, wind mixing, precipitation and nutrient input. According to Wetzel (2001), algal community structure and growth rates among species are likely to be limited by different resources, including differing nutrients. Therefore different species can survive on varying nutrient concentrations and others compete for the same nutrients (Sitoki, 2010).

The changes in the nutrient availability can lead to variations in phytoplankton diversity and species composition in aquatic systems. Elevated pH, dissolved oxygen, NH_4-N, NO_2-N and silica favored the growth of Cyanophyceae and Chrysophyceae in the tropical ponds of Pindamonhangaba Brazil (Beyruth and Tanaka, 2000) and the most represented species were *Microcystis* spp, *Anabaena solitaries, Crucigeneilla crucifera* and *Oocystis lacustris*. Furthermore, nutrient concentration level changes due to manuring or fertilizing the aquatic systems, affects the community structure of phytoplankton in the system. The increase in nutrient concentrations leads to differences in phytoplankton community structure.

Competition, exclusion, disturbance and grazing (Lung'ayia et al., 2001) dominate the phytoplankton dynamics too. The phytoplankton community in the small water bodies is affected by the grazers and the physical chemical characteristics. The presence or the feeding behavior of the fish, result in size selection of the phytoplankton in ponds (Beyruth and Tanaka, 2000). For instance, usually *Oreochromis niloticus* ponds were dominated by chlorophyceae and cyanobacteria. In a situation of many fish, phytoplankton feeders will reduce and thus the phytoplankton will increase in case other essential requirements are present. In this way invertebrates and fish possessing size-selective feeding habits on zooplankton can influence zooplankton grazing effectiveness and in turn algal succession (Wetzel, 2001). During rainy season, the water surface temperature is reduced and the water becomes more turbid, this favours the development of volvocales, especially *Eudorina* spp, *Microcystis flos-aquae* and *Aulocaseira granulata* (Beyruth and Tanka, 2000).

Phytoplankton, as primary producers, forms the vital energy source at the first trophic tier. It is reported that over 90% of Kenya's total fish production comes from the Kenya waters of Lake Victoria and its basin (Maithya et al., 2002). The Basin is endowed with numerous perennial rivers that drain into the lake. Along the river drainage basins are important natural (small lakes and swamps) and man-made (reservoirs and dams) small water bodies (SWBs) built during the pre-independence era and stocked with various fish species. As they also serve as food to many aquatic animals, they also have an important role in the material circulation in aquatic ecosystems by controlling the growth, reproductive capacity and population characteristics of aquatic biota. Furthermore, their standing crops exhibit variations that depend on several factors. The supply of major nutrients (mainly phosphorus and nitrogen); light availability; grazing by zooplankton; water mixing regimes; and basin morphometry (Reynolds et al., 2001; Gurung et al., 2006).

Evaluation of phytoplankton community structure is essential and useful as an indicator of the water quality. To this end, recent studies in Kenya on phyto-plankton community structure, dynamics and productivity, have been conducted in a variety of freshwater lakes, enriching our knowledge and understanding of the phytoplankton ecology and community structures in such lakes. Similar studies in small water bodies of Kenya, however, are rare and sporadic. Several surveys of the phytoplankton populations have been conducted in open waters Lake Victoria (Okely et al., 2010; Krienitz et al., 2001; Lung'ayia et al., 2001, 2000), but no systematic attempt has been made to relate phytoplankton community composition and ambient water quality conditions to land use within the SWBs within the basin, despite the significant implications for integrated ecosystem management. Generalized ecoregion design nations are assumed to reflect larger scale similarities and differences than individual watersheds in geology physiography, vegetation, climate, soils, and management goals for nonpoint-source pollution. The present study is an absolutely one of the pioneering work for Kenya, and it aims to evaluate both the spatial and temporal scales of the distribution of different phytoplankton community structure and to relate phytoplankton community composition and ambient water quality conditions to land use within the northern Lake Victoria

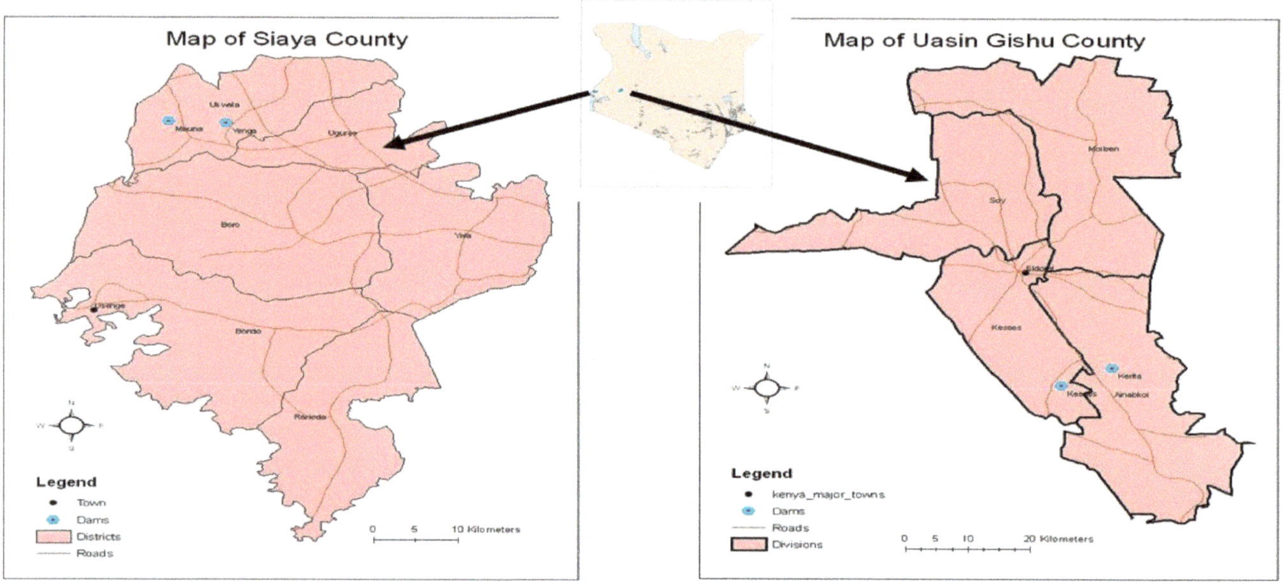

Figure 1. Map of Siaya and Uasin-Gishu counties showing the locations of small water bodies within the northern Lake Victoria basin, Kenya.

basin, Kenya.

MATERIALS AND METHODS

The study was carried out in SWBs within the northern Lake Victoria basin Kenya which was stratified in terms of altitude difference, their sizes and continuous availability of water. The low altitude sites were represented by Yenga and Mauna dams in Siaya county and high altitude by Kesses and Kerita Dams in Uasin-Gishu County (Figure 1).

Kesses dam is located at N00^0 17.263'and E035^0 19.852', at an altitude of about 2, 750 m above sea level (m.a.s.l). The littoral zone is dominated by *Typha latifolia* and *Cyperus papyrus* species. In the shallow littoral areas, water lilies, *Nyphae lotus*, water fern, *Azolla* spp., water cabbage, *Pistia stratoites*, *Salvinia*, *Lemna*, *Ceratophylum dermersum, Potamogeton, Ultricularia* and *Najas* species are found. The dam is drained by two rivers from the east including Endaragwa and Endaragweta. It has a surface area of approximately 2 km^2 and a maximum depth of 4.48 m while Kerita lies at N00^0 19.263'and E035^0 24.329', at an altitude of about 2, 800 m.a.s.l. The most noticeable emergent macrophytes community in the littoral zone is dominated by *Typhae latifolia* and *Cyperus papyrus*. It is also fed by two rivers that is, River Chebolol entering in the south-east direction and Kabiyemit which enters the dam in the south-west direction. It has a surface area of approximately 0.15 km^2 and a maximum depth of 3 m. The main human activities in the drainage basin of Kesses and Kerita dam includes majorly subsistence-crop farming, horticulture, agro-forestry, forestry and livestock rearing.

Yenga is located at N00^0 13' 03'' and E034^0 12' 44', at an elevation of about 1, 251 m.a.s.l. It has steep sided edges. The littoral zones are composed of sandy and rocky bottoms with loose macrophytic detrital and animal manure deposits brought in by surface runoff from the catchment characterize the littoral zones. The dam is dendritic in shape, has a seasonal feeder stream called Ugege and a permanent spillway. It has a surface area of approximately 0.08 km^2 and a maximum depth of 4.5 m. Mauna situated at N00^0 12' 358'' and E034^0 09' 433'', at an elevation of about 1, 217 m.a.s.l. It has steep sided edges. The littoral zones are composed of sandy and rocky bottoms with emergent macrophytes such as *Typha* sp., *papyrus sp*, reeds, and sedges dominated the zone. The main human activities in the drainage basin of Yenga and Mauna are subsistence crop farming and livestock keeping. It has a surface area of approximately 0.2 km^2 and a maximum depth of 4 m. These SWBs become over flooded during heavy down pour and rainy season as they have no direct outlets. The volume capacity depends on the dam's size. Sampling was done in each dam monthly on various sampling sites (Figure 2).

Sampling period covered both dry (November 2010 to February 2011) and rainy (March to June 2011) seasons. Triplicate samples were collected at 10 cm below the water surface in each dam at monthly intervals from November 2010 to June 2011 for the determination of phytoplankton species composition and water quality parameters. Phytoplankton was collected by plankton net (10 µm mesh), transferred into 500 mL polythene bottle and fixed with Lugol's solution and transported in cool box to the laboratory. About 300 mL quantitative samples were fixed in iodine, sedimented for 24 h, and concentrated to 50 ml. Numerical phytoplankton identification was carried out up to genus level with an inverted microscope 400X and 1000X magnifications (IMT-2, Model), using Needham Needham (1962) identification key and illustrations.

The surface water temperature and pH in the different microhabitats were measured *in situ* by a combined pH-and-temperature-meter, (OAKTON^R, Model pH/Mv/ºC METER, Singapore). Two sets of triplicate water samples were collected in glass stoppered bottles at each sampling station for dissolved oxygen (D.O) and biological oxygen demand (BOD_5) using Winkler's method (APHA 1998). The first set used to determine D.O was fixed using 2 ml manganous sulphate followed by 2 ml of Winkler's reagent.

Samples for determination of total phosphorus (TP) and total nitrogen (TN)) were collected using plastic sampling bottles of 500 ml. The samples were fixed with 3 drops of concentrated sulphuric

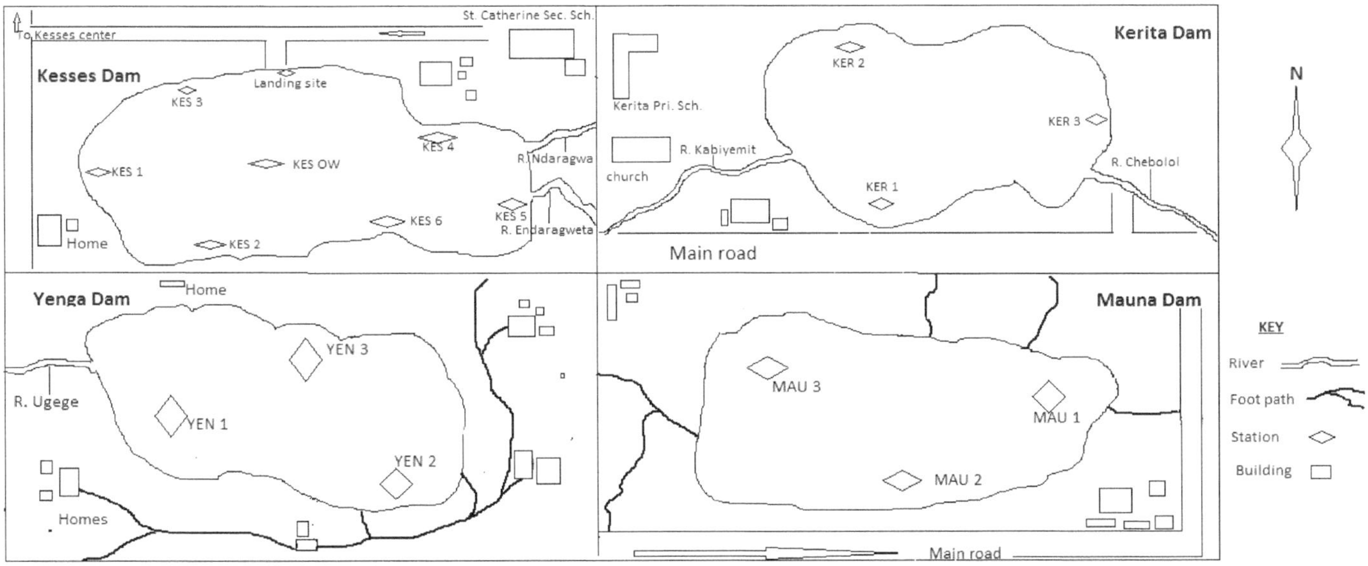

Figure 2. A sketch of the small water bodies showing the sampling sites (KES 1=Kesses station 1, KES 2=Kesses station 2, KES 3=Kesses station 3, KES 4=Kesses station 4, KES 5=Kesses station 5, KES 6=Kesses station 6, KES OW=Kesses open waters, KER 1=Kerita station 1, KER 2=Kerita station 2, KER 3=Kerita station 3, YEN 1=Yenga station 1, YEN 2=Yenga station 2, YEN 3=Yenga station 3, MAU 1=Mauna station 1, MAU 2=Mauna station 2 and MAU 3=Mauna station 3); **NB**- Not drawn to scale

acid in the field and transported in a cool-box to maintain the nitrogen balance to the laboratory for further analyses. Total nitrogen and phosphorus were determined using the Kjeldahl and Persulfate digestion methods, respectively in the laboratory (APHA, 2000).

Data analysis

The mean and standard errors of the physico-chemical parameters were determined. Species diversity (Shannon-Weinner index, H') was calculated for comparison and prior analysis of any uniqueness or difference in the sampled biological community at different space and time (Microsoft excel 2013 and Minitab 16). General linear model (GLM) was used to determine if there were any significant difference in surface water temperature, D.O, BOD_5, TN, TP and pH among the sampling stations and months. Further analysis of the above six water quality parameters related to the species composition was done using Spearman's correlation. The Shannon-Weiner diversity index was calculated as (Roy et al., 2001);

$H' = -\Sigma p_i \ln * p_i$;

Where, H' = Shannon diversity,
ln = natural logarithm
p_i = Proportion of the ith species (n/N)
n = Number of individuals of one taxon
N= Total number of individuals in a station of all taxa

RESULTS

Water quality

The mean monthly values of physico-chemical parameters and nutrients recorded among the dams during this study are summarized in Figures 3 to 8. Mean pH change fluctuated at the four dams and at different months. Kesses dam reached its peak during January 2011 and April 2011 but recorded low values during December 2010 and June 2011. In Kerita dam the highest value was registered during the months of March and April 2011 with small variation during other months. Yenga dam also fluctuates during the sampling periods with the peak during January 2011 and the same trend was also seen in Mauna dam with the highest value registered during November 2010 and lowest in April and May 2011. General linear model did not show significant difference in pH between and within the dams.

The temperature values were relatively high in the low altitude dams (Yenga and Mauna). The temperature values were recorded during the last three months (April, May and June, 2011) in Yenga dam while it was highest and lowest during the months of February 2011 and December 2010 respectively in Mauna dam. In the high altitude dams (Kesses and Kerita), there was no clear trend with high values recorded during November 2010 and May 2011 in Kerita while in Kesses the highest was registered during the months of March and June 2011. There was significant difference between the low and high altitude dams (F=31.24; p=0.002).

Mean D.O concentration change also revealed a similar trend in all dams and at different months. There was a similar trend in Kerita and Yenga with slight differences in values recorded in all months sampled. Mauna dam was different from other three dams (F=42.24; p=0.00). It registered the lowest D.O concentrations throughout the study period with its highest recorded during the month of

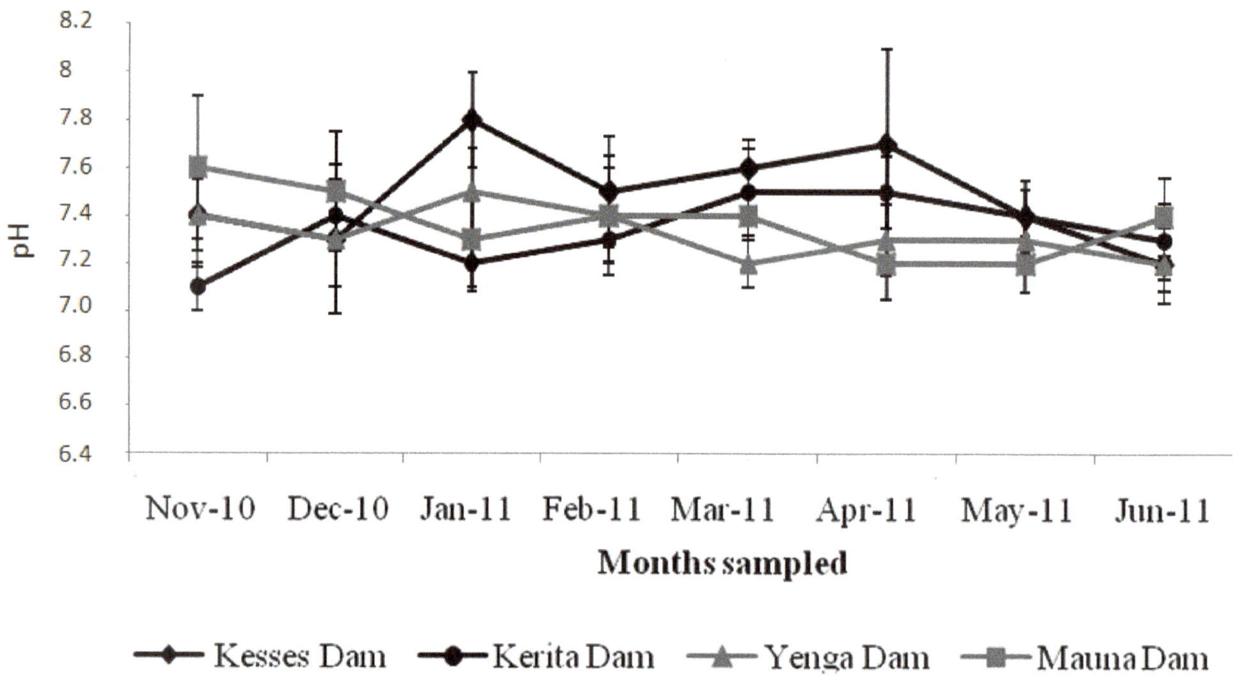

Figure 3. Mean monthly variation of surface water pH values in small water bodies during the study period (Kesses, n=56 while Kerita, Yenga and Mauna, n=24).

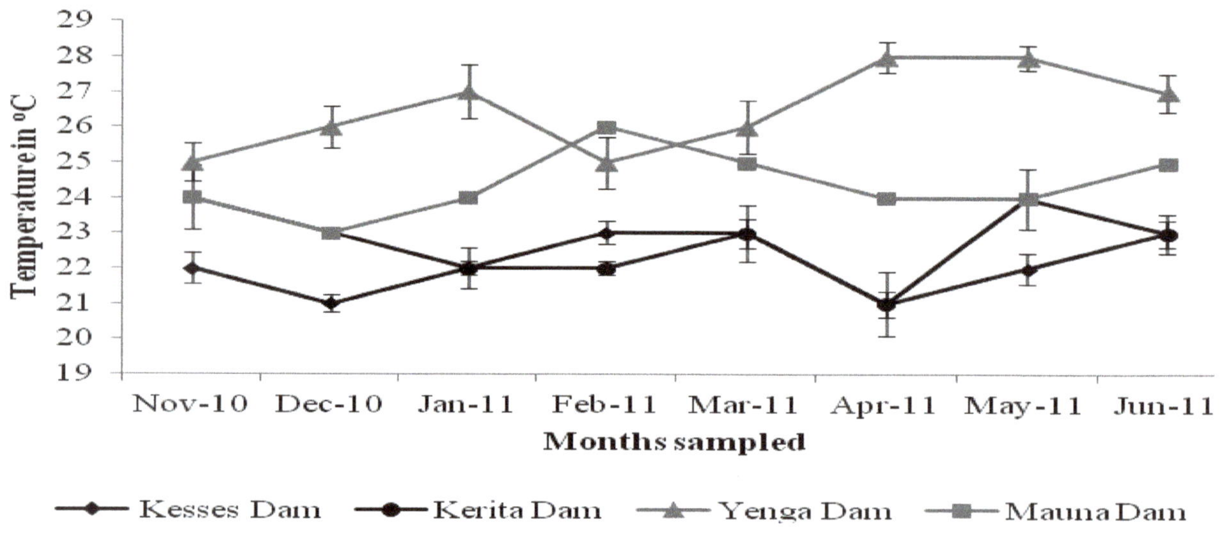

Figure 4. Mean monthly variation of surface water temperature values in small water bodies during the study period (Kesses, n=56 while Kerita, Yenga and Mauna, n=24).

January 2011 with almost the same value during the other months (Figure 5).

Kesses and Yenga dams registered the highest BOD_5 values and reached their peak during the month of May 2011 while their lowest values were recorded during the months of December 2010 and February 2011 (Figure 6). On the other hand Kerita dam had intermediate values with the highest value registered during the months of January 2011 and May 2011 while Mauna dam had the lowest BOD_5 concentration in all of the studied months except during the month of February and April 2011 where the value was above what was seen in Kerita dam. Both Mauna and Kerita registered the same BOD_5 during the month of March 2011.

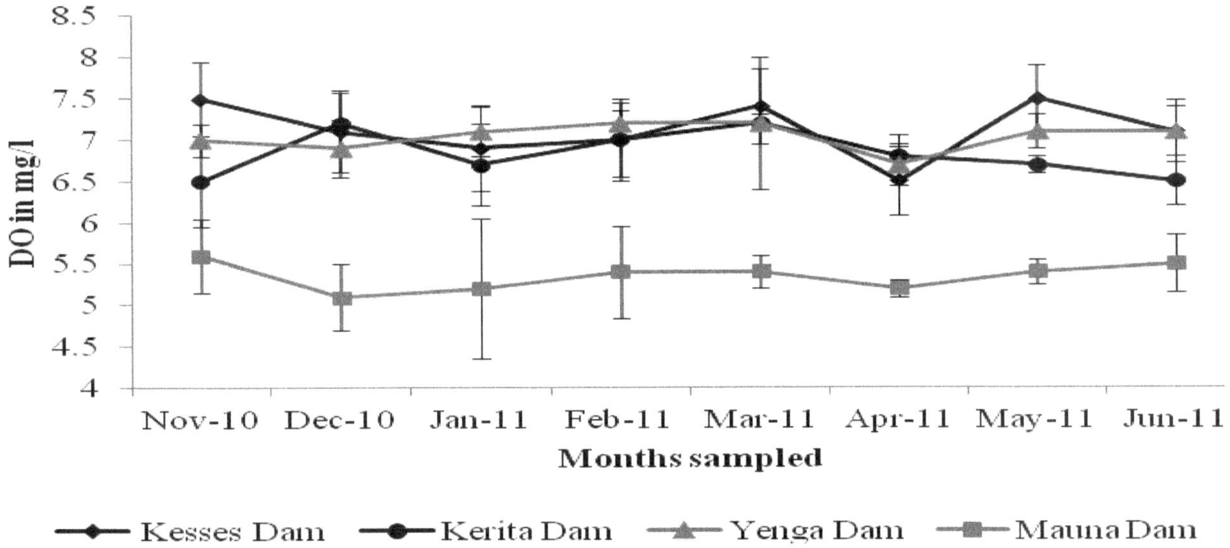

Figure 5. Mean monthly variation of surface water D.O concentration in small water bodies during the study period (Kesses, n=56 while Kerita, Yenga and Mauna, n=24).

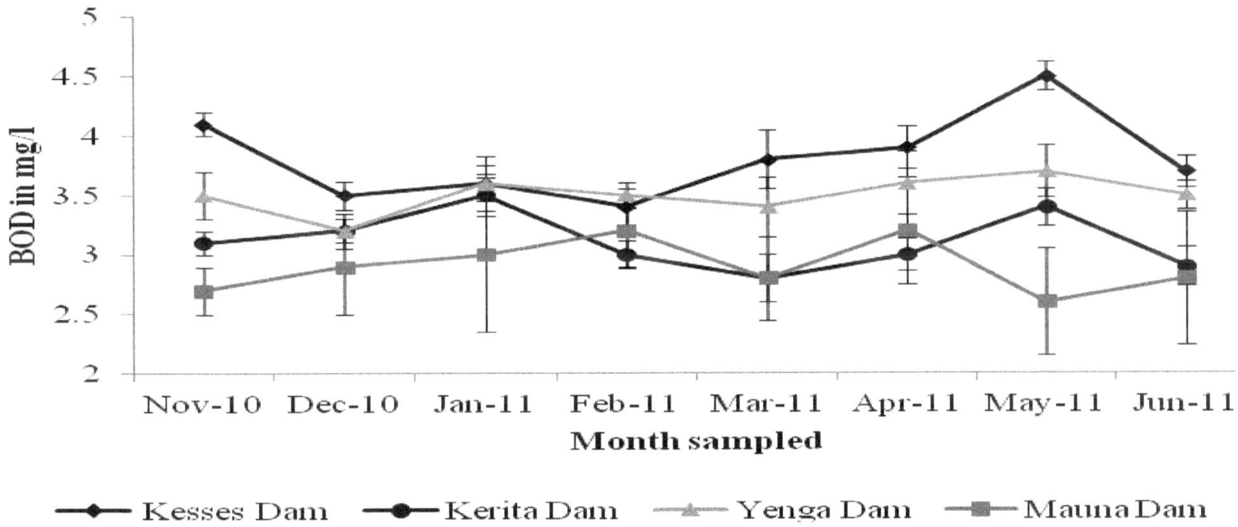

Figure 6. Mean monthly variation of surface water BOD_5 concentration in small water bodies during the study period (Kesses, n=56 while Kerita, Yenga and Mauna, n=24).

Monthly total nitrogen (TN) concentrations were high in high altitude dams with the highest values recorded in Kerita during the months of December 2010 and February 2011 while the low altitude dams registered moderately high figures with a similar trend (Figure 7). There was a significant difference between the dams (F=12; p=0.003). Total phosphorus (TP) concentration did not follow the same trend as the values were highest in Kesses and Mauna dams, Yenga dam recorded the lowest concentrations during the months of January and May 201. The GLM did not reveal significant difference between dams (Figure 8).

Phytoplankton species composition

A total of 1392 phytoplankton species belonging to four families and 20 genera were identified in Kesses dam while in Kerita dam a total of 376 phytoplankton species belonging to four families and 10 genera were sampled

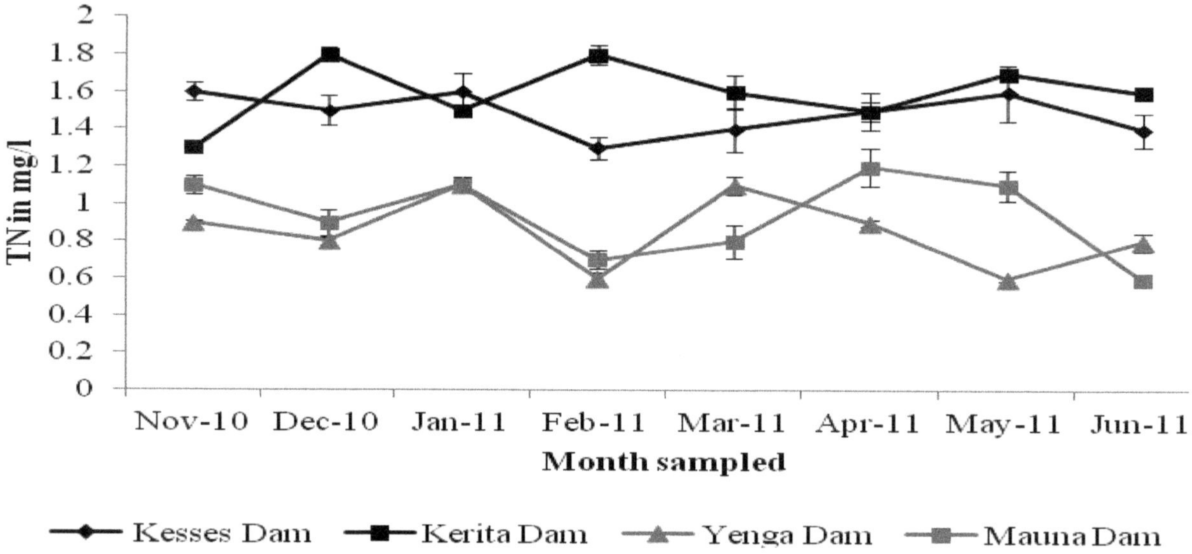

Figure 7. Mean monthly variation of surface water total nitrogen (TN) concentration in small water bodies during the study period (Kesses, n=56 while Kerita, Yenga and Mauna, n=24).

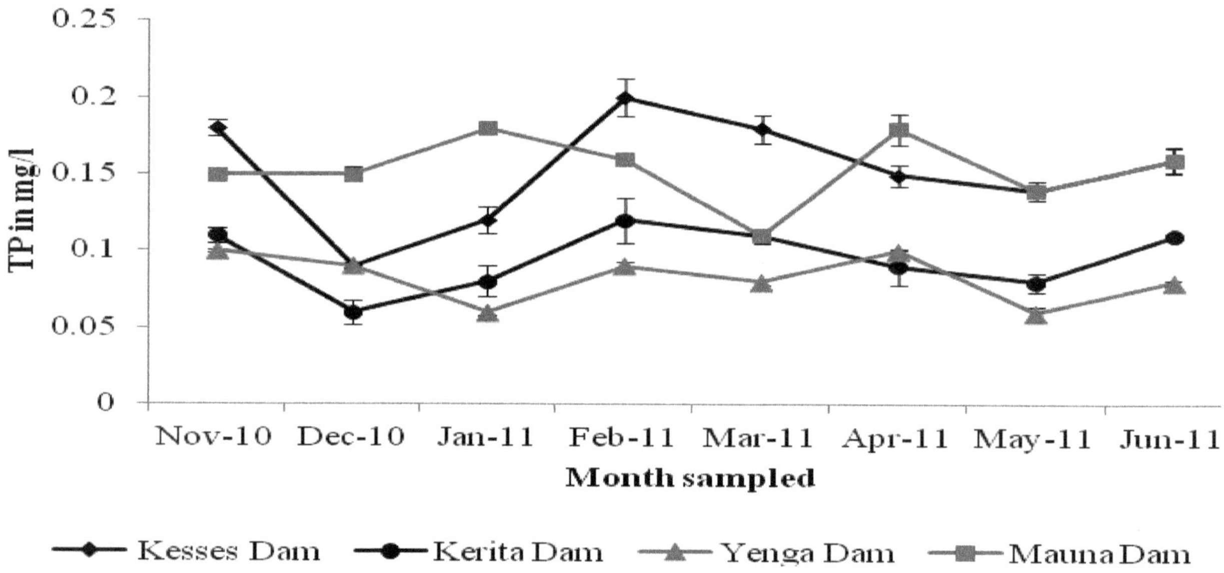

Figure 8. Mean monthly variation of surface water total phosphorus (TP) concentration in small water bodies during the study period (Kesses, n=56 while Kerita, Yenga and Mauna, n=24)

(Table 1).

In Kesses dam a total of four families of phytoplankton were found. They were Bacillariophyceae, Chlorophyceae, Cyanophyceae and Desmidiaceae. There were seven genera under Bacillariophyceae, one under Cyanophyceae, while Desmidiaceae composed by three with Chlorophyceae being the richest family with nine genera. In Kerita Dam, the same number of families (four) was sampled (Bacillariophyceae, Chlorophyceae, Cyanophyceae and Euglenophyceae). Unlike Kesses, Bacillariophyceae was the richest consisting of six genera. The family Cyanophyceae and Euglenophyceae had only one genus each (*Phormidium* sp., and *phecus* sp., respectively), Chlorophyceae had two genera (*Botryoccoccus* sp. and *Scenedesmus sp.).*

In the Siaya dams, Yenga dam had three families of phytoplankton; Chlorophyceae, Euglenophyceae and Cyanophyceae. Chlorophyceae had three genera. The

Table 1. List of phytoplankton taxa identified in Kesses, Kerita, Yenga and Mauna Dams during the study period (Blank space=Genus missing; x=present).

Family	Taxa	Kesses	Kerita	Yenga	Mauna
Bacillariophyceae	*Synedra* sp.	x	x		
	Navicula sp.	x	x		x
	Aulacoseira sp.	x	x		x
	Frustilia sp.	x	x		x
	Diatoma sp.	x	x		x
	Cyclotella sp.	x	x		
	Tabellaria sp.	x			
Desmidiaceae	*Cosmarion* sp.	x			x
	Gonatozygon sp.	x			
	Closterium sp.	x			
Chlorophyceae	*Botryoccoccus* sp.	x	x	x	x
	Crucigenia sp.	x			
	Cladophora sp.	x			
	Coelastrum sp.	x		x	
	Tetraspora sp.	x		x	
	Spirogyra sp.	x			x
	Pediastrum sp.	x			
	Scenedra sp.	x			
	Scenedesmus sp.	x	x		x
Cyanophyceae	*Phormidium* sp.	x	x	x	x
	Coelosphaerium sp.				x
Euglenophyceae	*Phecus* sp.		x	x	
	Euglena sp.			x	

family Euglenophyceae composed of two genera. The Cyanophyceae had a single genus. The family Desmidiaceae which was missing in Yenga was an extra taxon in Mauna dam which registered a total of four families of phytoplankton was found, that is, Chlorophyceae, Euglenophyceae, Cyanophyceae and. The family Chlorophyceae had three genera. The Botryococcus was the richest family with four). Desmidiaphyceae had a single taxon (*Cosmarion sp.)* while the family Cyanophyceae had two genera (Table 1).

Species monthly diversity index

The mean Shannon Weiner diversity index (H') in the small water bodies is given in Figure 9. The highest value of H' was obtained during the rainy season (March to June) in all small water bodies while dry season (November to February) had low species diversity. Kesses dam registered the highest diversity index in all sampled months, followed closely with Yenga dam then Mauna and lastly the lowest was recorded in Kerita dam in most of the sampled months. The diversity indices in all dams were generally low (Figure 9).

Spearman's correlations between phytoplankton and physico-chemical parameters in SWBs (Kesses, n=56, Kerita, Yenga and Kerita, n=24)

Phytoplankton community relative abundance in Kesses dam revealed different relationship with most physico-chemical parameters and nutrients (Table 2). The family Bacillariophyceae was negatively related to temperature, BOD_5, D.O and a significant negative and positive correlation to TP and pH (r=-0.29; p=0.000; r=0.372; p=0.013) respectively. Cyanophyceae was negatively related to all parameters except pH and BOD_5 with chlorophyceae showing significant negative relationship with temperature, D.O and TN r=-0.15, -0.18, -0.07; p=0.000) respectively. Desmidiaceae had an significant negative relationship with the D.O (r=-0.26; p=0.025).

In Kerita dam the relative abundance of phytoplankton families showed some significant negative and positive

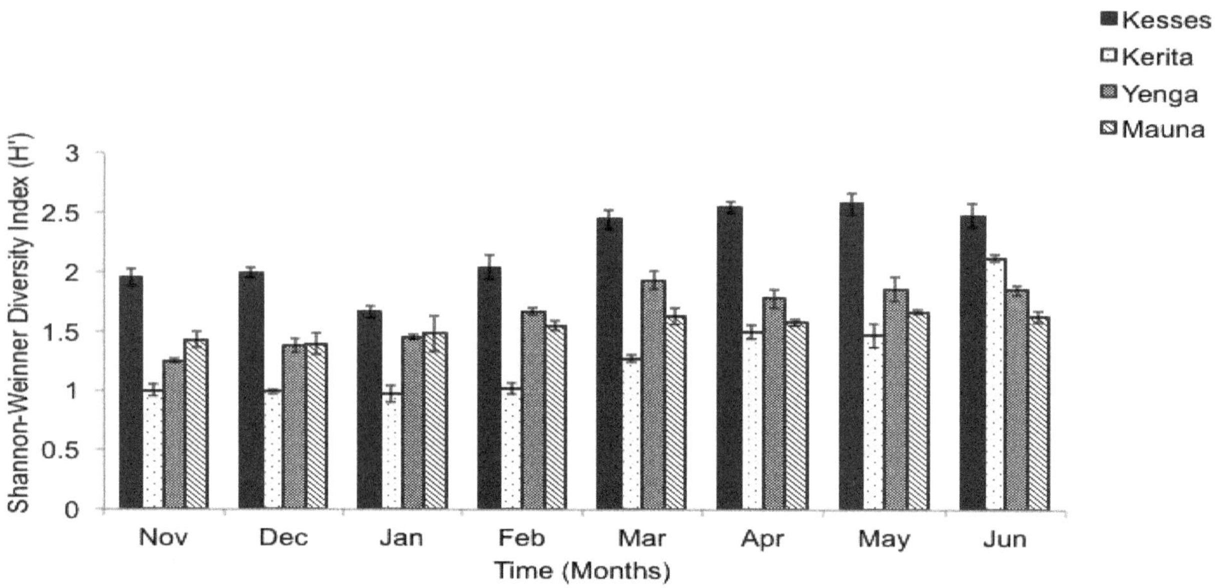

Figure 9. Spatio-temporal variations in phytoplankton diversity indices in SWBs during November 2010 to June 2011 (Kesses, n=56 while Kerita, Yenga and Mauna, n=24).

Table 2. Spearman rank order correlation coefficients between physicochemical parameters and taxa community structure attributes (* = significant correlation at α = 0.05) (Kesses, n = 56 while Kerita, Yenga and Mauna, n = 24).

County	Dam	Physico-chemical parameter						
		Taxa	pH	Temp	BOD_5	D.O	TN	TP
Uasin-Gishu	Kesses	Bacillariophyceae	0.372*	-0.114	-0.052	-0.074	0.117	-0.29*
		Chlorophyceae	0.095	-0.013	0.051	0.024	-0.013	-0.289
		Cyanophyceae	0.188	-0.15*	0.03*	-0.18*	-0.07*	-0.131
		Desmidiaceae	-0.108	0.053	-0.235	-0.26*	0.265	-0.227
		Euglenophyceae	-	-	-	-	-	-
	Kerita	Bacillariophycaea	0.079	-0.284	-0.11*	0.075	0.533*	-0.309
		Chlorophyceae	0.20*	-0.006	0.389	0.17	-0.502	-0.364
		Cyanophyceae	0.609	-0.026	0.135	0.024	-0.336	-0.342
		Desmidiaceae	-	-	-	-	-	-
		Euglenophyceae	0.125	0.11*	0.064	-0.271	0.28*	-0.284
Siaya	Yenga	Bacillariophyceae	0	0	0	0	0	0
		Cyanophyceae	-0.135	-0.209	0.03	-0.199	-0.168	0.256
		Chlorophyceae	0.374	-0.358	0.2*	0.323	-0.065	-0.472
		Euglenophyceae	-0.089	0.155	-0.09	0.016	-0.102	0.32*
	Mauna	Bacillariophyceae	0.34	0.163	0.024*	0.115	-0.188	-0.223
		Chlorophyceae	0.1*	0.316	-0.205	-0.179	-0.198	0.258
		Cyanophyceae	0.17*	-0.282	-0.337	-0.497	0.005*	-0.02*
		Desmidiaceae	-0.462	0.198	-0.13*	-0.225	0.241*	0.619*

relationship with physico-chemical parameters and nutrients. The family Bacillariophyceae was positively related to only pH, D.O and a significant relationship with TN (r=0.533, p=0.0042) with an insignificant negative relationship with all other parameters except BOD_5 (r=-0.11, p=0.000). Cyanophyceae showed an insignificant positive related to pH, BOD_5 and D.O (r=0.609, 0.135 and 0.024) respectively. Chlorophyceae was negatively related to temperature and both nutrients and positively related to the other parameters. It revealed a significant positive relationship with pH (r=0.20, p=0.035). Euglenophyceae had a weak significant positive relationship with temperature and TN (r=0.11, P=0.029 and r=0.28, p=0.001) respectively and a weak insignificant positive relationship with BOD_5 and pH and weak negative relationship with D.O and TP.

Phytoplankton relative community abundance in Siaya dams revealed different significant relationship with physico-chemical parameters and nutrients. In Yenga the family Bacillariophyceae revealed no relationship between phytoplankton, physico-chemical parameters and nutrients. Chlorophyceae revealed a significant positive relationship with BOD_5 (r=0.2, p=0.002) respectively, pH and D.O showed a weak positive relationship. The family Cyanophyceae showed no significant relationship with pH, temperature, D.O, and TN. The family Euglenophyceae revealed a positive relationship with temperature, D.O and TP with a negative relationship with the rest of the parameters. There was a significant positive relationship with TN (r=0.32, p=0.000). In Mauna the trend was the same with different phytoplankton families showing different relationships with physico-chemical parameters and nutrients. The family Chlorophyceae and Bacillariophyceae showed a significant relationship with pH and BOD_5 (r=0.100, p=0.045 and r=0.024, p=0.014), respectively. Cyanophyceae was positively related to pH and TN while negatively correlated to other parameters while Desmidiaceae had both weak negative and positive relationship with physico-chemical parameters.

DISCUSSION

The physicochemical environment of SWBs within Lake Victoria Basin, Kenya displayed considerable spatial and temporal variation in relation to prevailing environmental conditions. Some physicochemical parameters examined in this study were considerably outside the ranges reported for other similar SWBs (Maithya, 2008) this is a clear indication that these ecosystems have undergone fundamental shifts over time. Possible explanation for this involves variations in the frequency of lotic nutrient loading and the rates of mineralization of organic matter for lentic self-loading. High altitude regions generally recorded higher nutrient concentrations as compared to low altitude regions. This could be due to high nutrient concentration that is eroded from the catchment of high altitude regions which is a prime agricultural zone as opposed to low regions with no fertilizer-intensive agricultural activities located in its catchment and this accounts for moderately low total nitrogen and total phosphorus level recorded in Siaya dams. These extreme variations in SWBs physicochemical parameters give these water bodies the distinction of displaying an extreme physical and chemical environment, which seem to interact in determining the nature and structure of the autotrophic assemblages of organisms in them. This is in line with Carpenter et al. (1998) who stated that nutrient availability in an aquatic environment is primarily influenced by seasonal application of fertilizers among other physical factors.

Although there was optimal temperature, pH, D.O concentration and nutrient availability in SWBs, offering good conditions for high algal composition and diversity, only five algal families were observed in the samples collected during the Eight months sampling period, indicating that the SWBs were poor in species richness, and suggesting that its extreme water quality together with the presence of phytophagous fish species fundamentally shapes the ecological functioning of these dams. The SWBs' taxa composition also indicates a low species number which, although exhibiting some degree of similarity to other studies such as Oyoo et al. (2009). The observed low species number in SWBs, particularly for the more tolerant Cyanophyceae and Bacillariophyceae taxa, could be attributed to changes in habitat quality resulting from water quality changes originating from catchment areas that have undergone more land conversion to agriculture. Cyanophyceae have particular preferences for polluted environments because they grow in competitive interactions, with their enhanced ability to utilize carbon (mainly in polluted environments) at high pH levels, giving them an advantage over other phytoplankton species (Atici and Olcay, 2006). This observation seems to suggest that the water draining into Lake Victoria, Kenya from its catchments has some degree of pollution resulting from both point and non point sources. This possibility could explain the decline of sp., *Crucigenia* sp., *Cladophora* sp., *Coelastrum* sp., *Tetraspora* sp., *Spirogyra* sp., *Pediastrum* sp., *Scenedra* sp. and *Scenedesmus* sp with species from the family Euglenophyceae and Desmidiaceae missing completely from the phytoplankton populations during the rainy seasons, it is possible the population pressures in the catchment is leading to massive water quality changes that could alter the quality of the water draining into the SWBs, with subsequent impacts (alterations) on the species assemblage in these SWBs. This observation is supported by (Lung'ayia et al., 2000) who stated that phytoplankton species composition, numerical abundance, spatial distribution and total biomass are directly affected by the environmental factors.

The phytoplankton species variation observed in this study can therefore be attributed to a combination of these unique physiological attributes under different environmental conditions. The presence of large numbers of diatoms at some sampled stations indicates that these

SWBs are constantly changing, putting a recurring ecosystem balance between eutrophism and meso-trophism. This was reflected in this study by the alternate seasonality of occurrence of some few genera or species of Bacillariophyceae. The seasonality of occurrence is a function of nutrients availability or resurgence in the water bodies (Reynolds et al., 2001).

Seasonal variations in phytoplankton composition and diversity were significant. The peak composition and diversity corresponded to the onset of the rainy seasons. Composition of *Synedra* sp., *Navicula* sp., *Aulacoseira* sp., *Frustilia* sp., *Diatoma* sp. and *Cyclotella* sp increased during the rainy seasons. This observation is in line with Oyoo et al. (2009) who also reported increased phytoplankton composition and diversity during rainy season. Thus, current changes in the phytoplankton composition appear to be directly related to the prevailing environmental conditions. Precipitation can play a critical role in determining limnological properties and plankton dynamics of water bodies (Kalff, 2002) by washing a substantial quantity of allochthonous materials and nutrients from the catchment into the SWBs that can stimulate the growth and sustenance of phytoplankton populations, thereby also increasing phytoplankton composition. In contrast, the increased water inflow into the SWBs from the streams feeding it during the rainy season could decrease the water transparency, thereby reducing the availability of light in the water column to stimulate phytoplanktonic growth, especially for phytoplankton species residing some distance below the water surface.

Spatial variability is a structural character of an ecosystem. Spatial distributions allow for complex population interactions involving energy transfer, competition and niche apportionment (Brower et al. 1990). It is generally expected that less-disturbed sites exhibit higher species diversity (Jones et al., 2002). Anthropogenic impacts from changes in land-use practices are known to affect phytoplankton diversity patterns in lakes receiving drainage from modified stream water quality from the catchments (Death, 2000), often leading to decreased abundance of taxa, as well as shifts to a more unevenly distributed community containing only one or two numerically dominant taxa (Jones et al., 2002). The high level of dominance by a few taxa at sampling sites that experience water inflows compared to the more evenly distributed taxa in inshore sites, suggests that human activities already are shaping SWB's phytoplankton communities. Sitoki (2010) reported a wide range of land-use activities in the upper catchments of L.Victoria. These human activities can ultimately affect spatial variations of species assemblage in SWBs.

The observed correlation variations could be attributed to variations in the degree of anthropogenic activities resulting into different concentrations of chemical discharge into the different SWBs. Other studies support this present result, such as Sabater et al. (2004) who noted that changes in the structure of the microbial components did not significantly affect the organisms community (where only a few taxa showed changes because of the nutrient addition).

Although most of the phytoplankton families have not been found to produce toxic secondary compounds, other cyanobacteria species (for example, *Microcystis*; *Anabaena*) have produced cyanotoxins under conditions of changed water quality that can cause death, when consumed by the SWB's biodiversity, for example, birds. This underscores an urgent need for continuing monitoring of the SWBs' phytoplankton populations, as well as a closer collaboration between the government and the communities along the rivers draining to these water bodies to avoid polluting it. The urbanization process characterizing the nearby towns also should be better studied as a means of avoiding direct pollution effects that could influence water bodies' phytoplankton community structure. Furthermore, nutrient reduction strategies should be used to control the excessive phosphorus load to the SWBs, especially during the rainy seasons. In order to manage other water bodies, ecological approaches concerned with the phytoplankton community structure, and prevention of activities or measures through which this structure can be modified, should be used. With the use of such initiatives, water quality changes likely to cause adverse changes to the biotic assemblages can be discerned sufficiently early to allow corrective measures to be developed and implemented before irreparable ecological damages to the aquatic system occur.

ACKNOWLEDGEMENT

The authors wish to thank National Council of Science and Technology (NCST) for funding this study and also Department of Fisheries and Aquatic Sciences of University of Eldoret for provision of both laboratory space and technical support during the entire study period.

REFERENCES

Adame MF, Alcocer J, Escobar E (2008). Size-fractioned phytoplankton biomass and its implications for the dynamics of an oligotrophic tropical lake. Freshwater Biol. 53:22-31.

APHA (1998) (American Public Health Association) Standard methods for the examination of water and wastewater, 20th edn Washington, D.C USA.

APHA (2000). (American Public Health Association) Standard methods for the analysis of water and wastewater, 15th Ed. American Public Health Association and Water Pollution Control Federation. Washington DC, USA. p. 44.

Atici T, Olcay O (2006). Seasonal variations of phytoplankton and value of chlorophyll a in the Sariyar Dam Reservoir (Ankara, Turkey). Turk.. J. Bot. 30:349-57

Beyruth Z, Tanka M (2000). Biovolume of the phytoplankton in aquaculture tropical ponds. Verh. Internat. Verein. Limnol. 27:689-695. Brower JE, Zar JH, von Ende CN (1990). Field and Laboratory

Methods for General Ecology. William C. Brown Publishers, New York.
Carpenter SR, Caraco NF, Correll DL, Howarth RW, Sharply AN, Smith VH (1998). Non-point Pollution of Surface Waters with Phosphorus and Nitrogen. Ecological Applications. 8:559-568.
Cecchi P (2007). L'eau en partage: les petits barrages de Côte d'Ivoire. IRD (éd.). Collection latitudes 23. Paris: 2007, p. 295.
Death RG (2000). The effect of land use on the speciesarea relationships in benthic stream invertebrates. Verh. Int. Ver. Theor. Angew. Limnol. 27:2519-2522.
Descy JP, Hardy MA, Stenuite S, Pirlot S, Leporcq B, Kimirei I., Sekadende B, Mwaitega SR, Sinyenza D (2005). Phytoplankton pigments and community composition in Lake Tanganyika. In: Freshwater Biology. Vol. 50, 2005. pp. 668-684.
Descy JP, Sarmento H (2008). Microorgansms of the East African Great Lakes and their response to environmental changes. In Freshwater Reviews. Vol. 1, 2008. pp. 59-73.
Gurung TG, Dhakal RP, Bista JD (2006). Phytoplankton primary production, chlorophyll-a and nutrient concentration in water column of mountainous Lake Phewa, Nepal. Lake Reserv. Manage. 11:141-148.
Jones C, Palmer RM, Matkaluk S, Walters M (2002). Watershed Health Monitoring: Emerging Technologies. CRC Press, Boca Raton, FL.
Kalff J (2002). Limnology: Inland Water Ecosystems. Prentice Hall Inc. Upper Saddle River, NJ.
Krienitz L, Ballot AA, Wiegand C, Kotut K, Codd CA, Phlugmacher S (2001). Cyanotoxin-producing Bloom of Anabaena flos-aquae, Anabaena discoidea and Microcystis aeruginosa (Cyanobacteria) in Nyanza Gulf Lake Victoria, Kenya. J. Appl. Bot. Angewandte Botanik 76. pp. 179-193.
Lafond MB, Pinel A, Ross P (1990). Biomass and photosynthesis of sizefractionated phytoplankton in Canadian Shield lakes. Hydrobiologia. 196:25-38.
Levandowsky WM (1972). Tropical Limnology. Ann. Rev. Ecol. Syst. 18:159-184.
Lewis WM (1976). Surface/volume ratio: implications for phytoplankton morphology. Science 192:885-887.
Likens GE (1975). Primary production of inland aquatic ecosystems. In: Primary productivity of the biosphere. Lieth RH, Lieth & RH Whittaker (eds.): 185-202. Springer Verlag, New York.
Lung'ayia H, Sitoki L, Kenyanya M (2001). The nutrient enrichment of Lake Victoria (Kenyan waters). Hydrobiologia. 458:75-82.
Lung'ayia, HBO, M'Harzi A, Tackx M, Gichuki J, Symoens JJ (2000). Phytoplankton community structure and environment in the Kenyan waters of Lake Victoria. Freshwater Biol. 43:529-543.
Maithya J, Charo H, Ouma H (2002). Aquaculture potential of small water bodies in Lake Victoria basin, Kenya. Kenya Marine and Fisheries Research Institute Annual Report, 2001/2002.
Maithya JM (2008). Ecology, Biology and Aquaculture potential of (Oreochromis variabilis) (Boulenger 1906) in Lake Victoria basin, Kenya PhD thesis Moi University, Kenya.
Needham JG, Needham PR (1962). A guide to the study of freshwater biology.
Okely P, Imberger J, Antenucci JP (2010). Processes affecting horizontal mixing and dispersion in Winam Gulf, Lake Victoria. Limnol. Oceanogr. 55:1865-1880.
Oyoo EO, Muchai M, Shivoga WA, Miller NS, Rasowo J, Ngugi CC (2009). Spatial and seasonal variations in phytoplankton community structure in alkaline-saline Lake Nakuru, Kenya. Lakes & Reservoirs: Res. Manage. 14:57-69.
Reynolds CS (1999). Non-determinism to probability, or N:P in the community ecology of phytoplankton: Nutrient ratios. In Archiv für Hydrobiologie, Vol. 146, 1999. pp. 23-35.
Reynolds CS, Irish AE, Eliot JA (2001). The ecological basis for stimulating phytoplankton responses to environmental change (PROTECH). Ecol. Model. 140:271-91.
Roy AH, Rosemond DA, Leigh DS, Paul MJ, Wallace JB (2001). Efforts of changing land use on macroinvertebrate integrity. Georgia Water Resource Conference. University of Georgia, Athens, G.A. March 2001. p. 47.
Sabater S (2004). Diatom communities as indicators of environmental stress in the Gaudiamar river, S-W. Spain, following a major mine tailing spill. J. App. Phycol. 12:113-124.
Simciv T (2005). The role of plankton, zoobenthos, and sediment in organic matter degradation in oligotrophic and eutrophic mountain lakes. In Hydrobiologia, vol. 532, 2005, pp.69-79.
Sitoki LM (2010). Phytoplankton and eutrophication in Nyanza Gulf (with special reference to toxic cyanobacteria and planktonic diatoms). Ph.D. thesis, University of Innsbruck, Faculty of Biology, Institute of Botany p.106.
Stockner JG, Suttle CA, Harrison PJ (1987). Effects of nutrient pulses on community structure and cell size of a freshwater phytoplankton assemblage in culture. Can. J. Fish. Aqua. Sci. 44:1768-1774.
Talling JF (1966). Photosynthetic behavior in stratified and unstratified lake populations of a plankton diatom, Freshwater Biology. 54:99-127.
Walsh CJ, Sharpe AK., Breen PF, Sonneman JA (2001). Effects of urbanization on streams of the Melbourne region, victoria, Australia. I. Benthic macroinvertebrate communities. In Freshwater Biol. Vol. 46, 2001. pp. 535-551.
Wetzel RG (2001). Limnology: Lakes and River Ecosystems. Academic Press, San Diego. p. 1006

Mortality assessment of *Oreochromis niloticus* fingerlings in varying salinity and influence of salinity changes on acute toxicity of lead

Osuala Fidelia Ijeoma and Bawa- Allah Kafilat Adebola

Department of Zoology, Faculty of Science, University of Lagos, P.M.B. 101017, Lagos, Nigeria.

The setting of safe limits for lead (Pb) into the lagoon and protection of fishes such as *Oreochromis niloticus* has been based on its toxicity ignoring the influence of salinity, an important parameter in the lagoon. The study therefore investigates the salinity tolerance and relative acute toxicity of lead nitrate [$Pb(NO_3)_2$] under fresh water and varying salinity against *O. niloticus* fingerlings. A total of 280 fingerlings were used as test animals and $Pb(NO_3)_2$ as test chemical. Concentrations were 1, 2, 4, 6, 8, 10, 30, 40, 50, 60, 80, 120, 160 and 180mg/l at salinity levels of 2, 12, 18, 22, 32, 35 parts per thousand (ppt) and fresh water. Dose response data were analyzed by Probit. Mortality assessment showed that the fingerlings could not survive at salinities of 22, 32 and 35 ppt within 24 h but survived well in 2 ppt. The 96 h LC_{50} was 130.094 mg/l at 12 ppt but below 12 ppt there was a steady increase in toxicity (3.255 and 6.243 mg/l) in fresh water and 2 ppt respectively. Similarly, the toxicity also increased with an increase in salinity from 12.1 to 18 ppt (113.191 mg/l). The significance of this study shows the need for inclusion of varying salinity in setting of safe limits to confer protection on the delicate biotic components of the rich lagoon biodiversity.

Key words: Lead nitrate, *Oreochromis niloticus*, salinity.

INTRODUCTION

There is paucity of data on the setting of safe limits for discharge of heavy metals such as lead into the lagoon ecosystems, aimed at protecting brackish water organisms at varying salinity conditions. Salinity is one of the major factors (Lawson and Anetekhai, 2011) affecting organisms including fishes such as Tilapia in aquatic medium (Lagoon) and also has the ability to influence the toxicity of pollutants such as heavy metals (Oyewo, 1998; Lawson, 2011). There have been a lot of studies on the influence on the changing salinity on various physicochemical parameters, pollutant toxicity (heavy metals and pesticides) and their effects on fishes (Oyewo, 1998; Chukwu and Okpe, 2006; Breves et al., 2010; Lawson, 2011; Samy et al., 2011). The contamination, persistence, bioaccumulation and biomagnification of heavy metals (manganese, nickel, lead and zinc) in natural brackish are of global concern (Bhupander et al., 2011; Iulia et al., 2012; Marleen et al.,

2013). Researchers have established high levels of heavy metals in the Lagos lagoon (Oyewo, 1998; Aderinola et al., 2009). Lead (Pb) is one of the most common heavy metals detected in the Lagos lagoon (Oyewo, 1998) which is not useful in living systems and extremely toxic at low concentration thus an indicator of pollution. In Nigeria, the recent increase in the environmental Pb has been attributed to industrial discharges and mining activities. The toxicity of Pb to aquatic organisms and its accumulation in the aquatic biota are well documented. Leblond and Hontela (1999) reported that the 96 h median tolerance limit value for fathead minnows and brook trout were 40 to 70 mg/L in soft water and 440 to 480 mg/L in hard water. The 96-h LC_{50} for Lead acetate in bluegill was found to be approximately 400 mg/L (McKim, 1985).

The principal toxic effects of chronic Pb exposure to fish cause hematological (Małgorzata et al., 2010; Usama et al., 2013), neurological (Talia et al., 2009) and renal (Patel et al., 2006) disorders. In water bodies contaminated with Pb, it has been observed in fishes that it causes the disruption of Na^+, Cl^- and Ca^{2+} regulation during acute exposure, development of black tails and spinal curvature during chronic exposure as well as disruption in hemoglobin synthesis (Sippel et al., 1983; Rogers et al., 2003; Rogers et al., 2005). Concentration of Pb in pond water ranged from 0.03 - 0.16 g mL^{-1} (Himadri and Anilava, 2000). In polluted areas dissolved Pb has been observed in concentration of 50400 $nmolL^{-1}$ (Bowles et al., 2006) while Pb toxicity has been reported for some freshwater species at concentrations as low as 50 nmol L^{-1} (Grosell et al., 2006).

A typical species that inhabit the brackish water is *Oreochromis niloticus.* It is among some of the edible commercial fish species that inhabit the Lagos lagoon and is also widely bred in commercial fish farms. *O. niloticus* can be cultured in fresh water coupled with slight challenges due to fluctuations in salinity and this makes it not so easy to be cultured in the laboratory. Since *O. niloticus is* sensitive to heavy metals (Taweel et al., 2013) it is therefore of interest to study the influence salinity changes will have on the effect of heavy metals such as lead which could be a biological constraint and crucial in setting of safe limits for this fish.

MATERIALS AND METHODS

Test animals

O. niloticus (Nile tilapia, Gnathastomato, Cichlidae) fingerlings of similar sizes (total length; 11 ± 6 cm, mean weight; 20 ± 4 g and age (4 - 6 weeks old) were purchased from Agboola farms and transported to the laboratory in 25 L container into holding tanks (50 X 30 X 35 cm), which contained fresh water from fish pond (opened at the top for aeration) in the laboratory. The fingerlings were kept in the holding plastic tanks, half filled (15 L) with dechlorinated tap water, to acclimatize to laboratory conditions (28 ± 2°C, R.H 70 ± 2%) for a period of seven days before they were used in the bioassays. The fingerlings were fed with fish food (Coppens) at 3% of their body weight and water was changed once every 48 h, aerated continuously with circular plastic bowls (volume = 0 5 L, bottom diameter = 18 cm) were used as bioassay container.

Test chemicals

The heavy metal, lead as [$Pb(NO_3)_2$] investigated in this work was obtained as metallic salts of Fisons laboratory reagents, Analar grades of molecular weight 331.21 g) with 98% purity manufactured by British Drug House (BDH). The choice of Pb as the heavy metal for this study was based on the available and common metal from the results of a chemical survey of industrial effluents that empty into the Lagos lagoon (Oyewo, 1998).

Preparation of treatments at varying salinities

Seawater was obtained in 25 L container, Lagos State and taken to the laboratory, where the salinity (35 ppt) was measured with a salinometer. On the basis of salinity level, computed volumes of the sea water were measured out and mixed with dechlorinated tap water to obtain sea water solutions at pre-determined salinity levels. The salinity of the prepared media was verified with a salinometer for more accurate determination of the salinity was recorded and used in appropriate experiments.

Preparation of treatments for acute toxicity test

A pre-determined amount of Pb compound was weighed (using an oertling 30TD top loading balance) and diluted with given volume of dechlorinated tapwater to obtain a stock solution of known strength. The resultant stock solution was serially diluted to obtain solutions of required concentrations. Actual concentration of Pb ions in each solution of known strength was computed based on molecular weight of test compound. Test media were always made up to 2 L because preliminary studies showed that 10 fingerlings survived well in 2 L of media for a period of 7 days without aeration. In range finding experiments, the fishes were exposed to a wide range of concentrations (600, 300, 100, 50, 25, 5 and 1 mg/l) of the test compound to obtain an effective range of activity, initially relying on trial and error techniques.

Assessment quantal response (mortality)

Fingerlings were taken to be dead if no body movements including the operculum were observed, even when prodded with a blunt glass rod and submerged in water.

Salinity tolerance of *O. niloticus*

Five active fingerlings of similar age and size (as described above) were taken from plastic holding tanks with a sieve (200 µm) and randomly assigned to bioassay containers already holding treatments at varying salinities as described below. Each treatment was replicated twice, giving a total of 10 fingerlings that were exposed per salinity level (35, 32, 22, 18, 12, 2 ppt) and fresh water. Mortality was assessed as described above, once every 24 h for a period of 7 days.

Acute toxicity of ($Pb(NO_3)_2$) against *O. niloticus* in varying salinity treatments

A similar set up as described above in the case of salinity tolerance

Table 1. Acute toxicity concentration at fresh water, 2, 12 and 18 ppt.

Fresh water (mg/l)	2 ppt	12 ppt	18 ppt
1.0	1.0	30.0	40.0
2.0	2.0	50.0	60.0
4.0	4.0	80.0	80.0
6.0	6.0	120.0	120.0
8.0	8.0	160.0	180.0
10.0	10.0	200.0	240.0

Table 2. Salinity tolerance of *O. niloticus* under varying salinity (mortality assessment).

Salinity (ppt)	Time (h)					% Mortality		
	24	48	72	96	120	144	168	
35	10	10	10	10	10	10	10	100
32	10	10	10	10	10	10	10	100
22	9	10	10	10	10	10	10	100
18	5	6	7	7	7	7	7	70
12	2	2	3	3	3	3	3	30
2	0	1	1	1	1	1	1	10
Fresh water	0	0	0	0	0	0	0	0

of *O. niloticus* was carried out in this case but mortality was assessed once every 24 h for a period of four days.

Acute Toxicity concentrations

Acute toxicity concentrations under which relative acute toxicity of test heavy metal against *O. niloticus* was investigated were as shown in Table 1.

Statistics

The dose response (mortality) data for freshwater and varying salinities were analyzed with a computer program using Probit analysis after Finney (1971) as adopted by Don Pedro (1989) and used as indices for measuring toxicity of the test organisms to the toxicant. The Indices of measuring toxicity derived from these analyses were Lethal Concentration that caused 95, 50 and 5% response [mortality] of exposed organisms (LC_{95}, LC_{50}, LC_5, respectively) and their 95% confidence limits.

RESULTS

Salinity tolerance of *O. niloticus*

The results of salinity tolerance of *O. niloticus* based on mortality assessment, showed that the fishes could not survive for any appreciable period in treatments with salinity at 22 ppt and above (Table 2). In treatments with salinity of 18 ppt, between 50 - 60 % mortality was recorded in 48 h but from 72 - 168 h, 70 % mortality was observed. At 2 and 12 ppt, mortality occurred at low levels (10 - 20 %) after 24 - 48 h respectively but stabilized at 30% after 72 h at 12 ppt (Table 1). No mortality was recorded in fingerlings exposed to media with fresh water.

Acute toxicity of $Pb(NO_3)_2$ against *O. niloticus* in varying salinity treatments

The acute toxicity of $Pb(NO_3)_2$, based on 96 h LC_{50} values against *O. niloticus* fingerlings at salinity of 12 ppt, was 130.094 mg/l whereas at 2 ppt and freshwater, it was 6.243 and 3.255 mg/l respectively which was significantly (no overlap in 95 % CL of 96 h LC_{50}) lower than the toxicity values at 12 and 18 ppt (113.191 mg/l) treatments evaluated (Figure 1).

Additionally, the relative toxicity of $Pb(NO_3)_2$ based on 24, 48 and 72 h LC_{50} values followed a similar trend of relative toxicity against the fingerlings as 96 h LC_{50} values in all test media (freshwater, 2,12 and 18 ppt) and there was no overlap in 95% confidence limit for 96 h LC_{50} values for all media. Figures 2, displays the Probit line graphs of the toxicity data for the test freshwater fish species under varying salinity conditions indicating increasing toxicity with exposure time.

DISCUSSION

The established salinity tolerance and salinity influence of Pb against *O. niloticus* at varying salinity range reported

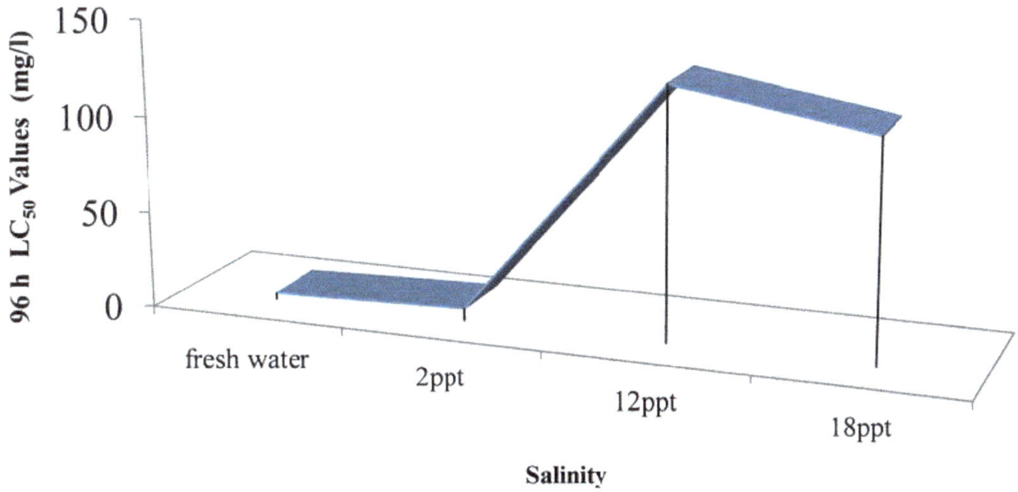

Figure 1. Relationship Between 96 h Lc_{50} values of $Pb(NO_3)_2$ against *O. niloticus* (n = 10) at different salinities.

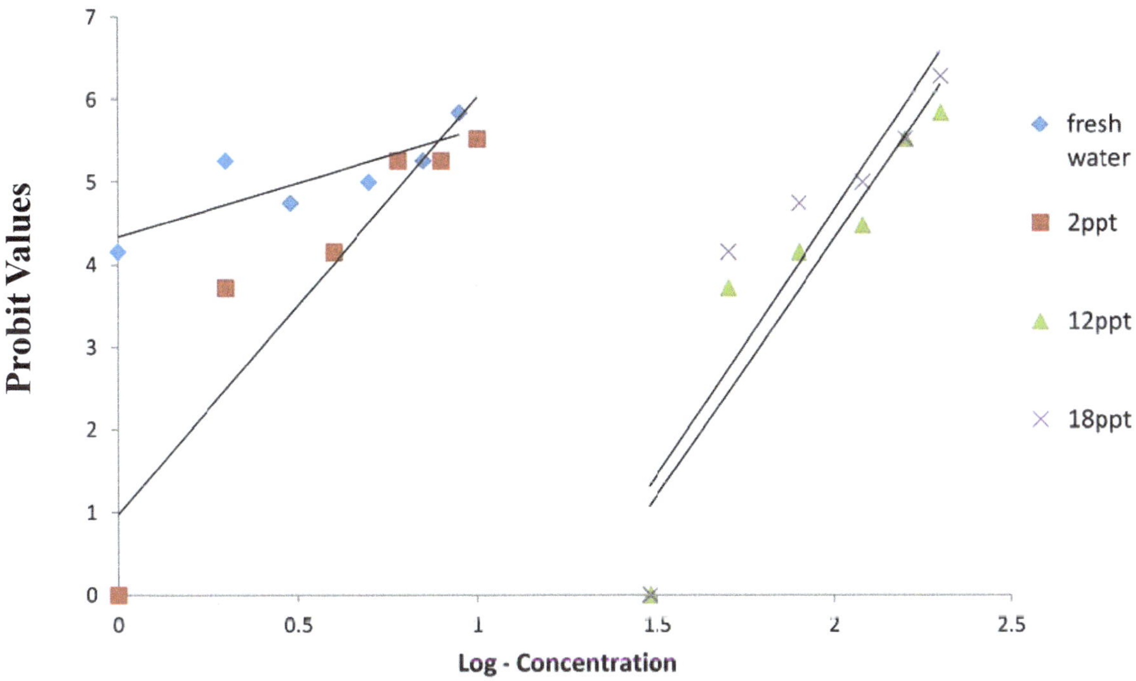

Figure 2. Probit transformed response of *O. niloticus* Log-concentration of $Pb(NO3)_2$ at fresh water, 2, 12 and 18 ppt conditions.

in this study showed that based on mortality assessment, *O. niloticus* fingerlings could not survive for any appreciable period in media with salinity equal to and higher than 22 ppt. This is not surprising because *O. niloticus* is a brackish water species and optimum survival is usually at salinity range between 10 - 15 ppt though can survive in waters with salinity between 0-15 ppt (Kurata, 1959, Thomas and Masser, 1999; Lawson and Anetekhai, 2011). Although the study conducted by Nugon (2003) showed that *O. niloticus* exhibited good survival (81%) in salinity regimes up to 20 ppt, which is in close agreement with the findings of this study and that of Osuala and Bawa-allah (2013).

However in fresh water and at 2 ppt the toxicity of Pb was higher than that observed at 12-18 ppt as depicted by the 96 h LC_{50} values suggesting that the toxicity of Pb against *O. niloticus* fingerlings may have synergized osmotic stress at lower salinities. Moreover, salinity played a significant role in influencing the acute toxicity of Pb. This probably resulted in the fish being less resistant and unable to regulate the body fluid to restore levels of osmotic pressure to near normal. Additionally it could be due to other toxicity modifying factors such as exposure duration and concentrations of Pb. Chemicals may have multiple effects on populations of organisms such as in mortality, reproductive failure, and productivity (Chukwu and Okpe, 2006). Sensitivity of populations depends upon such factors as age groups and temporal patterns of exposure. Passage of toxins or toxicants into an organism is also highly dependent on the specific physical-chemical characteristics of a given toxicant (Maheswaran et al., 2008; Ololade and Oginni, 2010). $Pb(NO_3)_2$, being the test chemical could be regarded as one of the major reasons for induced mortality due to its lipophilic and surfactant-containing nature (Ezemonye et al., 2007). Most of the test organisms survived initial attack at the early stage of test initiation and at the lower test concentrations of the toxicants. This could be attributed to protective adaptations as well as individual physiological nature of *O. niloticus* (Olalade and Oginni, 2010). However, as exposure progressed to 96 h, inevitable mortality could be due individual physiology and cumulative impact of the chemical toxicity. The results observed in this study are in agreement with other related studies (Omoregie et al., 1990; George and Clark, 2000; Johnson et al., 2005; Scarlett et al., 2005; Ezemonye et al., 2007). It is interesting to note that $Pb(NO_3)_2$ was most toxic to the fish at freshwater conditions, contrary to expectations that since the fish has the ability to survive at extremely low salinity such as in freshwater conditions, they should show more tolerance to the metal in freshwater. This is in agreement with findings of Oyewo (1998) who showed that similar brackish water adapted bony fishes such as *Tilapia guinensis* and *Nerite senegalensis* were known to be most susceptible to heavy metal pollutants including $CuSO_4$ at salinities tending towards the extremities (below 5 ppt and above 25 ppt), but were several folds more tolerant at salinity of up to 15ppt which falls within typical brackish water salinity (10-20 ppt).

The results reported for mean percentage mortality and 96 h LC_{50} values for the fresh water test could probably have been due to the physiology of *O. niloticus* under freshwater conditions, which have a body fluid concentration (about one-third) their surrounding environment, thus constantly taking in water by diffusion through their gills and skin for osmotic balance (Delbeek, 1987). Thus in a situation where there is damage to the skin and other tissues as is the case in exposure to high concentrations of surfactant-containing chemical ($Pb(NO_3)_2$), there is an influx of not only water but also the test chemical leading to a high lethal toxicity of the chemical and death rate of *O. niloticus* under freshwater conditions (Bury et al., 1999; Abel, 2006).

The authors' findings suggested that *O. niloticus* remains a brackish water fish as it retains its innate characteristics which allows it to do well in brackish water as its original natural habitat. Though, it can also strive relatively well in freshwater media because they are euryhaline (being able to adapt to a wide range of salinities due to their ability to osmoregulate especially in habitats where salinity changes regularly).

O. niloticus carry out hyper-hypo osmoregulation switching from hyper osmoregulation to hypo osmoregulation and vice versa during variations in salt content of water in their habitat. However, variations tending towards the extremities may result in stress or death, as was demonstrated with *O. niloticus* in extremely high salinities in this study.

Therefore, the practical significance of this study could be utilized in setting ecologically sound, safe limits for heavy metal (such as Pb) containing discharges into the lagoon ecosystems, aimed at protecting brackish water organisms. It is therefore suggested that data used for setting of safe limits should be based on those obtained from studies carried out over a wide range of salinity conditions.

Additionally this study provides data and information to already existing reviews on the influence of heavy metal pollutant in waters with varying salinity range against *O. niloticus*

Conclusion

This study has shown that salinity affects the acute toxicity of lead on fingerlings of *O. niloticus*. In the freshwater, 2, 12 and 18 ppt experimental media, death caused to *O. niloticus* may be detrimental since this could lead to elimination of potentially reproductive organisms in its natural brackish water habitat. The inclusion of salinity factor in the setting of safe limits would ensure amongst others that the delicate biotic components of the rich Lagoon biodiversity are prudently protected. The findings will assist our relevant agencies on the need to stipulate and implement stiffer regulations to control inflow of nutrient and chemical contaminants into brackish habitats coupled with enforcement of penalties imposed for illegal and unsustainable development that impacts these habitats.

Conflict of Interests

The author(s) have not declared any conflict of interests.

ACKNOWLEDGEMENT

The authors are grateful to the staff of the analytical Laboratory, University of Lagos for the analysis of samples and to Dr Igwo - Ezikpe for reviewing the manuscript.

REFERENCES

Abel PD (2006). Toxicity of synthetic detergents to fish and aquatic invertebrates. J. Fish Biol. 6:279-298.

Aderinola OJ, Clarke EO, Olarinmoye OM, Kusemiju V, Anatekhai MA (2009). Heavy metals in surface water, sediments, fish and perwinkles of Lagos lagoon. Am. Eurasian J. Agric. Environ. Sci. 5:609-617.

Bhupander K, Mukherjee DP, Sanjay K, Meenu M, Dev P, Singh SK, Sharma CS (2011). Bioaccumulation of heavy metals in muscle tissue of fishes from selected aquaculture ponds in East Kolkata wetlands. Ann. Biol. Res. 2:125-134.

Bowles KC, Apte SC, Batley GE, Hales LT, Rogers NJ (2006). A rapid Chelex column method for the determination of metal speciation in natural waters. Anal. Chim. Acta. 558:237-245.

Breves JP, Hasegawa S, Yoshioka M, Fox BK, Davis LK, Lerner DT, Takei Y, Hirano T Grau EG (2010). Acute salinity challenges In Mozambique and Nile Tilapia: differential responses of plasma prolactin, growth hormone and brachial expression of ion transporters. Gen. Comp. Endocrinol. 167(1):135-142.

Bury NR, McGeer JC, Wood CM (1999). Effects of Aatering freshwater chemistry on physiological responses of Rainbow Trout to silver exposure. Environ. Toxicol. Chem. 18:49-55.

Chukwu LO, Okpe HA (2006). Differential response of *Tilapia guineensis* fingerlings to inorganic fertilizer under various salinity regimes. J. Environ. Biol. 27(4):687-690.

Delbeek CJ (1987). Osmoregulation: Please give me a drink. J. Fish Biol. 1:3.

Ezemonye LIN, Ogeleka DF, Okieimen FE (2007). Acute Toxicity of industrial detergent (Neatex) and corrosion inhibitor (Norust CR 486) to early stages of Cichlids; *Tilapia guineensis*. Chem. Ecol. 23:1-8.

Finney DJ (1971). Probit Analysis. 3rd Edition University Press, London. 276pp.

George A, Clark JR (2000). Acute toxicity of two Corexit Dispersants. Chemosphere. 40: 897-906.

Grosell M, Gerdes RM, Brix KV (2006). Chronic toxicity of lead to three freshwater invertebrates - *Brachionus calyciflorus, Chironomus tentans, and Lymnaea stagnalis*. Environ. Toxicol. Chem. 25:97-104.

Himadri G, Anilava K (2000). Heavy metal concentration in water, sediment, Shrimp (*Penaeus monodon*) and Mullet (*Liza parsia*) in some brackish water ponds of Sunderban, India Mar. Pollut. Bull. 40 (11): 914-920.

Iulia GD, Marius LM, Alin T, Gabriel C, Laurentiu R, Gabriel LR (2012). Food chain biomagnification of heavy metals in samples from the lower Prut floodplain natural park. Environ. Eng. Manag. J. 11: 69-73.

Johnson TA, Miller LM, Whittle DM, Brown SB, Wieg DM, Kapuscinnsk AR, Leggett WC (2005). Effects of materially transferred Organochlorine contaminants of early life survival in a freshwater fish. Environ. Toxicol. Chem. 24:2594-2602.

Kurata H (1959). Preliminary report on the rearing of herring larvae. Bull. Hokkaido Resour. Fish Res. Lab. 20: 117-138.

Lawson EO (2011). Physico-chemical parameters and heavy metal contents of water from the mangrove swamps of Lagos lagoon, Lagos, Nigeria. Adv. Biol. Res. 5:08-21.

Lawson EO, Anetekhai MA (2011). Salinity tolerance and preference of hatchery reared Nile Tilapia, *Oreochromis niloticus* (Linneaus 1758). Asian J. Agric. Sci. 3(2): 104-110.

Leblond VS, Hontela A (1999). Effects of *in vitro* Exposures to cadmium, mercury, zinc and1-(2-chlorophenyl)-1-(4chlorophenyl)-2, 2-Dichloroethane on Steridogenesis by dispersed interrenal cells of Rainbow Trout (*Oncorhynchus mykiss*). Toxicol. Appl. Phamacol. 157(1):10-22.

Maheswaran R, Devapanl A, Muralidharan S, Velmurugan B, Ignaeimuthu S (2008).Haematological studies of fresh water fish, *Clariasbatradrus* (L) exposed to mercuric chloride. Int. J. Integr. Biol. 2(1): 49-54.

Małgorzata W, Elżbieta K, Monika S, Małgorzata, O (2010). Hematological changes in common Carp (*Cyprinus carpio* L.) after short-term lead (Pb) Exposure. Pol. J. Environ. Stud. 19:825-831.

Marleen E, Isabelle L, Chris MW, Gudrun DB (2013). Physiological effects of waterborne lead exposure in Spiny Dogfish (*Squalus acanthias*). Aquat. Toxicol. 126:373-381.

McKim JM (1985). Early life toxicity tests. In: Fundamentals of aquatic toxicology, hemisphere and New York. pp. 58-59.

Nugon RW (2003). Salinity tolerance of juveniles of four varieties of Tilapia. M.Sc. Thesis, Louisiana state University, USA. p. 76.

Ololade IA, Oginni O (2010). Toxic stress and hematological effects of Nickel on African catfish, *Clarias gariepinus* fingerlings. J. Environ. Chem. Ecotox. 2:14-19.

Omoregie E, Udodike EBC, Keke IR (1990). Tissue chemistry of *O. niloticus* exposed to sublethal concentrations of gammalin-20 and Actellic- 25EC. J. Appl. Sci. 5:33-36.

Osuala FI, Bawa-Allah KA (2013). Influence of salinity changes on acute toxicity of copper sulphate against *Oreochromis niloticus* fingerlings, in laboratory bioassays. Int. J. Environ. Sci. Toxicol. Food Technol. 6:22-27.

Oyewo EO (1998). Industrial Sources and distribution of heavy metal in Lagos lagoon and their biological effects on estuarine animals. Ph.D Thesis, University of Lagos. 247pp.

Patel M, Rogers JT, Pane EF, Wood CM (2006). Renal responses to acute Lead waterborne exposure in the freshwater Rainbow Trout (*Oncorhynchus mykiss*) Aquat. Toxicol. 80:362-378.

Rogers JT, Patel M, Gilmour KM, Wood CM (2005). Mechanisms behind Pb induced disruptions of Na^+ and Cl^- balance in Rainbow Trout (*Oncorhynchus mykiss*). Am. J. Physiol. Regul. Integr. Comp. Physiol. 289(2):R463-R472.

Rogers JT, Richards JG, Wood CM (2003). Ionoregulatory disruption as the acute toxic mechanism for lead in Rainbow Trout (*Oncorhynchus mykiss*). Aquat. Toxicol. 64:215-234.

Samy YE, Mohamed MMA, Mohamed ES, Hamad AE (2011). Production of salinity tolerant Nile Tilapia, *Oreochromis niloticus* through traditional and modern breeding methods: II. application of genetically modified breeding by introducing foreign DNA into fish gonads. Afr. J. Biotechnol. 10(4):684-695.

Scarlett A, Galloway TC, Canty M, Smith EL, Nilsson J (2005). Toxicity of two oil dispersants, Superdispersants - 25 and Corexit 9527, to a range of coastal species. Environ. Toxicol. Chem. 15:1219-1227.

Sippel AJA, Geraci JR, Hodson PV (1983). Histopathological and physiological responses of rainbow Trout (*Salmo Gairdneri Richardson*) to sublethal levels of Lead. Water Res. 17: 1115-1118.

Talia S, Yiming L, Virginia B, Paul BT (2009). Neurotoxic effects and biomarkers of lead exposure. Rev. Environ. Health 24:15-45.

Taweel A, Shuhaimi-Othman M, Ahmad AK (2013). In vivo acute toxicity tests of some heavy metals to Tilapia fish (*Oreochromis niloticus*). J. Biol. Sci. 13(5):365.

Thomas P, Masser M (1999). Tilapia life history and biology. Southern Regional Aquaculture Center. Publication No. 283

Usama MM, Abdel-Basset ME, Salsa MM (2013). Effect of lead on some haematological and biochemical characteristics of *Clarias gariepinus* dietary supplemented with lycopene and vitamin E. Egypt Acad. J. Biol. Sci. 5:67-89.

13

Spatial distribution, chemistry and subsurface temperatures of geothermal springs in Nkhata bay, Malawi

Burnett J. W. Msika[1] , John D. Kalenga Saka[2] and Zuze Dulanya[3]

[1]Mines Department, P.O Box 251, Lilongwe, Malawi.
[2]Department of Chemistry, University of Malawi, Chancellor College, P.O Box 280, Zomba, Malawi.
[3]Department of Geography and Earth Sciences, University of Malawi, Chancellor College, P.O Box 280, Zomba, Malawi.

This study investigated the spatial distribution, chemistry and subsurface temperatures of geothermal springs in Nkhata Bay, Malawi, which lies in the Western Branch of the East African Rift System. Seventeen geothermal springs were identified and mapped in the District. Eleven of these geothermal springs were purposively sampled for chemistry study on the basis of their spatial distribution representativeness and accessibility for reassessment. Two water types were hydrochemically delineated: sodium-sulfate (Na-SO_4) and sodium-sulfate-bicarbonate (Na-SO_4-HCO_3). Additionally, several mineral species were inferred to have been in contact with the geothermal spring waters in the district: calcite, aragonite, dolomite, gypsum, anhydrite, halite, fluorite, goethite, siderite and haematite. The thermal spring waters were undersaturated with respect to aragonite, calcite, dolomite, gypsum, anhydrite, halite and fluorite. However, four geothermal springs were supersaturated with respect to goethite, haematite and siderite. Although all the geothermal springs were undersaturated with fluorite (CaF_2), all contained higher F^- concentrations (from 2.21±0.06 to 10.27 ± 0.25 mg/L) than the World Health Organization (WHO) recommended limit for drinking water (1.5 mg/L). This suggests that fluoride does not precipitate as CaF_2 in the springs but mobilizes freely in the thermal waters without forming other complexes. The cadmium levels in four thermal springs exceeded the Malawi Bureau of Standards (MBS) and WHO safe limit for human consumption (0.005 mg/L). Furthermore, the iron levels in three springs exceeded the WHO and MBS safe limit (0.20 mg/L). Evidently, the chemistry of geothermal springs in Nkhata Bay shows that their utilization for drinking is not suitable. The Giggenbach Triangle revealed that all but one of the springs had fully equilibrated waters. Hence, chemical geothermometers were used to estimate their original subsurface temperatures. Information about these temperatures is required for the assessment of possible geothermal power stations. Amongst the geothermometers used, the Na/K/Ca and Na/K geothermometers offered better and more robust estimates (P>0.01) of geothermal subsurface temperatures. These geothermometers indicated that the original subsurface temperatures were in the ranges of 85 - 209 and 112 - 280°C, respectively.

Key words: Geothermal springs, spatial distribution, chemistry, subsurface temperatures.

INTRODUCTION

Most geothermal springs in the world occur in rift systems, more particularly, in areas along the tectonic plate boundaries where magma tends to rise towards the earth's surface (GEO, 2007). Malawi, which lies in the

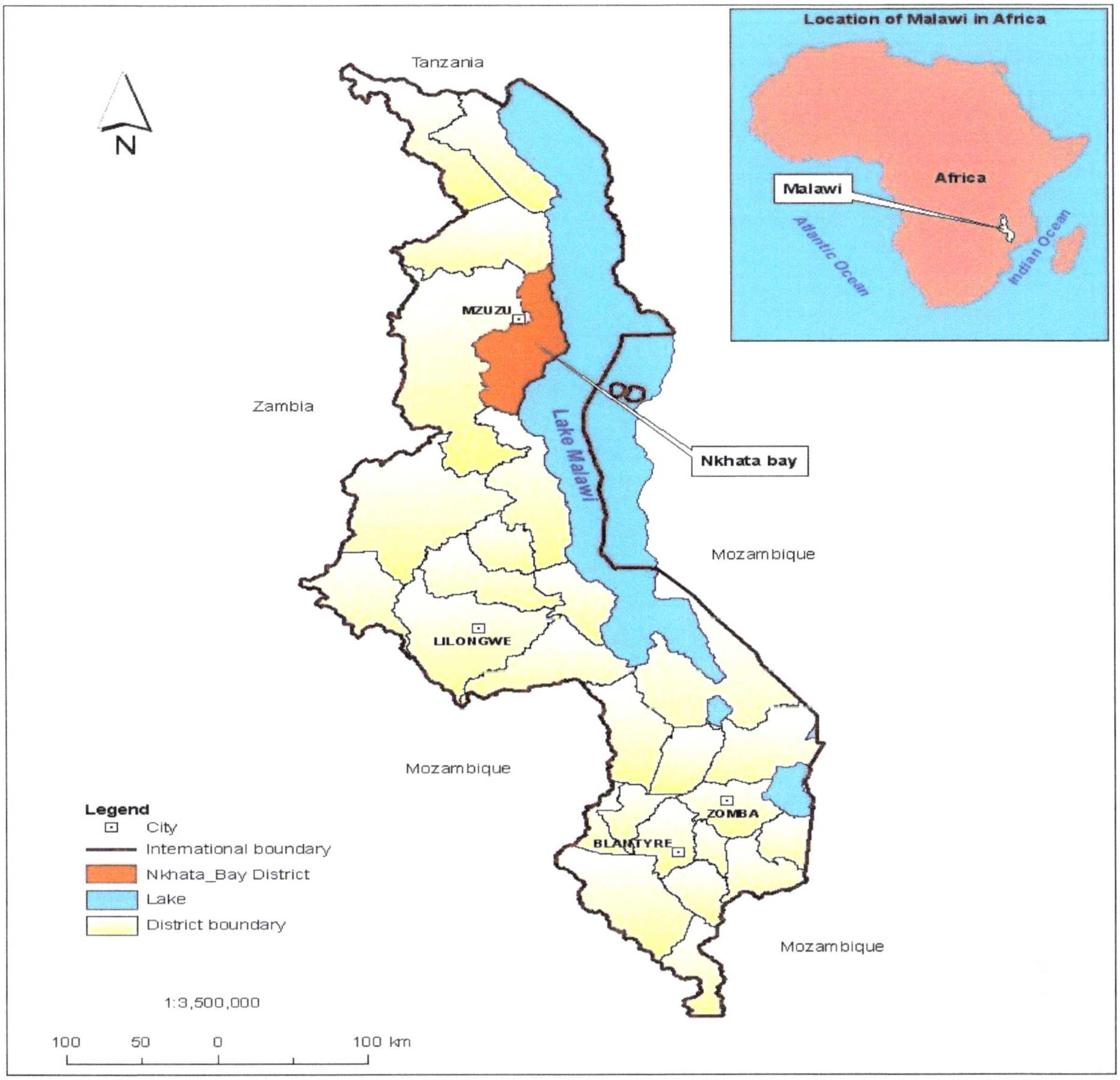

Figure 1. Location of Nkhata Bay, Malawi.

Western Branch of the East African Rift System, has thus many geothermal springs across the country from north to south (Dulanya, 2006).

The presence and distribution of geothermal springs in some parts of Malawi such as Chitipa, Karonga, Rumphi, Nkhotakota, Machinga, Mulanje, Zomba and Chikwawa Districts have been documented and their chemistry described by Harrison and Chapusa (1975), Ray (1975), Kalindekafe (2003) and Dulanya (2006), among others. Nkhata Bay located in the rift-valley floor in the north-eastern part of the country, along Lake Malawi (Figure 1), has numerous geothermal springs but lack research details of their spatial (Appendix 1) distribution and chemistry.

This study was therefore undertaken to investigate the spatial distribution, chemistry and subsurface temperatures of geothermal springs in Nkhata Bay, Malawi, recognizing the important roles similar geothermal resources play in many different parts of the world including boosting tourism in the country, mineral extraction, balneology and production of geothermal electricity.

MATERIALS AND METHODS

Locations of geothermal springs

Geothermal springs were identified in Nkhata Bay using indigenous information. The Universal Transverse Mercator (UTM) coordinates of the thermal springs were recorded using Global Positioning System (GPS) receiver. These UTM coordinates were superimposed on the electronic (scanned) map of Nkhata Bay obtained from Malawi National spatial Data Center, geo-referenced and digitized in the ArcView 3.3.Software.

Thermal springs chemistry

Water sampling

Eleven thermal springs were purposively sampled in order to study their chemistry. Sampling was done in December, 2007. Grab sampling was used in collecting water samples from geothermal spring-heads. Samples were collected in triplicates using 2 L polyethylene bottles and labelled accordingly. Samples meant for metal or cation analyses were acidified on the spot with concentrated nitric acid (1.5 mL HNO_3 per litre of collected water samples) according to APHA (1989).

Analytical procedures

Physico-chemical and chemical parameters

The physico-chemical parameters such as temperature, pH and electrical conductivity were measured in the field and chemical parameters, Na, K, Ca, Mg, Fe, Al, Cd, SiO_2, Pb, Cr, F^-, Cl^-, NO_3, PO_4^{3-}, HCO_3^- and SO_4^{2-} were determined in the laboratory according to APHA (1989).

1. Quantitative analysis of physico-chemical and chemical parameters

Determination of electrical conductivity (EC): The electrical conductivity of water samples was determined in the field using WPA - CM 35 conductivity meter. The electrode of the meter was rinsed before dipping into subsequent water samples to prevent inter-sample contamination (APHA, 1989).

Determination of temperature: A silver thermometer (graduated up to 300°C) was used to determine temperature at the spring-heads of the geothermal springs *in situ*. The thermometer was immersed in the thermal water and readings were taken on the spot according to APHA (1989).

Determination of pH: The pH of water samples was determined in the field at sampling points using a glass electrode Kent EIL 7020 pH meter. Prior to the pH determination, the meter was calibrated using pH 7 and 4 buffering solutions, respectively. The meter's electrode was rinsed with distilled water before determining the pH of any subsequent sample to prevent inter-sample contamination (APHA, 1989).

2. Analysis of metals/cations

The nitric acid digestion method was used for metal/cation determination (APHA, 1989). A well-homogenized unfiltered sample (100 ml) was transferred to a 250 ml flask and 5 ml of concentrated nitric acid was added to it. The mixture was then brought to slow boiling and evaporated on a hot plate until the volume was reduced to 15-20 ml. Concentrated nitric acid (5 ml) was added to the solution and the flask covered with a watch glass before heating it to obtain a gentle refluxing. The heating continued whilst adding 5 ml portion of nitric acid until the solution became light coloured clear solution.

Finally, concentrated nitric acid (2 ml) was added to dissolve any remaining residues and later the solution was quan-titatively filtered into 100 ml volumetric flask and diluted to the mark (APHA, 1989). The preparation of standard solutions for analysis of metals in water samples were done briefly as follows: For each of the metals, an aliquot of 10 mL taken from 1000 ppm standard stock solution were diluted with distilled water in a 100 mL volumetric flask to prepare an intermediate stock solution (100 ppm). Then, appropriate volumes of the intermediate stock solutions (0.0, 2.0, 4.0, 6.0, 8.0, 10.0, and 20 mL) were diluted with distilled water in 100 mL volumetric flasks to prepare working standard solutions of 0.0, 2.0, 4.0, 6.0, 8.0, 10.0 and 20.0 ppm, respectively. Absorbances of working standard solutions and water samples were determined using Atomic Absorption Spectrophotometer (AAS), Buck Scientific model no. 200A.

Cadmium (Cd): A standard stock solution (1000 ppm) was prepared by dissolving cadmium metal (1.00g) in concentrated nitric acid (4ml) before diluting the resultant solution with distilled water in a 1-L volumetric flask. Absorbances of the working standard solutions and samples were read at 228.8 nm (APHA, 1989).

Chromium (Cr): A standard stock solution (1000 ppm) was prepared by dissolving anhydrous potassium dichromate (2.83 g) in distilled water (200mL). Concentrated nitric acid (2 mL) was added to the solution before diluting with distilled water in a 1-L volumetric flask. Absorbances of the working standard solutions and samples were read at 357.9 nm (APHA, 1989).

Lead (Pb): A standard stock solution (1000 ppm) was prepared by dissolving lead nitrate (1.60 g) in distilled water (200 mL). Concentrated nitric acid (2 mL) was added to the solution before diluting with deionised water in a 1-Litre volumetric flask. Absorbances of the working standard solutions and samples were obtained at 283.3 nm (APHA, 1989).

Iron (Fe): A standard stock solution (1000 ppm) was prepared by dissolving iron wire (1.00 g) in 1+1 nitric acid water (50 mL). The standard solutions used in the final determination were 0.00, 0.5, 1.0, 2.0, 5.0 and 10.0 ppm. Absorbance of the working standard solutions and samples were read at 248.3 nm (APHA, 1989).

Sodium (Na): A standard stock solution (1000 mg/L) was prepared by dissolving sodium chloride (2.542 g) in 1 L volumetric flask and diluted to volume with the distilled water. The standard solutions used in the final determination were 0.0, 2.0, 4.0, 6.0, 8.0, 10.0, and 20 mg/L. Absorbances of the working standard solutions and samples were read at 589.0 nm (APHA, 1989).

Potassium (K): A standard stock solution (1000 ppm) was prepared by dissolving potassium chloride (1.907 g) dried at 110°C and later diluted to 1 L with distilled water. The standard solutions used in the final determination were 0.00, 0.5, 1.0, 2.0, 5.0 and 10.0 ppm. Absorbances of the working standard solutions and samples were read at 769.9 nm (APHA, 1989).

Calcium (Ca): The standard calcium (Ca) stock solution (1000 mg/L) was prepared by adding distilled water (50 mL) to calcium carbonate, $CaCO_3$, (2.497 g) and adding drop-wise concentrated hydrochloric acid, HCl, (10 mL) to complete dissolution. The solution was subsequently diluted to the mark in 1000-ml volumetric flask. The stock solution was used to prepare an intermediate stock solution (100 mg/L) from which the working standard solutions were prepared. The standard solutions used in the final determination were 0.00, 0.5, 1.0, 2.0, 5.0 and 10.0 mg/L. Absorbances of the standard solutions and samples were read at 422.7 nm on the AAS (APHA, 1989).

Determination of magnesium (Mg): A standard stock solution of Mg was prepared as follows: 1.000 g of magnesium ribbon was dissolved in a minimum volume (5mL) of (1+1) HCl. The solution was diluted to 1 L with 1% (v/v) HCl. The stock solution was used to prepare an intermediate stock solution (100 mg/L) from which the working standard solutions were prepared. The standard solutions used in the final determination were 0.00, 0.5, 1.0, 2.0, 5.0 and 10.0 mg/L. Absorbances of the working standard solutions and samples were read at 285.2 nm on the AAS (APHA, 1989).

Analysis of aluminum (Al): Aluminum concentrations in geothermal spring water samples were determined colorimetrically using UV-VIS spectrophotometer (APHA, 1989). Samples (2 mL each) and aluminum (Al) standards (10 mL each) were separately pipetted into 100-mL flasks. The following were added into each of the standards and samples: 1+199 thioglycolic acid (10mL) (prepared from 5 mL of the acid diluted to 1000 mL), solochrome cynine solution (10 ml, 0.075%) and ammonium acetate solution (pH 6). Solochrome cynine solution (0.075%) was prepared as follows: solochrome cynine analar (0.75 g) was dissolved in 200 mL of deionised water; sodium chloride (25 g), ammonium nitrate (25 g) and concentrated nitric acid (2 mL) were added into the solochrome cynine solution. This solution was then filtered through 542 Whatman paper (without washing) and was left to stand for overnight before use. Ammonium acetate solution (pH6) was prepared as follows: acetic acid (570 mL) was mixed with distilled water (1130 mL); then ammonium solution (570 mL) was carefully added; the resultant solution was mixed, cooled and adjusted to pH 6 with ammonia using a pH meter. The samples and the standards were topped to the 100-ml mark with distilled water. The aluminum (Al) standards were made from intermediate Al standard solution (100 mg/L) prepared from Al stock solution (1000 mg/L) as follows: Al wire (1.000 g) was dissolved in a minimum amount of (1+1) HCl, a small drop of mercury was added as a catalyst. The solution was filtered to remove mercury and then diluted to 1000 mL with 1% (v/v) HCl. Aluminum standard stock solution (10 ml, 1000 mg/L) was pipetted into a 100-mL volumetric flask and diluted to the mark to make intermediate aluminum standard solution (100 mg/L). The aluminum standard concentrations (0, 2, 5, 10, 15, 20 and 25 mg/L) were prepared from intermediate aluminum standard solution (100 mg/L).

This was done by pipetting 0, 2, 5, 10, 15, 20 and 25 mL from the intermediate standard aluminum solution into 100-mL volumetric flasks and diluting to the mark. The standards were run on UV-visible spectrophotometer and a concentration versus absorbance calibration curve for the standards was generated. Then sample absorbances and concentration were read from the colorimeter (UV-Visible spectrophotometer).

3. Determination of anions: Sulfate, nitrate, chloride, fluoride, bicarbonate, phosphate and silica

Analysis of sulfate (SO_4^{2-}): Sulfate (SO_4) in geothermal spring water samples was determined colorimetrically on UV-visible spectrophotometer (APHA, 1989). A calibration curve was generated from standard stock sulphate solution (100 mg/l) which was prepared by dissolving sodium sulphate (1.479 g) in distilled water and diluted to the mark in 1000-ml volumetric flask. Sulfate standard solution series for calibration were prepared from this stock solution by diluting 0, 2.0, 5.0, 10.0, 20.0 and 25.0 ml to 100 mL to provide the following sulfate concentrations: 0, 2.0, 5.0, 10.0, 20.0, and 25.0 mg/L, respectively.

Then the standard sulfate solutions (100 mL) of concentrations 0, 2.0, 5.0, 10.0, 20.0, and 25.0 mg/L were transferred into 250-mL Erlenmeyer flask. A buffer solution (25.0 mL) was added and, while stirring, a spatulaful of barium chloride ($BaCl_2$) was added. After 60 seconds, absorbance was measured at 420 nm on UV-Visible Spectrophotometer. A calibration curve was prepared by relating absorbance of standards with their concentrations. Sulfate concentration in geothermal spring water samples was determined as follows: samples (100 mL) were each transferred into 250-ml Erlenmeyer flasks. A buffer solution (20.0 mL) was added into each sample and, while stirring, a spatulaful of $BaCl_2$ was added.

After 60 s, absorbance was measured at 420 nm on UV-Visible spectrophotometer. In order to make correction for sample colour and turbidity present in the original sample, blanks from which $BaCl_2$ was withheld were prepared and run. The sulfate concentrations in the samples were determined directly by referring to the calibration curve after subtracting blank absorbance from sample absorbance.

Determination of nitrate (NO_3^-): A nitrate Ion-Selective Electrode (ELIT 8021), on a Mettler Toledo Seven Multimeter, was used to determine nitrate concentration in the geothermal spring water samples (APHA, 1989). Prior to sample reading, the electrode and meter were calibrated using sodium nitrate ($NaNO_3$) standard concentrations (1.0, 2.0, 5.0, 7.0 and 10.0 mg/L) prepared from the intermediate $NaNO_3$ standard (100 mg/L). This intermediate standard was prepared from 1000mg/L stock solution, which was again prepared by dissolving $NaNO_3$ (1.371g) which was dried at 105°C for 4 h. The standard concentrations were prepared by pipetting the following volumes from the intermediate $NaNO_3$ standard solution (100 mg/L) respectively: 1.0, 2.0, 5.0, 7.0, and 10 mL. These volumes were diluted to the mark in 100-ml volumetric flasks. Then the standards and samples (50 mL each) were carefully pipetted into 250-ml beakers separately. Ammonium sulfate ($(NH_4)_2SO_4$) (1.0 mL, 2.0 M) was added to these standards and samples as ionic strength adjustment buffer (ISAB) in order to ensure stable conditions before reading on the meter. Then after 20 min, $NaNO_3$ standards (1.0, 2.0, 5.0, 7.0 and 10.0 mg/L) were read from the meter as a means of electrode/ meter calibration. The nitrate concentrations of the samples were read from the meter.

Determination of chloride (Cl^-): A chloride Ion-Selective Electrode (ELIT 8261) on a Mettler Toledo Seven Multi meter was used to determine chloride concentrations in the geothermal spring water samples (APHA, 1989). Before sample reading, the electrode and meter were calibrated using NaCl standard concentrations (1.0, 2.0, 5.0, 7.0, 10.0, 15.0 and 20.0 mg/L) prepared from the intermediate NaCl standard (100mg/L). The intermediate standard was prepared from 1000mg/L stock solution, which was again prepared by dissolving NaCl (1.649 g) dried at 110°C for 4 h. The standard concentrations were prepared by pipetting the following volumes from the intermediate NaCl standard solution (100 mg/L) respectively: 0, 2.0, 5.0, 7.0, 10.0, 15.0 and 20.0 ml. These volumes were diluted to the mark in 100-ml volumetric flasks. Then the standards and samples (50 ml each) were carefully pipetted into 250-ml beakers separately. Sodium nitrate (1 ml, 2% v/v, 5M) was added to 50ml of individual samples and standards, as Ionic Strength Adjustment Buffer (ISAB), to stabilize conditions before taking readings. Then after 20 min, NaCl standards (0, 2.0, 5.0, 7.0, 10.0, 15.0 and 20.0 mg/L) were read from the meter as a means of electrode/ meter calibration. Thereafter, the nitrate concentrations in the samples were read from the meter.

Determination of fluoride (F^-): The fluoride ion-selective electrode Thermo Orion Ion Plus (9609BN) on a Mettler Toledo Seven Multimeter was used to determine fluoride concentration in the geothermal spring water samples against the sodium fluoride (NaF) standard curve (APHA, 1989). The NaF standards (0, 2.0, 5.0, 10.0, 20.0 and 25 mg/L), prepared from intermediate NaF standard (100 mg/L), were used to calibrate the electrode and meter. This intermediate standard was prepared from NaF stock solution (1000 mg/L), which was again prepared by dissolving NaF (2.210 g) dried at 105°C for 6 h. The standard NaF concentrations were prepared by pipetting the following volumes from the intermediate NaF standard solution (100 mg/L) respectively: 0, 2.0, 5.0, 7.0, 10.0, 15.0 and 20.0 ml.

These volumes were diluted to the mark in 100-ml volumetric flasks. Total Ionic Strength Adjustment Buffer (TISAB) (25 mL) was added to 25 mL of individual samples and standards to adjust pH and reduce complex interferences before fluoride concentration reading in both standards and samples on the meter. Then after 20 min, NaF standards (0, 2.0, 5.0, 7.0, 10.0, 15.0 and 20.0 mg/L) were read from the meter as a means of electrode/ meter calibration. The nitrate concentrations of the samples were obtained directly from the meter.

Determination of bicarbonate (HCO_3^-): Bicarbonates were determined using titrimetric method with hydrochloric acid (APHA, 1989). Hydrochloric acid (HCl) (0.02 M) was prepared by pipetting concentrated HCl (8.3 mL, 1.18 S.G) into 1000-mL volumetric flask and diluted to the mark with distilled water. Exactly 200mL of this stock solution was diluted to 1000 mL with distilled water which was standardized against sodium carbonate (Na_2CO_3, 0.0100 M).
Na_2CO_3 (0.0100 M) was prepared as follows: anhydrous sodium carbonate (1.0599 g) previously dried at 250°C for one hour was dissolved in 300ml of distilled water, diluted to 1000-mL mark and stored in a polyethylene bottle. Then the prepared Na_2CO_3 solution (25.0 mL, 0.0100 M) was transferred into a 250-mL beaker, distilled water (25 mL) was added, and titrated with hydrochloric acid (0.02 M) to phenolphthalein end point. The resultant solution was boiled gently for 3-5 min, cooled to room temperature and titrated to bromocresol green/methyl red end-point.

The volume of the hydrochloric acid titrant used was (S), and concentration, M, of the hydrochloric acid was calculated as follows:

$$\text{Molarity (M)} = \frac{0.01 \times 25.0 \times 2}{S}$$

Samples were titrated in triplicates as follows: a sample of volume, V, (25 mL) was transferred into a 250-ml conical flask, phenolphthalein indicator solution (4 drops) were added and titrated with standard hydrochloric acid of concentration, M, until pink colour was just discharged.

The volume of titrant used was recorded as (P). Bromocresol green/ methyl red indicator (four drops) was added and titration was continued to a light pink colour. Total volume of titrant used was recorded as (T).
The bicarbonate concentration was calculated as follows:

$$HCO^{3-}\ \text{mg/L} = \frac{(T-2P) \times M \times 61000}{V}$$

Determination of phosphate (PO_4^{3-}): The phosphate (PO_4^{3-}) standards were used to generate a calibration curve in the UV-Visible Spectrophotometer from which the PO_4^{3-} concentrations in geothermal spring water samples were read (APHA, 1989).

The standard PO_4^{3-} solution (50 µg/mL) was prepared as follows: 3 g of potassium dihydrogen orthophosphate (KH_2PO_4) was dried on a watch glass at 105-110°C for overnight and cooled in a dessicator. KH_2PO_4 (0.9594 g) was weighed out and dissolved into a 500-Ml volumetric flask with distilled water. This was 1 mg/ml PO_4^{3-} solution (standard stock solution).

For analysis, 5 ml of 1 mg/mL PO_4^{3-} solution was diluted to 100 ml with distilled water. This was 50 µg/mL PO_4^{3-} solution (intermediate standard solution). The PO_4^{3-} standard concentrations (0, 0.05, 0.1, 0.2, 0.3, 0.4 mg/L) were prepared from this intermediate standard solution by pipetting the following volumes into 100-mL volumetric flasks respectively and diluting them to the mark: 0, 1, 2, 4, 6 and 8 mL. Then the geothermal spring water samples (1 mL) were each pipetted into 100-mL beakers and into each beaker, HNO_3 (2 mL) and 5 mL of perchloric acid ($HClO_4$) were added. The resultant solution was heated on a hot plate until 2mL of solution remained (about 30 min) and then allowed to cool. HNO_3 (2 mL) and $HClO_4$ (5 mL) were added and the solution was again heated on the hot plate until 1 mL of solution remained. The solution was then allowed to cool.

Distilled water (20 mL) was added and the solution was heated gently (for 10 min, but it was not allowed to boil). The solution was decanted into a 100-ml volumetric flask and distilled water (70 mL) and reducing solution (20 mL) were added and the solution was diluted to volume with distilled water. This was left overnight before reading. The absorbance of the blank, standards and samples was measured at 555 nm.

Determination of silica (SiO_2): The first step was preparation of silica (Si) standards for derivation of a calibration curve from which Si concentration in the geothermal spring water samples were to be read on the UV-Visible spectrophotometer (APHA, 1989). From standard stock Si solution (1000 mg/L) (purchased), 5ml were pipetted into 500-mL volumetric flask and diluted to volume with distilled water.

Si standard volumes of 0, 2, 5, 10, 15 and 20 mL were measured from standard Si stock solution (0.01 mg/mL) using a 50-ml burette into 100-ml volumetric flasks. These standards contained the following concentration respectively: 0, 0.02, 0.05, 0.10, 0.15, and 0.20 mg/mL. Samples (1 mL) each were pipetted into clean nickel crucibles. Sodium hydroxide (2 g) was added and dissolved. The solution was heated over an open flame, starting with a small flame and gradually increasing the flame until the bottom of crucible was just red. Heating was continued for 3 min. Then the contents were removed and cooled. The crucible was then put into a 500-mL crucible polypropylene beaker. Hot distilled water was added into the crucible to dissolve the melt. The crucible was then removed, rinsed with hot distilled water and allowed to cool. The solution was then transferred into a 250-ml volumetric flask and diluted to the mark with distilled water. The solution was filtered through filter paper into a 500-ml propylene beaker. (First 10-20 mL solution was discarded).

The filtered solution was transferred to a plastic bottle to serve as sample stock solution. To sample stock solutions (1 ml) pipetted into 100-mL volumetric flasks in triplicates, P-nitrophenol indicator (1 drop) was added and acidified with drops of HCl (1+4) until the colour changed from yellow to colourless. The solution was diluted to about 50mL. To the mixture, HCl (1+4) (1 mL) and 5% ammonium molybdate solution (4 mL) were added and mixed thoroughly, allowed to stand for 10 min and 10% v/v tartaric acid solution (5 mL) and ascorbic acid solution (2 mL) were added. The resultant solution was diluted to the mark with distilled water, allowed to stand for 30 min at room temperature and absorbance and concentration of Si measured at 650nm wavelength. The SiO_2 were calculated from Si results.

Subsurface temperatures

The subsurface temperatures of geothermal springs in Nkhata Bay were determined using the following cation geothermometers: 1) Na/K chemical geothermometers by Truesdell (1976), Fournier

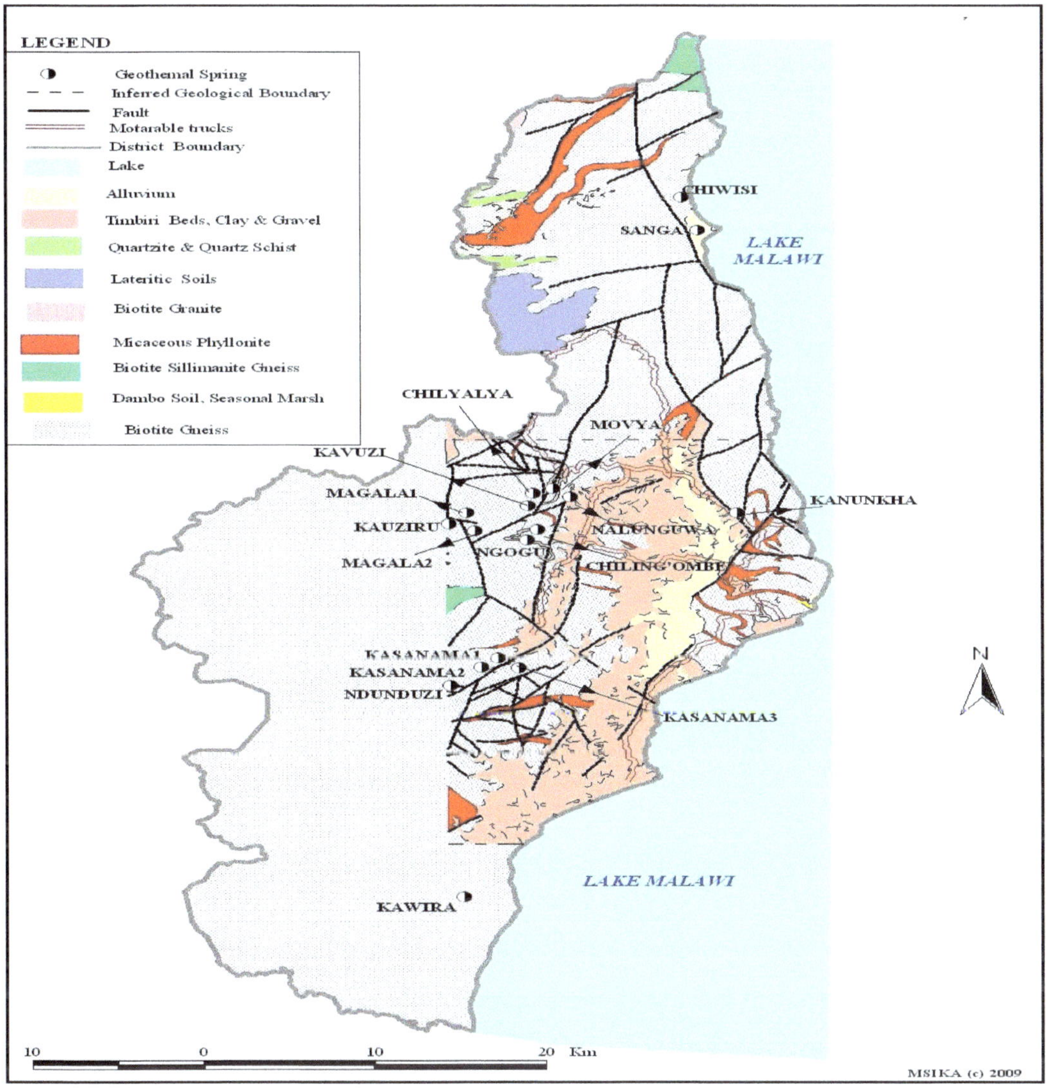

Figure 2. Spatial distribution of geothermal springs in Nkhata Bay, Malawi.

(1979), Arnorson (1983), Giggenbach (1988), and Verma and Santoyo (1997). These geothermometers have been coded as Na/K1, Na/K2, Na/K3, Na/K4 and Na/K5 respectively; 2) Na/K/Ca chemical geothermometer by Fournier and Truesdell (1973); and, 3) K/Mg chemical geothermometer by Giggenbach (1988).

RESULTS AND DISCUSSION

Spatial locations of geothermal springs in Nkhata Bay, Malawi

Seventeen geothermal springs, with surface temperatures ranging from 29±0.3 to 74±0.5°C, were identified and their distribution mapped in Nkhata Bay (Figure 2).

Spatially, 76% of the springs are clustered towards Nkhata Bay West and 94% of the springs characteristically occur along the major faults in the Basement Complex in the District. Ostensibly, the spatial distribution of geothermal springs in the district is controlled by faulting which possibly provides conduits for hydrothermal convection in the Basement Complex (Figure 2).

Chemistry of geothermal springs

A summary of the chemistry of the thermal springs is provided in Table 1.

Physico-chemical characteristics of the geothermal springs

The surface temperature values of the thermal springs in Nkhata Bay ranged from 74.0±0.5 to 29.0 ±0.3°C. The

Table 1. Some chemical and physico-chemical characteristics of geothermal springs in Nkhata Bay, Malawi.

Parameter (chemical species in mg/l)	Kawira	Kanunkha	Nalunguwa	Chilyalya	Kavuzi	Movya	Chiling'ombe	Ngogu	Magala1	Kasanama1	Kasanama3
Na	97.46±2.98	97.23±4.96	97.26±1.36	88.06±1.87	87.65±5.56	81.57±3.60	96.00±1.79	96.62±7.21	93.53±	96.90±1.70	107.50±3.00
K	3.18±0.037	17.25±0.371	1.76±0.038	1.47±0.011	1.63±0.015	1.54±0.027	1.34±0.003	1.59±0.008	1.73±0.064	2.51±0.057	4.64±0.016
Ca	0.65±0.004	11.09±0.03	1.14±0.007	1.97±0.035	1.63±0.024	1.98±0.025	0.87±0.012	0.86±0.009	1.05±0.003	0.94±0.054	1.89±0.012
Mg	0.33±0.006	9.26±0.049	0.002±0.0002	0.24±0.001	0.17±0.002	0.07±0.002	0.06±0.001	0.15±0.001	0.08±0.002	0.02±0.003	1.32±0.014
Fe	0.66±0.044	27.23±0.103	Not detected (Nd)	Nd	Nd	Nd	Nd	Nd	Nd	Nd	1.40±0.133
Cd	0.033±0.007	0.049±0.007	0.020±0.008	0.023±0.010	Nd	0.001±0.009	Nd	Nd	Nd	Nd	Nd
SiO_2	0.16±0.007	0.44±0.020	Nd	Nd	Nd	Nd	0.02±0.007	0.05±0.007	0.08±0.007	0.17±0.003	0.20±0.011
F^-	10.27±0.25	7.31±0.10	3.29±0.07	2.78±0.19	2.73±0.03	2.55±0.24	2.47±0.03	3.16±0.17	2.21±0.06	5.67±0.01	5.39±0.14
Cl^-	6.49±0.83	14.73±0.56	9.55±0.20	7.62±0.10	8.74±0.27	7.26±0.13	5.64±0.87	5.57±0.12	11.68±0.36	6.17±0.10	5.61±0.17
HCO_3^-	67.21±9.60	96.02±13.58	43.21±4.8	76.81±0.00	43.01±6.80	76.81±0.00	52.81±4.80	76.81±0.00	48.01±0.00	52.81±4.80	33.61±4.81
SO_4^{2-}	46.02±2.73	106.52±2.10	99.82±0.86	85.86±4.83	93.52±2.35	87.48±2.85	60.55±2.10	123.63±3.98	55.92±0.63	86.98±0.99	53.16±3.14
Ionic Strength	0.0041	0.0070	0.0046	0.0045	0.0045	0.0040	0.0055	0.0039	0.0049	0.0047	0.0041
Temp (° C)	74.0±0.5	53.2±0.5	39.5±1.8	46±0.0	50.3±1.0	41.2±0.3	40.4±0.3	39.0±1.0	48.3±0.6	61.5±1.3	65.0±0.0
pH	8.12±0.09	6.50±15	8.65±0.12	8.23±0.12	7.64±0.04	8.08±0.05	8.27±0.10	8.43±0.11	7.89±0.23	8.36±0.07	8.39±0.14
EC (µS/cm)	5267±153	7183±76	5300±0.0	4767±58	5183±104	4430±20	4917±176	4617±29	4800±0.0	6393±51	6810±193

electrical conductivities ranged from 7183- 4430 µS/cm (Table 1). The pH in the geothermal springs studied was in the range of 6.5-8.65. Water with a pH below 6.5 is considered acidic and a pH above 8.5 is considered basic (GNWT, 2008). Thus, practically all the studied thermal spring waters are of neutral type. The exception is Nalunguwa, which is slightly alkaline (Table 1).

Chemical characteristics

Sodium (Na) represented the highest concentration values in almost all the studied geothermal springs, this was followed by sulfate (SO_4^{2-}), bicarbonate (HCO_3^-), chloride (Cl^-), fluoride (F^-), potassium (K), calcium (Ca), nitrate (NO^{3-}), magnesium (Mg) and Cadmium (Cd) - in descending order of abundance. Thus, in terms of cations, Na is followed by K in abundance in the geothermal springs.

However, unlike Na (>80 mg/L), the concentration of K is considerably lower (<10 mg/L) in most thermal springs in Nkhata Bay. With regard to anions, the following is their descending order of abundance (mg/L): $SO_4^{2-} > HCO^{3-} > Cl^-, > F^- > NO^{3-}$.

Lead (Pb), chromium (Cr) and phos-phate (PO_4^{3-}) were below detection limits in all the studied geothermal springs. Trilinear Piper plot (Figure 3) was used to delineate the principal hydrochemical characteristics of the thermal springs.

The trilinear piper plot hydrochemically distinguished two water facies of geothermal springs: 1) sodium - sulfate ($Na\text{-}SO_4$) water type for Kasanama1 and Nalunguwa geothermal springs and 2) Sodium- sulfate-bicarbonate ($Na\text{-}SO_4\text{-}HCO_3$) water type for Kawira, Kanunkha, Movya, Chilyalya, Chiling'ombe, Ngogu, Magala1 and Kasanama3 springs.

Equilibrium conditions and water quality of geothermal springs in Nkhata Bay, Malawi

Several mineral species in contact with the thermal waters and equilibrium conditions (as reflected by the saturation indices (SI)) in the geothermal springs in Nkhata Bay were monitored and analyzed using AquaChem 5.1 software (SWS, 2008).

The following mineral species were inferred to have been in contact with the geothermal

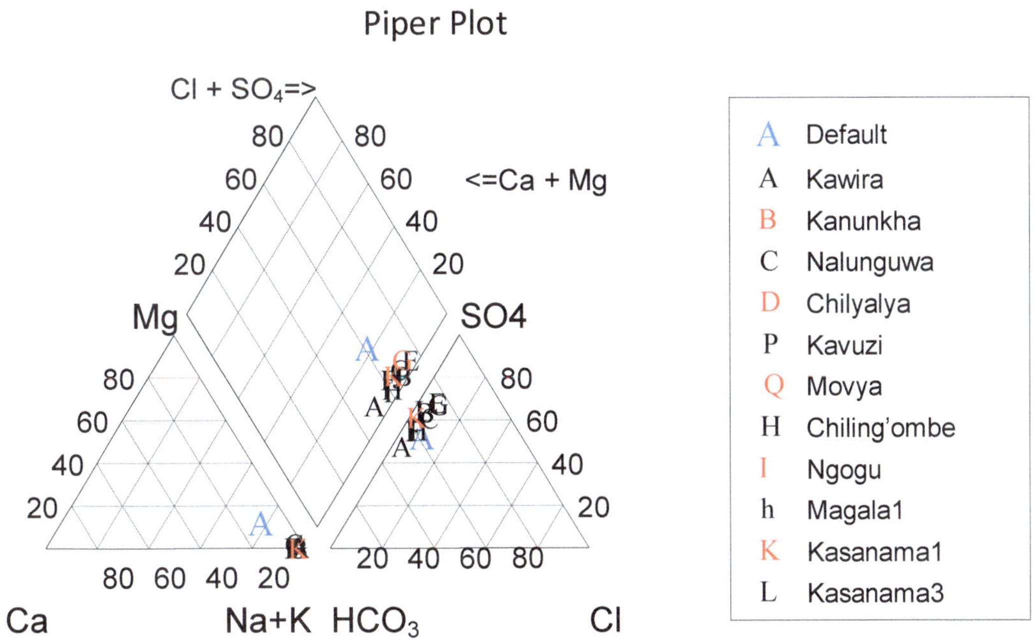

Figure 3. Principal chemistry of the water facies of geothermal springs in Nkhata Bay.

spring waters in the district: calcite, aragonite, dolomite, gypsum, anhydrite, halite, fluorite, goethite, siderite and haematite.

There was undersaturation (SI<0) in all of the springs with respect to the following mineral species: calcite ($CaCO_3$), aragonite ($CaCO_3$ in its rhombic form, harder and less common or stable than calcite), dolomite ($CaMg(CO_3)_2$), anhydrite ($CaSO_4$), halite or rock salt (NaCl), gypsum ($CaSO_4{\cdot}2H_2O$) and fluorite or fluorospar (CaF_2). However, there was supersaturation (SI>0) with respect to goethite (FeO(OH), haematite (Fe_2O_3) and siderite ($FeCO_3$) in Kanunkha, Kawira, Nalunguwa and Kasanama3 thermal waters.

Water, which is calcite-undersaturated (SI (calcite) <0), is corrosive to the pipes and on the other hand, if the water is calcite-oversaturated (SI (calcite>0), the calcite precipitates out and settles on the pipelines (Purschel, 2006).

Similarly, siderite, goethite and haematite (SI>0) precipitate and settle down in the springs in which they are inferred, unlike calcite, aragonite, halite, gypsum, anhydrite and fluorite (SI<0) which suggestively remain in thermal water column.

Despite fluorite undersaturation in the studied geothermal springs, fluoride concentration (Table 1) was high in all of them (>1.5 mg/L, WHO guidelines for drinking water). Moreover, due to the undersaturation conditions, fluoride cannot precipitate as CaF_2, and so mobilizes freely without forming other complexes in the thermal waters as also observed by Rango et al. (2008). Chronic ingestion of concentrations much greater than 1.5 mg/L is linked with development of dental fluorosis and, in extreme cases, skeletal fluorosis; high doses have also been linked to cancer (BGS, 2008).

Iron (Fe) levels observed in Kawira, Kanunkha and Kasanama3 thermal springs exceeded the safe limit of 0.20 mg/L set by MBS. High levels of iron oxides may result in development of a benign pneumoconiosis, called siderosis (lenntech, 2006).

The cadmium (Cd) levels observed in Kawira, Kanunkha, Nalunguwa and Chilyalya geothermal springs exceed the safe limit of 0.005 mg/L set by both MBS and WHO. Except Chilyalya which occurs on a hill where no farming takes place, the cadmium levels in other springs are likely to be exacerbated by phosphate fertilizer, which is commonly used in the farming field along the banks since fertilizer is a natural sink of cadmium.

Recent data indicate that adverse health effects of cadmium exposure may occur at lower exposure levels than previously anticipated, primarily in the form of kidney damage but possibly also bone effects and fractures (Järup, 2003).

The EC range of 7183- 4430 µS/cm in thermal waters in Nkhata Bay exceeds the WHO recommended EC value in surface water of 2500 µS/cm.

The pH of most of these thermal waters was within 6.5-8.5 (recommended limit, WHO water quality criteria) except Nalunguwa thermal waters which were slightly alkaline (pH = 8.65).

The extent of equilibrium conditions in the studied geotheothermal springs in Nkhata Bay is been represented in Figure 4. The Gibbenbach Triangle, the triangular K-Mg-Na representation by Giggenbach(1988), allows the verification of the extent to which water-rock

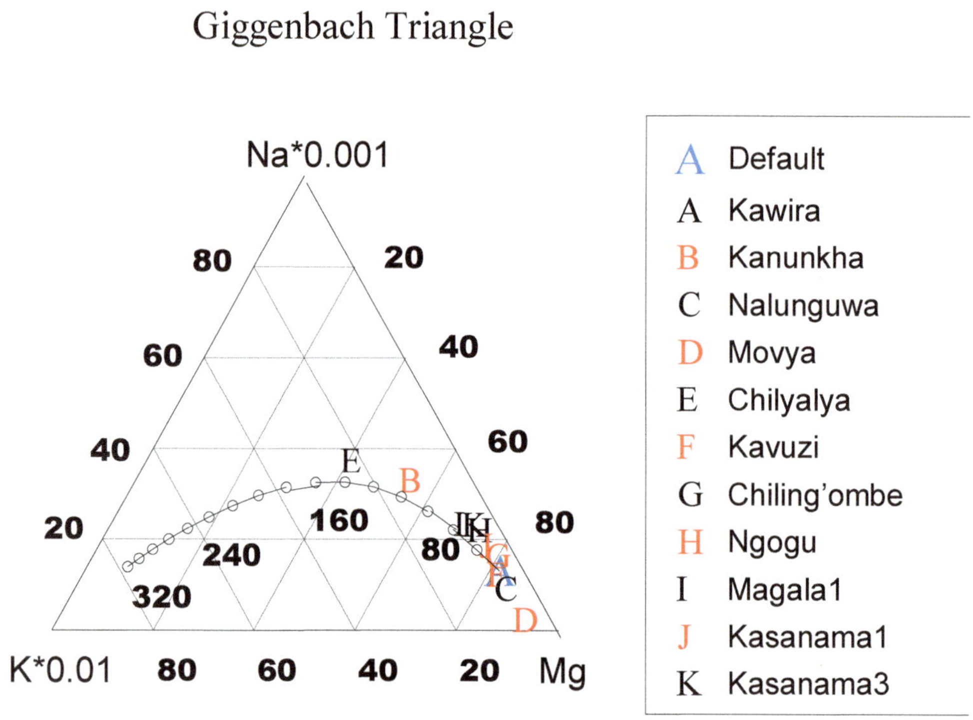

Figure 4. Extent of water-rock equilibrium in geothermal springs understudy.

equilibriums have been attained in the reservoirs (SWS, 2008). The triangle (Figure 4) comprises three zones: Immature waters (at the base); Partially equilibrated waters (in the middle); and Fully equilibrated waters (along the curve).

Except Movya which reflects immature water, Kavuzi, Chilyalya, Nalunguwa, Chiling'ombe, Ngogu, Magala1, Kawira, Kanunkha, Kasanama1 and Kasanama3 geothermal springs reflect fully equilibrated waters (Figure 4).

Since the Giggenbach Triangle (Giggenbach, 1988) indicates that most of the geothermal springs studied in Nkhata Bay have fully equilibrated waters, their chemistry can reliably be used to calculate the probable subsurface temperatures at which water-rock equilibriums were attained.

Subsurface temperatures of geothermal springs in Nkhata Bay

Three cation geothermometers (Na/K, Na/K/Ca and K/Mg) geothermometers were considered for estimating the sub-surface temperature of the geothermal springs in Nkhata Bay.

Na/K geothermometer usually offers the best chemical indication of high temperatures at depth since the Na/K ratio is low in hot areas and high in cool areas (Elder, 1981). These chemical constituents of the hot water are leached from geological sources in such a manner that at higher temperatures, there is an increased tendency for K to replace Na in feldspars (Elder, 1981).

The chemical reaction that occurs is reflected in the chemical Equation (1).

$$\text{Na-feldspar} + \text{K} \leftrightarrow \text{Na} + \text{K-feldspar} \quad (1)$$

Further, Na/K geothermometers are not affected by any concentration-dilution process since alkali metals are generally aquaphilic and their ratio is not affected by the kinetics of the reservoir (CER, 2007). They can, thus, be indicative of the highest temperature at which the waters have reached at chemical equilibrium with the reservoir rocks.

Evident from Table 2, only cation (Na/K, Na/K/Ca and K/Mg) geothermometers were considered for sub-surface temperature estimation of the geothermal springs in Nkhata Bay. Silica geothermometers were not considered relevant in this study because they gave subsurface temperature values which were too inconsistent to be plausible (Appendix 2).

The values were, thus, too low for most silica geothermometers and too high for one geothermometer. For example; silica (SiO_2) geothermometer models by Truesdell (1976), Fournier (1977) and Verma and Santoyo (1997) gave values which were too low (<0°C) while one model by Rimstidt (1979) gave temperatures values which were too high (> 500 ≤ 40922°C) - the reason being that silica was found in very low concentrations in geothermal springs understudy (Table 1). This was probably because of dilution that might have

Table 2. Sub-surface temperatures of Nkhata Bay geothermal springs estimated using some cation (Na/K, Na/K/Ca and K/Mg) geothermometers.

Geothermal Spring	Measured temperature (°C)	Na/K_1 (°C)	Na/K_2 (°C)	Na/K_3 (°C)	Na/K_4 (°C)	Na/K_5 (°C)	$Na/K/Ca_1$* (°C)	K/Mg (°C)
Kawira	74	92	137	103	156	142	144	26
Kanunkha	53	259	272	262	283	272	209	26
Nalunguwa	39.5	56	104	68	125	111	114	10
Movya	41	52	100	64	121	107	85	31
Chilyalya	46	58	106	69	126	112	93	29
Kavuzi	50	59	106	70	127	113	86	26
Chiling'ombe	40	42	91	54	112	98	106	26
Ngogu	39	51	99	63	120	106	113	29
Magala1	48	57	105	69	126	112	116	25
Kasanama1	61.5	77	123	89	143	129	131	16
Kasanama3	65	112	154	123	173	159	150	29

*No Mg-correction was done on geotemperature values for Na/K/Ca geothermometer.

occurred as a result of mixing between thermal and ambient ground waters. On the other hand, this could also be due to silica precipitation that might have happened prior to the collection of geothermal water samples for analysis as was also observed in one study conducted by Wahl (1977).

In this study, however, most Na/K geothermometers were used to estimate geotemperatures of the thermal springs, considering that the best chemical indications of high temperatures at depth are given by the Na/K ratio which, for example, is low in hot areas and high in cool areas (Elder, 1981). These chemical constituents of the hot water are leached from geological sources in such a manner that at higher temperatures, there is an increased tendency for K to replace Na in feldspars (Elder, ibid). The chemical reaction that occurs is reflected in the chemical equation:

Na-feldspar + K$\leftrightarrow$ Na + K-feldspar

CER (2007) also observes that Na/K geothermometers are not affected by any concentration-dilution process since alkali metals are generally aquaphilic and their ratio is not affected by the kinetics of the reservoir. Therefore, they can be indicative of the highest temperature at which the waters have reached at chemical equilibrium with the reservoir rocks (CER, 2007)

The deep reservoir temperatures in the geothermal springs in Nkhata Bay ranged from 112 to 283°C, estimated using Na/K geothermometers (Table 2). Similarly, Na/K/Ca geothermometer gave a subsurface temperature range of 85 - 209°C in the thermal springs. On the contrary, the K/Mg geothermometer gave a sub-surface temperature range of 10 to 29°C for geothermal springs in Nkhata Bay. The highest geotemperatures were associated with Kanunkha, Kawira and Kasanama3 geothermal springs; the lowest geotemperatures were associated with Chiling'ombe and Ngogu geothermal springs.

Pirlo (2004) indicated that the better the correlation between the emergence spring temperatures and subsurface temperatures estimated by the geothermometer, the better the geothermometer becomes for efficient use in a particular hydrothermal system.

From the correlation analysis conducted using Genstat 12.0 Software (NAG, 2008), Na/K/Ca geotemperatures showed the strongest positive correlation with measured temperatures of the geothermal springs ($r = 0.50$) and a geotemperature range of 85 - 209°C has been estimated for geothermal springs in Nkhata Bay.

Similarly, Na/K geotemperatures showed relatively strong positive correlation with the measured (emergence) temperatures of the geothermal springs ($r \geq 0.36$, $P<0.01$).

In particular, Na/K_4 geotemperatures displayed the strongest positive correlation with the measured temperatures of the thermal springs ($r \geq 0.39$, $P<0.01$). Using Na/K_4 geothermometer, the geotemperature range of 112 - 283°C was estimated for geothermal springs in Nkhata Bay. On the contrary, K/Mg geothermometer that gave the lowest geotemperatures (Table 2), had the weakest positive correlation with the measured temperatures of the geothermal springs ($r \geq 0.03$, $P<0.01$).

Hong et al. (2005) observe that K/Mg geothermometers can only be used for chloride waters as it is sensitive to mixing. The low geotemperatures given by K/Mg geothermometers can be attributed to very low Mg content (<1 mg/L) in most thermal waters understudy (Table 1) possibly due to dilution as a result of mixing between thermal waters and ambient groundwater.

To this end, it seems, therefore, that the Na/K/Ca and Na/K geothermometers offer better and more robust estimates of original subsurface temperatures of geothermal springs in Nkhata Bay; they both gave significant positive correlation with the emergence temperatures of the springs ($p<0.01$).

Conclusion

The chemistry of geothermal springs in Nkhata Bay reveals two spring water types: (1) sodium-sulfate (Na-SO4) springs, which include Nalunguwa and Kasanama1 geothermal springs; and, (2) sodium-sulphate-bicabornate (Na-SO4-HCO3) springs, which include Kawira, Kanunkha, Movya, Chilyalya, Ngogu, Chiling'ombe, Magala1 and Kasanama3 geothermal springs. Except Movya which has immature water, all these thermal springs have fully equilibrated waters.

The chemistry of geothermal springs in Nkhata Bay, Malawi, evidently shows that utilization of the thermal waters for drinking is not suitable due to high fluoride concentrations (>1.5 mg/L, WHO guidelines for drinking water) and high EC values (>2500 μScm^{-1}, WHO recommended EC value).

In addition, high concentrations of iron (Fe) (>0.20 mg/L set by MBS) and cadmium (>0.005 mg/L, safe limit set by both MBS and WHO) were determined in several thermal springs in the District, such as Kawira, Kanunkha, Kasanama3, Nalunguwa and Chilyalya. The original subsurface temperatures of Nkhata Bay geothermal springs, estimated by the Na/K/Ca and Na/K geothermometers, ranged from 85 to 209 and 112 to 283°C, respectively. The highest geotemperatures were associated with Kanunkha, Kawira and Kasanama3 geothermal springs; the lowest geotemperatures were associated with Chiling'ombe and Ngogu thermal springs.

The Na/K/Ca and Na/K geothermometers were found to provide better and more robust estimates for subsurface temperatures (geotemperatures) of geothermal springs in Nkhata Bay as these most significantly ($p<0.01$) correlated with the emergence temperatures of the thermal springs with correlation coefficient (r) =0.50 and $r \geq 0.36$ for Na/K/Ca and Na/K geotemperatures, respectively. The estimation of subsurface temperature in the geothermal springs in Nkhata Bay, however, needs re-evaluation by use of more extensive hydrogeochemical and geophysical surveys in order to establish whether they can be used for generation of geothermal electricity in Malawi.

Conflict of Interests

The author(s) have not declared any conflict of interests.

REFERENCES

APHA (American public Health Association) (1989). Standard Methods for Examination of Water and Wastewater. 15th ed. 1015. Eighteenth Street NW, Washington, DC 20036.

BGS (British Geological Survey) (2008). Water Quality Factsheet: Fluoride. http://www.wateraid.org/documents/plugin_documents/fluoride1. Accessed July 3, 2008.

CER (Comstock Energy Report) (2007)). Qualifying Report on: A High-grade Geothermal Resource in the Canadian Rockies; Canoe Hot Springs,Valemount, British Columbia. Retrieved from World Wide Web: http://www.comstockenergy.com/Qualifying%20Report.html on October 21, 2008.

Dulanya Z (2006). Geothermal Resources of Malawi - An Overview. PROCEEDINGS, Thirty-First Workshop on Geothermal Reservoir Engineering Stanford University, Stanford, California, January 30-February 1, 2006. SGP-TR-179.

Elder J (1981). Geothermal Systems. Academic Press Inc (London) Ltd. 24/28 Oval Road, London NW1. pp. 341-343.

Fournier RO (1979). A revised Equation for the Na-K Geothermometer. Geothermal Resources Council Transactions 3:221-224.

Fournier RO, Truesdell AH (1973). An Empirical Na-K-Ca Geothermometer. Geochim. Cosmochim. Acta 43:1543 -1550.

GEO (Geothermal Energy Office) (2007). Geothermal Energy. Retrieved from World Wide http://www.geo-energy.org/aboutGE/basics on October 10, 2007.

Giggenbach WF(1988). Geothermal Solute Equilibria: Derivation of Na-K-Mg-Ca Geoindicators.n Geochim. Cosmochim. Acta 52:2749-2765.

GNWT (Government of the North-West Territories) (2008). Operations-Drinking Water Quality. Retrieved from World Wide Web.Http://www.maca.gov.nt.ca/operations/water/homepage.asp on September 30, 2008.

Harrison DR, Chapusa FWP (1975). The Geology of the Nkhotakota - Benga Area. Bull. Geol. Surv. Mw, p. 32.

Hong H, Manchao H, Jichu Z (2005). Applications of Geothermometers to Hydrology Assessment. Proceedings World Geothermal Congress 2005, Antalya, Turkey.

Järup L (2003). Hazards of heavy metal contamination. British Medical Bulletin 68:167-182, The British council.

Kalindekafe LSN (2003). Geothermal Energy Potential for Malawi. Geological Survey, Malawi Government. Zomba, Malawi.

Lenntech (2006) Chemical properties, health and environmental effects of zinc, Rotterdamseweg 402 M, 2629 HH Delft, The Netherlands. Retrieved from World Wide Web. http://geothermal.marin.org/pwrheat.html#Q1 on October 10, 2007.

NAG (Numerical Algorithms Group) (2008). Genstat 12.0 Software. The Numerical Algorithms Group Ltd, Oxford, UK.

Pirlo MC (2004). Hydrochemistry and Geothermometry of Thermal Groundwaters from Birdsville Track Ridge, Great Artesian Basin, South Australia. Geothermics 33(2004):743-777.

Purschel M (2006). Hydrothermalism in Ethiopia: Interpretation of Chemical Composition of Hot Springs from Aluto-Langano, Dofan-Fantale and Gergedi-Sodere Geothermal Areas. TU Bergakademie Freiberg, Faculty of Geosciences, Geotechnology and Mining, Gustav-Zeuner- Street, 1209599 Freiberg, Germany.

Ray GE (1975). The Geology of the Chitipa-Karonga Area. Bull. Geol. Surv. Malawi.

SWS (Schlumberger Water Services) (2008). AquaChem Manual. U.S.A.

Verma SP, Santoyo E (1997). New Improved Equations for Na/K, Na/Li and SiO_2 Geothermometers by Outlier detection and Rejection. J. Volcanol. Geotherm. Res. 79:9-23.

Wahl FE (1977). Geothermal Energy Utilization. John Wiley & Sons, New York. U.S.A.

Appendix 1. Spatial data for geothermal springs in Nkhata Bay, Malawi.

Geothermal spring	utm coordinates	Village	Traditional authority
Kawira	0610504; 8665862	Chaphulika	Zilakoma
Kanunkha	0635870; 8718574	Chighachang'ombe	Timbiri
Nalunguwa	0620479; 8720782	Chilongozi	Kabunduli
Movya	0618818; 8721976	Chilongozi	Kabunduli
Chilyalya	0618291; 8721782	Chamawoya	Kabunduli
Kavuzi	0618187; 8721622	Chamawoya	Kabunduli
Chiling'ombe	0616070; 8715216	Kachiweya	Kabunduli
Ngogu	0616096; 8714990	Kachiweya	Kabunduli
Kasanama1	0614246; 8697772	Chilipeta II	Kabunduli
Kasanama2	0614163; 8697752	Chilipeta II	Kabunduli
Kasanama3	0615297; 8698416	Chilipeta II	Kabunduli
Ndunduzi	0609352; 8694896	Chilipeta II	Kabunduli
Magala1	0610111; 8716546	Vimaso	Kabunduli
Magala2	0610164; 8716592	Vimaso	Kabunduli
Kauziru	0609554; 8716788	Vimaso	Kabunduli
Chiwisi	0630933; 8761965	Siyalimba	Musisya
Sanga	0632384; 8757501	Kajizinge	Musisya

Appendix 2. Geotemperatures estimated by silica geothermometers for thermal springs in Nkhata Bay, Malawi.

Spring name	Measured Temperature (°C)	QTZ_1	QTZ_2	QTZ_3	QTZ_4	QTZ_5
Kawira	74	-41.37	-41.82	2089.63	-81.19	-73.09
Kanunkha	53	-21.81	-22.23	40921.95	-46.02	-54.36
Nalunguwa	39.5	-	-	-	-	-
Movya	41	-	-	-	-	-
Chilyalya	46	-	-	-	-	-
Kavuzi	50	-	-	-	-	-
Chiling'ombe	40	-74.00	-74.45	518.75	-154.97	-103.63
Ngogu	39	-59.80	-60.26	881.68	-120.11	-90.45
Magala1	48	-53.05	-53.50	1165.38	-105.11	-84.12
Kasanama1	61.5	-40.35	-40.80	2221.82	-79.22	-72.13
Kasanama3	65	-37.13	-37.57	2747.08	-73.08	-69.06

QTZ_1: Silica geothermometer by Trusdell (1976)

Formula: $$T_{SiO2} = \frac{1315 - 273.15}{5.205 - \log m}$$

QTZ_2: Silica geothermometer by Fournier (1977)

Formula: $$T_{SiO2} = \frac{1309 - 273.15}{5.19 - \log m}$$

QTZ3: Silica geothermometer by Rimstidt (1997)

Formula: $T_{SiO2} =$ (856/ log m) + 0.0254 - 273.15

QTZ4: Silica geothermometer by Verma and Santoyo (1997)

Formula: $T_{SiO2} = -44.119 + (0.24469m) + ((-1.7414 \times 10^{-4})\ m^2) + (79.305 \times \log m)$

QTZ5: Silica geothermometer by Fournier (1977)

Formula: $$T_{SiO2} = \frac{1032 - 273.15}{4.69 - \log m}$$

Note: m =concentration of SiO_2 in ppm or mg/L.

Heavy metal concentration in selected fish species from Eleyele reservoir Ibadan Oyo State South-western Nigeria

Ayeloja, A. A.[1] , George, F. O. A.[2], Shorinmade, A. Y.[3], Jimoh, W. A.[1], Afolabi, Q. O.[3]and Olawepo, K. D.[1]

[1]Department of Fisheries Technology, Federal College of Animal Health and Production Technology Moor Plantation, PMB 5029 Ibadan, Oyo State, Nigeria.
[2]Department of Aquaculture and Fisheries Management, Federal University of Agriculture, Abeokuta (FUNAAB) PO Box 2240 Abeokuta, Ogun State, Nigeria.
[3]Department of Science Laboratory Technology, Federal College of Animal Health and Production Technology Moor Plantation, PMB 5029 Ibadan, Oyo State, Nigeria.

Heavy metal concentrations (lead, cadmium, zinc, copper, chromium and manganese) in selected fish species from Eleyele reservoir Ibadan Oyo State South-Western Nigeria were investigated. A sample each of catfish (*Clarias gariepinus*), Africa pike *(Hepsetus odoe*) and tilapia fish (*Oreochromis niloticus*) from Eleyele lake Ibadan Oyo state were collected from fish farmers at the landing site early in the morning using standard procedure and transported to the laboratory within 30 minutes for laboratory analysis. Heavy metal concentration in the flesh of the fish species and water sample were analysed using atomic absorption spectrophotometer. There existed significant difference ($p<0.05$) in the concentration of heavy metals in fish flesh of various fish species with that of herbivorous fish (Tilapia fish) being the highest followed by omnivorous fish (catfish), followed by piscivorous fish (*H. odoe*), while that of the water body was the lowest. It was concluded that fish could be considered bio-indicator of environmental contamination within the aquatic ecosystem; it also indicates that fish could be useful in estimating bioavailability of metal to freshwater biota.

Key words: Heavy and trace metal, *Clarias gariepinus*, *Hepsetus odoe*, *Oreochromis niloticus*, Eleyele reservoir.

INTRODUCTION

Heavy metal comprises a number of elements which are manganese, chromium lead, cadmium, etc. (Mazvila, 2001), but these metals are also the most important source of necessary for living organisms: that is, iron, zinc, copper, pollution (Chovanec et al., 2003; Popek et al., 2008). The subject of heavy metal is receiving increasing popularity in food industry due to high incidence of contamination in agricultural and seafood products (Mohammad et al., 2011). Heavy metal affects freshwater organisms and induces certain harmful modifications at histological and morphological levels, also decreases the growth and developmental rates resulting in increase of death rate and the decrease of birth rate (Authman, 2008). Their potential toxic effects are given by the presence in water solution at concentrations exceeding certain threshold levels and their long persistence in the aquatic ecosystems and their bioaccumulation and biomagnification in the food webs (Naeem et al., 2011). Metal accumulation in fish tissues poses a direct threat for human being (Papagiannis et al., 2004). The distribution of metals varies between fish species, depending on age, development status and other physiological factors (Kagie and Schaffer, 1998). The ingestion of food is an obvious means of exposure to metals, not only because many metals are natural components of food stuffs, but also environmental contamination and contamination during processing (Voegborlo et al., 1999). Heavy metals having penetrated into human being through food chains might cause various disturbances or serious diseases (Idzelis et al., 2007). Once heavy metals are ingested, numerous health problems will take place. Lead may cause learning disabilities, impaired protein and hemoglobin synthesis and shorten the lifespan of red blood cells which leads to severe anemia (hypochromic microcytic anemia) in children (Sultana et al., 1998). The most common toxic effects of cadmium in human is renal failure, accumulation in the bone resulting in calcium loss and malfunctioning of peripheral and central nervous system (Schroeder et al., 1965; Castro et al., 2008). Nickel has different undesirable effects on human health such as impairing the biological activity of cells, lung and nasal cancer in long-term exposure; respiratory, nervous and digestive disorders and also psychological problems will be increased (Sultana and Rao, 1998). Zinc causes slow growth in children, reduced fertility, dry mouth, headache and nausea (Castro et al., 2008). Although fish can be effective in preventing cardiovascular disease (CVD), the fish found in waters with heavy metals may increase the incidence of some illnesses such as cancer (Capar and Yess, 1996). Therefore investigation of the heavy metals in fishes becomes important to estimate freshwater pollution and the risk potential of human consumption (Dural et al., 2007). This study therefore evaluates the presence of heavy metals in three commercial fishes from selected water bodies in Ibadan metropolis.

MATERIALS AND METHOD

The study area

Eleyele reservoir

The study location Eleyele Reservoir is located in north-eastern part of Ibadan, Southwestern Nigeria within longitude N07025'00" and N07027'00" and Latitude E03050'00" and E03053'00" (Figure 1). The study site is surrounded by Eleyele neighborhood in the south, Apete in the east and Awotan in the north. Eleyele wetland is a modified natural riverine wetland type with area of about 100 km2 including the catchment area. The elevation is relatively low ranging between 100- 150 m above sea level and surrounded by quartz-ridge hills toward the downstream section where the Eleyele dam barrage is located. A number of stream channels serve as feeding / recharge streams to the Eleyele wetland basin. In 1942, the quest to create a modern water supply system to meet the challenge of water scarcity for the emerging Ibadan metropolis led to the construction of Eleyele Dam on the main River Ona with a reservoir storage capacity of 29.5 million litres (Tijani et al., 2001).

Sample collection

A sample each of catfish (*C. gariepinus*), African pike *(Hepsetus odoe)* and tilapia fish (*O. niloticus*) (Figure 2) from Eleyele lake Ibadan Oyo state were bought from fish farmers at the landing site by 8.00 am, the fish were collected inside an ice box and transported to SMO laboratory Joyce B road Ibadan, Oyo State, Nigeria within 30 min for laboratory analysis. Sample of the water body in which these fish were sourced was also collected in water sampling bottle using a standard procedures described by Welz and Sperling (1999): the sampling bottles were conditioned by washing with detergent solution that is metal free and non-ionic and finally rinsing several times with distilled water. The sampling bottles were rinsed with the dam water first before the sample were finally collected. The heavy metal concentration of the fish species and the water sample were analyzed in triplicate using the following analytical procedure:

Digestion of fish samples

1 g of sample was weighed into a 50 ml digestion tube, 20 ml of acid mixture H_2SO_4, HNO_3, $HClO_4$ (2:1:1) were added to the sample and set in the appropriate hole of the digestion block to digest to a clear colourless solution. The digest was cooled down and carefully washed with deionized water into a 50 ml volumetric flask and made up to mark. The element digest was used to read for the level of the trace metals on the Atomic Absorption Spectrophotometer at a wavelength and allow cathode lamp related to each trace metal.

RESULTS AND DISCUSSION

Table 1 shows heavy metal concentration (mg/l) in fish species sampled from Eleyele Lake, Ibadan, Oyo State, Nigeria. The result of this study indicates that there was significant difference ($P<0.05$) in the concentration of lead (Pb) present in the different fish species and the water body; the lead concentration in the water body was the lowest with mean value of 0.2730 ± 0.05% while that of African pike was the highest with mean value 0.663 +0.02%. The levels of lead in these fish samples were less than those recommended by European commission (EC) 2001 guideline and FAO, as reported by Sivaperumal et al. (2007) that the allowable level of lead in fish, is 0.4 and 0.5 mg/kg respectively. Furthermore, there was significant difference ($p<0.05$) in the concentration of cadmium (Cd) present in different fish species with the cadmium concentration in tilapia body being the lowest with the mean value of 0.073 ±0.02%

Figure 1. Map of the study area. Source: Tijani et al. (2001).

Table 1. Heavy metal concentration (mg/l) in fish species sampled from Eleyele lake, Ibadan, Oyo State, Nigeria.

Fish	Pb	Cd	Zn	Cu	Cr	Mn
Tilapia	0.393 ± 0. 04^{c}	0.073 ± 0.02^{b}	0.442 ± 0.83^{b}	0.236 ± 0.11^{a}	0.273 ± 0.05^{c}	0.008 ± 0.13^{d}
Catfish	0.463 ± 0.12^{b}	0.123 ± 0.04^{a}	0.464 ± 0.03^{a}	0.108 ± 0.32^{b}	0.173 ± 0.02^{d}	0.012 ± 0.02^{c}
African pike	0.663 ± 0.02^{a}	0.093 ± 0.12^{b}	0.235 ± 0.13^{c}	0.042 ± 0.22^{d}	0.325 ± 0.22^{b}	0.162 ± 0.24^{a}
Water	0.273 ± 0.05^{d}	0.833 ± 0.02^{b}	0.093 ± 0.12^{d}	0.097 ± 0.02^{c}	0.113 ± 0.03^{a}	0.074 ± 0.33^{b}

Column means with different superscripts are significantly different (P<0.05) from one another.

while that of water is the highest with the mean value of 0.833 ± 0.02%. The values obtained for cadmium present in fish samples in this study is far less than those reported in Egypt and Saudi Arabia as reported by Voegborlo et al. (1999) when they determined the amount of cadmium in the tuna fish; the amount of cadmium in tuna was 0.32 and 0.35 $\mu g\ g^{-1}$ in each sample respectively.

However, prevention of cadmium in Eleyele water body should be encouraged so as to prevent its bioaccumulation because when cadmium accumulates in the human body, it may induce kidney dysfunction, skeletal damage and reproductive deficiencies (Mohammad et al., 2011). There was also significant difference (P<0.05) in the concentration of zinc (Zn) present in different fish species with the zinc concentration in water being the lowest with the main value of 0.09317±0.12%. There was significant difference (P<0.5) in the copper (Cu) concentration with that of *H. odoe* been the lowest with the mean value of 0.042±0.22% while the copper concentration in tilapia fish is the highest with the mean value of 0.236±0.11. This agrees with the opinion of Bordajandi et al. (2003) that diet has a remarkable role in the bioconcentration process for some metals, mainly for the Cu and Zn. Protasowicki et al. (1983) also reported that feeding strategy influenced the content of Cu and Zn in fish.

The result also indicates that there was significant different (P<0.5) in the chromium (Cr) concentration present in different fish species with the chromium concentration in water been the lowest (0.11273±0.03) while that of African pike was the highest (0.3253±0.22). Also, there exists significant difference (P<0.05) in the

Catfish (*Clarias gariepinus*)

African pike (*Hepsetus odoe*)

Tilapia fish (*Oreochromis niloticus*)

Figure 2. Fish samples used for the study.

manganese (Mn) concentration present in different fish species with the manganese concentration in tilapia fish been the lowest (0.008±0.13) while that of African pike was the highest (0.162±0.24). Similar manganese range of 0.14 - 3.36 mg/kg for muscle of fish was reported by Turkmen and Ciminli (2007). This study indicate that heavy metals concentration in different fish species differs; this in line with the opinion of Atuanya et al. (2011) that concentration of heavy metals varies with variation in fish species. This study also indicate that mineral content of herbivorous fish (Tillapia fish) is highest followed by ominiovorous fish (catfish) followed by piscivorous African pike (*H. odoe*), while that of the water body was lowest. This agrees with the report of Farkas et al. (2000) that the concentrations of element in fish body could be related primarily to their feeding habits.

Burger et al. (2002) also reported that fishes are good indicators for heavy metal contamination in aquatic systems because they occupy different trophic levels with different sizes and ages. Figure 3 shows heavy metal concentration in fish species sampled from Eleyele lake, Ibadan Oyo State Nigeria.

Transfer factor (TF)

The transfer factor in fish tissues from the aquatic ecosystem was calculated according to Kalfakakour and Akrida-Demertzi (2000), Rasheed (2001) and Anim-Gyampo et al. (2013):

$$TF = \frac{\text{Metal concentration in tissue}}{\text{Metal concentration in water}}$$

Table 2 shows heavy metal concentration (mg/l) in the water of Eleyele reservoir. The result of this study indicates that the mean heavy metal concentrations in raw water from Eleyele reservoir follow an increasing order of Mn< Zn<Cu<Cr < Pb < Cd. This result trend was similar to what was reported in the work of Anim-Gyampo et al. (2013).

The heavy metal concentrations obtained from the raw water with the exception of Zn and Cu exceeded the various raw water quality guidelines (WHO, 2003; USEPA, 1986). The concentrations of Pb, Cd and Cr (0.273, 0.833 and 0.113 mg/l) exceeded all the raw water guideline values, while Mn (0.074 mg/l) exceeded the guideline value of USEPA (1986).

The transfer factor is as presented in Table 3 which indicates that the concentration of heavy metals in the tissue was more than the concentration of the heavy metals in the water body with exceptions of Cd and in all the fish samples; Cu in African pike and Mn in tilapia and catfish. This result is in consonance with the report of Anim-Gyampo et al. (2013) that the mean heavy metal concentrations in tissues of the fish were higher than

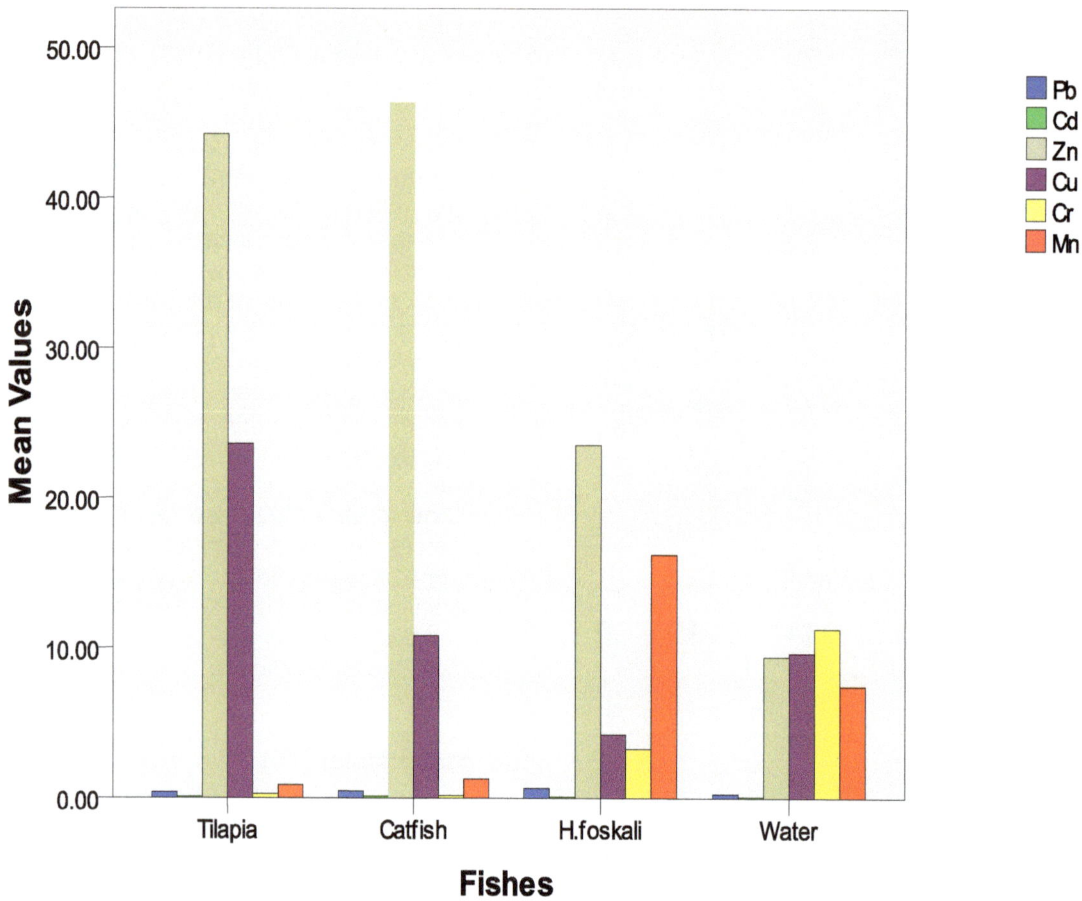

Figure 3. Chart showing heavy metal concentration in fish species sampled from Eleyele Lake, Ibadan Oyo State Nigeria.

Table 2. Mean heavy metal concentration (mg/l) in the Water of Eleyele reservoir.

Guideline	Pb	Cd	Zn	Cu	Cr	Mn	Reference
WHO	0.01	0.010	0.50	2.25	0.05	0.50	WHO,2003
USEPA	0.11	0.010	0.50	2.25	0.10	0.02	USEPA,1986
Water	0.273	0.833	0.093	0.097	0.113	0.074	This study

Table 3. Transfer factor (water/muscle) in different fish species.

Fish species	Pb	Cd	Zn	Cu	Cr	Mn
Tilapia	1.44	0.09	4.75	2.43	2.42	0.11
Catfish	1.70	0.15	4.99	1.11	1.53	0.16
African pike	2.43	0.12	2.53	0.43	2.96	2.19
WHO	0.005	0.003	5.0	2.00	0.15	0.50

metal concentration in raw water. Similarly, Chale (2002) also reported that heavy metal concentration in fish tissue is higher than that of water.

Conclusion

The result of this study supplied valuable information on

the heavy metal level of some fish species in Eleyele lake Ibadan Oyo State Nigeria. The values obtain for heavy and trace metals in this study were below the limits in muscle tissue for fish proposed by EU commission Regulation (2001). The study indicates that fish could be considered bio-indicator of environmental contamination within the aquatic ecosystem. It also indicates that fish could be useful in estimating bioavailability of metal to freshwater biota. It is therefore recommended that heavy metal pollution of the rivers and lakes by Industries in Ibadan should be checked by relevant authorities so as to guarantee food safety and minimize environmental pollution.

Conflict of Interests

The author(s) have not declared any conflict of interests.

REFERENCES

Atuanya EI, Edefetah MA, Nwogu NA (2011). Microbiological Qualities and some Heavy Metals (Mercury and Cadmium) Levels of fresh and Dry Fish Species sold in Benin City, Edo State, Nigeria. Bull. Environ. Pharmacol. Life Sci. 1(1):10-14.

Authman MMN (2008). *Oreochromis niloticus* as a biomonitor of heavy metal pollution with emphasis on potential risk and relation to some biological aspects. Global Veterinaria. 2(3):104-109.

Bordajandi IR, Gómez G, Fernández MA, Abad E, Rivera J, González, MJ (2003). Study on PCBs, PCDD/Fs, organochlorine pesticides, heavy metals and arsenic content in freshwater fish species from the River Turia (Spain). Chemosphere 53:163-171.

Burger J, Gaines KF, Shane BC, Step HWL, Snod GJ, Dixo NC, McMahon M, Shuk LS, Shuk LT, Goc HLM (2002). Metal levels in fish from the Savannah river: potential hazards to fish and other receptors. Environ. Res. 89(85):215-25.

Capar SG, Yess NJ (1996). US Food and Drug Administration survey of cadmium, lead and other elements in clams and oysters. Food Addit. Contam. A 13(5):553-560.

Castro-Gonzalez MI, Méndez-Armenta M (2008). "Heavy metals: Implications associated to fish consumption." Environ. Toxicol. Pharmacol. 26(3):263-271.

Chovanec A, Hofer R, Schiemer F (2003). Fishes as bioindicators. In: Principles, Concepts and Applications. Markert BA, Breure AM, Zechmeister HG (eds.), Elsevier, Amsterdam. pp. 639-676.

Dural M, Goksu MZL, Ozak AA (2007). Investigation of heavy metal levels in economically important fish species captured from Tuzla lagoon. Food Chem. 102:415-421

Farkas A, Salanki J, Varanka I (2000). Heavy metal concentrations in fish of Lake Balaton. Lakes Reserv. Res. Manage. 5:271-279.

Idzelis RL, Ladygien R, Sinkevicius S (2007). Radiological investigation of meat of game and dose estimation for hunters and members of their families. J. Environ. Eng. Landsc. Manage. 15(2):99-104.

Kagi JH, Schaffer A (1998). Biochemistry of metallothionein. Biochemistry. 27:8509-8515.

Mazvila J (2001). Heavy metals in Lithuanian soils and plants: Monograph. Kaunas. p. 343.

USEPA (1986). Quality Criteria for Water.EPA-440/5-86-001,Office of Water Regulations Standards, Washington DC, USA.

Mohammad M, Mahsa T, Mohammad D, Kiomars A (2011). Determination of Pb, Cd, Ni, and Zn concentrations in canned fish in Southern Iran. Sacha J. Environ. Studies. 1(1): 94-100.

Naeem M, Salam A, Khokhar MY, Nasir MF, Nouman MF, Ishtiaq A and Allah-Yar R (2011). Evaluation of metal contents in female *Oreochromis niloticus* with emphasis on potential risk of consumption and relation to some biological aspects. Afr. J. Biotechnol. 10 (53):11054-11057.

Papagiannis I, Kagalou I, Leonardos J, Petridis D, Kalfakakou V (2004). Copper and zinc in four freshwater fish species from Lake Pamvotis (Greece). Environ. Int. 30:357-362.

Popek W, Nowak M, Popek J, Deptuła S, Epler P (2008). Heavy metals concentration in tissues of the Eurasian minnow *Phoxinus phoxinus* from the Czarna Orawa River system, Poland. AACL Bioflux, 1(2):165-171.

Protasowicki M, Chodyniecki A, Ociepa A (1983). Heavy metals in fish caught in 1976-1980. Zesz. Nauk. AR Szczecin. Ryb. Mor. 103:181-197.

Sivaperumal PTV, Sankar TL (2007). Heavy metal concentrations in fish, shellfish and fish products from internal markets of India vis-a-vis international standards. Food chemistry 102(3):612-620.

Sultana R, Rao DP (1998). Bioaccumulation patterns of zinc, copper, lead, and India. Bull. Environ. Contam. toxicol. 60(6):949-955.

Turkmen M, Ciminli C (2007). Determination of metals in fish and Mussel species by inductively coupled plasma-atomic emission spectrometry. Food Chem. 103:670-675.

Voegborlo RB, El-Methnani AM, Abedin M Z (1999). Mercury, cadmium and lead content of canned tuna fish. Food Chem. 67(4):341-345.

WHO(2003), Malathion in cadmium in grey mullet, Mugil cephalus (L.), from harbour waters of Visakhapatnam, drinking water. Backgroung Document for Preparation of WHO Guidelines for Drinking Water Quality. World Health Organisation (WHO/SDE/WSH/03.04/103). Geneva.

Tijani MN, Olaleye AO, Olubanjo OO (2001).Impact of Urbanization of wetland Degradation: A case study of Eleyele wetland Ibadan, South-West, Nigeria. Proceeding of the Environmental Management Conference, Federal University of Agriculture, Abeokuta, Nigeria.

Anim-Gyampo M, Kumi M, Zango MS (2013). Heavy Metals Concentrations of some selected Fish Species in Tono Irrigation Reservoir in Navrongo, Ghana. J. Environ. Earth Sci. 3(1):109-119 pp. 435-457.

Variation of small scale wetland fishery in relation to land use within Mpologoma riverine marsh in Eastern Uganda

Grace Asiyo Ssanyu [1], Julius Kipkemboi [2], Jude M. Mathooko [3], Sylvester B. Wandera[4] and John Balirwa[4]

[1]Kyambogo University, P.O Box 1 Kyambogo, Uganda.
[2]Egerton University, P.O Box 536-20115, Nakuru, Kenya.
[3]Global Research Akademik and Mentoring Services (GRAMSs) P.O Box 14824-20100, Nakuru, Kenya.
[4]National Fisheries Resource and Research Institute, P.O Box 343 Jinja, Uganda.

In densely populated areas such as those in Eastern Uganda where livelihoods demands create immense pressure on environmental resources, small scale wetland fisheries may be disturbed by agriculture practices. The study carried out between 2011 and 2012, investigated the variation of fish catch at the different wetland sites in relation to land use in Mpologoma riverine marsh. Four sites were identified to represent different land uses; intact wetland, minimally disturbed, highly disturbed with small scale farmers and one with a large scale irrigation scheme. Data was collected on water quality, wetland fish species catch and catch per unit effort from the different sites. Conductivity and dissolved oxygen levels significantly differed between sites and explained 72.03% of the variance among sites. Seven fish taxa dominated the wetland fishery. Large sized fish species catch, *Clarias gariepinus* and *Protopterus aethiopicus* (range of 0.45 to 38 and 0.25 to 20 kg/day, respectively) was higher at the less disturbed sites than at highly disturbed sites which accounted for over 91.5% of total wetland catch. *Tilapia zillii* and *Oreochomis leucostictus* catch were also higher at the less disturbed sites while the small fish species (*Haplochomis* sp, *Clarias liocephalus and C. alluaudi*) did not vary with site. Conductivity and dissolved oxygen significantly correlated with the two large fish species' catch but did not correlate with small fish species catch. Agricultural activities in the wetland negatively affected the life history strategies of large fish species, leading to low catch rates at the highly disturbed site. Therefore, there is need to control land use changes to secure high productivity of small scale fisheries in the riverine wetland.

Key words: Papyrus wetland, fish, catch, water quality, disturbance.

INTRODUCTION

Small-scale fisheries play a significant role in human and socio-economic development and they are an entry point for poverty reduction though their role in generating revenues and creating employment, and their contribution to food security (Heck et al., 2007; Bene et al., 2010). These are wetland fisheries which often provide a 'safety valve' for people who cannot access other sources of livelihood (Bene, 2004). The small scale wetland fisheries provide nutritional security in remote areas that lack adequate supplies of animal protein and sustain the livelihood of landless fishers who can no longer survive by fishing in depleted freshwater bodies (Vass et al., 2009). Although wetlands provide habitat for 40% of all fish species (Arthington et al., 2004), 20% of their biota are amongst the most theatened components of global biodiversity (Smith et al., 2005). This is largely due to human induced environmental degradation. The demand for increased food production to cater for the rising human population and large numbers of undernourished or starving people, especially in the developing countries (Okechi, 2004); have led to widespread conversion of wetlands into farmland. Climate variability which has emerging signals within the River Nile Basin (Di Baldassarre et al., 2011), is likely to have a profound impact on land management practices leading to more changes in land use decisions (Verburg et al., 2011). Consequently, land requirements, reclamation potential and general environmental management are affected (Carvalho et al., 2004).

Many permanently flooded wetlands are open to fishing by anyone (open access) or subject to exclusive communal or individual use rights (Garaway et al., 2006; Martin et al., 2011). Often, wetlands under exclusive access arrangements are exploited less intensely and maintain higher standing stocks of fish than open access wetlands (Lorenzen et al., 1998). The open access wetlands may also be subject to very intensive exploitation of both water and fisheries resources (Garaway et al., 2006). In eastern Uganda, over 10% of Mpologoma River wetland is under cultivation of rice, maize and vegetable to cater for food demand of the increasing population (NEMA, 2004). Simultaneously, there appears to be reduction in the large preferred fish species, *Clarias gariepinus* and *Protopterus aethiopicus* in the Lake Kyoga basin (Muhoozi, 2003). *C. gariepinus* catch reduced from more than 1600 metric tonnes in the 1980s to 1.35 metric tonnes in 2000 (Muhoozi, 2003).
Understanding the effects of land use changes on the fisheries is essential for the management of exploited wetlands (Rientjes et al., 2011). Although there are a few existing models based on observed pattern in temperate areas, scarcity of data for most tropical riverine wetlands limits prediction of the effects of human-induced perturbations on the fish assemblages and catch (Ibanez et al., 2007). Therefore Mpologoma wetland fisheries that support local communities with no access to open lake fisheries and yet it is under intensive exploitation need evaluation. The aim of the study was the assessment of the small scale fishery dynamics in relation to land use in Mpologoma river wetland. We used comparison of the wetland fish species catch at the wetland sites with different land use patterns. It was hypothesized that the different land use at the wetland sites would affect the fish habitat though water quality variations, leading variation in the wetland fish species catch. The spatial-temporal variation of the wetland fish species catch in Mpologoma riverine wetland was expected.

MATERIALS AND METHODS

Study area and sampling sites

Mpologoma river wetland in Uganda (latitude 1°12' N and longitude 34°40' E) extending up to 102 km, discharges 610 million m^3 of water annually into Lake Kyoga complex (Ramsar, 2008). The climate is tropical with rainfall ranging between 1470 - 2300 mm in the longer wet season. The maximum temperature range is 27 – 32°C and the minimum is 16 – 18°C. This permanent wetland is dominated by papyrus (*Cyperus papyrus*) and hippo grass (*Vossia cuspidata*). The fish species of economic importance in the Lake Kyoga complex include *Clarias gariepinus, Protopterus aethiopicus Oreochomis leucostictus, O. niloticus, Bagrus docmak Rastrineobola argentea* and *Tilapia spp* (Vanden-Bossche and Bernacsek, 1990; NaFIRRI, 2007).

Using the Digital Elevation Model of the AVSwat, a map of Mpologoma river wetland with the four differently disturbed study sites was delineated (Figure 1).The main depression which is Mpologoma riverine wetland system consists of a network of small vegetated valley bottoms, fed by 38 sub-catchments. Four study sites with different levels of disturbance were selected along the wetland. Land use cover at each site was derived from Landsat ETM images of July 2011, verified by field survey data. Seven land use classes were identified in 200 hectare polygons area about each site (Table 1). Budumba and Mazuba sites were minimally disturbed with small farms at the edge of the wetland. Kapyani site was highly disturbed with small scale rice and maize farms of acreage of 2 – 6 acres inside the wetland, using low implement agriculture and rain-fed lowland systems (MAAIF, 2009). Nsango site was highly disturbed with a large scale rice scheme occupying over 3,000 hectares of the wetland. Artificial fertilisers, rainfed lowland and irrigation systems were used at the rice scheme.

Wetland fish assemblage

Fish samples from several fishing techniques were used to determine the wetland fish species composition and fishery production. Experimental fish sampling was done with gill nets hanged in the lagoons, while traps were placed inside the wetland. Gill nets of 2, 2.5 and 3 cm mesh sizes were set in parallel in the lagoon. Ten local basket traps were randomly deployed to catch fish inside the flooded vegetation zone of the wetland. All gear were set at around 7:00 h and retrieved before 14.00 h. This was done to reduce loss of fishing gear to the local fishermen during overnight fishing. On landing, fish samples were counted, total length and weight measured, and records of the gear and effort used at every

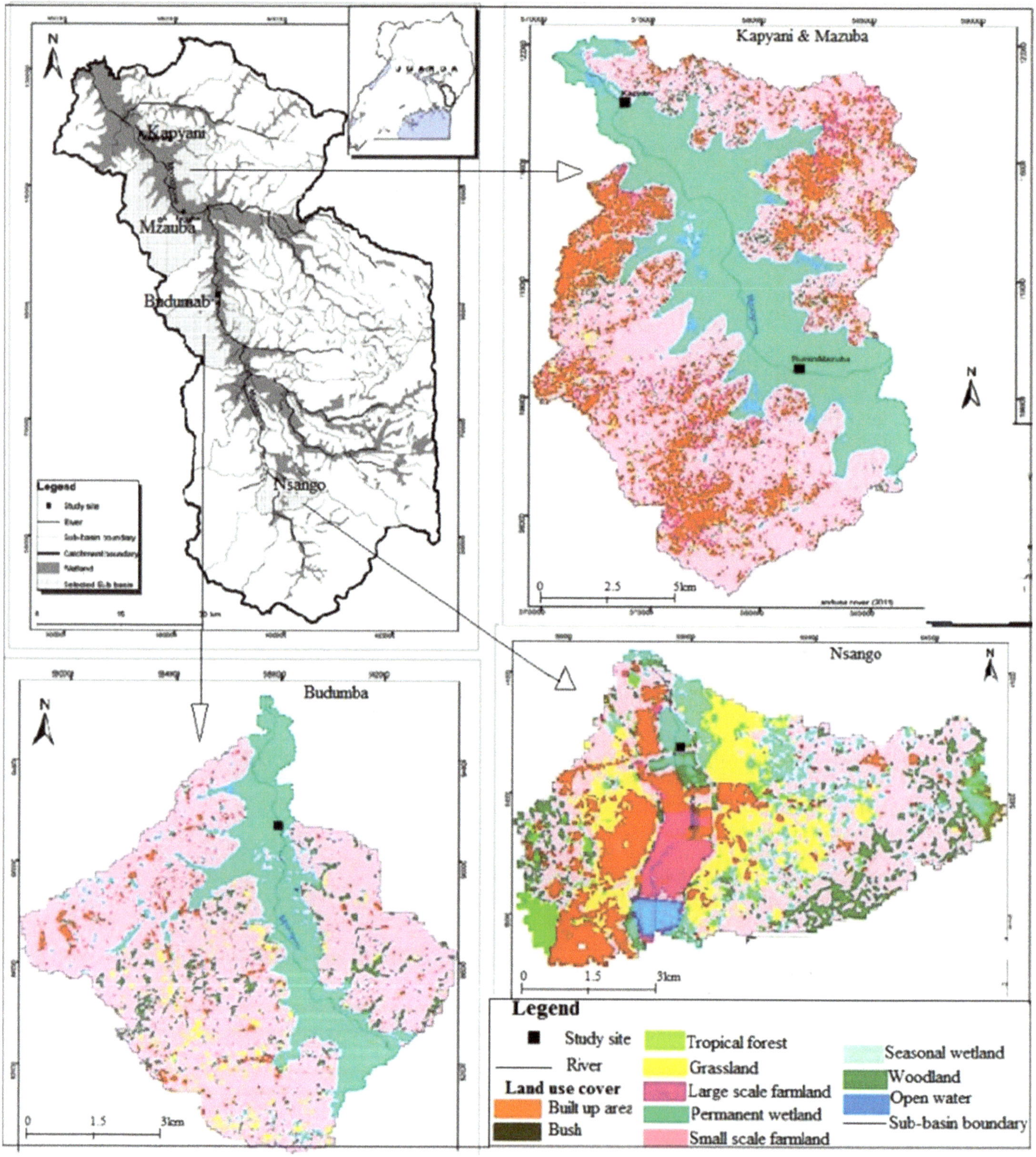

Figure 1. Land-use cover of three sub-basins surrounding study sites in the Mpologoma River catchment, Uganda, in July 2011.

site. Some fish were preserved in 40% ethanol and retained for identification using keys (Greenwood, 1974; Witte and van Densen, 1994). To cater for the seasonal changes in fishery, each site was sampled once a month for 12 months between September

Table 1. Land use percentage cover of 200 ha area around study sites in Mpologoma River wetland, Uganda, 2011.

Land use	Mazuba	Budumba	Kapyani	Nsango
	Intact wetland	Less disturbed	Highly disturbed (with small scale farms)	Highly disturbed (Large scale rice scheme)
Man made type				
Built up area	0	0.48	1.52	18.13
Subsistence	0	8.67	55.25	16.18
Large scale farmland	0	0	0.00	16.16
Subtotal cover	0	9.15	56.77	50.47
Natural type				
Permanent wetland	100	90.63	31.21	17.30
Temporary wetland	0	0	10.15	5.58
Woodland	0	0.23	1.38	21.09
Grassland	0	0	0.48	4.31
Subtotal cover	100	90.86	43.22	48.28

2011 and August 2012.

More data on wetland fish species and catch was collected from fishermen landing catches at each site for every sampling day. The local fishermen distributed their fishing effort thoughout the study site each trip to sample each portion of the study site each day. At each site, fishermen had their own culture of who fishes where, and regulated the number of fishermen per landing site. Therefore the decision of about where and how to fish was left to the fishermen. The local fishermen dominantly used gill nets (2.5 cm and 3.0 cm mesh sizes) of about 45 m long by 1.5 m deep, fishing lines with varying number of hooks and local basket traps (0.4 by 0.4 by 0.3 m, on average) as fishing gears. Fishing lines consisted of a weight (approximately 2 kg) with varying number of baited hooks mainly set to catch *Protopterus* sp. and *C. gariepinus* within the wetland. Bait materials included pieces of small fish species such as small *Clarias* species, *Tilapia* species, frogs and large insects for hooks, while dead earthworms were for basket traps. Most fishing gear were set at 18 hours and retrieved at 6:00 hours to prepare for the markets. Data on captured species brought to the landing sites were recorded including fish taxa, total length (TL, in centimeters), weight (Wt, in grams), type of gear and effort used. Total length and weight were measured as quickly as possible after the fishermen returned with their catch. For small species such as small *Clarias* sp., *Haplochomis* spp., *Tilapia* spp. and *Barbus* sp., random sub samples were taken to measure standard lengths but their total number in the catch recorded. For large species such as *C. gariepinus* and *Protopterus* sp. groups of similar sizes were counted and sub sampled for measuring their total length and weight to get as much data as possible in the limited time allowed by the fishermen.

Physical habitat

To characterise water quality of the wetland sites with different land uses, thee water samples were collected from different points (within the wetland, middle of lagoon and edge of the lagoon) at each site. Within the wetland, thee water samples were randomly collected close to where traps were placed at each site. Conductivity and pH were determined *in-situ* using a Hanna Instruments HI 9813-6 N Waterproof pH/Ec/TDS/°C meter. Dissolved oxygen and temperature were determined *in-situ* by Oakton DO 110 meter. Chemical analysis was performed on 0.45 µm membrane filtrate for alkalinity, nitrate-nitrogen and orthophosphate and unfiltered samples for total phosphorus within 48 h of collection according to standard procedures (APHA, 1995).

Data analysis

Fish data were summarised into the species composition catch and catch per unit effort (CPUE) for spatial and temporal variation along the wetland. The fishermen usually fished only once a day and therefore species catch was considered as total species catch in kilogram per day. Since the fishery was a multispecies type and uses non-selective fishing gear, for gill nets and traps fishing gear, CPUE was calculated by dividing the catch in grammes of the most dominant fish species caught by the individual gear by the number of hours. Most gillnets were set for less than 5 h and most fisherman had only one net, therefore the units of CPUE of gill nets derived from the mean sizes of gillnets and fishing hours used along the wetland. Most fishermen used locally made basket traps of similar sizes and shape. Therefore the units of CPUE of traps was derived from number of traps and hours used per fisherman. CPUE of lines with hooks was calculated by dividing the catch in grammes of most dominant fish species by the hours spent in fishing considering the mean number of hooks per fisherman. Most fishermen using line hooks in the evening and check on the fish caught early morning. Therefore 12 hwere used in the calculation of the hooks CPUE. Due to differences in mean number of hooks per sites, the hooks' CPUE was standardized to 100 hooks per hour. Species richness was tested using Shannon Wiener's index (H') of diversity (Thiebaut, 2006) in the equation;

$$H' = -\sum p_i \ln(p_i),$$

Where, p_i is the relative abundance of individual fish species.

Statistical analysis on all data was performed using SPSS version16 (IBM ©). All the wetland fish species catch and CPUE were treated as separate dependent variables. Data was log transformed since it was not normally distributed (Kolmogorov-Simonov test (p = 0.00) and Skewness of 4.523). One-way ANOVA followed by Tukey's HSD post hoc test was used to test differences

Table 2. Mean number of fishermen, fishing gear type per fisherman and boats at one major landing site of each of the differently disturbed sites along Mpologoma river wetland in 2012.

Parameter	Budumba	Mazuba	Kapyani	Nsango
Number of fishermen	22 ± 5	22 ± 4	30 ± 9	17 ± 5
Number of hooks per fisherman	85 ± 34	50 ± 35	64 ± 41	30 ± 17
Number of gill nets per fisherman	1	1	1	1
Number of traps per fisherman	8 ± 3	6 ± 3	5 ± 2	7 ± 3
Number of boats per site	16 ± 4	17 ± 2	13 ± 2	7 ± 2
Mean Catch per day ($Kgday^{-1}$)	29.03 ± 9.66[a]	21.28 ± 10.46[a]	24.10 ± 11.70[a]	2.07 ± 1.62[b]

The catch values with same superscript are not significantly different at $P < 0.05$.

in daily catch among fishing gears, species CPUE, catch and fish length between sites. Using R-statistic software (*R version* 3.0.2; R Development Core Team 2008), non-metric multidimensional scaling (NMDS) was more appropriate in measuring community dissimilarities between study sites along the wetland. The relationship between study sites and water quality parameters was evaluated to ordinate the similarity between sites. Water quality parameters were further fitted to the ordination as vectors to identify the parameters that significantly differentiate the study sites. Stepwise multiple linear regressions with forward selection of variables were used to analyze the relationship between fish catch and water quality parameters.

RESULTS

Fish species composition and catch

Data on a total 5137 fish specimens were collected from all study sites over the sampling period. Six fish species dominated the wetland fish community. *C. gariepinus, C. liocephalus, C. alluaudi, P. aethiopicus*, *O. leucostictus, T. zillii* and several unidentified *Haplochomis* species were observed at all site along the wetland. *Synodontis afrofischeri* was observed only at highly disturbed Kapyani site close to Lake Lemwa. The number of species ranged from seven at the highly disturbed sites to nine at the less disturbed sites and the mean Shannon Wiener index of diversity was 5.4 ± 1.2 per site. During the wet season months, a higher number of species was recorded than during the dry season months. For instance, the species richness index of 4.94 and 4.81 were realized in January and February, while 6.77 and 6.69 were realized in September and October respectively at less disturbed Nsango site.

Wetland fishery production was estimated to be 57 kg per ha per year. The number of fishermen ranged from 12 at the highly disturbed Nsango site to 27 at the less disturbed Budumba site (Table 2). Thee fishing gear (gill nets, line hooks and basket traps) were dominantly used at all sites. Many fishermen used line hooks with a range of 13 hooks at the highly disturbed Nsango site to over 119 hooks at the less disturbed Budumba site. At the less disturbed Budumba and Mazuba sites, the fishermen had a high number of hooks to catch as much fish as possible within in the intact wetland. Traps and gill nets were used minimally at all sites. During the rainy season, gillnets with 3.0 cm mesh size harvested more fish than other gears. Hooks were dominantly used during the dry season. More than 50% of the catch comes from the hooks fishery. Gill nets with 3.0 cm mesh size was the second most important, followed by the gill nets with 2.5 cm mesh size and traps (Figure 2). The number of hooks, gill nets of 2.5 cm mesh size and traps catch did not vary between season ($p = 0.001$). But among sites, the number of hooks were significantly higher at the less disturbed Budumba site than at the highly disturbed Nsango site ($p = 0.023$). The catch varied among the dominant fishing gears used. Significantly higher catch was realized by the use of line hooks than all other fishing gears ($p = 0.001$). The line hooks' mean catch of 11423 ± 6936 g/day was observed in the wet season and 14863 ± 9558 g/day observed during the dry season which high compared to 3294 ± 3108 and 4860 ± 4392 g/day the wet and dry season gill nets' catch respectively. There were temporal and spatial differences in the catch of individual fish species. Relatively higher catch was observed in the dry season than in the wet season and more at the less disturbed sites than at the highly disturbed sites (Figure 3). *C. gariepinus* and *P. aethiopicus* dominated the fishery by catch weight with a range of 0.45 kg per day at the highly disturbed Nsango site to 38.67 kg per day at the less disturbed Budumba site and 0.245 to 20.18 kg per day, respectively (Table 2). Catch at Nsango was significantly lower than that of other sites (Tukey's HSD test; $p < 0.05$). Higher catch of both *C. gariepinus* and *P. aethiopicus* species was recorded during the dry months (January, February, July and August, Figure 3). Significantly higher catch of both species was recorded at the less disturbed sites (Budumba and Mazuba) than at the highly disturbed site (Nsango) at $p = 0.05$ (Table 3). The disturbed Kapyani site had relatively high *C. gariepinus* catch almost the whole sampling period. The catch of small *C. liocephalus* and *C. alluaudi* was higher at the highly disturbed site than at the less disturbed sites (Figure 4).

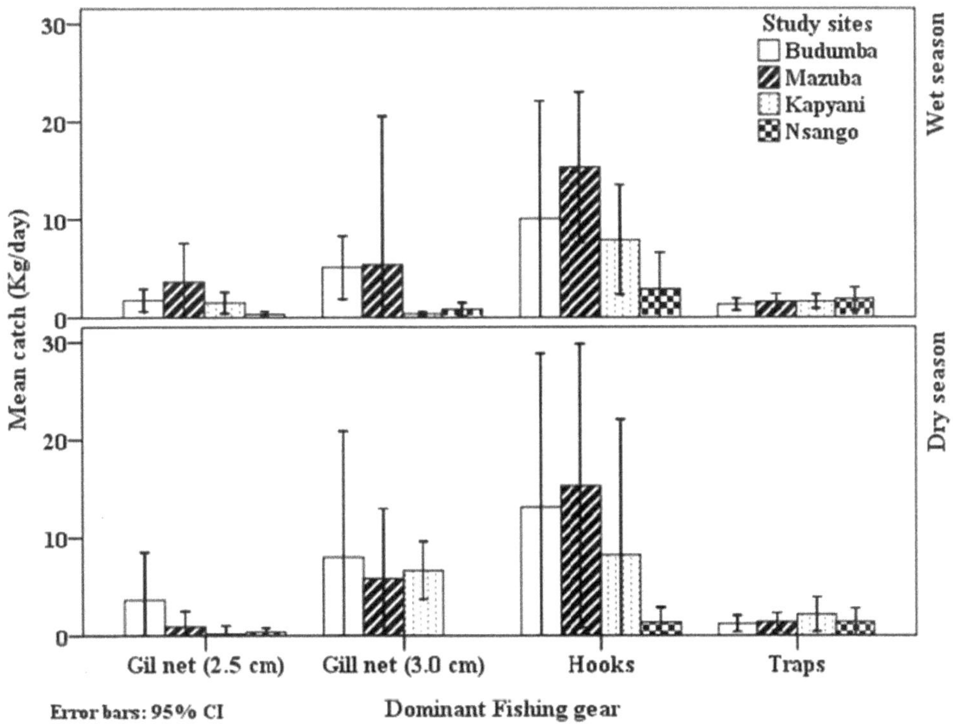

Figure 2. Seasonal mean catch of the major fishing gear at the four sites along Mpologoma wetland between September 2011 to August 2012.

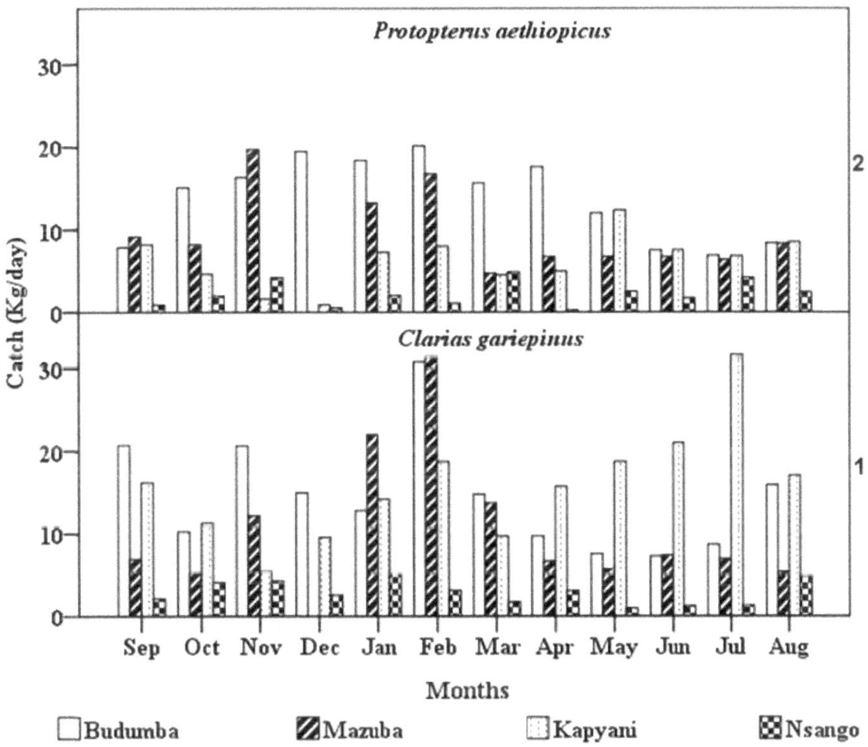

Figure 3. Mean catch of the big fish species in Mpologoma wetland fishery between September 2011 to August 2012.

Table 3. Mean catch (Kg/day, ± SD) of major fish species at differently disturbed sites along Mpologoma river wetland during the sampling period.

Fish species	N	Budumba	Mazuba	Kapyani	Nsango
Oreochomis leucostictus (Trewavas, 1933)	565	1.56 ± 1.07[a]	1.13 ± 1.51[a]	0.61 ± 0.41[a]	0.29 ± 0.15
Tilapia zilli (Garvais, 1848)	703	1.92 ± 1.28	0.86 ± 0.58[a]	0.93 ± 1.14[a]	0.32 ± 0.18
Haplochomis spp	1115	0.26 ± 0.25[a]	0.19 ± 0.09[a]	0.33 ± 0.42[a]	0.19 ± 0.13[a]
Clarias liocephalus (Boulenger, 1902)	788	0.50 ± 0.32[a]	0.42 ± 0.21[a]	0.66 ± 0.70[a]	0.75 ± 0.52[a]
Clarias alluaudi (Boulenger, 1906)	419	0.37 ± 0.24[a]	0.30 ± 0.25[a]	0.27 ± 0.19[a]	0.42 ± 0.36[a]
Clarias gariepinus (Burchell, 1815)	1062	14.64 ± 6.77[a]	11.28 ± 8.39[a]	15.88 ± 6.56	3.05 ± 1.62
Protopterus aethiopicus (Heckel, 1851)	598	13.38 ± 5.04[a]	9.72 ± 4.78[a]	6.47 ± 3.09[a]	2.25 ± 1.46

Values in the same row with the same superscript are not significantly different ($p < 0.05$).

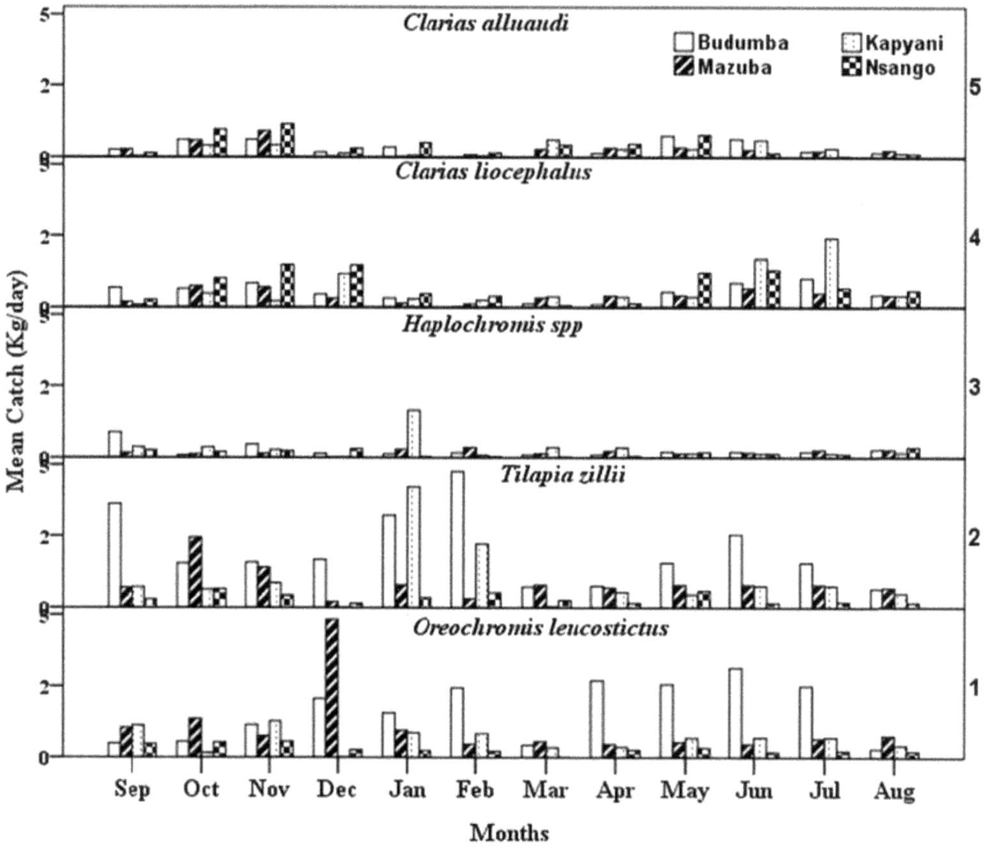

Figure 4. Fish catch of the small fish species at diferent sites along Mpologoma river wetland between September 2011 to August 2012.

The mean CPUE for the major commercial fish species varied with fishing gear, season and study site. The CPUE for *C. gariepinus* and *Protopterus* sp was higher with the use of hooks, at the less disturbed sites than that of other fishing gear,particularly the highly disturbed Nsango site (Table 4). At the less disturbed Budumba site, mean CPUE for *C. gariepinus* was higher (1.31 Kg100 hooks^{-1}h^{-1}) in the dry season as compared to 0.90 Kg 100hooks^{-1}h^{-1} recorded during the wet season. CPUE for the two large fish species at highly disturbed Nsango site was significantly different from that of the other sites (Tukey's HSD test; $p < 0.05$). However, the highly disturbed Kapyani site CPUE (4.50 Kg 100hooks^{-1}h^{-1}) was higher than the other highly disturbed Nsango site

Table 4. Mean catch per unit effort (CPUE) for the major commercial fish species/groups at each differently disturbed site along Mpologoma wetland during both the dry and wet season between September 2011 to August 2012.

Gear type	Species	Units of CPUE	Dry season				Wet season			
			Budumba	Mazuba	Kapyani	Nsango	Budumba	Mazuba	Kapyani	Nsango
GN 2.5	*Haplochomis* spp	Kg 45 $m^{-1} h^{-1}$	0.03	0.00	0.00	0.10	0.13	0.04	0.01	0.12
	C. gariepinus	Kg 45 $m^{-1} h^{-1}$	0.20	0.00	0.53	0.08		0.92	0.20	
	Protopterus sp	Kg 45 $m^{-1} h^{-1}$	0.40	0.00				0.29		
	Small *Clarias* spp	Kg 45 $m^{-1} h^{-1}$			0.45					
	Tilapia spp	Kg 45 $m^{-1} h^{-1}$	0.85	0.22	0.59	0.23	0.45	0.54	0.20	0.15
GN 3.0	*C. gariepinus*	Kg 45 $m^{-1} h^{-1}$	0.35	0.40	1.60		0.46	0.49	0.20	0.36
	Haplochomis spp	Kg 45 $m^{-1} h^{-1}$					0.07			
	Protopterus sp	Kg 45 $m^{-1} h^{-1}$	0.52	0.57	0.90		0.38	0.20	0.06	
	Tilapia spp	Kg 45 $m^{-1} h^{-1}$	0.13	0.14			0.46	0.17		0.47
Hooks	*C. gariepinus*	Kg 100hooks^{-1} h^{-1}	1.31	1.10	4.50	1.20	0.90	1.85	0.76	1.20
	Protopterus sp	Kg 100hooks^{-1} h^{-1}	0.86	1.05	1.28	0.54	1.32	1.88	0.81	1.09
	Small *Clarias* spp	Kg 100hooks^{-1} h^{-1}	0.16	0.20	0.33	0.27	0.15	0.16	0.11	0.26
Traps	*Barbus* sp	Kg trap^{-1} h^{-1}	0.04	0.02		0.09		0.02		0.03
	C. gariepinus	Kg trap^{-1} h^{-1}	0.04	0.06	0.11	0.06	0.09	-	0.11	0.08
	Haplochomis spp.	Kg trap^{-1} h^{-1}	0.01	0.04	0.09	0.23	0.03	0.14	0.05	0.07
	Protopterus sp.	Kg trap^{-1} h^{-1}	0.05	0.12	0.12	0.05	0.06	0.12	0.12	0.07
	Small *Clarias* spp.	Kg trap^{-1} h^{-1}	0.16	0.11	0.18	0.19	0.20	0.15	0.15	0.24

(1.20 100hooks^{-1}h^{-1}) in the dry season. The CPUE for *C. gariepinus* and *Protopterus* sp. showed an increasing trend during the dry months (Figure 5). While the CPUE of small fish species (*C. alluaudi*, *C. liocephalus*, *Haplochomis* spp and *Barbus* sp) did not vary
significantly between season and sites ($p = 0.05$).

Fish size

There were variations in *C. gariepinus* and *P. aethiopicus* total length in the wetland with a range of 12 to 135 cm and 16 to 151 cm respectively. The mean total length of *C. gariepinus* at Budumba and Mazuba was 47.27 ± 20.47 and 45.32 ± 16.76 cm, respectively. While at the highly disturbed sites (Kapyani and Nsango), the mean total length was 41.19 ± 21.85 and 27.33 ± 10.98 cm, respectively. Larger fish individuals of both species were caught during the wet season (April, May and October; Figure 6). The length of these two species was significantly lower at highly disturbed Nsango site than that of the less disturbed sites ($p < 0.01$; Figure 7). The less disturbed sites had larger fish even among the *Oreochomis* and *Tilapia* species than highly disturbed sites. However, the small fish species behaved differently. *C. liocephalus* total length ranged from 6.0 to 26.4 cm at Budumba and from 9.5 to 28.5 cm at Nsango.

Environmental data

The water quality parameters varied in space and

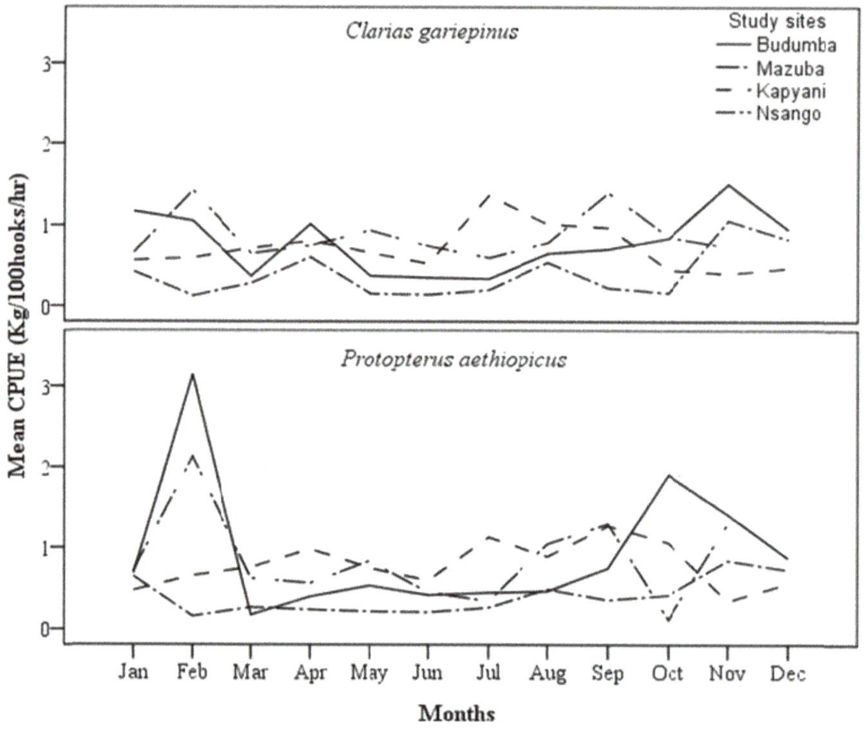

Figure 5. Monthly mean catch per unit effort (CPUE) of the two large fish species at the four differently disturbed sites along Mpologoma wetland between September 2011 to August 2012

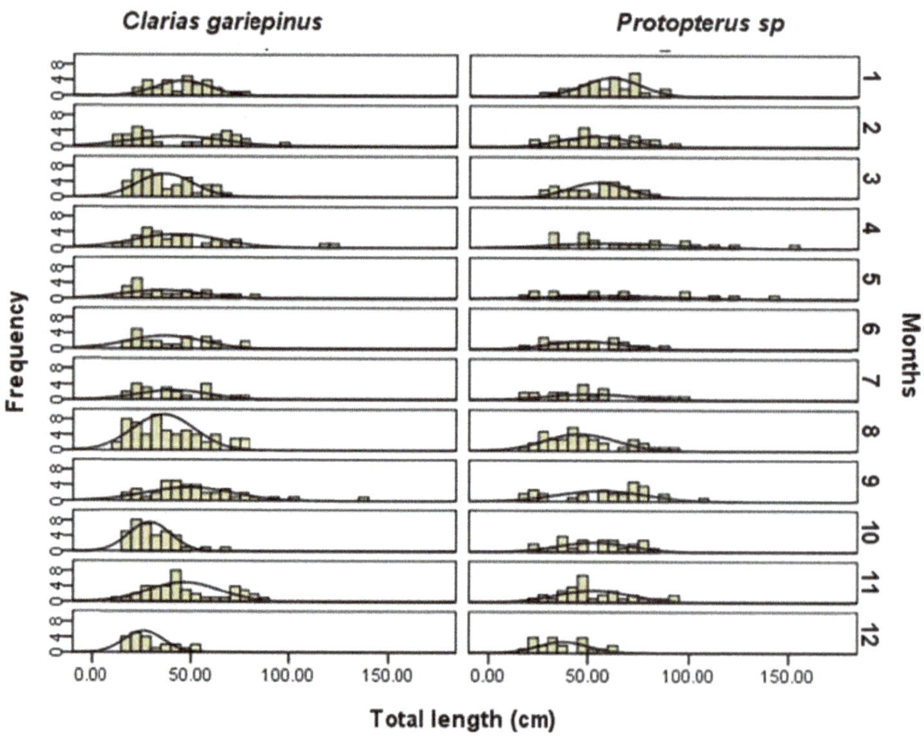

Figure 6. Length frequency distribution of two large fish species of Mpologoma river wetland fishery between September 2011 to August 2012.

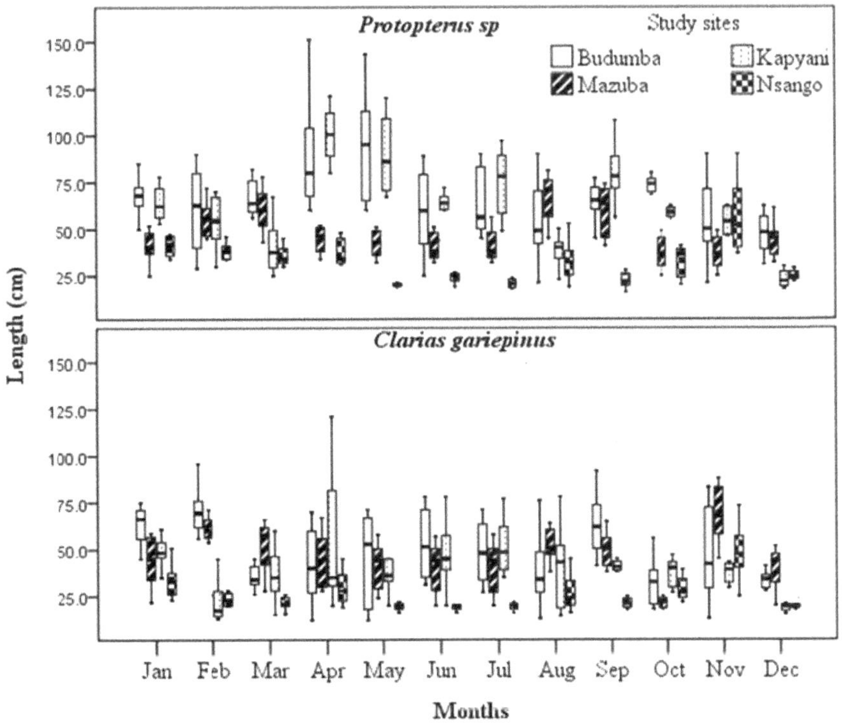

Figure 7. Mean total length of the large fish species caught at the differently disturbed sites along Mpologoma river wetland between September 2011 and August 2012.

Table 5. Mean (± STD) of water quality parameters of fish habitats (inside the wetland) at the different study sites along Mpologoma River wetland (September 2011 - August 2012).

Parameter	Budumba	Mazuba	Kapyani	Nsango
DO (mgl^{-1})	1.35 ± 1.03[a]	1.98 ± 1.26[a]	1.58 ± 0.67[a]	0.96 ± 0.76
Temp (°C)	23.75 ± 1.72	24.53 ± 0.81	25.35 ± 1.31	24.18 ± 1.49
Conductivity (µS/m)	152 ± 32.1[a]	161 ± 35.2[a]	218 ± 53.1[b]	351 ± 55.4[b]
OPO_4^{3-} (mgl^{-1})	0.28 ± 0.26[a]	0.27 ± 0.23[a]	0.13 ± 0.12[b]	0.14 ± 0.34[b]
TP (mgl^{-1})	0.67 ± 0.44[a]	0.62 ± 0.39[a]	0.29 ± 0.27[b]	0.23 ± 0.19[b]
Water depth (m) during the dry season	0.52 ± 0.27	0.45 ± 0.14	0.30 ± 0.15	0.35 ± 0.14

Values in the same row with the same superscript are not significantly different.

time along the wetland. Conductivity ranged from 119.9 µS/m at less disturbed Budumba site to 406.4 µS/m at the highly disturbed Nsango site. Conductivity at the less disturbed sites was significantly lower than that of disturbed sites and higher during the dry months. Dissolved oxygen (DO) within the papyrus ranged between 0.20 mgl^{-1} at highly disturbed site to 3.24 mgl^{-1} at less disturbed site (Table 5). DO was high during the wet season months (April, May, October and November) with a range of 2.2 to 4.4 mgl^{-1} in the mid river. DO was low in the dry season months (January, February, June and July) with a range of 0.6 to 1.8 mgl^{-1} in the mid river section. Nitrate levels were very low to undetectable levels at all sites during the wet season while in the dry season, it ranged from 0.01 to 0.12 mgl^{-1} at the less disturbed sites and 0.03 to 0.18 mgl^{-1} at the disturbed sites. Orthophosphate and total phosphorus ranged from 0.041 to 1.15 mgl^{-1} and 0.093 to 1.80 mgl^{-1} respectively in the wetland. From the NMDS analysis, the resulting ordination was two dimensional with final stress of 17.7 at $p < 0.05$. The less disturbed Budumba and Mazuba sites were identical as the distance between their circles was

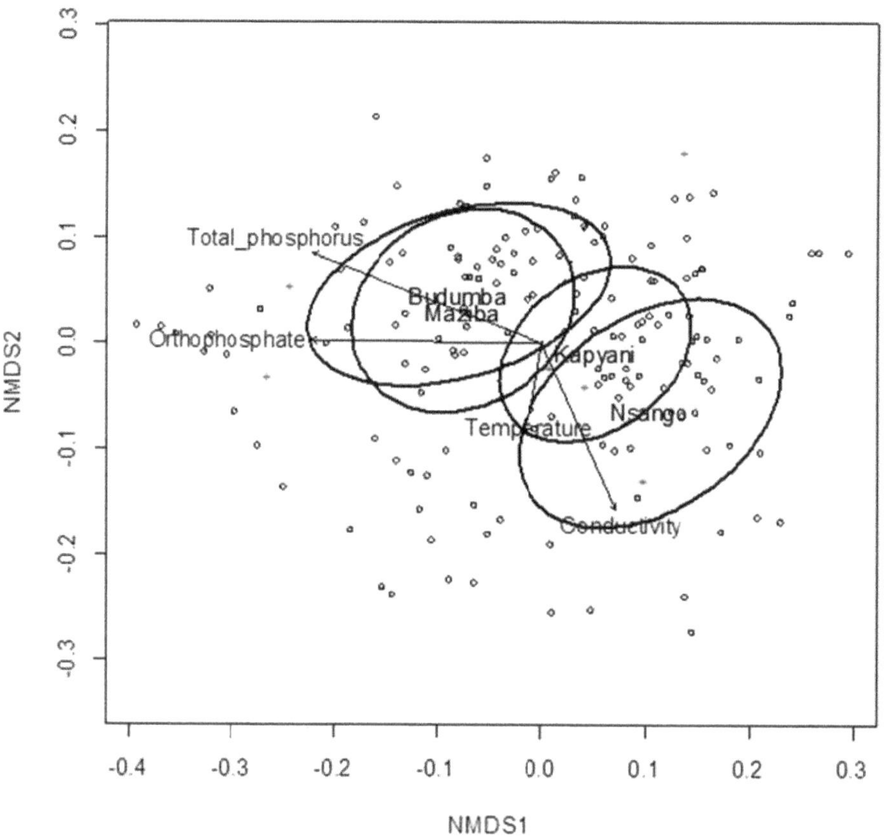

Figure 8. Non-metric multidimensional scaling plot of study sites along Mpologoma river wetland. Visually fit ellipses were drawn around sites that had similar levels of water quality parameters that were significantly different between sites.

Table 6. Environmental vector fitting of the NMDS analysis of the water quality parameters of the different study sites.

Water quality parameters	NMDS1	NMDS2	r^2	Pr(>r)
Dissolved oxygen	-0.95645	-0.29191	0.0150	0.316
Temperature	-0.15069	-0.98858	0.0651	0.07 *
Conductivity	0.41252	-0.91095	0.4269	0.001 ***
Orthophosphate	-0.99996	0.00869	0.7077	0.001 ***
Total phosphorus	-0.93377	0.35788	0.8128	0.001 ***

Significant codes: '***, 0'; **0.001; *0.01. P values were based on 999 permutations.

close to zero, while the highly disturbed Nsango site associated with a large scale rice scheme was different from the less disturbed sites as the distance between their circles was larger than zero (Figure 8). The vector analysis showed that conductivity, orthophosphate and total phosphorus were significantly important in differentiating the study sites all at r^2 of 0.43, 0.71 and 0.81 respectively, all at $P < 0.001$ (Table 6).

There was a significant relationship between fish catch and water quality parameters. A strong spearman rank correlation (rho' = 0.501 and 0.348; $p < 0.05$) between *C. gariepinus* catch with conductivity and orthophosphate respectively was realised. *P. aethiopicus* catch was significantly correlating with conductivity, orthophosphate and total phosphorus (rho' = 0.510, 0.465 and 0.441; $p < 0.01$ respectively). With stepwise multivariate linear regression, *P. aethiopicus* catch was significantly related to conductivity and orthophosphate ($R^2 = 0.544$; $p<0.01$;

catch = -13.214 - 0.023 conductivity + 16.69 orthophosphate). No significant relation was realized between water parameters and the small fish species catch and CPUE.

DISCUSSION

The wetland is a critical refuge for a subset of Lake Kyoga fish species given the high number of similar fish species recorded in the wetland compared to those reported in the open Lake Kyoga. *C. gariepinus, P. aethiopicus O. leucostictus* and *Tilapia* spp. which are important Lake Kyoga fish species (NaFIRRI, 2007), were also recorded in the wetland. High diversity, similarity and endemism stem largely from the fact that nearby fresh waters were once connected by heavy floods or drainage free of barriers that enabled dispersal of fish species (Olden et al., 2010). Barriers later maintained fish species in different parts of the riverine wetland. As level of wetland disturbance increase in certain areas of the wetland, the general similarity in species was still maintained in the wetland and this was due to the fact that some fish species are considered relatively tolerant to pollution. These Included *Oreochomis* species (Raburu and Masese, 2010), *P. aethiopicus, C. gariepinus* and *C. alluaudi* (Timmerman and Chapman, 2004). Normally reduction in species richness and abundance is expected along a degradation gradient (Raburu and Masese, 2010) and indeed at highly disturbed Nsango site, agricultural activities degraded the fish habitats that even tolerant species declined. During the relatively poor water quality periods (high conductivity and low dissolved oxygen), particularly in the dry months, at Nsango site the lowest number of fish species was recorded compared to other sites..

The Mpologoma wetland production estimate was in the order of magnitude of African floodplain production. Lightly exploited floodplain river systems produce about 40 - 60 kg ha^{1} yr^{1} of fish (Welcomme, 1985), while highly exploited tropical floodplains produce estimates more at 110 - 160 kg ha^{1} yr^{1} of fish (Bayley, 1988). The high wetland production estimated was attributed to the Increased exploitation rate to cater for the increasing need to diversify livelihoods along the wetland. The complex relationship between catch (fish abundance) and effort (function of fishermen's behaviour) is what controls the catch per unit of effort (Lopes, 2011). On average, the fishermen spent almost the same time span fishing both in wet and dry season, therefore the differences in CPUE resulted from the variation in fish abundance with time at the sites. Despite the few to no full time fishermen at all sites, the fishers were able to access distant less exploited areas with abundant fish resources particularly at the less disturbed sites leading to high catch and CPUE. The fishing gears used also contributed to the variation in catch and CPUE. The record of high number of hooks per fisherman at the less disturbed sites indicated the importance of this gear to achieving high catch and CPUE. The reliability of fishing gear CPUE as an index of fish density depends on fish activity, gear selectivity, avoidance and the morphology of the fishes (Olin et al., 2010). This explained the application various fishing gear that enabled exploitation of the various fish species at the different sites, during different seasons in the wetland.

Fish species abundance is influenced by physical and chemical composition of water (Randle and Chapman, 2004), habitat size and diversity (Budy et al., 2008), and water flow patterns into the wetland (Vorwerk et al., 2009). The differences in wetland fish species abundance and catch were governed primarily by water quality parameter variation among sites, given the strong correlation between catch and conductivity at the sites. This agreed with earlier studies on floodplain fisheries which highlighted dissolved oxygen, conductivity, pH and water depth as major determinant factors to fish abundance (Louca et al., 2009). Highly disturbed sites' conductivity and dissolved oxygen levels were similar to those of highly studied polluted Nakivubo wetland in Uganda (Kansiime et al., 2007). Wetland clearing for agriculture results in dramatic alteration of the river flood curve which, leads to decrease in both the amplitude and duration of flood regime. The wetland clearing also increases in evapotranspiration which, lead to increased salts in the water. These have pronounced impact such as high conductivity, alterations of vegetation species and cover, and decreased connectivity in wetland lagoons (Louca et al., 2009). Sustained increase in conductivity compounded by high sedimentation levels led to negative implications on the fish ecology (Chapman et al., 2003). The large rice scheme irrigation activities involve application fertilizers near Nsango site seem to have affected the water quality which, explain the reduced abundance of large fish species. Lungfish (*Protopterus* sp.) decline reflects the interaction of overexploitation and large-scale conversion of wetlands to agricultural land for the past few decades (Goudswaard et al., 2002). Such species with preferences for lower conductivities and high dissolved oxygen disappear from the wetland once these conditions persist (Goudswaard et al., 2002; Louca et al., 2009). This explains the low *C. gariepinus* and *Protopterus* species abundance at highly disturbed Nsango where those harsh conditions were observed.

Higher fish species composition abundance was recorded at the downstream disturbed Kapyani site and this was attributed to a number of factors. The interaction between the river and lake hydrology which modifies the fish habitats with modified vegetation and deeper waters (Cooper et al., 2007) allowing more fish to coexist, both wetland and open water dwelling species, despite the level of disturbance. The site was permanently connected to the nearby lakes Nakuwa and Lemwa. Also habitat recovery from disturbance during the flooding season is associated with increased food resources (Morris et al., 2007; Vorwerk et al., 2009), resulting in high fish abundance. Furthermore, wet season rice crop when cultivated as a largely rainfed crop and have little land engineering, causes less impacts on water quality and later on fish (Nguyen-Khao et al., 2005).

Temporal variation in fish species abundance and catch in the wetland was also attributed to their reproductive traits and predator avoidance factors. Low abundance of all *Clarias* species at beginning of the rainy season along the wetland was due to their breeding cycle (Offem et al., 2010). *T. zillii* spawn at the end of the dry season (El-Sayed and Moharram, 2007) and this could explain their low abundance during the dry season in the wetland. The observed high catch at Budumba and Kapyani during the dry season was due to the high water level maintained at these two sites which, offered refuge for fish avoiding harsh conditions. *Oreochomis* sp. spawn anytime but larger quantities are observed during the rainy season (Melcher et al., 2012). Variation of small fish species abundance which breed thoughout the year was attributed to avoidance of predators. High abundance of haplochomines in highly polluted areas is due to the non-avoidance of predators in turbid waters while their low abundance in less polluted areas is due to avoidance in clearer waters (Ogutu-Ohwayo, 1990).

Availability of abundant food during the rainy season is also an important factor responsible for the temporal variation in wetland fish (Offem et al., 2010). Rainy conditions lead to higher detritus, softer decomposing plant materials and low phosphorus favour in high abundance of benthic invertebrates (Hansson et al., 2005) which are important food for mainly the small fish species. Disruptions in the food base caused by alterations in water quality, particularly at highly disturbed sites, have been found to lead to higher percentage of omnivores (generalists) and a decrease in the proportions of insectivores and carnivores (Morris et al., 2007). *C. gariepinus* which is predatory (Raburu and Masese, 2010) was less abundant than *C. liocephalus*, an omnivore at highly degraded sites. Thus, feeding and reproductive strategies are among the divergent life-history characteristics that make riverine fish species respond to annual flood pulse and short-term environmental disturbances in different ways (Winemiller, 2005; Montaña et al., 2007).

The relationship between land use and small scale fishery was realized though strong relation between fish species catch and water quality parameters. The results should be of interest to resource managers because land use changes in the wetland have the potential to drastically affect the water quality, negatively impacting the production of large commercial fish species of such a small scale fishery. The wetland ecological shift phase need to be established in order for policy makers to regulate the land use change rate which may have potentially irreversible effects to both small and large fish species of this small scale fishery.

Conflict of Interests

The author(s) have not declared any conflict of interests.

ACKNOWLEDGEMENTS

The authors would like to thank the management at National Fisheries Resource and Research Institute, Jinja, who greatly assisted us with fishing equipment and space for the identification of wetland fish species, Mr. Peter Asimwe at Makerere University who assisted in the wetland land use characterization, Ms. Juliet Kutuusa and Mrs. Robinah Muhairwe at Bugolobi National Water and sewage corporation laboratory who assisted in water quality analysis and the fishermen who extended support during data collection in the wetland. We are most grateful for the financial support from Kyambogo University, the International Foundation for Science (IFS) and the Syngenta Foundation.

REFERENCES

APHA (1995). Standard methods for examination of water and wastewater. 19th edition. American Public Health Association, Washington, DC.

Arthington AH, Lorenzen K, Pusey BJ, Abell R, Halls A, Winemiller KO, Arrington DA, Baran E (2004). River fisheries: ecological basis for management and conservation. In Welcomme RL (Ed.) Proceedings of the 2nd International Large Rivers Symposium, Mekong River Commission, Phnom Penh, Cambodia. 1: 31- 60.

Bayley PB (1988). Factors affecting growth rates of young floodplain fishes: seasonality and density dependence. Environ. Biol. Fish. 21: 127-147.

Bene C (2004). Fisheries management and poverty. In: Neiland AE, Bene C (eds) Poverty and Small-scale Fisheries in Wet Africa. FAO and Kluwer Academic Publishers, Dordrecht/Boston/London. pp. 61-82.

Béné C, Hersoug B, Allison EH (2010). Not by rent alone: analysing the pro-poor functions of small-scale fisheries in developing countries. Dev. Policy Rev. 28:325-358.

Budy P, Thiede GP, McHugh P, Hansen ES, Wood J (2008). Exploring the relative influence of biotic interactions and environmental conditions on the abundance and distribution of exotic brown trout (*Salmo trutta*) in a high mountain stream. Ecol. Freshw. Fish 17:554 - 566.

Carvalho G, Moutinho P, Nepstad D, Mattos L, Santilli M (2004). An Amazon perspective on the forest–climate connection: opportunity for climate mitigation, conservation and development? Environ. Dev. Sustain. 6:163-174.

Chapman LJ, Chapman CA, Schofield PJ, Olowo JP, Kaufman L, Seehausen O, Ogutu-Ohwayo R (2003). Fish Faunal Resurgence in Lake Nabugabo, East Africa. Conserv. Biol. 17:500 - 511.

Cooper MJ, Uzarski DG, Burton TM (2007). Macroinvertebrate community composition in relation to anthopogenic disturbance, vegetation, and organic sediment depth in four Lake Michigan drowned

river-mouth wetlands. Wetlands 27(4):894-903.
Di Baldassarre G, Elshamy M, van Griensven A, Soliman E, Kigobe M, Ndomba P, Mutemi J, Mutua F, Moges S, Xuan JQ, Solomatine D, Uhlenbrook S (2011). Future hydrology and climate in the River Nile basin: a review. Hydrol. Sci. J. 56:199-211.
El-Sayed HKA, Moharram SG (2007). Reproductive biology of *Tilapia zillii* (Gerv, 1848) from Abu Qir bay, Egypt. Egypt. J. Aquat. Res. 33:379-394.
Garaway C, Arthur RA, Chamsingh B, Homekingkeo P, Lorenzen K, Saengvilaikham B, Sidavong K (2006). A social science perspective on stock enhancement outcomes: lessons learned from inland fisheries in southern Lao PDR. Fish Res. 80:37-45.
Goudswaard KPC, Witte F, Chapman LJ (2002). Decline of the African lungfish (*Protopterus aethiopicus*) in Lake Victoria (East Africa). Afr. J. Ecol. 40: 42-52.
Greenwood PH (1974). The cichlid fishes of Lake Victoria, East Africa: The biology and evolution of a species flock. Bull. Br. Mus. Nat. Hist. Zoo. Suppl. 6:11-34.
Hansson LA, Bronmark C, Nilsson PA, Abjornsson K (2005). Conflicting demands on wetland ecosystem services: nutrient retention, biodiversity or both? Freshw. Biol. 50:705-714.
Heck S, Béné C, Reyes-Gaskin R (2007). Investing in African Fisheries: Building Links to the Millennium Development Goals. Fish Fish. 8:211-226.
Ibanez C, Oberdorff T, Teugels G, Mamononekene V, Lavoue S, Fermon Y, Paugy D, Toham AK (2007). Fish assemblages structure and function along environmental gradients in rivers of Gabon (Africa). Ecol. Freshw. Fish 16:315-334.
Kansiime F, Kateyo E, Oryem-Origa H, Mucunguzi P (2007). Nutrient status and retention in pristine and disturbed wetlands in Uganda: management implications. Wetlands Ecol. Manage. 15:453-467.
Lopes PFM (2011). Decision-making processes by small-scale fishermen on the southeast coast of Brazil. Fish. Manage. Ecol. 18:400-410.
Lorenzen K, Garaway CJ, Chamsingh B, Warren TJ (1998). Effects of access restrictions and stocking on small water body fisheries in Laos. J. Fish Biol. 53:345-357.
Louca V, Lindsay SW, Majambere S, Lucas MC (2009). Fish community characteristics of the lower Gambia River floodplains: a study in the last major undisturbed West African river. Freshw. Biol. 54:254 - 27.
MAAIF (2009). Uganda national rice development strategy (UNRDS) 2nd draft, Ministry of Agriculture Animal Industry and Fisheries, Government of Uganda, Entebbe.
Martin SM, Lorenzen K, Arthur RI, Kaisone P, Souvannalangsy K (2011). Impacts of fishing by dewatering on fish assemblages of tropical floodplain wetlands: A matter of frequency and context. Biol. Conserv. 144:633-640.
Melcher AH, Ouedraogo R, Schmutz S (2012). Spatial and seasonal fish community patterns in impacted and protected semi-arid rivers of Burkina Faso. Ecol. Eng. 48:117-129.
Montaña CG, Choudhary SK, Dey S, Winemiller KO (2011). Compositional trends of fisheries in the River Gangoo, India. Fish. Manag. Ecol. 18:282-296.
Morris C, Green JM, Simms A (2007). Population components in a small Atlantic cod population in Labrador and their management implications for an MPA. Proceedings of the Sixth International Conference on Science and Management of Protected Areas, Wolfville, NS, May 21-25, 2007.
Muhoozi L (2003). Aspects of the Commercial Exploitation of Fish Stocks in Lake Victoria. PhD Thesis, Hull: University of Hull.
NaFIRRI (2007). Standard Operating Procedures (SOPs) for collecting biological information from fishes of Lake Victoria. National Fisheries Research and Resource Institute, vol. 1, Jinja.
National Environment Management Authority (NEMA) (2004). District State of Environment Report for Pallisa District, report. http://www.nemaug.org/district_reports/DSOER 2004_Pallisa.pd. Accessed 08 September, 2012.
Nguyen-Khao S, Lorenzen K, Garaway C, Chamsingh B, Siebert DJ, Rondone M (2005). Impacts of irrigation on fisheries in rain-fed rice farming landscapes. J. Appl. Ecol. 43:892-900.
Offem BO, Akegbejo S, Omoniyi Y (2010). Aspects of Ecology of *Clarias anguillaris* (Teleostei: Clariidae) in the Cross River, Nigeria. Turk. J. Fish. Aqua. Sci. 10:101-110.
Ogutu-Ohwayo R (1990). The decline of the native fishes of Lakes Victoria and Kyoga (East Africa) and the impact of introduced species, especially the Nile perch, *Lates niloticus* and the Nile tilapia, *Oreochomis niloticus*. Environ. Biol. Fish. 27:81-96.
Okechi JK (2004). Profitability assessment: A case study of African catfish (*Clarias gariepinus*) farming in the Lake Victoria basin, Kenya. United Nations University-Fisheries Training Program. wttp://www.unuftp.is/static/fellows/document/ okechiprf04.pdf. Accessed 16 May 2010.
Olden JD, Kennard MK, Leprieur F, Tedesco PA, Winemiller KO, Garcı´a-Berthou E (2010). Conservation biogeography of freshwater fishes: recent progress and future challenges. Divers. Distrib. 16:496 - 513.
Olin M, Vinni M, Lehtonen H, Rask M, Ruuhijarvi J, Saulamo K, Ala-Opas P (2010). Environmental factors regulate the effects of roach *Rutilus rutilus* and pike *Esox lucius* on perch *Perca fluviatilis* populations in small boreal forest lakes. J. Fish Biol. 76:1277 - 1293
R Development Core Team (2008). R: A language and environment for statistical computing. R Foundation for Statistical Computing. Vienna, Austria. ISBN 3-900051-07-0, URL http://www.R-project.org. Accessed 20 January 2014.
Raburu PO, Masese FO (2010). Development of a fish-based index of biotic integrity (FIBI) for monitoring riverine ecosystems in the Lake Victoria drainage basin, Kenya. River Res. Appl. 28:23 - 38.
Ramsar (2008). Uganda, Ramsar sites Information service from http://ramsar.wetlands.org/Portals/15/UGANDA.pdf, pp 18. Accessed 14 May 2014.
Randle AR, Chapman LJ (2004). Habitat use by the air-breathing fish *Ctenopoma muriei*: implications for costs of breathing. Ecol. Freshw. Fish. 13:37-45.
Rientjes THM, Haile AT, Kebede E, Mannaerts CMM, Habib E, Steenhuis TS (2011). Changes in land cover, rainfall and stream flow in Upper Gilgel Abbay catchment, Blue Nile basin Ethiopia. Hydrol. Earth Syst. Sci. 15:1979-1989.
Smith LED, Nguyen-Khoa S, Lorenzen K (2005). Livelihood functions of inland fisheries: policy implications in developing countries. Water Policy 7:359-383.
Thiebaut G (2006). Aquatic macrophyte approach to assess the impact of disturbances on the diversity of the ecosystem and on river quality. Int. Rev. Hydrobiol. 91:483-497.
Timmerman CM, Chapman LJ (2004). Behavioral and physiological compensation for chonic hypoxia in the sailfin molly (*Poecilia latipinna*). Physiol. Biochem. Zool. 77:601-610.
Vanden-Bossche JP, Bernacsek GM (1990). Source book for the inland fishery resources of Africa. CIFA Tech Pap, FAO, Rome, 18/1.
Vass KK, Shivastava NP, Katiha PK, Das AK (2009). Enhancing fishery productivity in small reservoir in India. A Technical Manual. WorldFish Center Technical Manual No. 949. The WorldFish Center, Penang, Malaysia. pp. 9.
Verburg PH, Neumannw K, Nol L (2011). Challenges in using land use and land cover data for global change studies. Glob. Chan. Biol. 17: 974-989.
Vorwerk PD, Paterson AWP, Froneman W, Whitfield AK (2009). Increased abundance of two important sport fishery species following renewed river flow into a freshwater-deprived South African estuary. Fish. Man. Ecol. 16:420-423.
Welcomme RL (1985). River fisheries. FAO Tech. Pap., 262, 330 pp.
Winemiller KO (2005). Life history strategies, population regulation, and their implications for fisheries management. Can. J. Fish. Aquat. Sci. 62:872-885.
Witte F, van Densen WLT (1994). Fish stocks and fisheries of Lake Victoria, a handbook for field observations. Samara Publishing Cardigan Wales UK.

Copepoda parasites in economically important fish, Mugilidae (*Mugil cephalus* and *Liza falcipinnis*) from Lac Nokoue Lagoon in Republic of Benin, West Africa

Nike F. Aladetohun[1,2] , Nestor G. Sakiti[2] and Emmanuel E. Babatunde[3]

[1]Federal College of Fisheries and Marine Technology, Ahmadu Bello Way, Victoria Island, Lagos State, Nigeria.
[2]Laboratory of Parasitological and Ecology of parasites, Department of Zoology and Genetics, Faculty of Sciences and Technology, University of Abomey Calavi, Republic of Benin.
[3]Department of Marine Sciences, University of Lagos, Akoka, Lagos State, Nigeria.

Fish parasitology is an indispensable tool in aquatic health studies and a basic understanding of richness of a community parasitism in many localities is essential for instituting control. Many copepod parasites are ecto-parasites which negatively affect the appearance and reduced production of species of economically important fish, both from the wild and fish farms, thus making them difficult to market. In this study, copepod parasitic investigation was carried out in gills of 856 fish mugilidae (*Mugil cephalus* and *Liza falcipinnis*) in three stations (Ganvie, Djdje and Zogbo) of Lac Nokoue lagoon. In the three stations, three species of parasitic copepod were found: *Nipergasilus bora, Ergasilus latus* and *Ergasilus lizae*. The highest total percentage copepod prevalence was found in Ganvie (86.23%) and the least in Djdje (63.14%). There is significant difference (P<0.05) in the rate of infestation of *L. falcipinnis* at Djdje than the other two stations, where as *M. cephalus* shows no significant difference in the rate of infestation in the three stations. The research reveals higher number of parasite copepods during the rainy season.

Key words: Copepod parasites, Mugilidae fish, Lac Nokoue Lagoon.

INTRODUCTION

Even though spawning can be induced on an environmental basis, mullet farming still depends on fry or fingerlings stocked from natural water. This fact combined with behaviour of mullets, promotes disease including parasite infestation (Paperna and Overstreet, 1981). This results in constant dissemination of disease-causing agent between cultured and natural population of mullet. Transmission can occur through water and intermediate host. Thus there is need for comprehensive study of parasites from natural stocks.

Lake Nokoue is the largest coastal lagoons and the most productive brackish water body in Republic of Benin (Laleye and Moreau, 2004). Lake Nokoue was formally a lake until 1185 when it was artificially linked with the ocean through a channel (Lang and Paradis, 1977). Majority of fish production (fish mugilidae inclusive) come from the lagoon (Laleye et al., 2003; Niyonkuru et al., 2010).

The Ganvie area is located near the floating village of Ganvie, where the water is characterised by a high level of organic pollution (Niyonkuru et al., 2003). The fishing technique used by the fishermen from whom samples were obtained included gill nets, cast nets and acadjas (a traditional fish harvesting system using nets to encircle man-made brush packs of branches placed in the lagoon (Gnohossou et al., 2009).

Fish mugilidae are economically important and are important food fishes. The euryhalinity, eurythermality and their simple diet as well as their rapid growth, have made them the object of aquaculture in many parts of the world (Oren, 1981). In Republic of Benin, these *(Mugil cephalus* and *Liza falcipinnis*) constitute important proportion of catches by artisanal or subsistence fishermen (Laleye et al., 2003; Gnohossou, 2009). They are of economic importance to the country.

Parasites have recently been highlighted as serious pathogenic problems in cultured mullet fish in marine and brackish water. Among the parasites, copepode family is commonly found on fishes cultured in brackish water (Noor El- Deen et al., 2012), and therefore a threat to the developing industry of finfish mariculture (Ru¨ckert et al., 2008).

Copepods are aquatic crustaceans which are diverse and are the most numerous metazoans in the water community with habitats ranging from fresh water to hypersaline conditions. Some copepods parasites of fishes all over the world may cause lesions that negatively affect the fisheries and aquaculture economies (Barreiro and Francnete, 2003).

Some studies have been carried out to discover the richness of community parasitic in many localities (Poulen and Rode, 1997; Sures and Street, 2001; Kimpela et al., 2006; Boualleg et al., 2011)

In Republic of Benin, many researchers have worked on parasites of fish (Doussou, 1985; Sakiti et al., 1991; Sakiti, 1997; Gbankoto et al., 2003) but no work has been carried out on parasites of mugilidae fish in Repulic of Benin.

The objective of this study is to determine copepod parasites in economically important fish *Mugil cephalus* and *Liza falcipinnis* in some major fishing areas of the largest fishing lagoon in Republic of Benin.

In determining the copepods parasite species of this world economically important fish with high commercial and environmental attributes will provide better culture condition for them, especially in mariculture. This will also help to solve some of the problems of fish diseases that can perturb their health and productivity, both in the wild and fish culture.

MATERIALS AND METHODS

The study site, Lake Nokoue (Figure 1) is the largest lagoon (Moreau, 2004), is a shallow, sub-tropical coastal lagoon (6°25N, 2°36E) (Table 8) with surface of 150 km and stretches 20 km in its east-west direction by 11 km in the north-south direction (Laleye et al., 2003). Lake Nokoue opens directly into the Atlantic Ocean through channel at cotonou which is about 24.5 km long.

For the purpose of this study, three stations were considered as sampling areas within the lagoon: Ganvie, Djdje and Zogbo (Figure 1). The Ganvie area is located at the northern part of the lake Nokoue lagoon, near the floating village of Ganvie where the water is characterised by a high level of organic pollution (Laleye et al., 2003). The Djdje and Zogbo areas are located at the southern part of the lake and are partly influenced by Atlantic Ocean waters. Sampling was carried out at each station both in the dry (December-March, 2011) and the rainy season (April-July, 2012).

Collection and analysis of water samples

Water samples were collected with 1 dm 3 water samplers and stored in 1 L water bottles and analysed in the laboratory for pH, conductivity, salinity and turbidity using a multi-meter water checker (Horiba U- 12). Water sample was collected in 250 ml dissolved oxygen bottle and fixed with Winkler solution at each station. This was followed by dissolved oxygen estimation using Iodometric Winkler's method. Water temperature was measured *in situ* using mercury-in-glass thermometers.

Sampling of the fish specimen and parasitological examination

M. cephalus and *L. falcipinnis* from the three stations were obtained by the assistance of fishermen. A total number of 856 fish specimen, *M. cephalus* and *L. falcipinnis* were examined for copepods parasites. In Station 1 (Ganvie) number of fish examined was (MC - 115, L.F - 132), in Station 2 (Djdje) it was (MC - 181, LF - 226) and station 3 (Zogbo) it was (MC - 96 and LF - 106). The collected fish were transported in ice boxes to the laboratory and examined for parasites. The weight of the fish were taken using digital weighing balance and recorded. The standard length (SL), the total length (TL) and the fork length (FL.) were measured with the aid of meter rule.

After dissection, stereo-microscopic observation was made on gills, for the presence of copepod parasites. Checking of prepared slides for parasites for better observation was carried out using light microscope. The standard length (SL), the total length (TL) and the fork length (FL) were measured with the aid of meter rule. The host species was identified using Fisher et al. (1987). The standard length (SL), the total length (TL) and the fork length (FL) were measured with the aid of meter rule. The host species was identified using Fisher et al. (1987). Collected copepods were fixed and preserved in ethanol (70%). Copepods on gills were studied using stereo and light microscopy. Parasites species identification was based on morphological features according to Yamaguti (1963), Vassiliades, (1975), Kabata (1979) (Skryabin et al., 1982), Ben-Hasseine (1983), Kabré (1997), Kabre et al. (1997), Ho and Kim (2004) and Moravec (2007).

Data analysis

The parasitological terms follow Bush et al. (1997): prevalence (P)

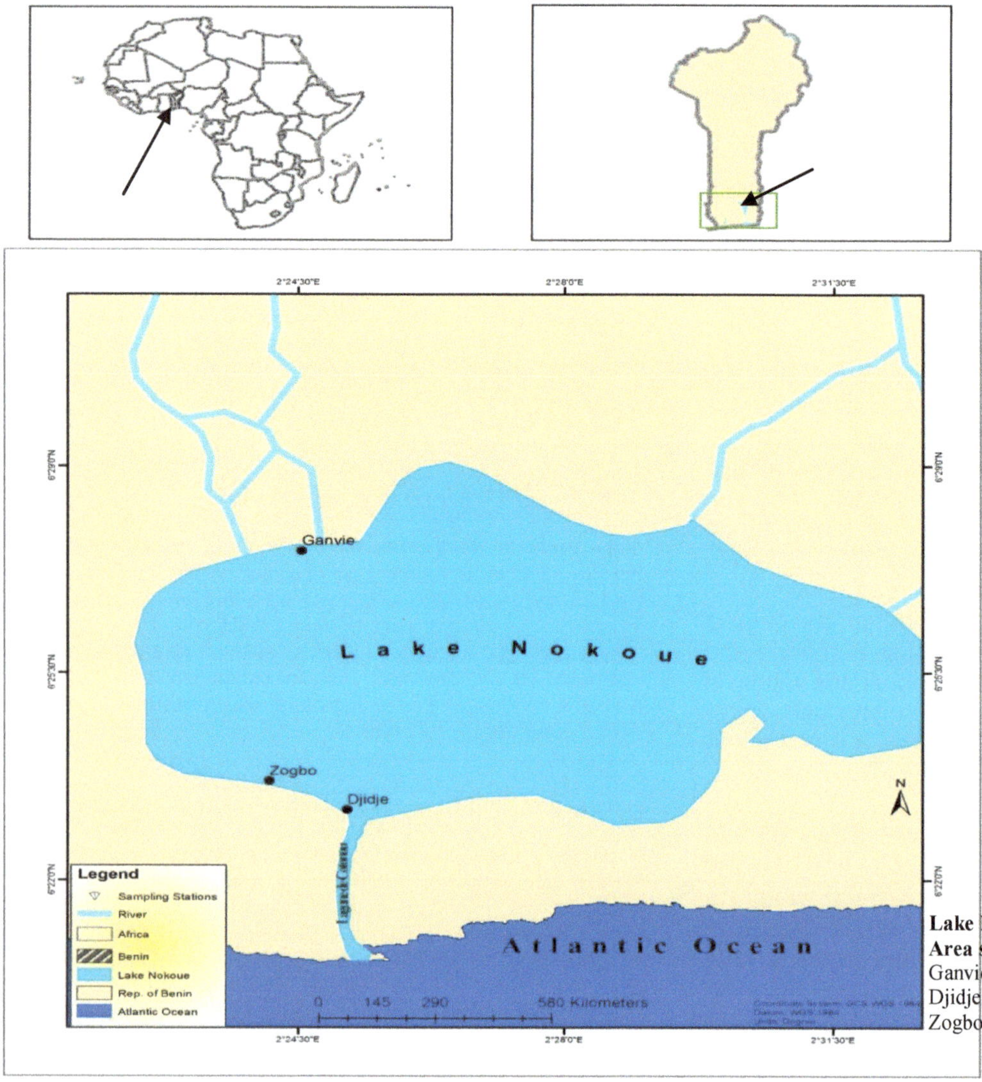

Lake Nokoue area

Area stations	Longitude	Latitude
Ganvie	20 24'32.615"E	6027'33.77"N
Djidje	20 25'8.393"E	6023'11.077"N
Zogbo	20 24'7.512"E	6023'40.162"N

Figure 1. Location of Lake Nokoue showing fish sampling stations.

is the number of fish infected with one or more individuals of a particular parasite species (or taxonomic group) divided by the number of hosts examined (expressed as a percentage):

$$Prevalence = \frac{No\ of\ hosts\ infested}{No\ of\ hosts\ examined} X\,100$$

Intensity (of infection, I) is the number of individuals of a particular parasite species in a single infected host (expressed as a numerical range); mean intensity (of infection, mI) is the average intensity, or the total number of parasites of a particular species found in a sample divided by the number of infected hosts:

$$\text{Mean intensity} = \frac{Total\ number\ of\ a\ particular\ parasite}{Number\ of\ infected\ hosts}$$

ANOVA test, using statistical software (SPSS) was also done on MC and LF to know the significant difference of copepod parasites infestation in each station.

RESULTS

Physicochemical parameters of the study area

The physicochemical parameters recorded at the three stations during rainy and dry seasons are presented in Table 1. Both in the rainy and dry season, the salinity levels were lowest in station 1 (Ganvie) with 0 and 11.4 parts per thousand, but highest in station 2 (Djdje) with 0.8 and 29.5 parts per thousand salinity levels respecttively. The highest salinity recorded at Djdje is because of its proximity to the sea. The lowest salinity value obtained

Table 1. Spatial and seasonal distributions of the physico-chemical parameters in some parts of Lake Nokoue.

Station	WT (°C)		PH		Cond. ($mScm^{-1}$)		Turb. (NTU)		Sal. (ppt)		DO (Mg/l)	
	RS	DS	RS	DS	RS	DS	RS	DS	RS	DS	RS	DS
Ganvie	29	28	8.1	8.5	0.18	19.5	78	10	0	11.4	11.2	8.0
Djdje	29	27	8.9	8.0	1.0	46.5	366	10	0.8	29.5	13.6	13.2
Zogbo	28	28.5	9.0	8.1	0.8	45.8	126	10	0.5	27	18.4	8.4

WT: Water temperature; Cond: conductivity; Turb: turbidity; Sal.: salinity; DO: dissolved oxygen; RS: rainy season; DS: dry season.

Table 2. Copepods for eight months in Benin.

Month	La Ganvie		Djdje		Zogbo		
	M.C	L.F	M.C	L.F	M.C	L.F	
Dec	7	19	10	26	10	11	
Jan	18	10	16	29	9	18	
Feb	14	22	12	28	10	5	
Mar	5	8	8	19	11	17	
Apr	16	20	7	18	6	10	
May	11	13	13	20	7	7	
June	10	11	16	15	13	13	
July	12	17	14	6	10	10	
Total	93	120	96	161	76	91	637

M.C: *M. cephalus,* L.F: *L. falcipinnis.*

Table 3. Percentage fish infested and percentage parasite prevalence in Benin.

Location	No. of fish examined		Total	No. of infested fish		Number of fishes infested (%)	No. of copepod parasite		Parasite prevalence (%)		Total parasite prevalence (%)
	MC	L.F		M.C	L.F		M.C	L.F	M.C	L.F	
Ganvie	115	132	247	90	110	80.97	93	120	43.66	56.34	86.23
Djdje	181	226	407	150	201	86.24	96	161	37.35	62.65	63.14
Zogbo	96	106	202	71	89	79.21	76	91	45.51	54.49	82.67
Total	392	464	856	311	400	246.42	265	372	126.52	173.48	232.04

can be explained by the proximity of the So River to the station (Laleye et al., 2003).

Generally low conductivity levels were recorded from all stations during rainy season, whereas relatively high levels which ranged from 19.5 $mScm^{-1}$ in station 1 to 46.5 $mScm^{-1}$ at station 2 were obtained in the dry season (Table 1). On the other hand, every station had higher turbidity, pH and DO levels during rainy than dry season (Table 1).

A total number of 637 copepods parasites were found in the three stations (Table 2): station 1 (MC- 93, LF - 120), station 2 (MC-96, Lf - 161 and station 3 (MC 76, L F- 91). Total % copepod prevalence was highest in Ganvie 86.23% followed by Zogbo (82.67%) and Djdje with 63.14% (Table 3).In all the three stations Liza falcipinnes (LF) has the highest percentage copepods parasite infestation. Station 1(MC-43.66%, LF-56.34%), Station2(MC-37.35%, LF-62.65%) and the third station (MC-45.51, L F-54.49).

In all the three stations, percentage of copepod para-

Table 4. Prevalence rate of copepod during dry and rainy season.

Season	Location	No of fish examined		No of copepods		Parasite prevalence (%)		Total parasite prevalence (%)
		M.C	L.F	M.C	L.F	M.C	L.F	
	Ganvie	63	71	44	59	0.4272	0.5728	0.7687
Dry	Djide	92	114	46	102	0.3108	0.6892	0.7185
	Zogbo	49	53	40	51	0.4396	0.5604	0.8922
	Ganvie	52	61	49	61	0.4455	0.5545	0.9735
Rainy	Djije	89	112	50	59	0.4587	0.5413	0.9732
	Zogbo	47	53	36	76	0.4737	0.5263	0.6184

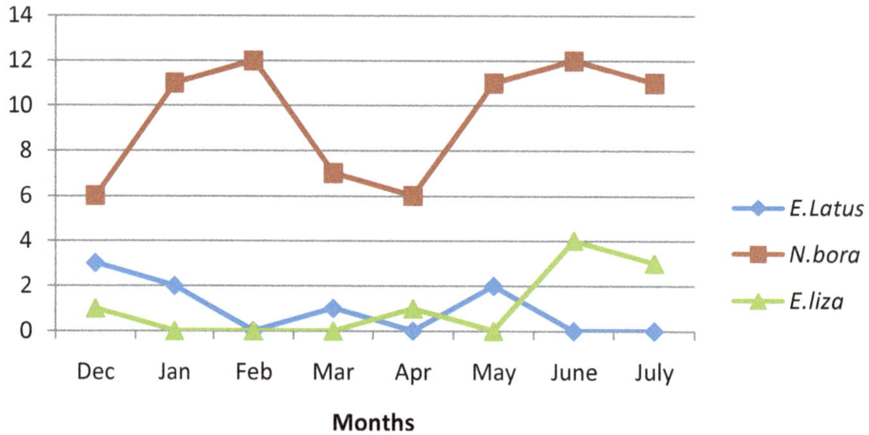

Figure 2. M.C copepod species at Djdje.

sites prevalence was much higher in *L. falcipinnis* than in *M. Cephalus*: In Ganvie (MC 43.66%, LF 56.34%), Djide (MC 37.35%), (LF 62.35%) and in Zogbo (MC 45.51%, LF 54.49%). Total percentage copepod parasites was highest in Ganvie (86.23%) and least in Djidje (63.14%) (Table 3).

Further analysis was carried out to investigate the prevalence rate of copepods during dry and raining season (Table 4). The prevalence rate during dry season; shows Ganvie (0.769), Djide (0.719) and Zogbo (0.892). The prevalence rate during the rainy season is Ganvie (0.974), Djide (0.973) and Zogbo (0.618). It is clearly shown in the analysis that the prevalence rate is higher in the rainy season than dry season. Zogbo shows highest percentage of parasite prevalence during dry season, while Ganvie shows highest copepod parasite prevalence during rainy season.

Figures 2 to 7 shows graphs of parasitic copepod species (*E. latus*, *N. bora* and *E. lizae*) found in both LF and MC in all the stations.

In Ganvie, *N. bora* shows the highest percentage prevalence both in the dry season and rainy season in both (MC 9.813%, LF 16.822%) and (MC 16.355, LF 28.972) (Table 5). Also in Djdje, *N. boras* shows highest percentage prevalence in dry season (MC 14.173, LF 21.2598) and at rainy season (MC 15.148, LF, 38.189 (Table 6), the same in Zogbo, *N. bora* also shows highest percentage rate both in the dry season MC (19.162 and LF, 15.569), and in rainy season (MC 16.767% and LF, 28.7437%) (Table 7).

DISCUSSION

Generally low conductivity levels were recorded from all stations during rainy season, whereas relatively high levels which ranged from 19.5 $mScm^{-1}$ in station 1 to 46.5 $mScm^{-1}$ in station 2 were obtained in the dry season (Table 1). On the other hand, every station had higher turbidity, pH and DO levels during rainy season than dry season (Table 1). This report coincides with that of Yakub et al. (2011).

Copepod parasites are common on fish hosts in

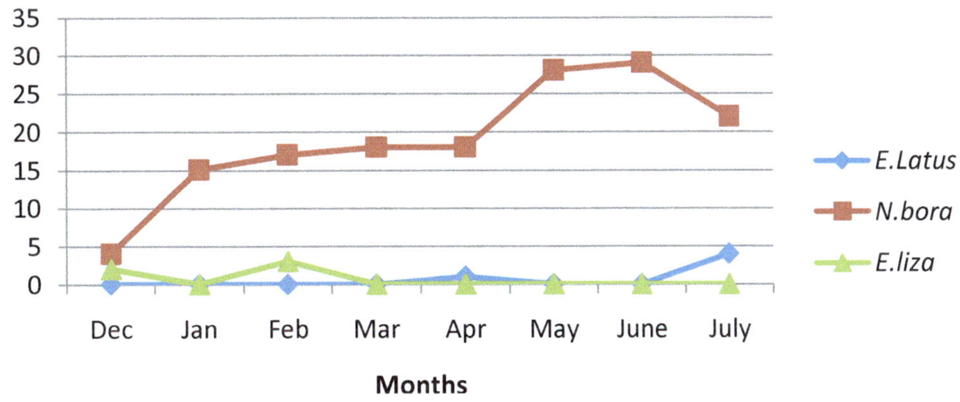

Figure 3. L.F copepod species at Djidje.

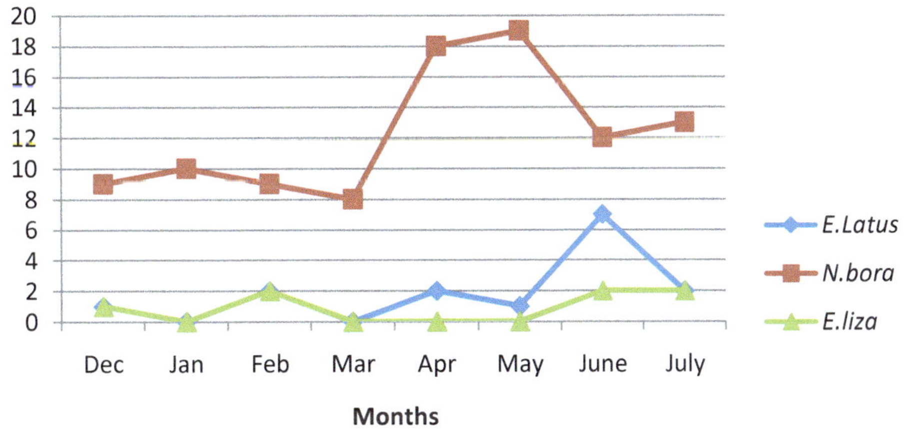

Figure 4. L.F copepod species at Ganvie.

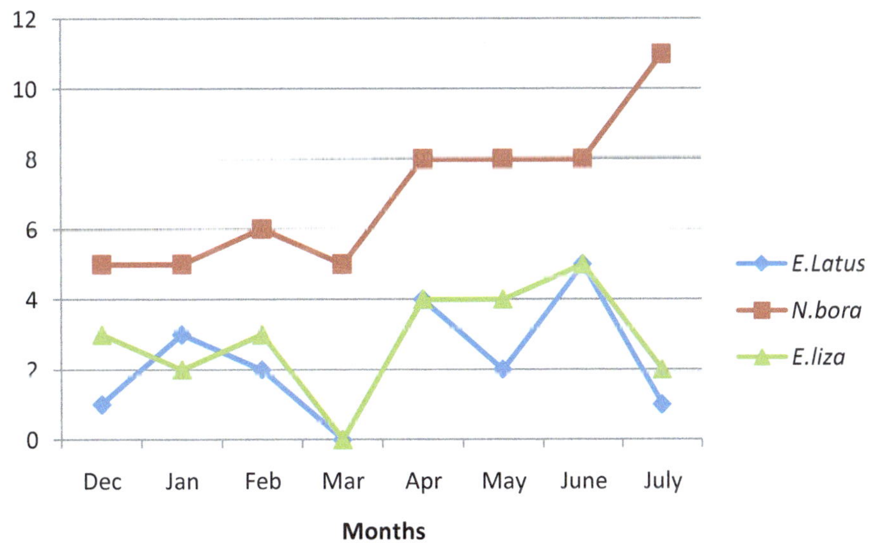

Figure 5. M.C copepod species at Ganvie.

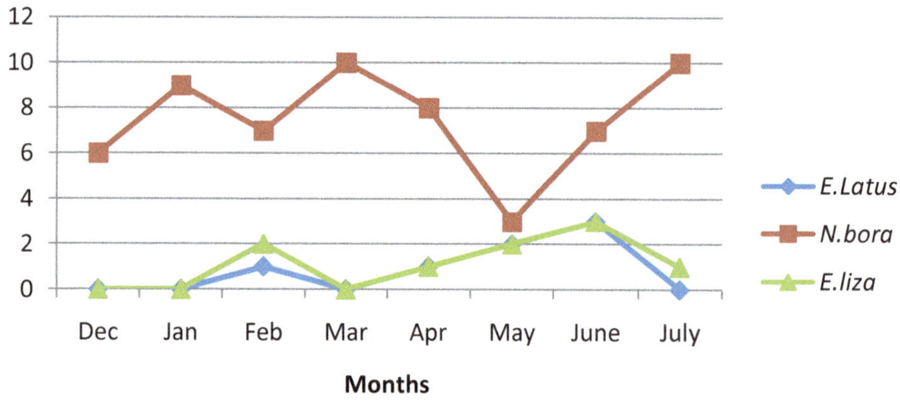

Figure 6. M.C copepod species at Zogbo.

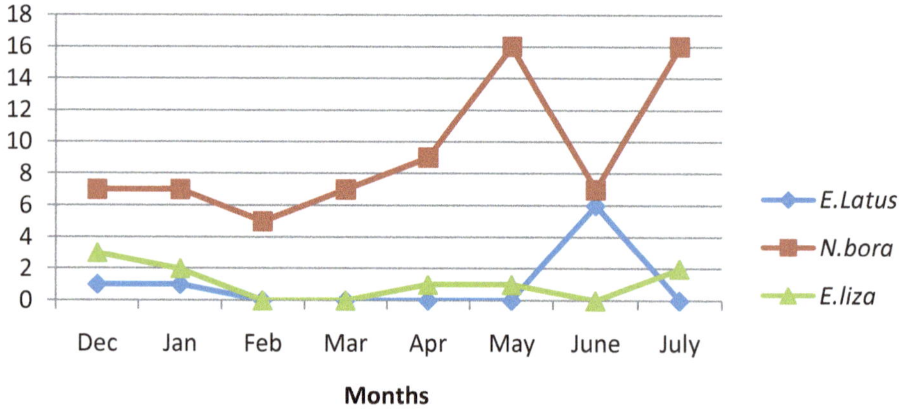

Figure 7. L.F copepod species at Zogbo.

Table 5. Prevalence rate of copepod species of Ganvie.

Ganvie	*E. latus*	*N. bora*	*E. lizae*	*E. Latus*	*N. bora*	*E. lizae*
	MC	MC	MC	LF	LF	LF
Dry	2.804	9.813	2.336	1.402	16.822	1.402
Rain	5.607	16.355	7.009	5.607	28.972	1.869

Table 6. Prevalence rate of copepods species of Djdje.

Djdje	*E. latus*	*N. bora*	*E. liza*	*E. latus*	*N. bora*	*E. liza*
	MC	MC	MC	LF	LF	LF
Dry	2.362	14.173	0.394	0	21.260	1.969
Rain	0.788	15.748	3.150	1.969	38.189	0

coastal, marine and brackish waters (Vinoth et al., 2010; Noor El - Deen et al., 2012). According to Vinoth et al., 2010 Infestation can result in serious loss and damage of gill rakers and gill lamella which hinders respiration and

Table 7. Prevalence rate of copepods species of Zogbo.

Zogbo	*E. latus*	*N. bora*	*E. liza*	*E. latus*	*N. bora*	*E. liza*
	MC	**MC**	**MC**	**LF**	**LF**	**LF**
Dry	0.599	19.162	1.198	1.198	15.569	2.994
Rain	3.593	16.767	4.192	3.593	28.742	2.400

N. bora: Nipergasilus bora, E. latus: Ergasilus latus and *E. lizae: Ergasilus lizae.*

Table 8. Geographical coordinate of sampling stations

Lake Nokoue area		
Area stations	**Longitude**	**Latitude**
Ganvie	2° 24'32.615"E	6°27'33.77"N
Djidje	2° 25'8.393"E	6°23'11.077"N
Zogbo	2° 24'7.512"E	6°23'40.162"N

and eventually result in the death of fish and significant economic loss. In his work he also reported maximum copepoda parasite infestation was noticed in *M. cephalus* and *Lates calcarifer* (P. Vinnobaba 2007). The result of this work agrees with that of Barreiro and Fancinete, (2003), who worked on parasite copepod from fishes of Santa Cruz Channel and Suape area of Pernathatmbico, Brazil. He discovered that, 66% family mugilidae were infested with copepod parasite

The highest total percentage copepod parasites recorded in Ganvie might have been caused by organic pollution resulting from the human settlement. Aladetohun et al. (2013, awating publication), in a similar work on copepod parasites of these species of mugilidae fish from Lagos lagoon, Nigeria also shows that ,*Liza falcipinnis* had more copepod parasites than *Mugil cephalus.*

In this work, *genuss Nipergasilus bora* copepods had the highest frequency of occurrence and relative abundance than genus ergasilus. This agrees with Morella and Garippa (2001) in his work on parasites of grey mullets from Mistras lagoon, Western meditarranean, also reported Caligus apolus and Nipergasilus bora to be the most prevalent crustaceans.

Ben-Hassin (1983) in his work on copepod parasites of mugilidae from the meditarrranean of France and Tunisia, reported *Nipergasilus bora.* Although *Ergasilus lizae* and *Ergasilus latus* were also found in all the three stations. Perperna and Lahav (1971), also discovered copepod parasite (*Ergasilus lizae*) in grey mullets in Israel.

Tores and Bareiro (2003) in his work also revealed higher no of copepod parasites during rainy season which also agrees with the result of this present work Yashouv, 1972, reported reduction in the number of harvested fish was about 50% of those stocked, rather than the expected 90% as a result of this infestation. Even though *M. capito, Tilapia aurea* and carp accompanied *M cephalus*, only *M cephalus* was involved in mortalities. Individual *M. capito* had about nine copepods compared with 100 to 120 on the emaciated *M. cephalus.*

In conclusion, high level of copepod parasites infestation of mugilidae fish *(Mugil cephalus* and *Liza falcipinnis*) in Lac Nokoue can cause serious health problems, fish mortalities and consequent production loss.

REFERENCES

Ben-Hassin OK (1983). Les copépodes parasites de poisson mugilidae en mediterrannéoccidentalale(cotes francaises et tunisiaennes).These de Doctorat d'Etat,U.S.T.L. pp. 452.

Doussou CT (1985) .Monogenous parasites de poissonsd'eaudounce au Benin (ouestafrican) ; These de doctoral d'etat. pp. 116.

Gbankoto A, Pampoulie C, Marques A, Sakiti N, Dramane K (2003). Infection pattern of *Myxobolus heterospora* in two tilapia species (Teleostea:Cichlidae) and its potential effects. Dis. Aquat. Org. 55:120 131.

Gnohossou P, Atachi P, Dauta C, Moreau J (2003). Influence of the salinity on the spatial distribution of the benthic fauna in lake Nokoue (Abstract). In : Picarts (ed.), proc. Intern. Conf. Pan Africa Fisheries Society, Coronou Benin. p. 23.

Kabré GB (1997). Parasites des poissons au Burkina Faso. Faunistique, *ultrastructure, biologie*. Thèse d'état, Univ. Ouagadougou, 265P. Kabata, Z. (1992): Copepods parasitic on Australian fish, XV, Family Ergasilidae (poecilastomatoida). J. Nat. Hist. 26:47- 66.

Laleye P, Niyonkuru C, Moreau J, Teugels GG (2003). Spatial and seasonal distribution on the ichtyofauna of lake Nokoue, Benin, West Afr. J. Aquat. Sci. 28:151-161.

Laleye P, Moreau J (2004). Resources and contains of West Africa coastal waters for fish production. In: Abban E. K., Casal C.M.V., Dunga P. and Falk T.M.(eds.), Biodiversity,Management and utilization of west Africa Fishes, Wordfish Center Conference, Proceedings Contribution 1718:28-30

Lang J, Paradis G (1977). Un exemple d'environement se'dimentaire biode'tritique n0n carbonate' marin et continental, h0l0cene, en climat intertropical [exDahomey]. Rev. Ge'ogr/ Phys. Ge'ol. Dyn., 19 (3)/ Paris: 295-312.

Moravec F (2007). Nematode parasites of fishes: recent advances and problems of their research. Parassitologia 49:155-160.

Niyonkuru C, Laleye P, Hounkpe C, Moreau J (2003). Impact de la peche sur les resources aquatiques des plans d'eau du sud-Benin. Astracts Third International Conference of the Pan African Fisheries Society, Cotonou, Benin. p. 221.

Niyonkuru C, Laleye P, Moreau J (2010). Impact of acadja fisheries on the population dynamics of Sarotherondon melanotheron and Hemichromic fasciatus in a lake Nokoue (Benin, West Africa). Knowl. Managt. Aquatic Ecosyst. 397:1

Noor El- Deen AE, Abdel Hady OK, Shalaby SI, Mona SZ (2012). Field Studies on Caligus Disease among Cultured Mugil Cephalus in Brackish Water Fish Farms. Life Sci. J. 9(3):733-737

Paperna I, Lahav M (1971). New records and further data on fish parasites in Israel. Bull. Fish Cult. Israel (Bamidgeh). 16: 77-86.

Paperna I, Overstreet RM (1981). Parasites and diseases of Mullets (Mugilidae). In: Oren, O.H. (ed.) Aquaculture of Grey Mullets. IBP 26, Cambridge University Press, U.K. pp. 1-19.

Ru¨ ckert S, Palm HW, Klimpel S (2008). Parasite fauna of seabass (Lates calcarifer) under mariculture conditions in Lampung Bay, Indonesia J. Appl. Ichthyol. 24:321-327

Sakiti NG (1997). *Myxosporidies et Microsporidies* de poissons du Sub-Benin faunistigue, ultrastucture, biologie. In These de Doctoratd'etatUAC. 1-300.

Sakiti NG, Blanc E, Marques A, Bouix.G (1991).(*Myxozoa,Myxosporea*) du genre *Myxobolus* Butschili,1882 parasites de poisons de Cichlidae du la lac Nokoué au Benin (Afrique de l'Quest). J. Afr. Zool.105:175-186.

Vassiliades G (1975). Nematodes parasites des poissons d'eau douce de la Republique de Senegal Bullet de l'I.F.A.N.T.37,ser A.N.3.605-618.

Vinoth R, Ajith Kumar TT, Ravichandran .S, Gopi M, Rameshkumar G (2010). Infestation of copepod parasites in the food fishes of Vellar estuary, Southeast coast of India. Acta Parasitologica Globalis 1(1):01-05.

Yakub AS, Balogun KJ, Ajani GE, Renner KO, Ariyo AA, Bello BO, Nkwoji JA, Igbo JK (2011). Distribution of phytoplankton in some parts of Lagos Lagoon, Southwest Nigeria. Int. J. Biol. Chem. Sci. 5(1):150-163.

Yamagute S (1963). Parasitic Coopepoda and Brachiura of fish. Inter science publishers, Inc. New York.

Spatial and temporal variations in selected heavy metals in water and sediment from the Mhlathuze Estuary, Richards Bay

H. M. Mzimela[1] , V. Wepener[2] and D. P. Cyrus[1]

[1]University of Zululand, Private Bag X1001, KwaDlangezwa 3886, South Africa.
[2]Department of Zoology, University of Johannesburg, P.O. Box 523, Auckland Park 2006, South Africa.

Spatial and temporal variations in water (total, dissolved and particulate), and sediment concentrations of seven heavy metals, Al, Cr, Cu, Mn, Pb and Zn from the Mhlathuze Estuary were analyzed. Effects of environmental factors on the metals in the estuary were also investigated. In water, metals concentrations varied spatially as well as seasonally with highest concentrations of Al Cr, Fe, Mn and Zn recorded in summer. Partitioning of metals was influenced by environmental factors including dissolved oxygen, turbidity and pH with two groups being identified, that is, metals that increased with increasing salinity and metals that increased with increasing turbidity. In sediment, metal concentrations showed little seasonal variation. There were, however, significant spatial differences in metal concentrations, with muddy, high organic areas of the estuary consistently having highest concentrations of metals as compared to other sites. These results suggest a high degree of heavy metal contamination in the Mhlathuze Estuary and also stress the importance of incorporating sediment metal analysis in any assessment of metal pollution in estuarine environments.

Key words: Estuaries, heavy metals, pollution.

INTRODUCTION

Rapid industrial and urban development has previously been linked to high concentrations of heavy metals recorded in the Richards Bay Harbour (Vermeulen and Wepener 1999). The Mhlathuze Estuary is located adjacent to Richards Bay harbour. The estuary and the harbour originally formed a single system called the Richards Bay estuary, but during the development of the deep water port, the original estuary was divided into half by a berm wall (Begg, 1978). The new mouth was created for the Mhlathuze Estuary on the southern side of the harbour. This area was reserved for environmental protection and declared a marine protected area (Begg, 1978).

Industrial activity in the Mhlathuze catchment includes manufacturing of metal products by some industries, while many others use metals in processing their products. The Mhlathuze estuarine system has a potential to be polluted by heavy metals from effluents discharged by these industries. Wepener and Vermeulen (1999) reported the estuary as having a low freshwater input in winter and floods occur during summer in the Richards Bay area. These floods initiate in the upper reaches and the freshwater may flush heavy metals into the estuary from the upper reaches.

The adjacent harbour is periodically dredged and the dredger spoil is either discharged on the beach or a few kilometres offshore. In both instances, there is a possibility of the dredger spoil finding its way into the estuary (Cyrus and Wepener, 1998).

The possibility of spoil ingression was predicted by the CSIR (1993) and observed by Begg (1978) who reported the ingression of the fine sediment from the dredger spoil south towards the mouth of the estuary during visual environmental audits. Cyrus and Wepener (1998) also reported the ingression of the fine sediment into the mouth of the estuary. A dramatic increase in the percentage of mud (<63 μm) was observed around the estuary in the area south of the Mhlathuze estuary (CSIR, 1993; Cyrus and Wepener, 1998). The mud fraction is known to be the component of sediment on to which most of the metals will adhere (Newman and Watling, 2007). This ingression brought about physical change and was reported as having impacted on certain components of the biota (CSIR, 1993).

Metals occur in aquatic environments either as dissolved ions or as metals bound to particular matter. The dissolved fraction is usually the bioavailable fraction however it is toxic to biota in estuarine environments (Silva et al., 2006). The particulate fraction is adsorbed to organic matter, organic fluvic and humic acids (Krupadam et al., 2006). A high amount of heavy metals may be locked onto sediments. These are metals that have been scavenged from the water column on to the particulate matter and later onto proximal sediments (Sarkar et al., 2004). In shallow estuaries such as the Mhlathuze Estuary, metals have a potential to be resuspended and thus become bio-available (Wepener and Vermeulen, 2005). They may also be available to benthic and bottom feeding organisms during feeding (Shirneshan et al., 2012). As the normal estuarine function rely on the mixing of both sea water and freshwater from the river, the type of metal species present in the particular system is determined by the chemistry of that freshwater/sea interface.

There is very little information available concerning the environmental and physicochemical processes that regulate the existence of metals in South African estuarine systems. Generally, in estuarine and marine environment, the behavior of heavy metals is governed by a number of physicochemical and environmental factors such as salinity, valence state and association with organic radicals, dissolved oxygen and pH (Cox and Micaela, 2005). These processes can remove metals dissolved in water by adsorption of metals onto particulate matter or cause resuspension of metals from sediment (Sarkar et al., 2004). Any physicochemical change that reduces the hydrophylic complexation of dissolved trace metals enhances the bioavailability of that metal by increasing the concentration of the free metal ions (Krupadam et al., 2006); thereby causing the bioavailable metal becomes toxic to biota. Since physical and chemical factors such as pH and salinity are dynamic in estuaries, metals behave differently across estuarine environments. Mhlathuze estuary is a marine dominated estuary with salinity normally ranging from 20-35 ppt.

In view of possible harbour expansion and the introduction of a metal mining plant in the catchment, the objectives of this study were to determine baseline information on the bioaccumulation of metals in water and sediment of the Mhlathuze Estuary.

MATERIALS AND METHODS

Quarterly water and sediment samples were collected in the Mhlathuze Estuary from April 1996 to December 1997. Samples were collected from the sampling sites (Figure 1).

Water samples

The surface water variables were determined *in situ* at each site using a Surveyor 3 Hydrolab connected to an H_2O water quality multiprobe: pH, water temperature, dissolved oxygen and percentage oxygen saturation, turbidity, and salinity. Water samples collected at each site were analysed by the analytical laboratory of Mhlathuze Water Scientific Services for orthophosphates, sulphates and fluorides. Two surface water samples were collected at each site, one was filtered through a 0.45 μm cellulose acetate filter and the other unfiltered.

The unfiltered sample represents total metal concentrations whereas the filtered sample represents dissolved metal concentrations. The samples were frozen until analysed for metal concentration in the laboratory. The samples were thawed and pre-concentration was carried out by acidifying 250 ml water with 10 ml of 55% nitric acid and 5 ml perchloric acid in Erlenmeyer flasks and evaporating to 5 ml on a hotplate (Standard Methods, 1998). Samples were made up to 50 ml with double distilled water.

Metal analyses

Aluminium, Cr, Cu, Fe, Mn and Zn in water and sediment were measured by flame furnace atomic absorption spectrophotometry using a Varian SpectrAA 50B spectrophotometer fitted with a deuterium arc background corrector. Calibration was carried out using matrix matched calibration standards. Analytical accuracy was determined using Standard Reference Material (SRM) of the National Bureau of Standards: standard for trace elements in water (SEM 1643c) and Buffalo River sediment (SRM 2704). Recoveries were within 10% of the certified values.

Statistical analyses

Statistical analyses of the data were performed using Tukey ANOVA multiple comparison test to measure ad hoc significant differences. Significance was regarded at the P-<0.05 significance level. Baseline normalization was performed using Fe as a normalizing metal [metal (μg/g)/Fe (μg/g)]. The principal component analysis (PCA) was used to determine the relationships between

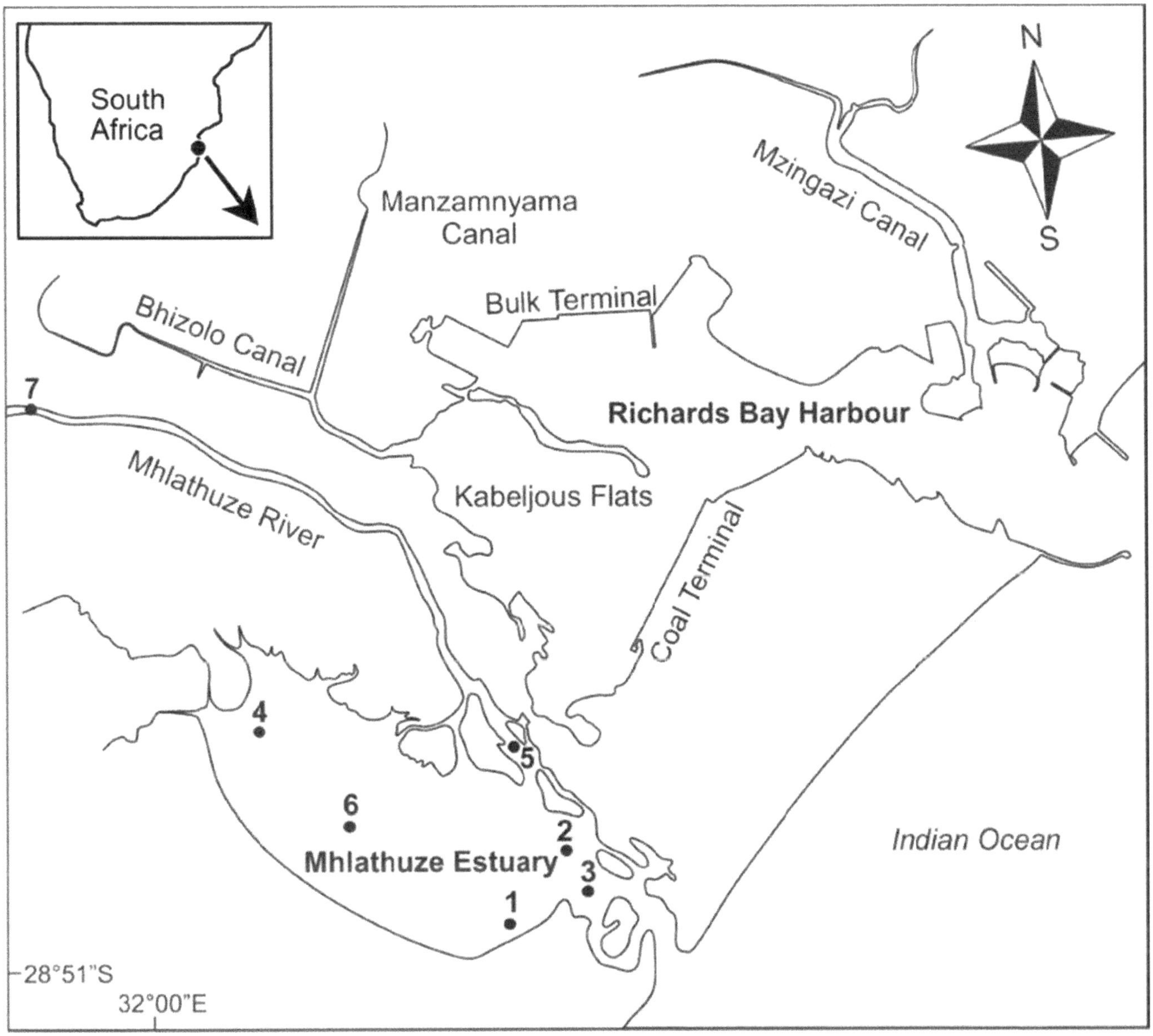

Figure 1. Map of Richards Bay Harbour and Mhlathuze Estuary showing sampling sites.

metals in water, sediment and environmental variables.

RESULTS AND DISCUSSION

Changes in the selected physico-chemical variables of the Mhlathuze Estuary are summarized in Table 1. Temperatures remained fairly constant at all sites during the quarterly surveys reflecting the natural seasonal temperature fluctuations in a subtropical environment. Only the water temperatures at site 4 seemed consistently different from those measured at the other sampling sites during the particular survey. Turbidity varied seasonally with elevated turbidities during Spring and Summer. Turbidities were elevated at site 4 when compared with the historical data. This could be attributed to the deposition of sediments from the catchment into the estuary. The deposition resulted in the buildup of very fine sediments on the bottom of the estuary. Due to its shallow nature, the strong tidal prism and strong prevailing north-easterly or south westerly winds, the soft bottom sediments can be brought into resuspension, with potential ensuing deleterious effectsto the aquatic biota. Although resuspension is a natural process, the spoil from dredging in the harbour has been shown to exacerbate the situation.

Recently, it was shown that ingression of fine sediments into the estuary from the marine environment was causing a build-up of very fine sediments in the

Table 1. Quarterly physico-chemical values recorded at five sites in the Mhlathuze Estuary for the sampling period from April 1996 to December 1997. Historical data in the original estuary before the construction of the harbour are represented by A after the construction of the harbour are presented by B.

Season	Site	Temp (°C)	Turbidity(NTU)	O_2 (mg/l)	O_2 (%)	Salinity (‰)	pH	PO_4 (mg/l)	NO_3 (mg/l)	SO_4 (mg/l)
Historical data A[1]		21.00	6.00	7.58	101.00	28.00	8.10	0.02	0.00	NA
Historical data B[2]		NA	NA	NA	NA	32.8	NA	0.017	0.067	NA
Autumn 1996	1	21.00	3.7	7.07	98.5	36.5	7.87	0.23	ND	2583
Autumn 1996	2	20.09	4.6	6.91	96.9	36.5	7.92	0.09	ND	2667
Autumn 1996	3	22.95	DL	6.84	94.3	36.7	7.79	0.24	ND	2639
Autumn 1996	4	19.91	31.7	6.34	88.2	29.2	7.72	0.16	ND	2222
Autumn 1996	5	19.86	BDL	8.00	89.4	0.8	7.91	0.25	0.80	107
Winter 1997	1	19.09	5.0	8.53	119.5	34.0	4.19	BDL	ND	2600
Winter 1997	2	20.15	19.6	7.42	98.8	30.4	5.40	BDL	ND	3100
Winter 1997	3	17.36	4.4	7.82	108.2	34.5	8.78	BDL	ND	3000
Winter 1997	4	16.31	6.0	7.41	96.8	30.5	7.56	BDL	ND	2600
Winter 1997	5	21.45	6.5	8.31	88.6	8.1	7.49	BDL	ND	410
Spring 1996	1	21.03	6.0	7.63	108.2	34.4	7.82	0.18	ND	3860
Spring 1996	2	20.96	26.0	7.66	109.5	34.4	7.86	0.12	ND	3700
Spring 1996	3	21.88	24.0	7.74	110.3	34.4	7.88	BDL	ND	3140
Spring 1996	4	24.12	5.0	7.48	107.2	34.2	7.78	BDL	ND	3860
Spring 1996	5	25.47	14.0	8.01	108.1	22.2	7.67	BDL	ND	2420
Summer 1996	1	23.62	8.0	7.74	115.3	34.6	8.58	0.10	ND	3300
Summer 1996	2	24.00	5.0	7.88	115.2	34.9	8.44	0.17	ND	2700
Summer 1996	3	25.84	4.0	8.23	12.2	34.9	8.49	0.17	ND	3150
Summer 1996	4	30.41	15.0	7.57	114.8	34.0	8.52	0.20	ND	2850
Summer 1996	5	24.58	28.0	7.94	114.7	14.2	8.44	ND	ND	ND
Autumn 1997	1	22.10	16.0	6.94	101.4	32.3	8.45	BDL	ND	4000
Autumn 1997	2	24.05	12.0	6.80	90.9	32.9	8.40	0.21	ND	3500
Autumn 1997	3	23.80	18.0	7.46	108.1	32.5	8.52	0.09	ND	2900
Autumn 1997	4	23.20	14.0	6.42	88.2	24.2	8.32	0.19	ND	2200
Autumn 1997	5	19.51	26.0	7.50	86.6	0.3	8.25	0.45	0.50	32
Winter 1997	1	19.18	15.0	5.84	105.6	34.2	8.40	0.21	0.07	2881
Winter 1997	2	19.16	22.0	8.01	105.7	35.1	8.40	0.21	0.07	2959
Winter 1997	3	19.04	20.0	8.12	105.1	35.1	8.40	0.20	0.06	3026
Winter 1997	4	14.76	24.0	8.22	94.8	6.5	7.74	0.18	0.06	189
Winter 1997	5	20.22	46.0	9.11	91.2	0.4	7.27	0.11	0.49	24
Spring 1997	1	21.15	14.0	7.62	103.8	35.0	8.82	0.17	BDL	2928
Spring 1997	2	20.03	28.0	7.45	98.4	36.5	8.64	0.13	0.14	1806
Spring 1997	3	18.71	17.5	20.30	102.0	35.3	8.80	0.22	0.12	3137
Spring 1997	4	21.72	10.0	7.03	89.8	29.4	8.36	0.13	0.11	2285
Spring 1997	5	21.42	29.0	21.42	78.4	0.3	8.68	0.06	0.21	32
Summer 1997	1	22.67	20.0	7.16	84.2	4.1	7.70	0.08	0.28	370
Summer 1997	2	22.66	25.0	6.58	76.1	0.2	7.43	0.06	0.46	185
Summer 1997	3	21.76	17.0	7.37	86.5	5.5	7.66	0.10	0.36	477
Summer 1997	4	22.48	45.5	7.36	85.4	0.7	7.78	BDL	0.35	46
Summer 1997	5	22.16	10.00	5.10	59.70	0.20	7.19	0.07	0.28	12

BDL = Below detection limit. ND = No data. [1]Hemens *et al.*.(1971) and [2]Hemens et al. (1976).

mouth of the estuary and embayment (Cyrus and Wepener, 1998). This ingression was attributed to dredging activities in Richards Bay Harbour. Salinity remained constant throughout the system during the study with the exception of summer 1997 where the salinity at all sites was below 6‰. The reduced salinity during summer 1997 was a result of the flood that occurred during the sampling trip.

Salinity measured at site 5 was consistently lower than other sites during a particular survey. Dissolved oxygen was also constant throughout the estuary. Site 3 which was the site located at the mouth of the estuary had consistently elevated dissolved oxygen concentrations.

This was also the case for sulphate and fluoride. Concentrations at site 5 were consistently different from those measured at any of the other sampling sites during the particular survey.

During summer 1997, sulphate concentrations were lower as compared to those measured in other seasons. Nutrients (nitrates and orthophosphates) showed little variations during the study (Table 1).

During the construction of the harbour, nutrients level increased due to redistribution of anaerobic silt dredge from several meters below the surface.

Hemens et al. (1976) speculated that this was due to the bottom consisting of redeposited silt that had been depleted of its soluble nitrogen, or that the release of nitrogen from dead and decaying plankton had decreased, together with plankton densities, due to the opening of the new mouth.

The present study showed substantially higher nutrient levels than during the 1970s. It is unlikely that runoff from the catchment could contribute to the high nitrate concentrations measured at Sites 5 and 4, since low nutrient levels were recorded in the lower reaches of the river and at Site 1 (Table 1).

According to Wepener and Vermeulen (1999), the source of the nutrients is possibly related to fixation of nitrogen by blue-green algae in the muds of the vast mangrove areas.

Another potential source of nutrients they mentioned was sewage runoff from the surrounding rural settlements was indicated by high faecal coliform counts the authors obtained (Wepener and Vermeulen, 1999).Seasonal and spatial organic content (mean ± standard error) of sediment samples from the Mhlathuze Estuary are presented in Table 2.

Mean organic content were highest during summer 1996 and lowest during autumn 1996. The highest mean organic content was recorded at Site 4 with the lowest levels measured at site 3.

Metals in water

Temporal and spatial variations of metals in water

Metals in the water column generally did not show any significant differences between seasons with the exception of December 1997 where concentrations were significantly different from the rest of the seasons. During Spring 1997, Cu and Pb concentrations were also significantly higher.

The increase in metals during summer 1997 was probably related to resorption of metals bound to suspended particles during river flood conditions that prevailed during the sampling trip. A general trend of a reduction in metals was observed during autumn 1997. This reduction was attributed to dilution of waterborne metals by freshwater due to floods. According to Krupadam (2006) wide variations in water-borne metal concentration within estuaries are normally related to the degree of fresh water contribution or the presence of industrial effluents.

Total metal concentrations in water samples showed little spatial variation (Table 2). Only Pb concentrations significantly differed between Site 3 at the mouth and Site 5 which is located up the Mhlathuze River. This is probably related to the short residence time of water in the estuary.

About 90% of water from the estuary is being drained during each tidal cycle (Begg, 1978). The residence time of water may be too short to allow for major spatial variations of metals in water. The highest Pb concentrations in water recorded at Site 3 suggests the mouth as the source of lead pollution into the estuary. Lead may have been discharged with the fine sediments during dredger spoil on the breach, which were transported into the estuary through the mouth.

The elevated concentrations of Fe, Al, Zn and Cr recorded in the water from the Mhlathuze Estuary are in agreement with the results reported by Vermeulen and Wepener (1999, 2005) for metals in water from the adjacent Richards Bay Harbour (Table 4). When compared with historical concentrations reported from the original Richards Bay Estuary, only Zn concentrations were higher during the present study. Hemens et al. (1976) commented on the high Zn concentrations recorded at that time as “reason for concern”. Their study was, however, conducted before the completion of the harbour.

The high metal concentrations measured in this study could therefore be related to subsequent activities in the catchment and adjacent harbour area. Wave borne metals could also enter the estuary from the harbour via tidal gates. These tidal gates that are no longer in use were built to serve as a connection between the water in the harbour and the estuary. During high tide water can be observed overtopping over the tidal gates (personal observation) and it is therefore highly likely that metals originating in the harbour will find its way into the estuary. The concen-trations of metals from this study were slightly elevated when compared with metal concentrations in the water of other estuaries on the east coast of South Africa (Table4). It must be borne in mind that surveys by Hemens et al. (1970, 1976) were undertaken between 20 and 30 years ago and that these concentrations may have increased in the intervening

Table 2. Comparisons of dissolved metal concentrations in water (A) and metals in sediment (B) in the Mhlathuze Estuary with corresponding metal concentrations in other estuaries on the eastern seaboard of South Africa and in other countries.

Estuary	Al (μg/l)	Cr (μg/l)	Cu Fe (μg/l)	Fe (μg/l)	Mn (μg/l)	Pb (μg/l)	Zn (μg/l)	Reference
Water								
Mhlathuze	990.0	48.0	39.1	907.0	48.2	130.15	66.5	This study
Kynsna	-	0.1	0.2	81	5	0.6	0.3	a
Gamtoos	-	-	0.6	372	41	0.6	1.1	c
Swartkops	-	-	3.9	275	41	1.5	3.6	d
Sundays	-	1.3	3.2	334	18	0.7	2.4	e
Bushmans	-	0.4	1.8	302	10.2	0.2	0.5	f
Kosi Bay	-	-	1.1	-	-	1.0	5.7	i
St Lucia E	-	-	3.3	68.8	-	39.2	2.3	j
St Lucia E	-	-	17	2000	-	-	11.7	i
Richards Bay	-	-	1.7	-	-	1.9	-	k
Richards Bay	-	-	4.0	-	-	4.2	3.8	i
Richards Bay	504.4	23.6	50.8	782.4	80.7	-	85.4	l
Durban	-	-	27	800	-	117	287	i
East London			42.4	183.0	16.3	23.9	27.6	Fatoki
Port Elizabeth			11.3	21.9	4.2	16.8	16.2	Fatoki
Sunderban				175.0		0.20	9.7	GandK
Sediment								
Mhlathuze	18677.4	64.4	12.2	20606.9	13.5	45.6	45.6	This study
Kynsna	-	21	5	-	14	17	17	a
Gamtoos	-	15	5	9180	7	16	16	c
Swartkops	-	5	18	20800	31	55	55	d
Sundays	-	38	16	72	18	57	57	e
Bushmans	-	22	3	7330	5	13	13	f
Kosi Bay	-	150	61	60000	19	72	72	i
St Lucia E	-	7	2	3000	0.8	3.4	3.4	i
St Lucia E	-		9.9	23640	24	98	98	k
Richards Bay	-	74.8	24.04	5814	17.47	87.16	87.16	j
Richards Bay	-	110.3	19.22	31762.7	-	95.54	95.54	l
Richards Bay	31323.4	388	57	40000	117	287	287	i
Durban	-		183.0	18114.0	549.0	332.0	332.0	m
East London			82.3	15182.0	499.0	126	126	m
Port Elizabeth				10068	25.15	3448.4	3448.4	n
Sunderban				16600	29	54	54	o
Victoria		16.0	9.0					

References: a, Watling and Watling, 1982a; c. Watling and Watling, 1982c; d, Watling and Watling, 1982d; e, Watling et al., 1982; f, Watling and Watling, 1983b; g, Watling et al.., 1985, 1985; I, Cloete et al.., 1979; j, Oliff et al.., 1986; k, Connell *et al.*., 1975 I; Vermeulen and Wepener, 1999; m., Fatoki and Mathabatha 2001; n. Guhathakurta and Kariva 2000 , Tanner et al., 2000.

period.

Metals such as Zn and Cr are important constituents of industrial mining and domestic effluents (Khan et al., 2005). High concentrations of these metals may be a direct consequence of anthropogenic contamination from these activities. It is not possible to comment on the high Al levels since there is no historical data (Hemens et al., 1976) to compare with it. However, it is highly likely that the current levels are due to a combination of anthropogenic sources (the nearby aluminium smelter complexes, dredging of sediments in Richards Bay Harbour and natural leaching and weathering processes in the catchment).

Metals in sediment

Metals concentrations in the sediment were in elevated

orders of magnitude above the concentrations recorded in water. According to Soumady and Asokan (2011), sediments represent the most concentrated pool of metals in aquatic environments, metals that can be assimilated by aquatic organisms resident in those sediments. Since those organisms form a primary food source for bottom feeding fish, the accumulation of metals by such fish potentially depends on uptake from food as well as from water (Shulkin et al., 2003; Silva et al., 2006). Various bioaccumulation studies also found concentrations of metals in the estuarine environments which were in orders of magnitude higher in sediments than in water (Allen, 1993; Lim et al., 2012).

No significant temporal differences were recorded for metals in the sediment. This may be due to most metals being present in sediment as precipitates or in an un-dissolved state.

In estuarine systems, these metals are not bio-available except to organisms that are resident in sediment (Williams et al., 1998). Metals would be taken up and absorbed through the skin, or be ingested with food particles in benthic feeders.

Highest metal concentrations in the sediment were recorded for Fe, Al and Zn (Table 4). These metals also displayed the highest accumulation in the particulate matter in the water column. Watling and Watling (1982a-d) also reported high concentrations of Fe and Zn from sediments in Gamtoos, Swartkops, and Sunday River estuaries (Table 4), although concentrations in these estuaries were lower than those in this study.

There were significant spatial variations in metals concentrations (Table 3). With the exception of Al and Pb, most metals differed significantly between Sites 3 and 4. Metals concentrations were highest at Site 4, and with the exception of Cu, all metals had lowest concentrations at Site 3. The reasons could be attributed to the differences in substrate types and the amount of organic material.

Site 4 is in the middle of the embayment and contains mainly fine sediment. It is also the site with the highest organic content when compared with the rest of the sites. Site 3 is at the mouth of the estuary and is dominated by coarse sand. It is also the lowest in terms of organic content (Table 3).

The metals in the Mhlathuze Estuary probably exist as particulate matter or precipitated metals. Their presence in sediment may be predominantly due to a result of anthropogenic activities and ingression with fine dredged spoil from the harbour (Van den Hurk et al., 1997).

The metals are then precipitated onto and into the sediments. According to Sarkar et al., 2004 and Magesh et al. (2011), high loads of heavy metals are normally concentrated in the fine sediments. It is likely that most of the particulate bound metals entered the estuary during the ingression of into the Mhlathuze Estuary which took place when dredger spoil was deposited on the beach north of the estuary (Wepener and Cyrus, 1997; Mackay and Cyrus, 1999). Wave action and the near-shore current would have resulted in contaminated fine silt drifting into the estuary. The very high levels of Al in sediment may either be from geological leaching, as the area is known for its high Al metal content (Kwazulu Natal Business, 2013), or from pollution effects such as dredger spoil (Wepener and Cyrus, 1997).

Concentrations of Cu and Pb in sediment were low, ranging from 20-50 μg/g. These metals exist mostly as dissolved ions in the water column hence their low concentrations in sediment.

Copper is also known to form complexes with organic matter (Sarkar et al., 2004). The highest Cu concentrations in the sediment (at site 4) could be attributed to the complexation of Cu with organic ligands since the highest organic content was also recorded at this site (Table 4). The presence of high organic content in estuaries can decrease toxicity of metals such as copper by binding copper to organic ligands (DePalmer et al., 2011).

Chromium concentrations were elevated in sediments of the Mhlathuze Estuary (Table 4). High concentrations of Cr may be a result of contamination from the harbour due to sediment bound metals entering the estuary as discussed in the previous paragraphs. The results suggest equal amounts of Cr being exchanged between dissolved fractions in pore water and particulate fractions in sediment. This is because the percentage of particulate and dissolved Cr reported in this study were almost equal and also both the fractions were significantly different between seasons. Studies in estuarine areas show that dissolved and particulate Cr are present in almost equal quantities with the flocculation processes increasing Cr^{3+} concentrations in the salinity that are below normal sea water (Zwolsman and van Eck, 1999).

Results of this study show that metals vary spatially and temporally in water at the Mhlathuze estuary. Although there are minor variations of metals in sediment these results, the concentrations are higher than those recorded in other estuaries along in the eastern seaboard (Table 2). Seasonal variation in the Mhlathuze Estuary may be as a result of the metal loads that may be delivered by the river or sea. Spatial differences may be as a result of behaviour of metals in response to gradients in the chemistry of the water body and the nature of the sediment.

The important factor that results in variation of metals between sites closer to the mouth in the Mhlathuze Estuary is the residence time of water in the estuary. The tidal cycle in the estuary affects the sites next to the mouth (Site 1, 2 and 3). Water from these sites is almost completely replaced during a tidal cycle. The quality of water brought in during each tidal cycle would be similar resulting in more or less similar values for sites next to the mouth (sites 1, 2 and 3). Salinity values form these sites are also similar and was close to sea water. Sites 4

Table 3. Spatial heavy metal concentrations (means ± standard error) in water (total dissolved and particulate) (μgL^{-1}) and sediment (μg.g^{-1}) in the Mhlathuze Estuary. All references to significant differences are made in the text and are not indicated in the table.

Parameter	Turbidity	Salinity	Organic	Dissolved 0_2	pH	Total	Suspended	Dissolved
Aluminium								
Turbidity	1							
Salinity	-0.64*	1						
Organic	0.02	-0.27	1					
Dissolved O_2	0.06	0.11	0.10	1				
pH	-0.18	0.45*	-0.04	0.22*	1			
Total	0.23	-0.24*	-0.19	-0.11	-0.58*	1		
Suspended	-0.28*	0.36*	-0.21*	-0.37*	-0.15	0.117	1	
Dissolved	0.05	0.03	-0.32*	0.13	-0.22*	0.44*	0.35*	1.00
Chromium								
Turbidity	1.00							
Salinity	-0.70*	1.00						
Organic	-0.02	-0.16	1.00					
Dissolved O_2	0.01	0.10	-0.04	1.00				
pH	-0.19*	0.40*	-0.01	0.11	1.00			
Total	-0.02	0.08	-0.28*	-0.17	0.19	1.00		
Suspended	0.24*	-0.29*	0.01	-0.34	0.33*	0.66*	1.00	
Dissolved	0.25	-0.05	-0.29*	-0.02	-0.23*	0.51*	0.26*	1.00
Copper								
Turbidity	1.00							
Salinity	-0.53*	1.00						
Organic	-0.10	-0.26*	1.00					
Dissolved O_2	-0.07	0.12	-0.05	1.00				
pH	-0.10	0.34*	-0.06	0.28*	1.00			
Total	-0.20*	0.44*	0.17	0.14	0.55*	100		
Suspended	-0.07	0.12	-0.46*	-0.16	-0.27*	-0.12	1.00	
Dissolved	0.07	0.30*	-0.06	-0.09	-0.11	0.29*	0.11	1.00
Manganese								
Turbidity	1.00							
Salinity	-0.61*	1.00						
Organic	-0.06	-0.26*	1.00					
Dissolved O_2	0.01	0.17	-0.10	1.00				
pH	-0.19*	0.46*	-0.09	0.19	1.00			
Total	0.11	0.03	-0.41*	-0.06	0.01	1.00		
Suspended	0.48*	-0.51*	0.17	-0.36*	-0.10	0.13	1.00	
Dissolved	0.11	0.03	-0.23*	-0.09	-00.03	.52*	-0.13	1.00
Iron								
Turbidity	1.00							
Salinity	-0.52*	1.00						
Organic	-0.22*	-0.22	1.00					
Dissolved O_2	-0.04	0.22*	-0.20*	1.00				
pH	-0.11	0.43*	-0.04	0.27*	1.00			
Total	0.11	0-.09	-0.30*	0.08	-0.50*	1.00		
Suspended	0.62*	0-.42	-0.23*	-0.13	-0.43*	0.508*	1.00	
Dissolved	-0.07	0.20*	-0.52*	-0.12	0.08	0.36*	0.05	1.00

Table 3. Contd.

Parameter	Turbidity	Salinity	Organic	Dissolved 0_2	pH	Total	Suspended	Dissolved
Lead								
Turbidity	1.00							
Salinity	-0.60*	1.00						
Organic	-0.05	0-.27*	1.00					
Dissolved O_2	0.06	0.18*	-0.13	1.00				
pH	-0.18*	0.44*	-0.02	0.16	1.00			
Total	-0.20*	0.44*	0.05	0.13	0.58*	1.00		
Suspended	-0.11	0.14	0.07	-0.28*	0.27*	-0.00	1.00	
Dissolved	-0.16	0.35*	-0.18*	0.19	-0.06	0.41*	-0.63*	1.00
Zinc								
Turbidity	1.00							
Salinity	-0.62*	1.00						
Organic	0.03	-0.29*	1.00					
Dissolved O_2	0.06	0.13	-0.12	1.00				
pH	-0.20*	0.43*	-0.02	0.24*	1.00			
Total	0.01	0.10	-0.22*	0.32*	-0.11	1.00		
Suspended	0.20*	-0.34*	0.08	-0.39*	-0.04	0.21*	1.00	
Dissolved	0.12	0.01	0.04	-0.23*	0.12	0.16	0.15	1.00

and 5 however were different. Site 4 is in the basin and site 5 up the river. There is limited tidal influence on these sites and consequently they have different levels of metals in comparison to the former sites. The low levels of metals consistently reported in water and sediment up river at site 5, when compared with the other sites lead to the conclusion that the ingression of sediment from the harbour as the source of most of the high metal concentrations found in sediments of the Mhlathuze Estuary. Harbour sediments have been reported by various authors as highly polluted environments, and a possible source of metal contamination in surrounding water bodies e.g. Victoria Harbour, (Tanner et al., 2000) East London and Port Elizabeth (Fatoki and Mathabatha, 2001), Tolo Harbour (Owen and Sandhu, 2000), Richards Bay Harbour (Vermeulen and Wepener, 2005) and Klaipeda Harbour (Galkus et al., 2012).

Temporal variations of metals in sediment could not be observed in the Mhlathuze Estuary. This is because of the relative stability of metals in sediments as well as variation of metals concentrations at different sites. Metals varied spatial because of the different substrates found in the Mhlathuze Estuary and their variability in accumulation of metals. According to Cox and Micaela (2005), heavy metal distribution in estuarine and marine deposits is influenced by sediment texture, clay content, organic carbon, iron hydrous oxides and carbonates. These components differ in the manner in which they adsorb metals (Newman and Watling, 2007). The clay-rich sediment and organic carbon rich sediment tend to contain higher metal content as compared to sand dominated regions. Mud has a very fine texture and it usually accumulates most of the metals. According to Herut and Sandler (2006), clay minerals readily absorb heavy metal and consequently clay-rich sediments tend to contain higher contamination levels than sand-dominated deposits. Fine sediments contribute by having large specific areas with many attachment sites for metals. Organic coating also occurs easily in fine sediments (Mortimer and Rae, 2000). Types of substrate sediments in the Mhlathuze Estuary are described in Cyrus and Wepener (1998). Sites 1, 2 and 4 have high mud and organic content (Table 3). They also had high metals concentration when compared with sandy sites such as sites 3 and 5, which had low concentrations of heavy metals in the sediment (Table 4).

One method used to determine the level of contamination of metals in sediment is the simple ratio between the concentration of a metal and the concentration of the normaliser. The conditions that are considered in selecting a normalizer is that the metal should be abundant, naturally occurring and should not be influenced by anthropogenic sources. Because of these conditions, metals that are considered as suitable for the normalization of metals in sediment are Al and Fe (Newman and Watling, 2007). For this study, Fe was considered more suitable due to the presence of aluminium smelters located around the harbour area with a possibility of estuarine sediment contamination from thesmelters' effluents. An attempt to normalize the sediment metals using Al showed results with very low R2 values, a high number of data points that behaved as

Table 4. Mean seasonal heavy metal concentrations in water (total, dissolved and particulate) (µg.g-1) in the Mhlathuze Estuary. All references to significant differences are made in the text and are not indicated in the table.

Sample	Site 1	Site 2	Site 3	Site 4	Site 5
Organic content	5.12±1.18	4.55±1.05	2.51±0.76	11.17±1.73	4.53±1.73
Aluminium					
Total	1995.8 ±1554.2	42705±3159.8	2060.0 ± 1554.3	5814.2 ±4973.1	2630.0±2029.6
Dissolved	632.8 ±281.9	721.0±403.6	990.3 ± 580.4	1546.5±1090.4	939.3±522.0
Particulate	513.5 ±347.9	564.3±356.6	537.3 ± 234.2	837.0±408.0	332.8±143.5
Sediment	22044.9 ±7119.2	21202.1 ±6081.8	10303.3 ± 3514.6	29118.4±7631.0	17230.9±4012.4
Chromium					
Total	37.5±13.0	61.8±26.7	32.8±13.8	47.0±27.3	26.3±16.2
Dissolved	46.0±10.7	65.3±10.8	46.3±12.1	48.8±17.7	33.5±15.1
Particulate	44.8±14.3	51.0±15.2	50.5±16.8	53.8±20.6	40.0±16.5
Sediment	84.1±15.1	42.4±12.2	41.7±8.9	107.9±16.5	55.6±10.8
Copper					
Total	53.1±10.2	51.3±7.2	48.3±6.6	38.3±9.0	21.8±6.0
Dissolved	42.0±13.4	52.3±16.2	50.8±15.9	38.3±15.9	14.5±4.4
Particulate	2.3±1.2	19.0±11.2	9.3±11.2	9.3±5.9	5.2±3.0
Sediment	12.1±1.8	8.4±2.5	8.8±2.3	22.72.0	11.7±2.7
Iron					
Total	1732.5±1054.9	3900.8±2817.1	1872.5±1241.4	4119.5±3286.1	2585.0±2077.8
Dissolved	655.0±218.8	787.1±292.4	961.8±584.7	1317.5±864.6	730.0±406.7
Particulate	1542.5±869.0	5598.3±4224.3	2067.5±954.3	1142.5±315.4	1797.5±857.7
Sediment	23132.3±2303.6	18715.9±4456.7	12585.8±33473	31280.7±3428.7	21048.5±5641.2
Manganese					
Total	77.8±11.7	110.5±24.7	69.8±12.0	109.0±46.9	82.8±20.2
Dissolved	42.8±8.2	88.0±39.6	45.8±9.1	40.0±6.4	61.3±20.2
Particulate	36.8±8.5	60.8±13.0	43.8±14.9	121.7±58.74	70.5±18.8
Sediment	274.2±44.2	308.2±89.4	165.3±38.7	731.5±141.2	335.5±66.6
Lead					
Total	168.8±22.9	233.5±37.8	240.0±41.5	193.0±55.0	78.8±19.7
Dissolved	167.5±72.2	129.5±35.8	186.0±62.8	130.3±65.8	37.5±12.5
Particulate	59.0±27.76	57.8±23.47	67.0±35.6	62.5±25.97	39.5±19.5
Sediment	18.0±3.8	14.1±3.5	9.1±2.1	18.7±4.2	12.3±2.8
Zinc					
Total	66.8±3.8	76.2±6.4	63.0±4.3	119.5±59.1	43.8±5.2
Dissolved	68.8±8.3	83.5±13.1	74.5±6.5	64.5±18.7	59.8±7.3
Particulate	41.8±10.4	69.0±14.3	47.0±12.5	114.2±59.8	47.0±9.9
Sediment	54.7±4.6	35.0±9.0	33.1±7.1	74.1±4.7	37.7±7.0

outliers and metals Cr, and Zn were not significantly different ($P<0.05$). Normalization plots of metals using Fe are presented in Figure 2. The ratios of metals to Fe were calculated and linear regressions were represented for metals Al, Cr, Mn, Pb and Zn. The R^2 values recorded are displayed in the graph. The regression recorded was significantly different for all metals. The outliers observed in the regression graphs for metals Al, Cr and Cu indicated increased concentrations of metals due to anthropogenic influence. Regression of Al and Fe was the lowest as compared to other metals suggesting an increase of Al from background concentrations. In the comparison of metals and physicochemical parameters using principal component analysis, the two axes retained 60.8% of the variances from the sample data. The multivariate PCA analysis, based on metals in water,

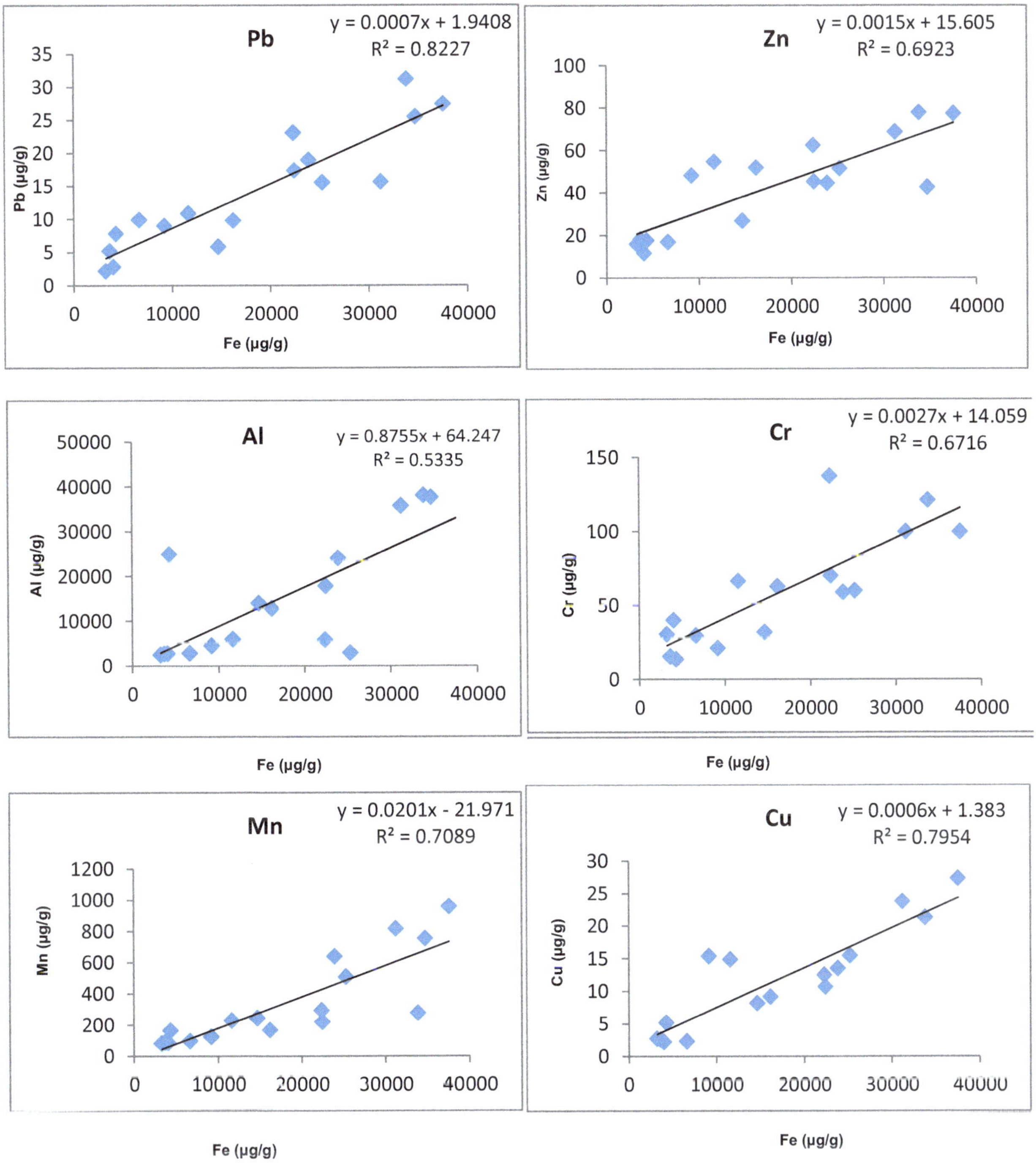

Figure 2. Normalised linear regression plots of Al, Cr, Mn, Pb and Zn concentrations (µg/g) to Fe (µg/g).

sediment and physicochemical data is plotted in Figure 3. The data for metals from the different phases and the physicochemical parameters influencing them formed three distinct groups. A fourth group consisted of nutrients and water quality parameters.

The first group consisted of sediment metals and the organic content. The correlation of metals in sediment suggested the increase of metals in sediment with an increase in the sediment organic content. Sediment metals in the estuary increased and were found in high concentration in the areas such as the basin at Site 4 that had high organic matter. The mouth area, however, had less organic content and low concentrations of metals.

The particulate concentrations of Al, Pb Cu and Zn formed the second group that increased with an increase in dissolved oxygen, pH and temperature. These

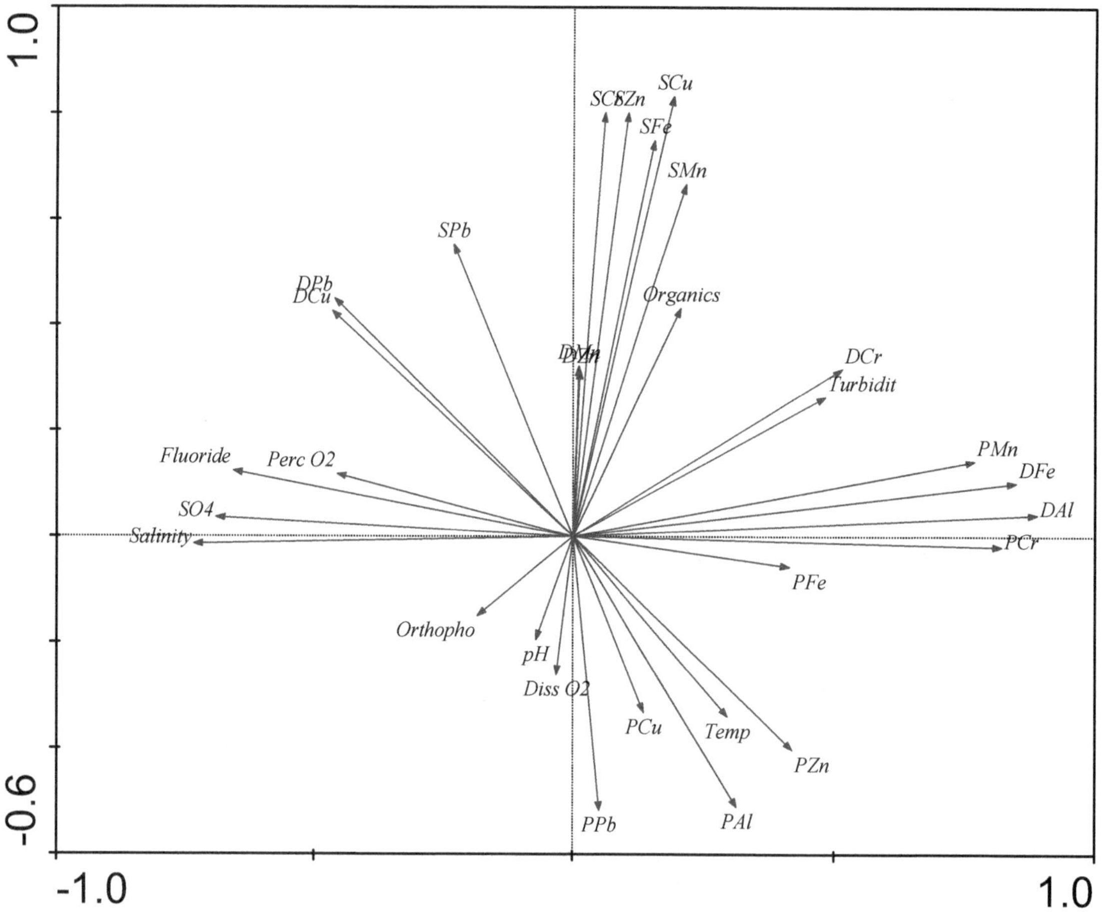

Figure 3. Principal component analysis biplots of particulate (P), dissolved (D) and sediment (S) concentrations (60.8% of variance explained)

conditions were found in the shallow waters towards the mouth of the estuary. This suggests that precipitation is the important process for the increased of these metals in the particulate phase. In the third group, the dissolved concentrations of Al, Cr and Fe and the particulate concentrations of Cr, Mn and Fe were correlated with turbidity, suggesting increasing concentrations of these metals in the basin where turbidity was high due to re-suspension as a result of the shallow nature of the estuary and the very fine bottom sediment. Dissolved Zn and Mn concentrations were both influenced by different factors to the other dissolved metal concentrations. This was also observed in the behaviour of dissolved Mn and Zn in other estuaries on the KZN eastern seaboard (unpublished data) where dissolved concentrations of both Mn and Zn increased in areas of high salinity and high turbidity. The correlations of dissolved Cu and Pb concentrations were probably related to the very low dissolved concentrations of these metals recorded in the Mhlathuze Estuary. The increase in salinity and percentage oxygen at the mouth of the estuary coincided with increases in sulphates and fluoride concentrations as identified by the water quality group. This suggests the sea as the source of the fluorides and sulphates in the estuary as these chemicals are observed in very low concentrations in the upstream sites.

Heavy metals discharged directly from industrial effluents, as well as those diffusing from activities in the harbour and from runoff storm water; result in contamination of water and harbour sediments (Vermeulen and Wepener,1999). Material that was dredged from the harbour was deposited on to the beach north of the estuary mouth. The dredge spoil is then taken up by wind and currents to the estuary mouth thereby transferring metal contamination into the estuary. The harbour authority has been looking at various ways of disposing of the spoil as an environmental exercise to mitigate the potential pollution of the estuary. The method that was recommended by CRUZ (Wepener and Cyrus, 1997; Cyrus and Wepener, 1998) of disposal a few kilometres offshore is environmentally sustainable. This would result in the pollution from the spoil becoming more diffuse and not drifting towards the Mhlathuze Estuary.

The value of an estuary for resource harvest, habitat for fish and other invertebrates and for aesthetic beauty can never be overestimated (Breen and McKenzie, 2001).

The idea of keeping the southern side of the original Richards Bay Estuary for resource protection proved to be a good environmental exercise. It was to ensure perpetuation of biodiversity and resource protection. It is however being affected by activity in surrounding areas. This is shown by the findings of this study in terms of metal pollution in the Mhlathuze Estuary.

Further studies are needed to clearly point to the actual sources of pollutants in the Mhlathuze Estuary. Activities such as mining on the boundary waters (Armah et al., 2010), and periodic dredging of the harbour have been found to have a marked effect on the accumulation of metals in estuarine environments (Van der Hurk et al., 1997; Galkus et al., 2012). A high concentration of these metals has an effect on biota of estuaries. This could be made worse by a number of factors. Plans are underway to expand the Port if Richards Bay. The operation of EXARRO Sands, a heavy metal company that has recently built on the catchment of the Mhlathuze River, has been in operation for about 10 years. Both these and other future developments may impact negatively on the estuary in terms of metal pollution.

Conflict of Interests

The author(s) have not declared any conflict of interests.

ACKNOWLEDGEMENTS

Dr Gavin Tweddle is thanked for reviewing the script. The fund for this project was provided by the Coastal Research Unit of Zululand (CRUZ) of the Department of Zooogy, University of Zululand.

REFERENCES

Allen HE (1993). The significance of trace metal speciation for water, sediment and soil quality standards. Sci. Total Environ. 134(Suppl. Part 1):23-45.

Armah FA, Obiri S, Yawson DO, Pappoe AN, Moi BA (2010). Mining and Heavy Metal Pollution: Assessment of Aquatic Environments in Tarkwa (Ghana) using Multivariate Statistical Analysis. J. Environ. Stat. 1(4): 1-13.

Begg G (1978). Estuaries of Natal. Town and regional planning. Report V41.

Breen CN, McKenzie M (2001). Managing Estuaries in South Africa: An Introduction. Institute of Natural Resources, Pietermaritzburg.

Cloete CE (Ed.) (1979) The transfer of pollutants in two southern hemispheric oceanic systems. Proceedings of a workshop held at Plettenberg Bay, South Africa, 23-26 April 1975. South African Scientific Programmes Report. No. 39.

Connell AD, Turner WD, Gardner BD, McClurg TP, Livingstone DJ, Carter JE Geternbach WJN (1975) National marine pollution monitoring programme, East Coast section. Progress Report. No. 2: 118.

Cox, ME, Micaela P (2005). Trace metal distribution within Marine and Estuarine sediments of Western Moreton, Bay Queesland, Australia: Relation to land use and settling. Geogr. Res. 43(2):173-193.

CSIR (1993). Physical impacts of the disposal of dredger spoil at Richards Bay. CSIR Report EMAS-C 93028.

Cyrus DP, Wepener V (1998). The Impact of Dredger Spoil on Selected Components of the Mhlathuze Estuary. Cruz Investigational Report No 52.

DePalmer SGS, Anorld WR, McGeer JC, Dixon DG, Smith DS (2011). Effects of dissolved organic matter and reduced sulphur on copper bioavailability in coastal marine environments Ecotox. Environ. Saf. 74(3): 230-237.

Fatoki OS, Mathabatha S (2001). An assessment of heavy metal pollution in the East London and Port Elizabeth Harbours. Water SA 27:233-240.

Galkus A, Joksas K, Stakeniene R, Lagunaviciene L (2012). Heavy metal contamination of habor bottom sediments. Pol. J. Environ. Stud. 6:1583-1594.

Hemens J, Connell A D, McClurg TP, Simpson DE, Warwick RJ, Muller MS (1976). Richards Bay Sanctuary Area: Report on field surveys in October 1975 and February 1976. Progress Report No 31 Thirty Second Steering Committee Meeting CSIR/NIWR.

Hemens J, Connell AD (1975). Richards Bay: Southern Bay Conservation Area. Progress Report No 29. Thirty Second Steering Committee Meeting CSIR/NIWR.

Hemens J, Simpson DE McClurg TP (1970). Detailed survey of Richards Bay. Progress Report, CSIRINIWR. 20th Steering Committee Meeting: Marine disposal of effluents, 20: 1-28.

Herut B, Sandler A (2006). Normalization method for pollutants in marine sediments: review and recommendations for the Mediterranean IORL Report H18/2006 Submitted to UNEP/MAP 23 pp.

Khan R, Israili SH, Ahman H, Mohan A (2005). Heavy metal pollution assessment in surface water bodies and its suitability for irrigation around the Neyevli Lignitemines and Associated industrial Complex Tamin Nadu India. Mine Water Environ. 24: 155-161.

Krupadam RJ, Smita P, Wate SR (2006). Geochemical fractionation of heavy metals in sediments of the Tapi estuary. Geochem. J. 40: 513-522.

Kwazulu Natal Business(2013). The guide to business and investment in Kwazulu natal.Mining and metals. Downloaded 13 August 2013. http://www.kwazulunatalbusiness.co.za/pls/cms/ti_secout.secout_prov?p_sid=19&p_site_id=130

Lim WY, Aries AZ, Zakaria MP (2012) Spatial variability of metals in surface water and sediment in the Langat river and geochemical factors that influence their water-sediment interactions. Sci. World J. 2012:652150.

Mackay CF, Cyrus DP (1999). A review of the macrobenthic fauna of the Mhlathuze Estuary: Setting the ecological Reserve. S. Afr. J. Aquat. Sci. 24: 111-129.

Magesh NS, Chandrasekar N, Roy VD (2011). Spatial analysis of trace element contamination in sediments from Tamiraparani estuary, South East coast of India. Est. Coast. Shelf Sci. 92: 618-628.

Mortimer RJG, Rae JE (2000). Metal speciation (Cu, Zn, Pb and Cd) and organic matter in oxic and suboxic salt marsh sediments, Servern Estuary. Mar. Pollut. Bull. 40: 377-386.

Newman, BK Walling, RJ (2007). Definition of baseline metal concentrations for assessing metal enrichment of sediment form south-eastern Cape coastline of South Africa. Water SA 33. 675-691.

Owen RB, Sandhu, N (2000). Heavy metal accumulation and anthropogenic impacts on Tolo Habour, Hong Kong. Mar. Pollut. Bull. 40: 174-180.

Sarkar SK, Franciskovic S, Bhattachrrya A, Saha M, Bilinski H (2004). Levels of elements in the surficial estuarine sediments of the Hugli River, northeast India and their environmental implications. Environ. Int. 30: 1089-1098.

Shirneshan G, Bachtiari AR, Kazemi A, Mohamadi M, Kheirabadi N (2012). Oyster *Saccostrea cucullata* as a biomonitor for Hg contamination and the risk for human on the coast of Qeshm Island, Persian Gulf, Iran. Bull. Contam. Toxicol. 88:962-966.

Shulkin VM, Presley BJ, Kavun VI (2003). Metal concentrations in mussel *Crenomytilis grayanus* and oyster *Crassostra gigas* in relation to contamination of ambient sediments. Environ. Int. 29(4): 493-502.

Silva CAR, Smith BD, Rainbow PS (2006). Comparative biomonitors of coastal trace metal contamination in tropical South America (N. Brazil). Mar. Environ. Res. 61: 439-455.

Soumady D, Asokan S (2011). A study on trace metal accumulation in selected organs of Perna viridis in Nagapattinam coastal waters, Tamil Nadu, India. Int. Pharm. Sci. 1(2):42-47.

Standard Methods (1998). Standard Methods for the Examination of Water Waste Water. (17th ed) American Public Health Association Port City Press, Maryland U.S.A.

Tanner PA, Shing-leong L, Ming pan S (2000). Contamination of heavy metals in marine sediment cores from Victoria Harbour, Hong Kong. Mar. Pollut. Bull. 40:769-779.

van der Hurk P, Eertman RHM, Stronkhorst J (1997). Toxicity of harbour canal sediment before dredging and after offshore disposal. Mar. Pollut. Bull. 34:244-249.

Watling HR, Watling RJ (1982). Metal concentrations in oyster from South African coast. Bull. Contam. Toxicol. 28:460-466.

Watling R,Talbot MMJ-F, Branch EB, Marsland SA (1982e). Metal Survey in South African Estuaries VI. Sundays River. Water SA 8:92-95.

Watling R, Talbot MMJ-F, Branch EB, Marsland SA (1985). Metal Survey in South African Estuaries IX, Buffalo River. Water SA 11:61-66.

Watling RJ, Watling HR (1982a). Metal Surveys in South African Estuaries II. Knysna River. Water SA 8:36-44.

Watling RJ, Watling HR (1982b). Metal Surveys in South African Estuaries Keurbooms and Bietou Rivers (Plattenberg Bay). Water SA 8:14-19.

Watling RJ, Watling HR (1982c). Metal Surveys in South African Estuaries V Kroome and Gamtoos (St Francis Bay). Water SA 8:87-91.

Watling RJ, Watling HR (1982d). Metal Surveys in South African Estuaries I. Swartkop River. Water SA 8:26-35.

Wepener V, Vermeulen L (1999). Comments on the water quality of the Mhlathuze Estuary un relation to determinining the ecological integrity class. S. Afr. J. Aquat. Sci. 24:86-98.

Wepener V, Cyrus DP (1997). The effects of Inter-basin Transfer on the Hydrochemistry, Benthic Invertebrates and Ichthyofauna of the Mhlathuze Estuary and Lake Nsezi: Second Interim Report. CRUZ Investigational Report 43.

Wepener V, Vermeulen LA (2005). A note on the concentrations and bioavailability of the selected metals in sediments of Richards Bay Harbour, South Africa. Water SA 31(4):589-596.

Williams PR, Attrill MJ, Nimmo M (1998). Heavy metal concentrations and bioaccumulation within Fal Estuary A Reappraisal. Mar. Pollut. Bull. 36:463-467.

Zwolsman JJG, van Eck GTM (1999). Geochemistry of major elements and trace metals in suspended matter of the Scheldt Estuary, southwest Netherlands. Mar. Chem. 66:91-111.

18

Variation of small scale wetland fishery in relation to land use along Mpologoma riverine marsh in Eastern Uganda

Ssanyu Grace Asiyo[1] , Julius Kipkemboi [2], Jude M. Mathooko [3] and John Balirwa[4]

[1]Kyambogo University, P. O Box 1 Kyambogo, Uganda.
[2]Egerton University, P. O Box 536-20115, Nakuru, Kenya.
[3]Global Research Akademik and Mentoring Services (GRAMSs) P.O Box 14824-20100, Nakuru, Kenya.
[4]National Fisheries Resource and Research Institute, P. O Box 343 Jinja, Uganda.

In densely populated areas such as those in eastern Uganda where livelihoods demands create immense pressure on environmental resources, small scale wetland fisheries may be disturbed by agriculture practices. The study carried out between 2011 and 2012, investigated the variation of fish catch at the different wetland sites in relation to land use in Mpologoma riverine marsh. Four sites were identified to represent different land uses; intact wetland, minimally disturbed, highly disturbed with small scale farmers and one with a large scale irrigation scheme. Data was collected on water quality, wetland fish species catch and catch per unit effort from the different sites. Conductivity and dissolved oxygen levels significantly differed between sites and explained 72.03% of the variance among sites. Seven fish taxa dominated the wetland fishery. Large sized fish species catch, *Clarias gariepinus* and *Protopterus aethiopicus* (range of 0.45 to 38 kg/day and 0.25 to 20 kg/day respectively), was higher at the less disturbed sites than at highly disturbed sites which accounted for over 91.5% of total wetland catch. *Tilapia zillii* and *Oreochomis leucostictus* catch were also higher at the less disturbed sites while the small fish species (*Haplochomis* sp, *Clarias liocephalus and C. alluaudi*) did not vary with site. Conductivity and dissolved oxygen significantly correlated with the two large fish species' catch but did not correlate with small fish species catch. Agricultural activities in the wetland negatively affected the life history strategies of large fish species, leading to low catch rates at the highly disturbed site. Therefore there is need to control land use changes to secure high productivity of small scale fisheries in the riverine wetland.

Key words: Papyrus wetland, fish, catch, water quality, disturbance.

INTRODUCTION

Small-scale fisheries play a significant role in human and socio-economic development and they are an entry point for poverty reduction though their role in generating revenues and creating employment, and their contribution to food security (Heck et al., 2007; Bene et al., 2010). These are wetland fisheries which often provide a 'safety valve' for people who cannot access other sources of livelihood (Bene, 2004; 2004). The small scale wetland fisheries provide nutritional security in remote areas that lack adequate supplies of animal protein and sustain the livelihood of landless fishers who can no longer survive by fishing in depleted freshwater bodies (Vass et al., 2009). Although wetlands provide habitat for 40% of all fish species (Arthington et al., 2004), 20% of their biota are amongst the most theatened components of global biodiversity (Smith et al., 2005). This is largely due to human induced environmental degradation. The demand for increased food production to cater for the rising human population and large numbers of undernourished or starving people, especially in the developing countries (Okechi, 2004), have led to widespread conversion of wetlands into farmland. Climate variability which has emerging signals within the River Nile Basin (Di Baldassarre et al., 2011), is likely to have a profound impact on land management practices leading to more changes in land use decisions (Verburg et al., 2011). Consequently, land requirements, reclamation potential and general environmental management are affected (Carvalho et al., 2004).

Many permanently flooded wetlands are open to fishing by anyone (open access) or subject to exclusive communal or individual use rights (Garaway et al., 2006; Martin et al., 2011). Often, wetlands under exclusive access arrangements are exploited less intensely and maintain higher standing stocks of fish than open access wetlands (Lorenzen et al., 1998). The open access wetlands may also be subject to very intensive exploitation of both water and fisheries resources (Garaway et al., 2006). In eastern Uganda, over 10% of Mpologoma River wetland is under cultivation of rice, maize and vegetable to cater for food demand of the increasing population (NEMA, 2004). Simultaneously, there appears to be reduction in the large preferred fish species, *Clarias gariepinus* and *Protopterus aethiopicus* in the Lake Kyoga basin (Muhoozi, 2003). *C. gariepinus* catch reduced from more than 1600 metric tonnes in the 1980s to 1.35 metric tonnes in 2000 (Muhoozi, 2003).

Understanding the effects of land use changes on the fisheries is essential for the management of exploited wetlands (Rientjes et al., 2011). Although there are a few existing models based on observed pattern in temperate areas, scarcity of data for most tropical riverine wetlands limits prediction of the effects of human-induced perturbbations on the fish assemblages and catch (Ibanez et al., 2007). Therefore Mpologoma wetland fisheries that support local communities with no access to open lake fisheries and yet it is under intensive exploitation need evaluation. The aim of the study was the assessment of the small scale fishery dynamics in relation to land use in Mpologoma river wetland. We used comparison of the wetland fish species catch at the wetland sites with different land use patterns. It was hypothesized that the different land use at the wetland sites would affect the fish habitat though water quality variations, leading variation in the wetland fish species catch. The spatial-temporal variation of the wetland fish species catch in Mpologoma riverine wetland was expected.

MATERIALS AND METHODS

Study area and sampling sites

Mpologoma river wetland in Uganda (latitude 1°12' N and longitude 34°40' E) extending up to 102 km, discharges 610 million m^3 of water annually into Lake Kyoga complex (Ramsar, 2008). The climate is tropical with rainfall ranging between 1470 - 2300 mm in the longer wet season. The maximum temperature range is 27 – 32°C and the minimum is 16 – 18°C. This permanent wetland is dominated by papyrus (*Cyperus papyrus*) and hippo grass (*Vossia cuspidata*). The fish species of economic importance in the Lake Kyoga complex include *Clarias gariepinus, Protopterus aethiopicus Oreochomis leucostictus, O. niloticus, Bagrus docmak Rastrineobola argentea* and *Tilapia spp* (Vanden-Bossche and Bernacsek, 1990; NaFIRRI, 2007).

Using the Digital Elevation Model of the AVSwat, a map of Mpologoma river wetland with the four differently disturbed study sites was delineated (Figure 1).The main depression which is Mpologoma riverine wetland system consists of a network of small vegetated valley bottoms, fed by 38 sub-catchments. Four study sites with different levels of disturbance were selected along the wetland. Land use cover at each site was derived from Landsat ETM images of July 2011, verified by field survey data. Seven land use classes were identified in 200 hectare polygons area about each site (Table 1). Budumba and Mazuba sites were minimally disturbed with small farms at the edge of the wetland. Kapyani site was highly disturbed with small scale rice and maize farms of acreage of 2 – 6 acres inside the wetland, using low implement agriculture and rain-fed lowland systems (MAAIF, 2009). Nsango site was highly disturbed with a large scale rice scheme occupying over 3,000 hectares of the wetland. Artificial fertilisers, rainfed lowland and irrigation systems were used at the rice scheme.

Wetland fish assemblage

Fish samples from several fishing techniques were used to determine the wetland fish species composition and fishery production. Experimental fish sampling was done with gill nets hanged in the lagoons, while traps were placed inside the wetland. Gill nets of 2, 2.5 and 3 cm mesh sizes were set in parallel in the lagoon. Ten local basket traps were randomly deployed to catch fish inside the flooded vegetation zone of the wetland. All gear were set at around 7:00 h and retrieved before 14:00 h. This was done to reduce loss of fishing gear to the local fishermen during overnight fishing. On landing, fish samples were counted, total length and weight measured, and records of the gear and effort used at every

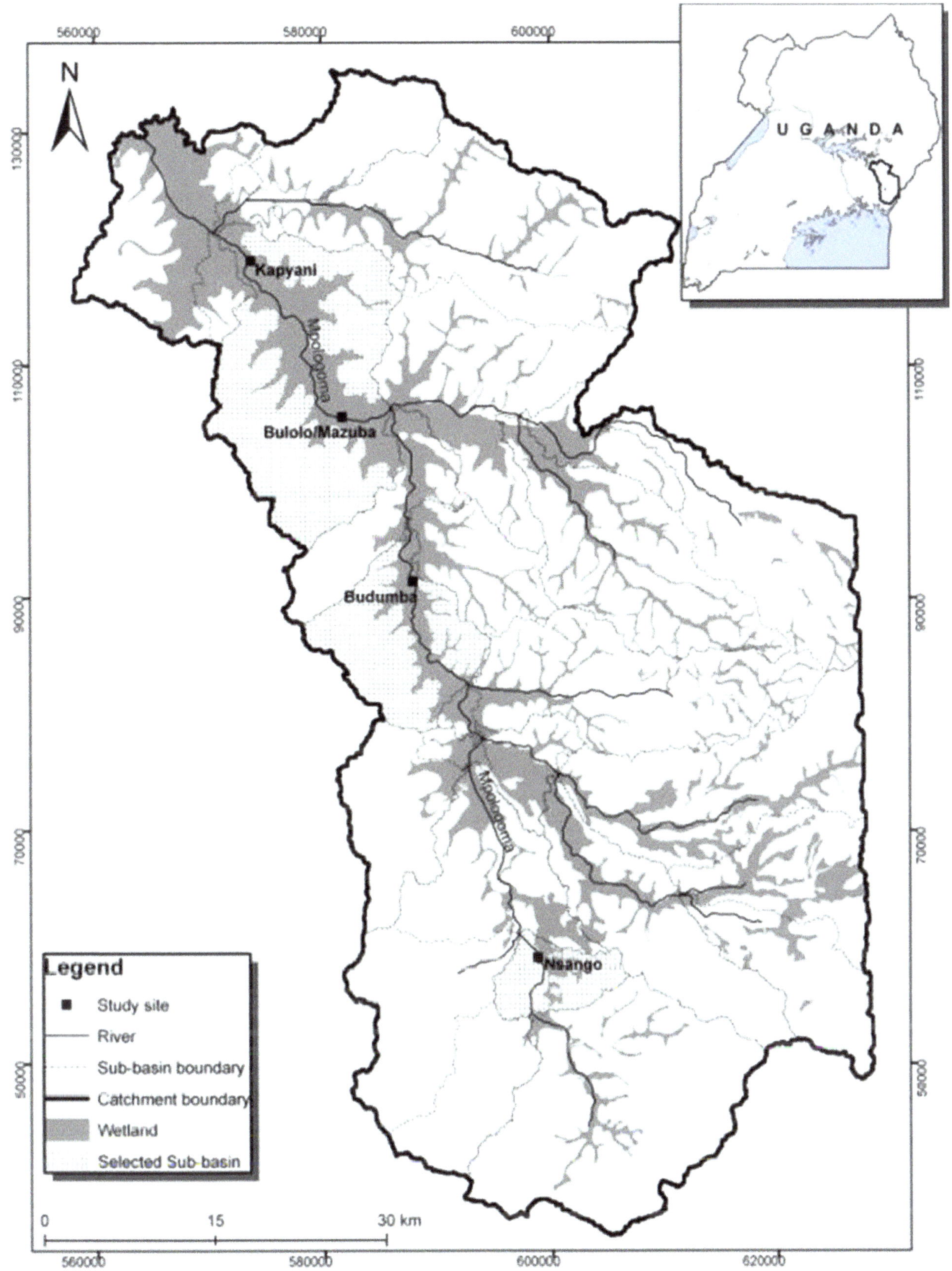

Figure 1. Location of the study sites along Mpologoma river wetland.

Table 1. Land use percentage cover of 200 ha area around study sites in Mpologoma River wetland, Uganda, 2011.

Land use	Mazuba	Budumba	Kapyani	Nsango
	Intact wetland	Less disturbed	Highly disturbed (with small scale farms)	Highly disturbed (Large scale rice scheme)
Man made type				
Built up area	0	0.48	1.52	18.13
Subsistence	0	8.67	55.25	16.18
Large scale farmland	0	0	0.00	16.16
Subtotal cover	0	9.15	56.77	50.47
Natural type				
Permanent wetland	100	90.63	31.21	17.30
Temporary wetland	0	0	10.15	5.58
Woodland	0	0.23	1.38	21.09
Grassland	0	0	0.48	4.31
Subtotal cover	100	90.86	43.22	48.28

site. Some fish were preserved in 40% ethanol and retained for identification using keys (Greenwood, 1974; Witte and van Densen, 1994). To cater for the seasonal changes in fishery, each site was sampled once a month for 12 months between September 2011 and August 2012.

More data on wetland fish species and catch was collected from fishermen landing catches at each site for every sampling day. The local fishermen distributed their fishing effort thoughout the study site each trip to sample each portion of the study site each day. At each site, fishermen had their own culture of who fishes where, and regulated the number of fishermen per landing site. Therefore the decision of about where and how to fish was left to the fishermen. The local fishermen dominantly used gill nets (2.5 cm and 3.0 cm mesh sizes) of about 45 m long by 1.5 m deep, fishing lines with varying number of hooks and local basket traps (0.4 by 0.4 by 0.3 m, on average) as fishing gears. Fishing lines consisted of a weight (approximately 2 kg) with varying number of baited hooks mainly set to catch *Protopterus* sp. and *C. gariepinus* within the wetland. Bait materials included pieces of small fish species such as small *Clarias* species, *Tilapia* species, frogs and large insects for hooks, while dead earthworms were for basket traps. Most fishing gear were set at 18 hours and retrieved at 6:00 hours to prepare for the markets. Data on captured species brought to the landing sites were recorded including fish taxa, total length (TL, in centimeters), weight (Wt, in grams), type of gear and effort used. Total length and weight were measured as quickly as possible after the fishermen returned with their catch. For small species such as small *Clarias* sp., *Haplochomis* spp., *Tilapia* spp. and *Barbus* sp., random sub samples were taken to measure standard lengths but their total number in the catch recorded. For large species such as *C. gariepinus* and *Protopterus* sp. groups of similar sizes were counted and sub sampled for measuring their total length and weight to get as much data as possible in the limited time allowed by the fishermen.

Physical habitat

To characterise water quality of the wetland sites with different land uses, thee water samples were collected from different points (within the wetland, middle of lagoon and edge of the lagoon) at each site. Within the wetland, thee water samples were randomly collected close to where traps were placed at each site. Conductivity and pH were determined *in-situ* using a Hanna Instruments HI 9813-6 N Waterproof pH/Ec/TDS/°C meter. Dissolved oxygen and temperature were determined *in-situ* by Oakton DO 110 meter. Chemical analysis was performed on 0.45 µm membrane filtrate for alkalinity, nitrate-nitrogen and orthophosphate and unfiltered samples for total phosphorus within 48 h of collection according to standard procedures (APHA, 1995).

Data analysis

Fish data were summarised into the species composition catch and catch per unit effort (CPUE) for spatial and temporal variation along the wetland. The fishermen usually fished only once a day and therefore species catch was considered as total species catch in kilogram per day. Since the fishery was a multispecies type and uses non-selective fishing gear, for gill nets and traps fishing gear, CPUE was calculated by dividing the catch in grammes of the most dominant fish species caught by the individual gear by the number of hours. Most gillnets were set for less than 5 h and most fisherman had only one net, therefore the units of CPUE of gill nets derived from the mean sizes of gillnets and fishing hours used along the wetland. Most fishermen used locally made basket traps of similar sizes and shape. Therefore the units of CPUE of traps was derived from number of traps and hours used per fisherman. CPUE of lines with hooks was calculated by dividing the catch in grammes of most dominant fish species by the hours spent in fishing considering the mean number of hooks per fisherman. Most fishermen using line hooks in the evening and check on the fish caught early morning. Therefore 12 hwere used in the calculation of the hooks CPUE. Due to differences in mean number of hooks per sites, the hooks' CPUE was standardized to 100 hooks per hour. Species richness was tested using Shannon Wiener's index (H') of diversity (Thiebaut, 2006) in the equation;

$$H' = - \sum p_i \ln(p_i),$$

Where, p_i is the relative abundance of individual fish species.

Statistical analysis on all data was performed using SPSS version16 (IBM ©). All the wetland fish species catch and CPUE

Table 2. Mean number of fishermen, fishing gear type per fisherman and boats at one major landing site of each of the differently disturbed sites along Mpologoma river wetland in 2012.

Parameter	Budumba	Mazuba	Kapyani	Nsango
Number of fishermen	22 ± 5	22 ± 4	30 ± 9	17 ± 5
Number of hooks per fisherman	85 ± 34	50 ± 35	64 ± 41	30 ± 17
Number of gill nets per fisherman	1	1	1	1
Number of traps per fisherman	8 ± 3	6 ± 3	5 ± 2	7 ± 3
Number of boats per site	16 ± 4	17 ± 2	13 ± 2	7 ± 2
Mean Catch per day ($Kgday^{-1}$)	29.03 ± 9.66^a	21.28 ± 10.46^a	24.10 ± 11.70^a	2.07 ± 1.62^b

were treated as separate dependent variables. Data was log transformed since it was not normally distributed (Kolmogorov-Simonov test (p = 0.00) and Skewness of 4.523). One-way ANOVA followed by Tukey's HSD post hoc test was used to test differences in daily catch among fishing gears, species CPUE, catch and fish length between sites. Using R-statistic software (*R version* 3.0.2; R Development Core Team 2008), non-metric multidimensional scaling (NMDS) was more appropriate in measuring community dissimilarities between study sites along the wetland. The relationship between study sites and water quality parameters was evaluated to ordinate the similarity between sites. Water quality parameters were further fitted to the ordination as vectors to identify the parameters that significantly differentiate the study sites. Stepwise multiple linear regressions with forward selection of variables were used to analyze the relationship between fish catch and water quality parameters.

RESULTS

Fish species composition and catch

Data on a total 5137 fish specimens were collected from all study sites over the sampling period. Six fish species dominated the wetland fish community. *C. gariepinus, C. liocephalus, C. alluaudi, P. aethiopicus, O. leucostictus, T. zillii* and several unidentified *Haplochomis* species were observed at all site along the wetland. *Synodontis afrofischeri* was observed only at highly disturbed Kapyani site close to Lake Lemwa. The number of species ranged from seven at the highly disturbed sites to nine at the less disturbed sites and the mean Shannon Wiener index of diversity was 5.4 ± 1.2 per site. During the wet season months, a higher number of species was recorded than during the dry season months. For instance, the species richness index of 4.94 and 4.81 were realized in January and February, while 0.77 and 0.69 were realized in September and October respectively at less disturbed Nsango site.

Wetland fishery production was estimated to be 57 kg per ha per year. The number of fishermen ranged from 12 at the highly disturbed Nsango site to 27 at the less disturbed Budumba site (Table 2). Thee fishing gear (gill nets, line hooks and basket traps) were dominantly used at all sites. Many fishermen used line hooks with a range of 13 hooks at the highly disturbed Nsango site to over 119 hooks at the less disturbed Budumba site. At the less disturbed Budumba and Mazuba sites, the fishermen had a high number of hooks to catch as much fish as possible within in the intact wetland. Traps and gill nets were used minimally at all sites. During the rainy season, gillnets with 3.0 cm mesh size harvested more fish than other gears. Hooks were dominantly used during the dry season. More than 50% of the catch comes from the hooks fishery. Gill nets with 3.0 cm mesh size was the second most important, followed by the gill nets with 2.5 cm mesh size and traps (Figure 2). The number of hooks, gill nets of 2.5 cm mesh size and traps catch did not vary between season (p = 0.001). But among sites, the number of hooks were significantly higher at the less disturbed Budumba site than at the highly disturbed Nsango site (p = 0.023). The catch varied among the dominant fishing gears used. Significantly higher catch was realized by the use of line hooks than all other fishing gears (p = 0.001). The line hooks' mean catch of 11423 ± 6936 g/day was observed in the wet season and 14863 ± 9558 g/day observed during the dry season which high compared to 3294 ± 3108 and 4860 ± 4392 g/day the wet and dry season gill nets' catch respectively.

There were temporal and spatial differences in the catch of individual fish species. Relatively higher catch was observed in the dry season than in the wet season and more at the less disturbed sites than at the highly disturbed sites (Figure 3). *C. gariepinus* and *P. aethiopicus* dominated the fishery by catch weight with a range of 0.45 kg per day at the highly disturbed Nsango site to 38.67 kg per day at the less disturbed Budumba site and 0.245 to 20.18 kg per day, respectively (Table 2). Catch at Nsango was significantly lower than that of other sites (Tukey's HSD test; $p < 0.05$). Higher catch of both *C. gariepinus* and *P. aethiopicus* species was recorded during the dry months (January, February, July and August, Figure 3). Significantly higher catch of both species was recorded at the less disturbed sites (Budumba and Mazuba) than at the highly disturbed site (Nsango) at p = 0.05 (Table 3). The disturbed Kapyani site had relatively high *C. gariepinus* catch almost the whole sampling period. The catch of small *C. liocephalus* and *C. alluaudi* was higher at the highly disturbed site than at the less disturbed sites

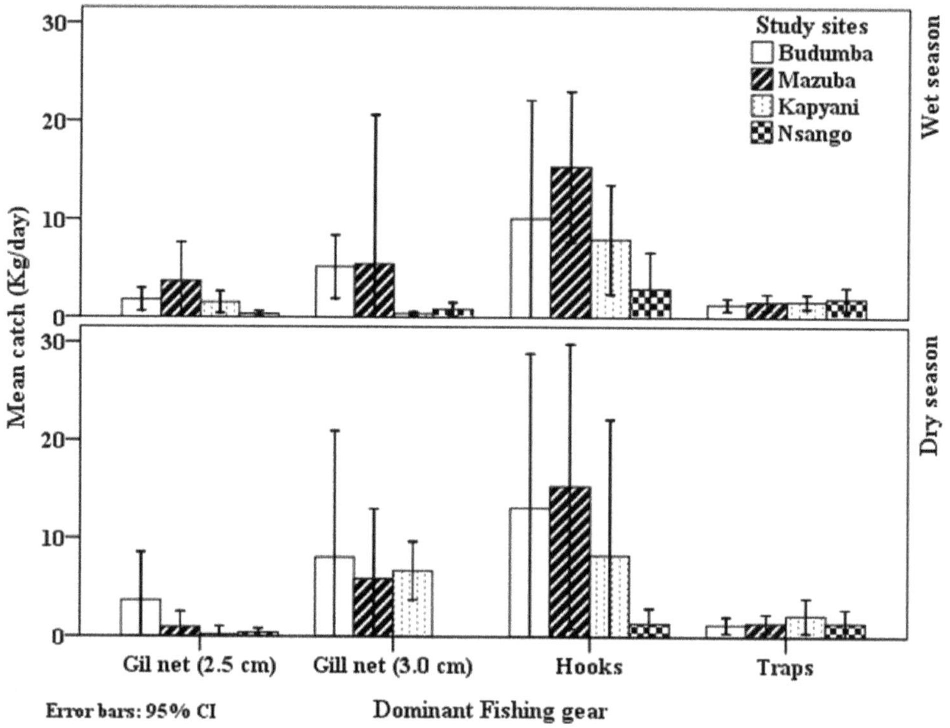

Figure 2. Seasonal mean catch of the major fishing gear at the four sites along Mpologoma wetland between September 2011 to August 2012.

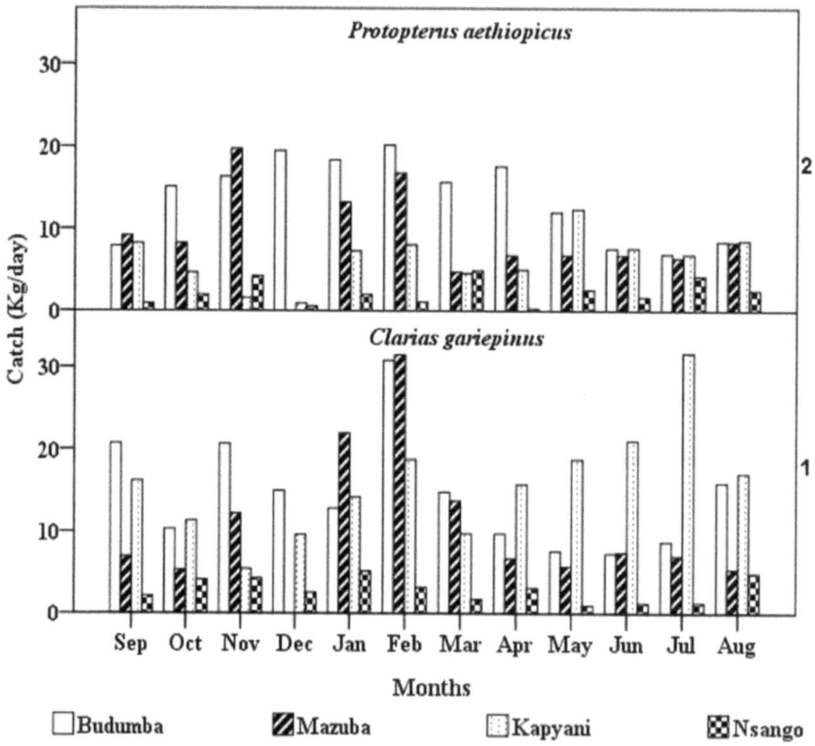

Figure 3. Mean catch of the big fish species in Mpologoma wetland fishery between September 2011 to August 2012.

Table 3. Mean catch (Kg/day, ± SD) of major fish species at differently disturbed sites along Mpologoma river wetland during the sampling period.

Fish species	N	Budumba	Mazuba	Kapyani	Nsango
Oreochomis leucostictus Trewavas 1933	565	1.56 ± 1.07[a]	1.13 ± 1.51[a]	0.61 ± 0.41[a]	0.29 ± 0.15
Tilapia zilli Garvais 1848	703	1.92 ± 1.28	0.86 ± 0.58[a]	0.93 ± 1.14[a]	0.32 ± 0.18
Haplochomis spp	1115	0.26 ± 0.25[a]	0.19 ± 0.09[a]	0.33 ± 0.42[a]	0.19 ± 0.13[a]
Clarias liocephalu Boulenger 1902	788	0.50 ± 0.32[a]	0.42 ± 0.21[a]	0.66 ± 0.70[a]	0.75 ± 0.52[a]
Clarias alluaudi Boulenger 1906	419	0.37 ± 0.24[a]	0.30 ± 0.25[a]	0.27 ± 0.19[a]	0.42 ± 0.36[a]
Clarias gariepinus Burchell 1815	1062	14.64 ± 6.77[a]	11.28 ± 8.39[a]	15.88 ± 6.56	3.05 ± 1.62
Protopterus aethiopicus Heckel 1851	598	13.38 ± 5.04[a]	9.72 ± 4.78[a]	6.47 ± 3.09[a]	2.25 ± 1.46

Values in the same row with the same superscript are not significantly different ($p < 0.05$).

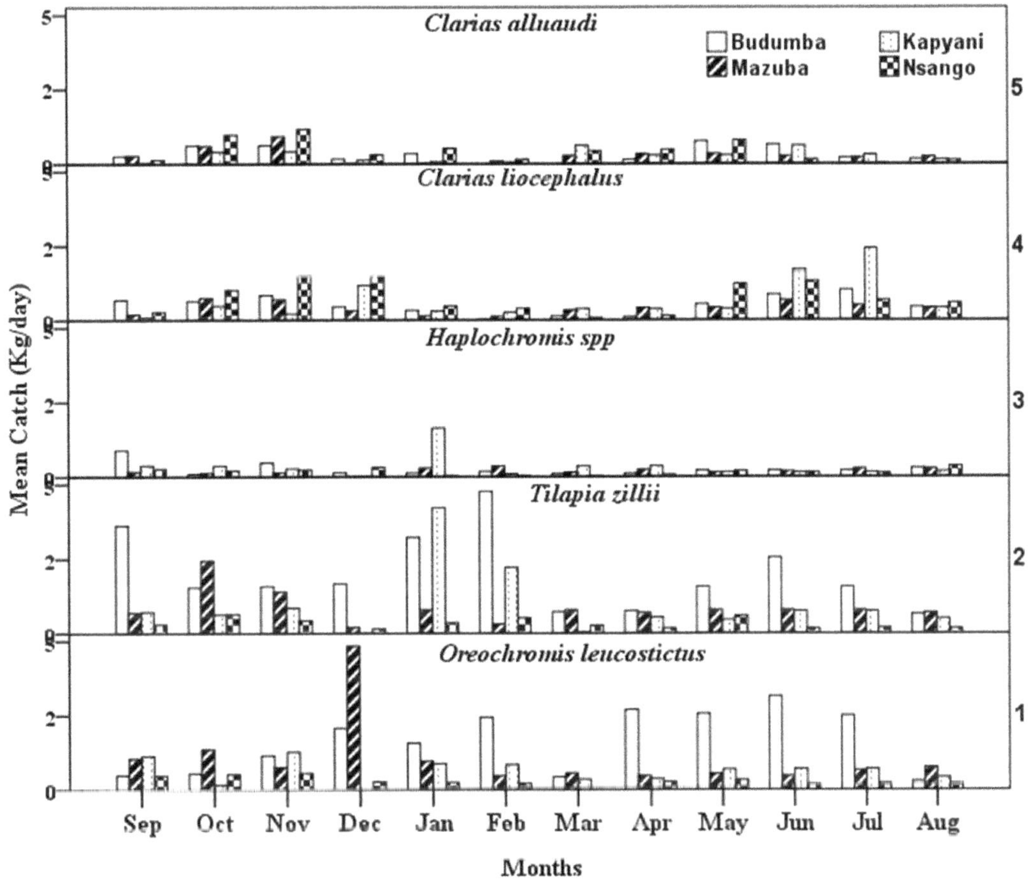

Figure 4. Fish catch of the small fish species at diferent sites along Mpologoma river wetland between September 2011 to August 2012.

(Figure 4).

The mean CPUE for the major commercial fish species varied with fishing gear, season and study site. The CPUE for *C. gariepinus* and *Protopterus* sp was higher with the use of hooks, at the less disturbed sites than that of other fishing gear, particularly the highly disturbed Nsango site (Table 4). At the less disturbed Budumba site, mean CPUE for *C. gariepinus* was higher (1.31 Kg100 hooks^{-1}h^{-1}) in the dry season as compared to 0.90 Kg 100hooks^{-1}h^{-1} recorded during the wet season. CPUE for the two large fish species at highly disturbed Nsango site was significantly different from that of the other sites (Tukey's

Table 4. Mean catch per unit effort (CPUE) for the major commercial fish species/groups at each differently disturbed site along Mpologoma wetland during both the dry and wet season between September 2011 to August 2012.

Gear type	Species	Units of CPUE	Dry season				Wet season			
			Budumba	Mazuba	Kapyani	Nsango	Budumba	Mazuba	Kapyani	Nsango
GN 2.5	*Haplochomis* spp	Kg 45 m^{-1} h^{-1}	0.03	0.00	0.00	0.10	0.13	0.04	0.01	0.12
	C. gariepinus	Kg 45 m^{-1} h^{-1}	0.20	0.00	0.53	0.08		0.92	0.20	
	Protopterus sp	Kg 45 m-1 h-1	0.40	0.00				0.29		
	Small *Clarias* spp	Kg 45 m^{-1} h^{-1}			0.45					
	Tilapia spp	Kg 45 m^{-1} h^{-1}	0.85	0.22	0.59	0.23	0.45	0.54	0.20	0.15
GN 3.0	*C. gariepinus*	Kg 45 m^{-1} h^{-1}	0.35	0.40	1.60		0.46	0.49	0.20	0.36
	Haplochomis spp	Kg 45 m^{-1} h^{-1}					0.07			
	Protopterus sp	Kg 45 m^{-1} h^{-1}	0.52	0.57	0.90		0.38	0.20	0.06	
	Tilapia spp	Kg 45 m^{-1} h^{-1}	0.13	0.14			0.46	0.17		0.47
Hooks	*C. gariepinus*	Kg 100hooks^{-1} h^{-1}	1.31	1.10	4.50	1.20	0.90	1.85	0.76	1.20
	Protopterus sp	Kg 100hooks^{-1} h^{-1}	0.86	1.05	1.28	0.54	1.32	1.88	0.81	1.09
	Small *Clarias* spp	Kg 100hooks^{-1} h^{-1}	0.16	0.20	0.33	0.27	0.15	0.16	0.11	0.26
Traps	*Barbus* sp	Kg trap^{-1} h^{-1}	0.04	0.02		0.09		0.02		0.03
	C. gariepinus	Kg trap^{-1} h^{-1}	0.04	0.06	0.11	0.06	0.09	-	0.11	0.08
	Haplochomis spp	Kg trap^{-1} h^{-1}	0.01	0.04	0.09	0.23	0.03	0.14	0.05	0.07
	Protopterus sp	Kg trap^{-1} h^{-1}	0.05	0.12	0.12	0.05	0.06	0.12	0.12	0.07
	Small *Clarias* spp	Kg trap^{-1} h^{-1}	0.16	0.11	0.18	0.19	0.20	0.15	0.15	0.24

HSD test; $p < 0.05$). However, the highly disturbed Kapyani site CPUE (4.50 Kg 100hooks^{-1}h^{-1}) was higher than the other highly disturbed Nsango site (1.20 100hooks^{-1}h^{-1}) in the dry season. The CPUE for *C. gariepinus* and *Protopterus* sp. showed an increasing trend during the dry months (Figure 5). While the CPUE of small fish species (*C. alluaudi*, *C. liocephalus*, *Haplochomis* spp and *Barbus* sp) did not vary significantly between season and sites ($p = 0.05$).

Fish size

There were variations in *C. gariepinus* and *P. aethiopicus* total length in the wetland with a range of 12 to 135 cm and 16 to 151 cm respectively. The mean total length of *C. gariepinus* at Budumba and Mazuba was 47.27 ± 20.47 cm and 45.32 ± 16.76 cm respectively. While at the highly disturbed sites (Kapyani and Nsango) the mean total length was 41.19 ± 21.85 cm and 27.33 ± 10.98 cm respectively. Larger fish individuals of both species were caught during the wet season (April, May and October; Figure 6). The length of these two species was significantly lower at highly disturbed Nsango site than that of the less disturbed sites ($p < 0.01$; Figure 7). The less disturbed sites had larger fish even among the *Oreochomis* and *Tilapia* species than highly disturbed sites. However, the small fish species behaved differently. *C. liocephalus* total length ranged from 6.0 to 26.4 cm at Budumba and from 9.5 to 28.5 cm at Nsango.

Environmental data

The water quality parameters varied in space and time along the wetland. Conductivity ranged from 119.9 μS/m at less disturbed Budumba site to 406.4 μS/m at the highly disturbed Nsango site. Conductivity at the less disturbed sites was significantly lower than that of disturbed sites and higher

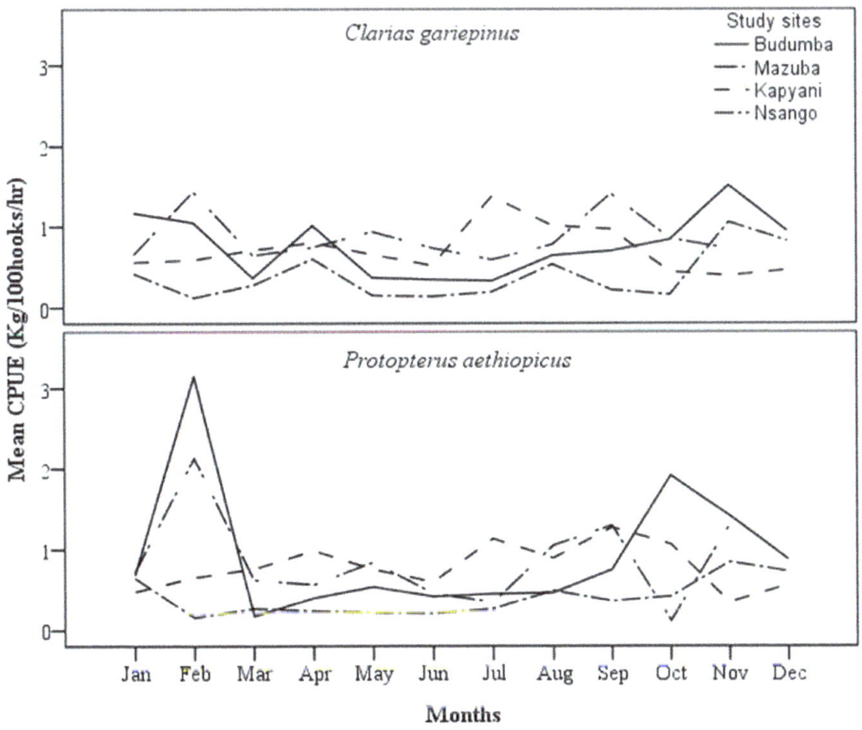

Figure 5. Monthly mean catch per unit effort (CPUE) of the two large fish species at the four differently disturbed sites along Mpologoma wetland between September 2011 to August 2012

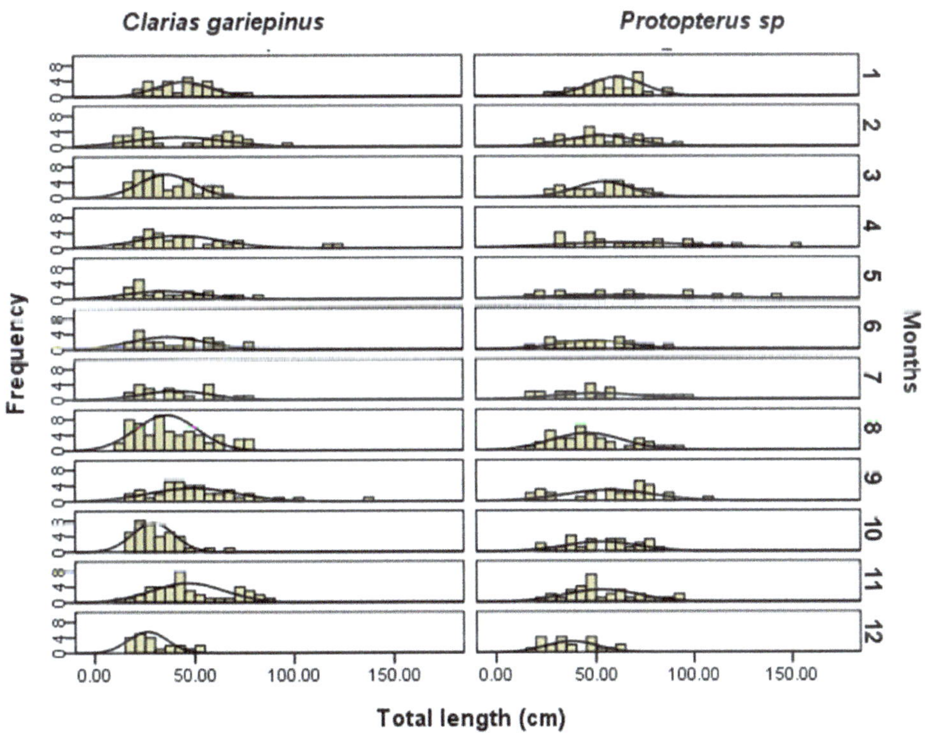

Figure 6. Length frequency distribution of two large fish species of Mpologoma river wetland fishery between September 2011 to August 2012.

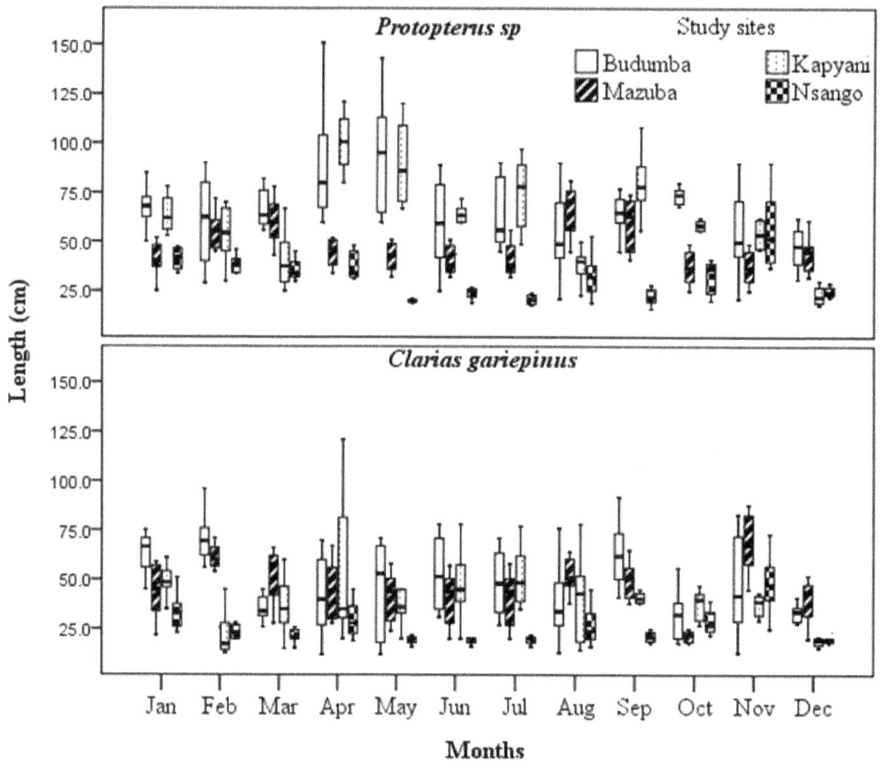

Figure 7. Mean total length of the large fish species caught at the differently disturbed sites along Mpologoma river wetland between September 2011 and August 2012.

Table 5. Mean (± STD) of water quality parameters of fish habitats (inside the wetland) at the different study sites along Mpologoma River wetland (September 2011 – August 2012).

Parameter	Budumba	Mazuba	Kapyani	Nsango
DO (mgl^{-1})	1.35 ± 1.03^{a}	1.98 ± 1.26^{a}	1.58 ± 0.67^{a}	0.96 ± 0.76
Temp (°C)	23.75 ± 1.72	24.53 ± 0.81	25.35 ± 1.31	24.18 ± 1.49
Conductivity (µS/m)	152 ± 32.1^{a}	161 ± 35.2^{a}	218 ± 53.1^{b}	351 ± 55.4^{b}
OPO_4^{3-} (mgl^{-1})	0.28 ± 0.26^{a}	0.27 ± 0.23^{a}	0.13 ± 0.12^{b}	0.14 ± 0.34^{b}
TP (mgl^{-1})	0.67 ± 0.44^{a}	0.62 ± 0.39^{a}	0.29 ± 0.27^{b}	0.23 ± 0.19^{b}
Water depth (m) during the dry season	0.52 ± 0.27	0.45 ± 0.14	0.30 ± 0.15	0.35 ± 0.14

Values in the same row with the same superscript are not significantly different.

during the dry months. Dissolved oxygen (DO) within the papyrus ranged between 0.20 mgl^{-1} at highly disturbed site to 3.24 mgl^{-1} at less disturbed site (Table 5). DO was high during the wet season months (April, May, October and November) with a range of 2.2 to 4.4 mgl^{-1} in the mid river. DO was low in the dry season months (January, February, June and July) with a range of 0.6 to 1.8 mgl^{-1} in the mid river section. Nitrate levels were very low to undetectable levels at all sites during the wet season while in the dry season, it ranged from 0.01 to 0.12 mgl^{-1} at the less disturbed sites and 0.03 to 0.18 mgl^{-1} at the disturbed sites. Orthophosphate and total phosphorus ranged from 0.041 to 1.15 mgl^{-1} and 0.093 to 1.80 mgl^{-1} respectively in the wetland. From the NMDS analysis, the resulting ordination was two dimensional with final stress of 17.7 at $p < 0.05$. The less disturbed Budumba and Mazuba sites were identical as the distance between their circles was close to zero, while the highly disturbed Nsango site associated with a large scale rice scheme was different from the less disturbed sites as the distance between their circles was larger than zero (Figure 8). The vector analysis showed that conductivity, orthophosphate

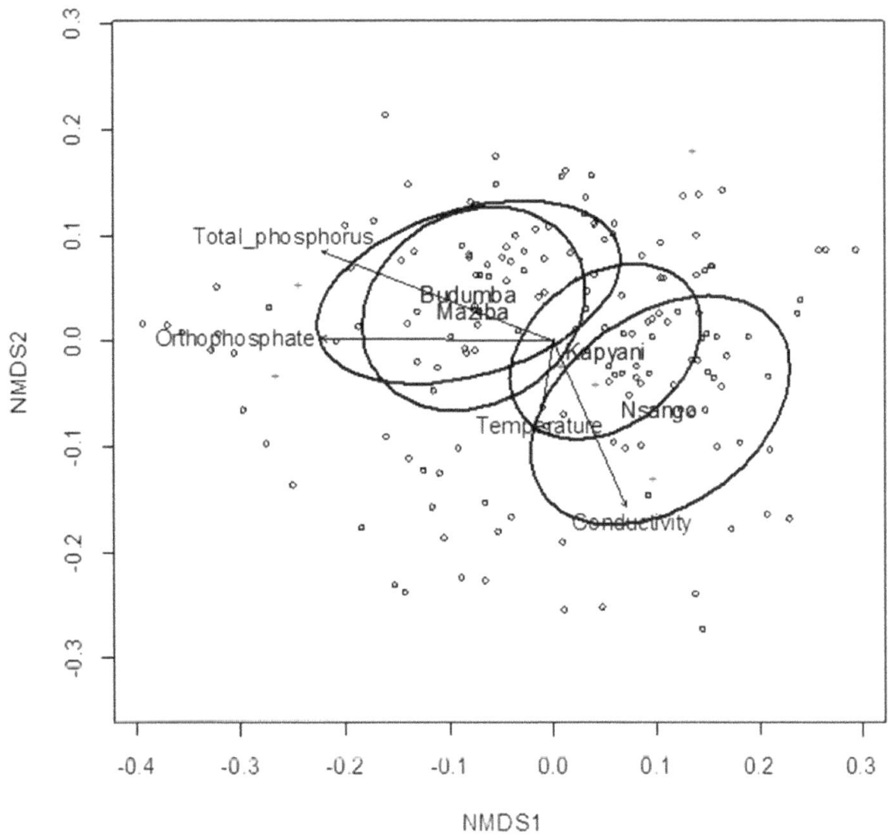

Figure 8. Non-metric multidimensional scaling plot of study sites along Mpologoma river wetland. Visually fit ellipses were drawn around sites that had similar levels of water quality parameters that were significantly different between sites.

Table 6. Environmental vector fitting of the NMDS analysis of the water quality parameters of the different study sites.

Water quality parameters	NMDS1	NMDS2	r^2	Pr(>r)
Dissolved oxygen	-0.95645	-0.29191	0.0150	0.316
Temperature	-0.15069	-0.98858	0.0651	0.07 *
Conductivity	0.41252	-0.91095	0.4269	0.001 ***
Orthophosphate	-0.99996	0.00869	0.7077	0.001 ***
Total phosphorus	-0.93377	0.35788	0.8128	0.001 ***

Significant codes: '***, 0', 0.001 '**'; '*' 0.01. P values were based on 999 permutations.

and total phosphorus were significantly important in differentiating the study sites all at r^2 of 0.43, 0.71 and 0.81 respectively, all at $P < 0.001$ (Table 6).

There was a significant relationship between fish catch and water quality parameters. A strong spearman rank correlation (rho' = 0.501 and 0.348; $p < 0.05$) between *C. gariepinus* catch with conductivity and orthophosphate respectively was realized. *P. aethiopicus* catch was significantly correlating with conductivity, orthophosphate and total phosphorus (rho' = 0.510, 0.465 and 0.441; $p < 0.01$ respectively). With stepwise multivariate linear regression, *P. aethiopicus* catch was significantly related to conductivity and orthophosphate ($R^2 = 0.544$; $p < 0.01$, catch = -13.214 – 0.023 conductivity + 16.69 orthophosphate). No significant relation was realized between water parameters and the small fish species catch and CPUE.

DISCUSSION

The wetland is a critical refuge for a subset of Lake Kyoga fish species given the high number of similar fish species recorded in the wetland compared to those reported in the open Lake Kyoga. *C. gariepinus, P. aethiopicus O. leucostictus* and *Tilapia* spp. which are important Lake Kyoga fish species (NaFIRRI, 2007), were also recorded in the wetland. High diversity, similarity and endemism stem largely from the fact that nearby fresh waters were once connected by heavy floods or drainage free of barriers that enabled dispersal of fish species (Olden et al., 2010). Barriers later maintained fish species in different parts of the riverine wetland. As level of wetland disturbance increase in certain areas of the wetland, the general similarity in species was still maintained in the wetland and this was due to the fact that some fish species are considered relatively tolerant to pollution. These included *Oreochomis* species (Raburu and Masese, 2010), *P. aethiopicus, C. gariepinus* and *C. alluaudi* (Timmerman and Chapman, 2004). Normally reduction in species richness and abundance is expected along a degradation gradient (Raburu and Masese, 2010) and indeed at highly disturbed Nsango site, agricultural activities degraded the fish habitats that even tolerant species declined. During the relatively poor water quality periods (high conductivity and low dissolved oxygen), particularly in the dry months, at Nsango site the lowest number of fish species was recorded compared to other sites..

The Mpologoma wetland production estimate was in the order of magnitude of African floodplain production. Lightly exploited floodplain river systems produce about 40 - 60 kg ha^{1} yr^{1} of fish (Welcomme, 1985), while highly exploited tropical floodplains produce estimates more at 110 - 160 kg ha^{1} yr^{1} of fish (Bayley, 1988). The high wetland production estimated was attributed to the increased exploitation rate to cater for the increasing need to diversify livelihoods along the wetland. The complex relationship between catch (fish abundance) and effort (function of fishermen's behaviour) is what controls the catch per unit of effort (Lopes, 2011). On average, the fishermen spent almost the same time span fishing both in wet and dry season, therefore the differences in CPUE resulted from the variation in fish abundance with time at the sites. Despite the few to no full time fishermen at all sites, the fishers were able to access distant less exploited areas with abundant fish resources particularly at the less disturbed sites leading to high catch and CPUE. The fishing gears used also contributed to the variation in catch and CPUE. The record of high number of hooks per fisherman at the less disturbed sites indicated the importance of this gear to achieving high catch and CPUE. The reliability of fishing gear CPUE as an index of fish density depends on fish activity, gear selectivity, avoidance and the morphology of the fishes (Olin et al., 2010). This explained the application various fishing gear that enabled exploitation of the various fish species at the different sites, during different seasons in the wetland.

Fish species abundance is influenced by physical and chemical composition of water (Randle and Chapman, 2004), habitat size and diversity (Budy et al., 2008), and water flow patterns into the wetland (Vorwerk et al., 2009). The differences in wetland fish species abundance and catch were governed primarily by water quality parameter variation among sites, given the strong correlation between catch and conductivity at the sites. This agreed with earlier studies on floodplain fisheries which highlighted dissolved oxygen, conductivity, pH and water depth as major determinant factors to fish abundance (Louca et al., 2009). Highly disturbed sites' conductivity and dissolved oxygen levels were similar to those of highly studied polluted Nakivubo wetland in Uganda (Kansiime et al., 2007). Wetland clearing for agriculture results in dramatic alteration of the river flood curve which, leads to decrease in both the amplitude and duration of flood regime.

The wetland clearing also increases in evapotranspiration which, lead to increased salts in the water. These have pronounced impact such as high conductivity, alterations of vegetation species and cover, and decreased connectivity in wetland lagoons (Louca et al., 2009). Sustained increase in conductivity compounded by high sedimentation levels led to negative implications on the fish ecology (Chapman et al., 2003). The large rice scheme irrigation activities involve application fertilizers near Nsango site seem to have affected the water quality which, explain the reduced abundance of large fish species. Lungfish (*Protopterus* sp.) decline reflects the interaction of overexploitation and large-scale conversion of wetlands to agricultural land for the past few decades (Goudswaard et al., 2002). Such species with preferences for lower conductivities and high dissolved oxygen disappear from the wetland once these conditions persist (Goudswaard et al., 2002; Louca et al., 2009). This explains the low *C. gariepinus* and *Protopterus* species abundance at highly disturbed Nsango where those harsh conditions were observed.

Higher fish species composition abundance was recorded at the downstream disturbed Kapyani site and this was attributed to a number of factors. The interaction between the river and lake hydrology which modifies the fish habitats with modified vegetation and deeper waters (Cooper et al., 2007) allowing more fish to coexist, both wetland and open water dwelling species, despite the level of disturbance. The site was permanently connected to the nearby lakes Nakuwa and Lemwa. Also habitat recovery from disturbance during the flooding season is associated with increased food resources (Morris et al., 2007; Vorwerk et al., 2009), resulting in high fish abundance. Furthermore, wet season rice crop when cultivated as a largely rainfed crop and have little land engineering, causes less impacts on water quality and later on fish (Nguyen-Khao et al., 2005).

Temporal variation in fish species abundance and catch

in the wetland was also attributed to their reproductive traits and predator avoidance factors. Low abundance of all *Clarias* species at beginning of the rainy season along the wetland was due to their breeding cycle (Offem et al., 2010). *T. zillii* spawn at the end of the dry season (El-Sayed and Moharram, 2007) and this could explain their low abundance during the dry season in the wetland. The observed high catch at Budumba and Kapyani during the dry season was due to the high water level maintained at these two sites which, offered refuge for fish avoiding harsh conditions. *Oreochomis* sp. spawn anytime but larger quantities are observed during the rainy season (Melcher et al., 2012). Variation of small fish species abundance which breed thoughout the year was attributed to avoidance of predators. High abundance of haplochomines in highly polluted areas is due to the non-avoidance of predators in turbid waters while their low abundance in less polluted areas is due to avoidance in clearer waters (Ogutu-Ohwayo, 1990).

Availability of abundant food during the rainy season is also an important factor responsible for the temporal variation in wetland fish (Offem et al., 2010). Rainy conditions lead to higher detritus, softer decomposing plant materials and low phosphorus favour in high abundance of benthic invertebrates (Hansson et al., 2005) which are important food for mainly the small fish species. Disruptions in the food base caused by alterations in water quality, particularly at highly disturbed sites, have been found to lead to higher percentage of omnivores (generalists) and a decrease in the proportions of insectivores and carnivores (Morris et al., 2007). *C. gariepinus* which is predatory (Raburu and Masese, 2010) was less abundant than *C. liocephalus*, an omnivore at highly degraded sites. Thus, feeding and reproductive strategies are among the divergent life-history characteristics that make riverine fish species respond to annual flood pulse and short-term environmental disturbances in different ways (Winemiller, 2005; Montaña et al., 2007).

The relationship between land use and small scale fishery was realized though strong relation between fish species catches and water quality parameters. The results should be of interest to resource managers because land use changes in the wetland have the potential to drastically affect the water quality, negatively impacting the production of large commercial fish species of such a small scale fishery. The wetland ecological shift phase need to be established in order for policy makers to regulate the land use change rate which may have potentially irreversible effects to both small and large fish species of this small scale fishery.

Conflict of Interests

The author(s) have not declared any conflict of interests.

ACKNOWLEDGEMENTS

The authors would like to thank the management at National Fisheries Resource and Research Institute, Jinja, who greatly assisted us with fishing equipment and space for the identification of wetland fish species, Mr. Peter Asimwe at Makerere University who assisted in the wetland land use characterization, Ms. Juliet Kutuusa and Mrs. Robinah Muhairwe at Bugolobi National Water and sewage corporation laboratory who assisted in water quality analysis and the fishermen who extended support during data collection in the wetland. We are most grateful for the financial support from Kyambogo University, the International Foundation for Science (IFS) and the Syngenta Foundation.

REFERENCES

APHA (1995). Standard methods for examination of water and wastewater. 19th edition. American Public Health Association, Washington, DC.

Arthington AH, Lorenzen K, Pusey BJ, Abell R, Halls A, Winemiller KO, Arrington DA, Baran E (2004). River fisheries: ecological basis for management and conservation. In Welcomme RL (Ed.) Proceedings of the 2nd International Large Rivers Symposium, Mekong River Commission, Phnom Penh, Cambodia. 1:31-60.

Bayley PB (1988). Factors affecting growth rates of young floodplain fishes: seasonality and density dependence. Environ. Biol. Fish. 21: 127-147.

Bene C (2004). Fisheries management and poverty. In: Neiland AE, Bene C (eds) Poverty and Small-scale Fisheries in Wet Africa. FAO and Kluwer Academic Publishers, Dordrecht/Boston/London. pp. 61-82.

Béné C, Hersoug B, Allison EH (2010). Not by rent alone: analysing the pro-poor functions of small-scale fisheries in developing countries. Dev. Policy Rev. 28:325-358.

Bene C, Neiland AE (2004). Empowerment yes ...but empowerment of whom? Fisheries decentralization reforms in developing countries: a critical assessment with specific reference to poverty reduction. Aquat. Resour. Cult. Dev. 1:35-49.

Budy P, Thiede GP, McHugh P, Hansen ES, Wood J (2008). Exploring the relative influence of biotic interactions and environmental conditions on the abundance and distribution of exotic brown trout (*Salmo trutta*) in a high mountain stream. Ecol. Freshw. Fish 17:554-566.

Carvalho G, Moutinho P, Nepstad D, Mattos L, Santilli M (2004). An Amazon perspective on the forest–climate connection: opportunity for climate mitigation, conservation and development? Environ. Dev. Sustain. 6:163-174.

Chapman LJ, Chapman CA, Schofield PJ, Olowo JP, Kaufman L, Seehausen O, Ogutu-Ohwayo R (2003). Fish Faunal Resurgence in Lake Nabugabo, East Africa. Conserv. Biol. 17:500-511.

Cooper MJ, Uzarski DG, Burton TM (2007). Macroinvertebrate community composition in relation to anthopogenic disturbance, vegetation, and organic sediment depth in four Lake Michigan drowned river-mouth wetlands. Wetlands 27(4):894 - 903.

Di Baldassarre G, Elshamy M, van Griensven A, Soliman E, Kigobe M, Ndomba P, Mutemi J, Mutua F, Moges S, Xuan JQ, Solomatine D, Uhlenbrook S (2011). Future hydrology and climate in the River Nile basin: a review. Hydrol. Sci. J. 56:199 - 211.

El-Sayed HKA, Moharram SG (2007). Reproductive biology of *Tilapia zillii* (Gerv, 1848) from Abu Qir bay, Egypt. Egypt. J. Aquat. Res. 33: 379-394.

Garaway C, Arthur RA, Chamsingh B, Homekingkeo P, Lorenzen K, Saengvilaikham B, Sidavong K (2006). A social science perspective on stock enhancement outcomes: lessons learned from inland fisheries in southern Lao PDR. Fish Res. 80:37-45.

Goudswaard KPC, Witte F, Chapman LJ (2002). Decline of the African lungfish (*Protopterus aethiopicus*) in Lake Victoria (East Africa). Afr. J. Ecol. 40:42-52.

Greenwood PH (1974). The cichlid fishes of Lake Victoria, East Africa: The biology and evolution of a species flock. Bull. Br. Mus. Nat. Hist. Zoo. Suppl. 6:11-34.

Hansson LA, Bronmark C, Nilsson PA, Abjornsson K (2005). Conflicting demands on wetland ecosystem services: nutrient retention, biodiversity or both? Freshw. Biol. 50:705-714.

Heck S, Béné C, Reyes-Gaskin R (2007). Investing in African Fisheries: Building Links to the Millennium Development Goals. Fish Fish. 8: 211-226.

Ibanez C, Oberdorff T, Teugels G, Mamononekene V, Lavoue S, Fermon Y, Paugy D, Toham AK (2007). Fish assemblages structure and function along environmental gradients in rivers of Gabon (Africa). Ecol. Freshw. Fish 16:315-334.

Kansiime F, Kateyo E, Oryem-Origa H, Mucunguzi P (2007). Nutrient status and retention in pristine and disturbed wetlands in Uganda: management implications. Wetlands Ecol. Manage. 15:453 - 467.

Lopes PFM (2011). Decision-making processes by small-scale fishermen on the southeast coast of Brazil. Fish. Manage. Ecol. 18: 400-410.

Lorenzen K, Garaway CJ, Chamsingh B, Warren TJ (1998). Effects of access restrictions and stocking on small water body fisheries in Laos. J. Fish Biol. 53:345 – 357.

Louca V, Lindsay SW, Majambere S, Lucas MC (2009). Fish community characteristics of the lower Gambia River floodplains: a study in the last major undisturbed West African river. Freshw. Biol. 54:254-27.

MAAIF (2009). Uganda national rice development strategy (UNRDS) 2nd draft, Ministry of Agriculture Animal Industry and Fisheries, Government of Uganda, Entebbe.

Martin SM, Lorenzen K, Arthur RI, Kaisone P, Souvannalangsy K (2011). Impacts of fishing by dewatering on fish assemblages of tropical floodplain wetlands: A matter of frequency and context. Biol. Conserv. 144:633 - 640.

Melcher AH, Ouedraogo R, Schmutz S (2012). Spatial and seasonal fish community patterns in impacted and protected semi-arid rivers of Burkina Faso. Ecol. Eng. 48:117-129.

Montaña CG, Choudhary SK, Dey S, Winemiller KO (2011). Compositional trends of fisheries in the River Ganges, India. Fish. Manag. Ecol. 18:282-296.

Morris C, Green JM, Simms A (2007). Population components in a small Atlantic cod population in Labrador and their management implications for an MPA. Proceedings of the Sixth International Conference on Science and Management of Protected Areas, Wolfville, NS, May 21-25, 2007.

Muhoozi L (2003). Aspects of the Commercial Exploitation of Fish Stocks in Lake Victoria. PhD Thesis, Hull: University of Hull.

NaFIRRI (2007). Standard Operating Procedures (SOPs) for collecting biological information from fishes of Lake Victoria. National Fisheries Research and Resource Institute, vol. 1, Jinja.

National Environment Management Authority (NEMA) (2004). District State of Environment Report for Pallisa District, report. http://www.nemaug.org/district_reports/DSOER 2004_Pallisa.pd. Accessed 08 September, 2012.

Nguyen-Khao S, Lorenzen K, Garaway C, Chamsingh B, Siebert DJ, Rondone M (2005). Impacts of irrigation on fisheries in rain-fed rice farming landscapes. J. Appl. Ecol. 43: 892-900.

Offem BO, Akegbejo S, Omoniyi Y (2010). Aspects of Ecology of *Clarias anguillaris* (Teleostei: Clariidae) in the Cross River, Nigeria. Turk. J. Fish. Aqua. Sci. 10:101-110.

Ogutu-Ohwayo R (1990). The decline of the native fishes of Lakes Victoria and Kyoga (East Africa) and the impact of introduced species, especially the Nile perch, *Lates niloticus* and the Nile tilapia, *Oreochomis niloticus*. Environ. Biol. Fish. 27:81 - 96.

Okechi JK (2004). Profitability assessment: A case study of African catfish (*Clarias gariepinus*) farming in the Lake Victoria basin, Kenya. United Nations University-Fisheries Training Program. wttp://www.unuftp.is/static/fellows/document/ okechiprf04.pdf. Accessed 16 May 2010.

Olden JD, Kennard MK, Leprieur F, Tedesco PA, Winemiller KO, Garcı´a-Berthou E (2010). Conservation biogeography of freshwater fishes: recent progress and future challenges. Divers. Distrib. 16:496-513.

Olin M, Vinni M, Lehtonen H, Rask M, Ruuhijarvi J, Saulamo K, Ala-Opas P (2010). Environmental factors regulate the effects of roach *Rutilus rutilus* and pike *Esox lucius* on perch *Perca fluviatilis* populations in small boreal forest lakes. J. Fish Biol. 76:1277-1293

R Development Core Team (2008). R: A language and environment for statistical computing. R Foundation for Statistical Computing. Vienna, Austria. ISBN 3-900051-07-0, URL http://www.R-project.org. Accessed 20 January 2014.

Raburu PO, Masese FO (2010). Development of a fish-based index of biotic integrity (FIBI) for monitoring riverine ecosystems in the Lake Victoria drainage basin, Kenya. River Res. Appl. 28:23-38.

Ramsar (2008). Uganda, Ramsar sites Information service from http://ramsar.wetlands.org/Portals/15/UGANDA.pdf, pp 18. Accessed 14 May 2014.

Randle AR, Chapman LJ (2004). Habitat use by the air-breathing fish *Ctenopoma muriei*: implications for costs of breathing. Ecol. Freshw. Fish. 13:37-45.

Rientjes THM, Haile AT, Kebede E, Mannaerts CMM, Habib E, Steenhuis TS (2011). Changes in land cover, rainfall and stream flow in Upper Gilgel Abbay catchment, Blue Nile basin Ethiopia. Hydrol. Earth Syst. Sci. 15:1979-1989.

Smith LED, Nguyen-Khoa S, Lorenzen K (2005). Livelihood functions of inland fisheries: policy implications in developing countries. Water Policy 7:359-383.

Thiebaut G (2006). Aquatic macrophyte approach to assess the impact of disturbances on the diversity of the ecosystem and on river quality. Int. Rev. Hydrobiol. 91:483 - 497.

Timmerman CM, Chapman LJ (2004). Behavioral and physiological compensation for chonic hypoxia in the sailfin molly (*Poecilia latipinna*). Physiol. Biochem. Zool. 77:601-610.

Vanden-Bossche JP, Bernacsek GM (1990). Source book for the inland fishery resources of Africa. CIFA Tech Pap, FAO, Rome, 18/1.

Vass KK, Shivastava NP, Katiha PK, Das AK (2009). Enhancing fishery productivity in small reservoir in India. A Technical Manual. WorldFish Center Technical Manual No. 949. The WorldFish Center, Penang, Malaysia. pp. 9.

Verburg PH, Neumannw K, Nol L (2011). Challenges in using land use and land cover data for global change studies. Glob. Chan. Biol. 17: 974-989.

Vorwerk PD, Paterson AWP, Froneman W, Whitfield AK (2009). Increased abundance of two important sport fishery species following renewed river flow into a freshwater-deprived South African estuary. Fish. Man. Ecol. 16:420-423.

Welcomme RL (1985). River fisheries. FAO Tech. Pap., 262:330 pp.

Winemiller KO (2005). Life history strategies, population regulation, and their implications for fisheries management. Can. J. Fish. Aquat. Sci. 62:872-885.

Witte F, van Densen WLT (1994). Fish stocks and fisheries of Lake Victoria, a handbook for field observations. Samara Publishing Cardigan Wales UK.

A comparative study of the defluoridation efficiency of synthetic dicalcium phosphate dihydrate (DCPD) and lacunar hydroxyapatite (L-HAp): An application of synthetic solution and Koundoumawa field water

A. S. Manzola[1,2] , M. S. Laouali[1] and M. Ben Amor[2]

[1]Laboratoire de Chimie Analytique et Minérale, Faculté des Sciences et Techniques, Université Abdou Moumouni BP 10662, Niamey, Niger.
[2]Laboratoire des Traitements des Eaux Naturelles, CERTE, Technopôle de Borj Cedria, BP 273, 8020, Soliman, Tunisia.

This paper deals with the comparison of defluoridation efficiency of two defluoridation agents by the use of dicalcium phosphate dihydrate (DCPD) and lacunar hydroxyapatite (L-Hap) as a fluoride sorbents. The DCPD and L-HAp are characterized by using XRD and FTIR techniques. Defluoridation of synthetic solution of sodium fluoride (NaF) and natural waters of Koudoumawa are studied. The fluoride removal capacity is as follows: DCPD: (26.37 mg.g^{-1}; 0.0174 g, 9.81 mg.g^{-1}; 0.1012 g) and L-Hap: (18.96 mg.g^{-1}; 0.0174 g, 8.00 mg.g^{-1}; 0.1012 g). The optimum 0.0623 g of DCPD/100 mL dosage of synthetic solution could bring down the level of fluoride within the tolerance limit, [F$^-$] = 0.38 mg/l (WHO guideline value = 0.8 mg/l), the pH rise is 5.10 and the defluoridation time is 72 h. For L-Hap, it is 0.1012 g of L-Hap/100 mL, [F$^-$] = 1.98 mg/l in the same conditions. For Koundoumawa natural waters, 0.0527 g of L-Hap/100 mL of solution could bring down the level of fluoride, [F$^-$] = 0.84 mg/l. New mechanisms of fluoride removal by DCPD and L-HAp are proposed from which it is established that this material removes fluoride by ion-exchange, adsorption process, dissolution, precipitation and co-precipitation.

Key words: Defluoridation, dicalcium phosphate dihydrate (DCPD), lacunar hydroxyapatite (L-Hap), adsorption, ion-exchange, dissolution-precipitation.

INTRODUCTION

To provide water requirements for the rural population and to fight against poverty, Niger government exploits waters from wells and drillings. In 1996, more than 6207 drillings and 10005 wells were done. These drillings play

Table 1. Physicochemical composition of Koundoumawa waters.

Parameter	pH	Conductivity ($\mu S.cm^{-1}$)	Concentrations (mg/l)								
			Ca^{2+}	Mg^{2+}	Na^{+}	K^{+}	SO_4^{2-}	HCO_3^{-}	NO_3^{-}	NO_2^{-}	F^{-}
Koundoumawa well water (summer)	7.1	384	42.1	24.45	87	5	25	216.08	8	0.02	4
Koundoumawa well water (rainy)	7	381	20.8	03	85	06	20	200.1	28	0.28	2.3
Koundoumawa drilled water	7.5	379	39.1	22.15	69	03	29	216.70	9	0.01	5

a primary role in water supply to the rural population. More than 90% of rural population in Niger depends on underground water as their drinking waters sources and fluoride excess is generally found in underground waters. A study carried out by the Ministry in charge of Waters, Environment and Fight against desertification (MHE/FAD) in 2005 on the physico-chemical analysis of raw water showed a high fluoride concentration according to World Health Organization (WHO) guidelines. WHO has set a guideline value of 0.8 mg/l for hot countries as the maximum permissible level of fluoride for drinking water to avoid health effects of fluoride water poisoning (WHO, 1996). In over a total of 211 studied drillings, 38% presented a fluoride concentration higher than 0.8 mg/l. The highest concentrations are recorded in Tibiri, Koundoumawa (in the Eastern part of Niger) and in Ingall (in the Northern part of Niger) with concentrations varying from 5 to 8 mg/l. Table 1 gives the chemical composition of Koundoumawa well and drilled waters during the summer rainy season. It has fluoride concentration of about 2 to 5 mg/l, its pH is about 7 to 7.5, and a bicarbonate concentration of about 167.7 to 216.08 mg/l.

A study carried out by UNICEF in 1998 classified Niger among the countries presenting fluorosis endemic contaminations. Several processes of defluoridation of drinking water have been reported in literature such as: ion-exchange, adsorption, and coagulation and precipitation process. Based on these processes, several defluoridation methods have been proposed by using nanosized hydroxyapatite (Yu et al., 2011), hydroxyapatite (Mourabet et al., 2011; Jiménez-Reyes and Solache-Ríos, 2010), surface coated hydroxyapatite powders (Subbaiah and Sankaran, 2014). Dicalcium phosphate dihydrate (DCPD) has been reported to be efficient for water defluoridation (Sekar et al., 2009; Masamoto and Tetsuji, 2004; Moubaret et al., 2011; Taewook et al., 2012). Several methods based on coconut shell carbon (Anirudhan et al., 2007; Amit et al., 2011), bone char (Medellin-Castillo et al., 2014), hybrid precipitation-microfiltration process (Nash and Liu, 2010), precipitated fluorhydroxyapatite nanoparticles (Kevin, 2014), Al (III) modified calcium hydroxyapatite, nano-hydroxyapatite/chitin composite (Yulun and Chun, 2012), alginate bioencapsulated nano-hydroxyapatite composite (Kalimuthu and Natrayasamy, 2014), and recently other defluoridation methods including membrane process based on reverse osmosis and nanofiltration (Simons, 1993; Rao et al., 1998; Mameri et al., 1998) and cellulose anhydroxyapatite nanocomposites (Xiaolin et al., 2013), glass derived hydroxyapatite (Wen et al., 2011), nano-hydroxyapatite/chitosan (Sundaram et al., 2008) were reported. Because of the socio-economic conditions in Niger, it is impossible to set up the defluoridation processes mentioned above. Thus, the rural populations are obliged to consume this fluoride poisoning water with its consequence, the appearance of the dental and skeletal fluorosis. The most dramatic example is the case of the children from Tibiri (Maradi, in the middle-east of Niger) where more than 450 children are affected by various forms of fluorosis, which include dental, skeletal and non-skeletal forms (Rapport Mission Internationale d'Enquête, 2002). However, for developing countries, precipitation and adsorption are the most accessible methods. Indeed the management and maintenance of current and proposed defluoridation technologies require expensive chemicals and/or a high level of technological skill and can be applied only in centralized water distribution systems (Rao et al., 1998). Precipitation methods are based on the addition of chemicals to water and removal of insoluble fluoride compounds as precipitates or co-precipitates or adsorbed onto the formed precipitates (Nash and Liu, 2010). In adsorption processes, fluoride is removed either by ion exchange, physical or surface chemical reactions with the adsorbent material. Hydroxyapatite (HAp) was used to remove cadmium, oxovanadium, cobalt, lead and zinc (Lusvardi et al., 2002; Vega et al., 2003; Hammari et al., 2004; Smiciklas et al., 2006; Sandrine et al., 2007). It appears that with hydroxyapatite (HAp), the adsorption and ion-exchange mechanisms are the most favorable mechanisms for fluoride removal (Meenakshi et al., 2007). The removal of fluoride using HAp has been reported earlier (Fan et al., 2003; Hammari et al., 2004). Sairam et al. (2008) have used nano-hydroxyapatites for water defluoridation. Lacunar hydroxyapatite or calcium-deficient hydroxyapatite (CDHA) nano-crystals incorporated with bovine serum albumin (BSA) to form BSA-loaded nano-carriers were synthesized via both *in situ* and *ex situ* processes (Tse-Ying et al., 2005). Spherical Ca-deficient hydroxyapatite (HA) granules are expected to be useful drug carriers in bony sites because of their bone regeneration and adsorption ability (Masanobu et al., 2013). To study the effects of nanocrystalline calcium deficient hydroxyapatite incorporation in glass ionomer cements, bioactive nanocrystalline calcium deficient hydroxyapatite (nCDHA) with improved mechanical and resorption properties was synthesized (Sumit et al., 2011). The objective of the present study was to investigate the performance of synthesized DCPD and Lacunar hydroxyapatite as feasible and suitable adsorbent. Synthesized L-HAp and DCPD are synthesized in the laboratory by precipitation method. Defluoridation studies are carried out under various equilibrating conditions like the amount of adsorbent, the effect of contact time and the pH. Detailed precisions during the defluoridation mechanism by L-HAp and DCPD and the kinetic studies are presented.

EXPERIMENTAL SET UP AND PROCEDURES

Synthesis of DCPD and L-Hap

The synthesis of L-HAp and DCPD involved adding variable volumes of 0.1 M of monosodium phosphate dihydrate to 100 mL of 0.025 M of calcium chloride monohydrate in a closed tricol balloon reaction vessel. The calcium phosphate precipitation is being controlled by the pH level adjustment with 1 M NaOH solution, using a pH-meter TACUSSEL giving an accuracy of 0.01 unit of pH. The synthesis of DCPD involves the reaction of calcium chloride monohydrate and monosodium phosphate dihydrate with a Ca/P ratio close to 1. pH level of the reaction should be maintained at pH ranging from 5.8 to 6.6 (Manzola et al., 2013), otherwise it may lead to the formation of OH-Apatite (Legeros et al., 1983; Casciani et al., 1980). From pH 7 to 7.8 it precipitates Lacunar hydroxyapatite (L-Hap) (Manzola et al., 2013). The medium agitation is carried out using a magnetic stick at the ambient temperature of 22 to 25°C. The precipitated solution is poured into a 250 mL bottle and kept for 1, 3 or 7 days, then filtered. The precipitated solid is dried from 60 and 70°C for 24 h to get DCPD ($CaHPO_4.2H_2O$) or L-Hap.

Characterisation of DCPD and L-Hap

The solid phases obtained at different pH are characterized by powder X-ray diffraction (XRD) and the phase identification is made by using a JCPDS cards (Joint Committee on Powder Diffraction Standards) and Fourier Transform Infrared Spectrometer (FTIR). The XRD patterns are obtained by using a PW 1050/37 diffractometer with a monochromatic radiation Kα1 of Cu (λ = 1.5418; 1.5405 A°). The FTIR spectra are performed by using the KBr pellet technique in a Shimadzu Fourier Transform-8300 in the range of 4500-400 cm^{-1} at a resolution of 4 cm^{-1}. Also, the results of FTIR spectrometer and XRD are used to confirm the fluoride uptake by the DCPD or L-Hap precipitates.

Adsorption experiments

Synthetic solution

The fluoride experimental solutions are prepared by a quantitative dilution of stock solution. The stock solution of 1000 mg/l fluoride is obtained by dissolving an appropriate amount of sodium fluoride in distilled water.

A 100 ml of the fluoride experimental solution (10 mg/l as initial fluoride concentration) is taken into a 150 mL of PVC conical flask and a known weight of adsorbent material (0.0174; 0.0318; 0.0527; 0.0623 and 0.1012 g) is added in this solution, and then kept for stirring at 150 rpm on a horizontal rotary shaker for 0; 2; 21; 48; 72 and 92 h in order to reach the equilibrium. After that, the solution is then filtered through Millipore filter paper 22 µm and the filtrate is analyzed for residual fluoride concentration by ion selective electrode (ISE) using field ion meter 340I/ION. In order to regulate the total ionic strength, the TISAB adjusting buffer is added to the sample and standard solutions. The TISAB buffer was added in order to maintain pH constant, to decomplex metal-F complexes contained in the sample during measurement and allowed us to get the total free fluoride concentration in the solution. The fluoride removal experiments are studied over different operational conditions including effect of adsorbent dose, the nature of adsorbent and the effect of contact time. All the experiments were carried out at 30°C, room temperature. During the experiments, the fluoride ion concentration and the pH solution were recorded.

Fluoride poisoning field water of Koundoumawa

In the second serial experiments, the fluoride removal ability of L-HAp and DCPD was tested in field water collected from KOUNDOUMAWA drilling (Zinder, in the far eastern part of Niger) at the same conditions as the synthetic water. The physicochemical characteristics of field water samples are determined before batch adsorption study. The detailed characteristics of field waters are given in Table 1.

RESULTS AND DISCUSSION

Characterization of DCPD sorbent

In order to characterize DCPD, XRD and FTIR were carried out on precipitated DCPD and DCPD samples mixed with fluoride solution. The XRD patterns of precipitated DCPD and the DCPD samples mixed with fluoride solution are respectively presented in Figures 1 and 2.

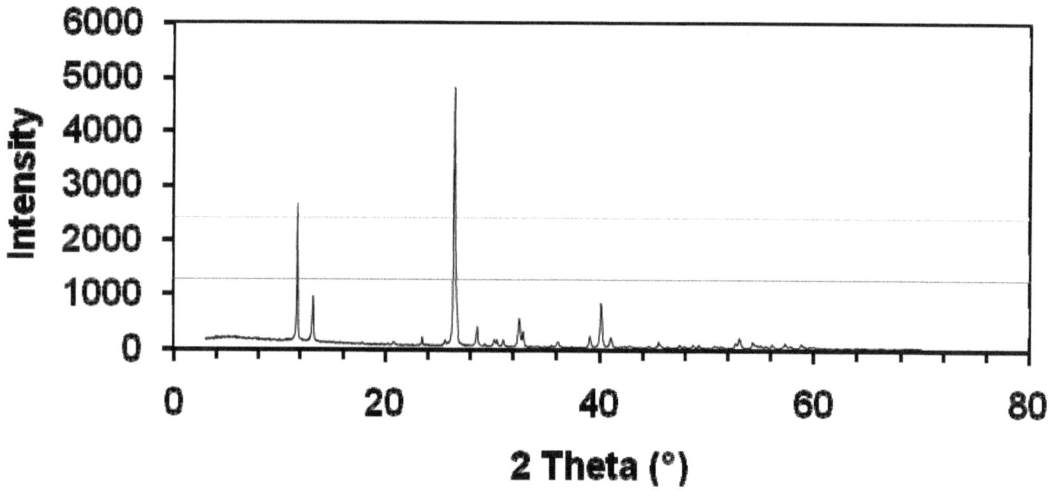

Figure 1. XRD patterns of precipitated DCPD.

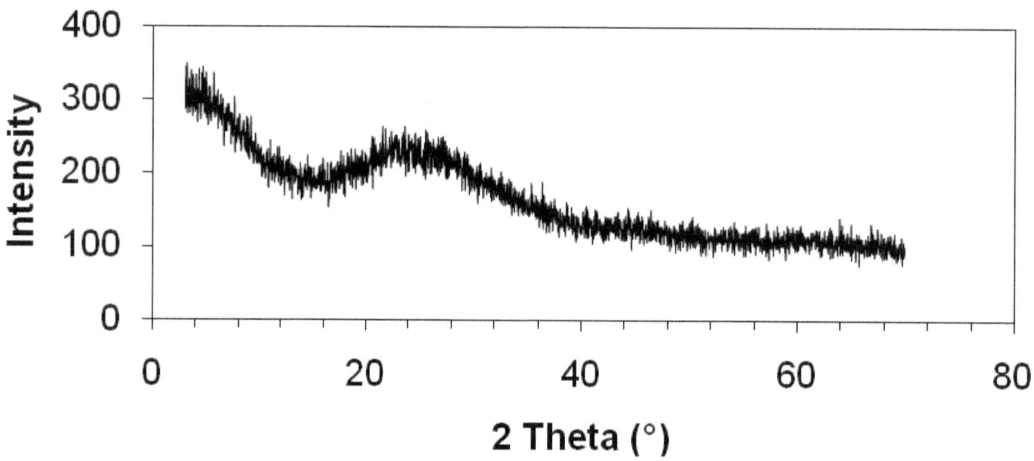

Figure 2. XRD patterns of fluoride sorbed precipitated DCPD.

The identification of crystal phases was done by using a JCPDS cards (Joint Committee on Powder Diffraction Standards). The XRD patterns shown in Figure 1 indicates the DCPD peaks at 2θ values of about 11.6°, 13.2°, 20.9°, 23.6°, 26.5°, 29.3° which are in the JCPDS cards (card number 09–0077). The maximum peak intensity is 26.5 degree theta. The spectra confirmed that the products obtained are mainly composed of the DCPD form of calcium phosphate (Figure 1). There is a meaningful change in the XRD patterns of DCPD after treatment with fluoride (Figure 2). The results of X-ray diffraction of treated product show amorphous product. Similar results are reported (Larsen et al., 1993) while studying the fluoride sorption on brushite, calcium hydroxide, and bone char. The fine powder obtained was poorly crystallized apatite.

Figure 3 represents FTIR spectra of the samples before and after treatment with fluoride. In Figure 3, the absorption peaks appeared at 3540 and 3488 cm^{-1}, characteristics of valence vibrations of free H_2O (ν_{H_2O}), at 3290 and 3161 cm^{-1}, characteristics of valence vibrations of associated H_20 (ν_{H_20}), at 2367 cm^{-1}, characteristic of valence vibrations $\nu_{O\text{-}H}$ of HPO_4^{2-} groups, at 1649 cm^{-1} characteristic of valence vibrations of H_20 of constitution (ν_{H_20}), at 1220 and 790 cm^{-1}, characteristics of bond elongation vibrations $\delta_{P\text{-}O\text{-}H}$ at 1135, 1059 and 985 cm^{-1}, characteristics of valence vibrations ν_{PO} at 873 cm^{-1},

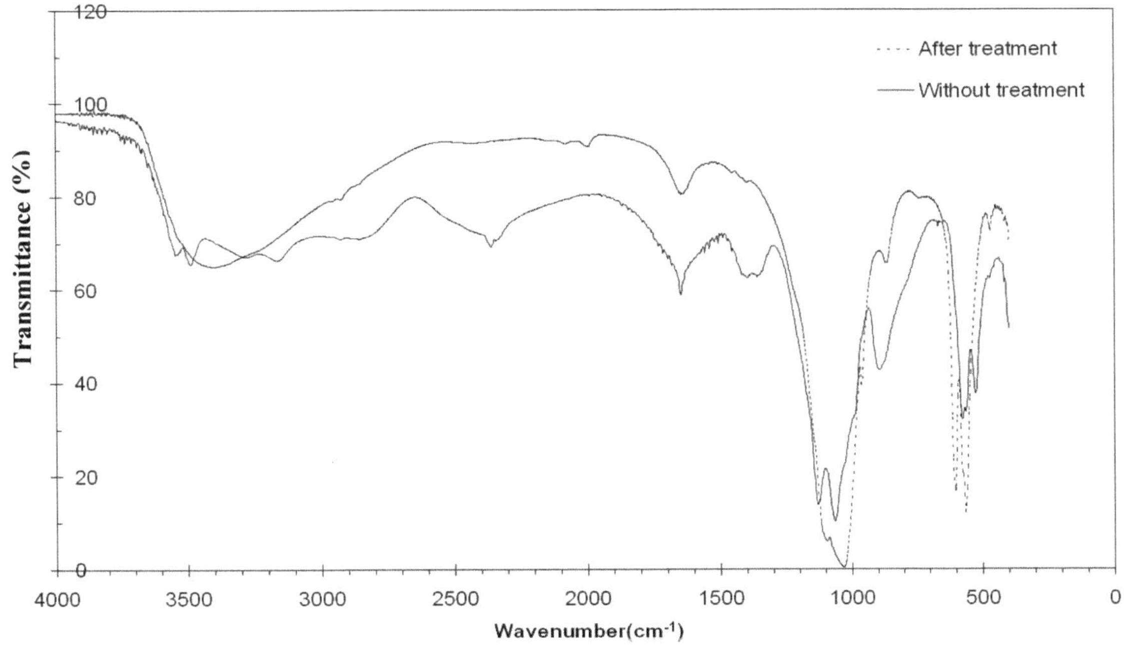

Figure 3. FTIR spectra of DCPD (without treatment and after treatment with fluoride).

characteristics of valence vibrations ν_{P-OH} and at 525 cm^{-1}, characteristics of valence vibrations ν_{O-P-O} (Legeros et al., 1983; Casciani et al., 1980). We can observe that the spectrum of precipitated solid is perfectly identical to infra-red spectrum of $CaHPO_4 \cdot 2H_2O$ (Legeros et al., 1983), and the characteristic bands are in conformity with those obtained (Legeros et al., 1983; Casciani et al., 1980).

There is a meaningful change in the FTIR spectrum of the sample after treatment with fluoride. In the presence of F^- ions, the DCPD is transformed into "OH-Apatite", then into fluorinated hydroxyapatite FAp (Maiti et al., 1981; Takahashi et al., 1978). Indeed for the infra-red spectrum of "OH-Apatite" (Legeros et al., 1983; Casciani et al., 1980), absorption peaks appeared at 3561 cm^{-1} characteristics of valence vibrations of ν_{OH^-}, at 3473 cm^{-1} characteristic of valence vibrations ν_{H-O-H} of free H_2O, at 1647 cm^{-1} characteristic of valence vibrations ν_{H-O-H}, at 1112, 1033 and 960 cm^{-1} characteristic of valence vibrations $\nu_{PO_4^{3-}}$, at 873 cm^{-1} characteristic of valence vibrations ν_{P-OH}, at 602 and 562 cm^{-1} characteristics of elongation vibrations $\delta_{PO_4^{3-}}$. The apparition of two new bands at 775 and 468 cm^{-1} may be due to the fluoride adsorption/exchange, indicating the incorporation of fluoride into the solid.

Characterization of L-Hap sorbent

The identification of crystal phases was done by using a JCPDS cards (Joint Committee on Powder Diffraction Standards). The crystalline peaks at 2θ = 25.9°, 32°, 33° and 40° (Figure 4) indicating the L-Hap peaks which are in the JCPDS cards (card number 9-432). The maximum peak intensity is 32 degree theta. The spectra confirmed that the products obtained are mainly composed of the L-Hap form of calcium phosphate (Figure 4). There is no significant change in the XRD pattern of L-HAp after treatment with fluoride (Figure 5). Similar results were reported (Diaz-Nava et al., 2002) while studying the fluoride sorption on zeolites.

In Figure 6, there is a meaningful change in the FTIR spectrum of the sample after treatment with fluoride. The presence of F^- ions makes the L-Hap transformed into fluorinated hydroxyapatite FAp (Maiti et al., 1981; Takahashi et al., 1978). For the infra-red spectrum of "OH Apatite" (Legeros et al., 1983; Casciani et al., 1980), absorption peaks appear at 3561 cm^{-1} characteristics of valence vibrations of ν_{OH^-}, at 3473 cm^{-1} characteristic of valence vibrations ν_{H-O-H} of free H_2O, at 1647 cm^{-1} characteristic of valence vibrations ν_{H-O-H}, at 1112, 1033 and 960 cm^{-1} characteristic of valence vibrations $\nu_{PO_4^{3-}}$, at 873 cm^{-1} characteristic of valence vibrations ν_{P-OH}, at 602

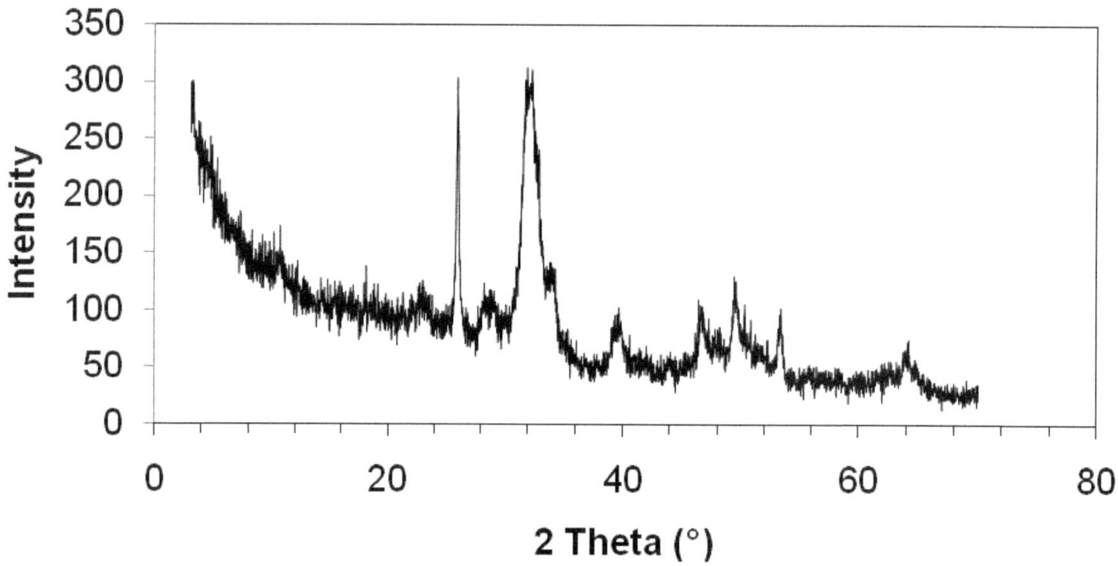

Figure 4. XRD patterns of precipitated L-HAp.

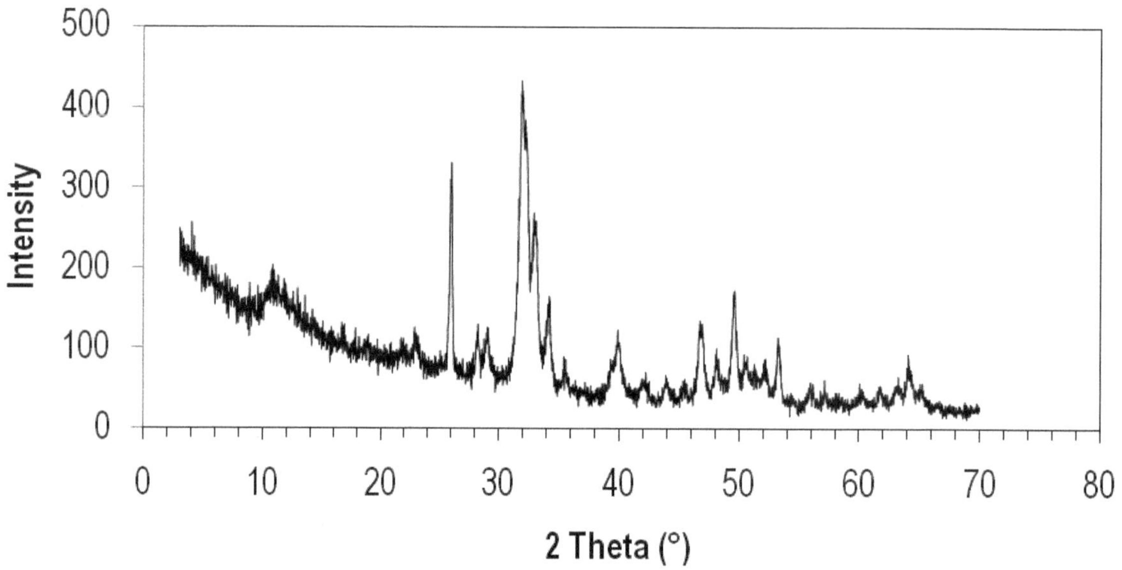

Figure 5. XRD patterns of fluoride sorbed precipitated L-HAp.

and 562 cm^{-1} characteristics of elongation vibrations $\delta_{PO4}{}^{3-}$. The presence of some OH^- with F^- disturbs the OH band frequencies of the pure calcium "OH-Apatite". There is a reduction in the intensity of OH bands at 3570 and 602 cm^{-1} with some displacement with lower frequencies in fluoride treated L-HAp which may be due to fluoride adsorption/exchange. The apparition of two new bands at 701 and 468 cm^{-1} for fluorinated hydroxyapatite FAp may be due to the fluoride adsorption/exchange, indicating the incorporation of fluoride into the solid. It is known that FAp is more stable in solution than "OH-Apatite" (Chow et al., 1997). For the initial fluoride treated "OH-Apatite", the new band appears at 701 cm^{-1}. The displacement of the new band from 775 to 701 cm^{-1} confirms the formation of O–H···F bond in agreement with other authors (Okazaki et al.,

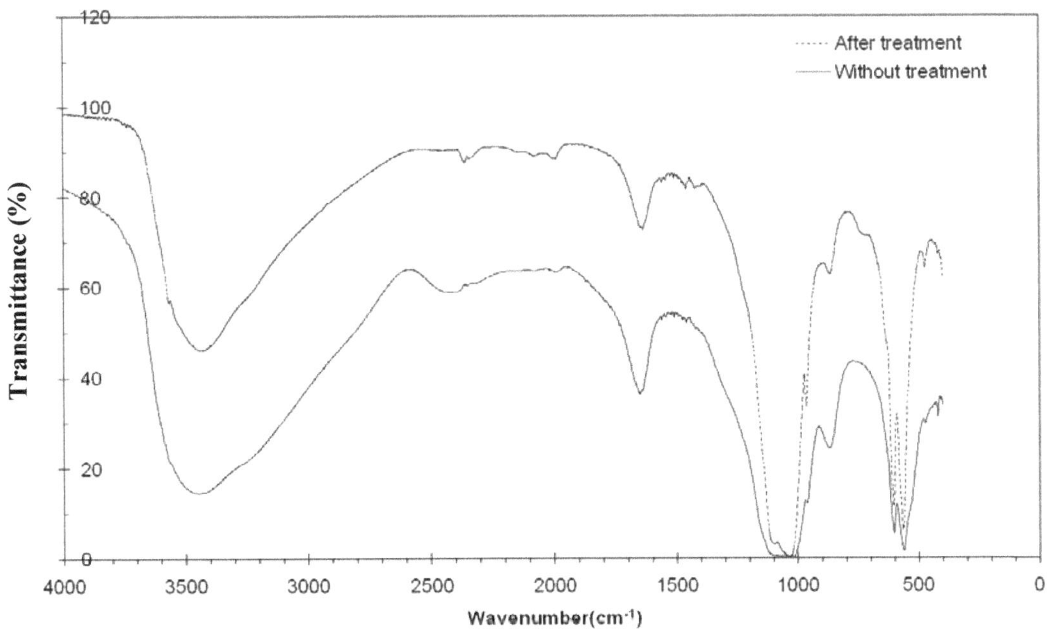

Figure 6. FTIR spectra of L-Hap (without treatment and after treatment with fluoride).

1981; Okazaki, 1992).

Adsorption experiments

Synthetic solution

DCPD sorbent - Effect of contact time and dose: In order to obtain the optimum DCPD sorbent dose defluoridation and contact time, experiments were carried out with various dosages of DCPD ranging from 0.0174 to 0.1012 g with 10 mg/l as initial fluoride concentration. We investigated the sorption of fluoride ion on DCPD as a function of contact time in the range of 0 to 92 h at room temperature and as a function of sorbent dose. The effect of fluoride removal capacity of DCPD with contact time and sorbent dose is shown in Figure 7. It was observed that fluoride removal capacity increases with contact time and contrary to the fluoride ion concentration (Figure 8), the fluoride removal capacity decreases with the raise of sorbent dose, where saturation is reached after 72 h for 0.0174; 0.0318; 0.0527 and 0.0623 g amount of DCPD sorbent dose. The maximum fluoride removal capacity is found to be 26.37 $mg.g^{-1}$ at a sorbent dose of 0.0174 g. For 0.1012 g amount of DCPD sorbent dose, the fluoride removal capacity kept increasing and reached over 92 h 9.81 $mg.g^{-1}$. Indeed at low sorbent dose, the fluoride uptake capacity is high because of the better utilization of the available active sites and at high sorbent dose, too many sites are available for limited quantity of fluoride and the lower equilibrium concentration of fluoride for sorption becomes negligible (Sanjay et al., 2009). Similar behavior has also been reported previously for other adsorbent (Kamble et al., 2007). The evaluation of fluoride ion concentration with contact time and sorbent dose is shown in Figure 8. We observed that the fluoride ion concentration decreased rapidly and continuously with the increase in the dose of the sorbent. This phenomenon was still observed for 72 h. From 72 to 92 h, it stayed steady for 0.0174; 0.0318; 0.0527 and 0.0623 g amount of DCPD sorbent dose. For 0.1012 g amount of DCPD sorbent dose, the fluoride ion concentration continued to decrease. For this amount, the concentration of fluoride ion reached for 92 h was $[F^-]$ = 0.15 mg/l. The optimum dosage can be fixed as 0.0623 g for further studies as this dosage was found to bring down the level of fluoride within the tolerance limit, $[F^-]$ = 0.38 mg/l for this amount (WHO guideline value = 0.8 mg/l). The steady state of DCPD sorbent was reached after 72 h for 0.0174; 0.0318; 0.0527 and 0.0623 g amount of DCPD sorbent dose. The uptake of fluoride can be controlled by adsorption or the dissolution-and-recrystallization mechanism. DCPD reached steady state only after 72 h, suggesting that the process is also governed by adsorption and mainly by dissolution-and-recrystallization mechanisms which are slow processes (Sairam et al., 2008; Meenakshi et al., 2007; Low et al., 1995; Taewook et al., 2012). For 0.1012 g amount of

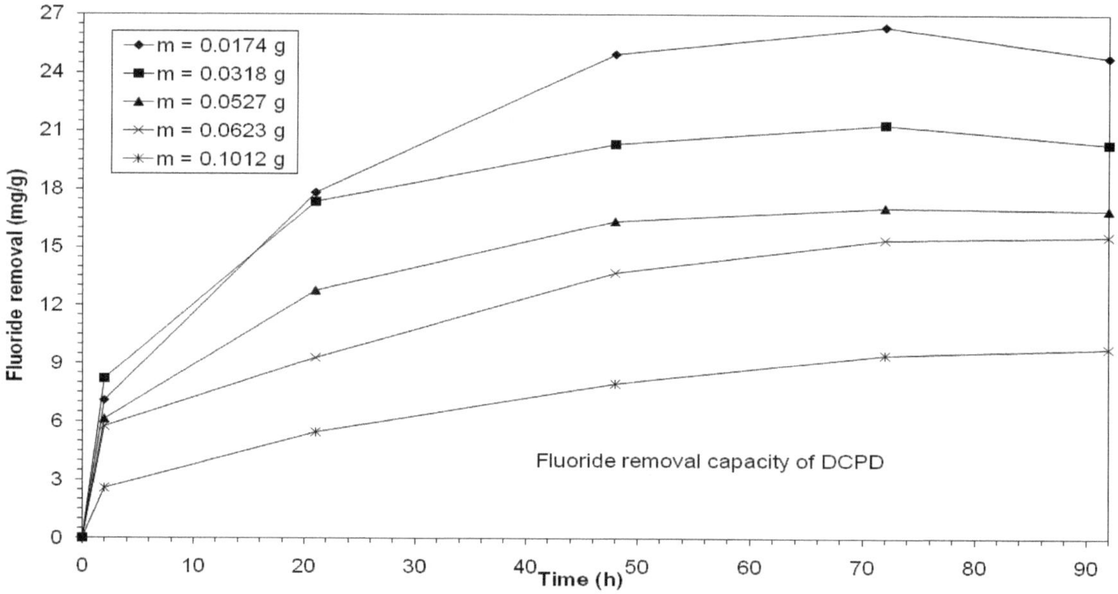

Figure 7. Effect of contact time and sorbent dose on the fluoride removal capacity of precipitated DCPD.

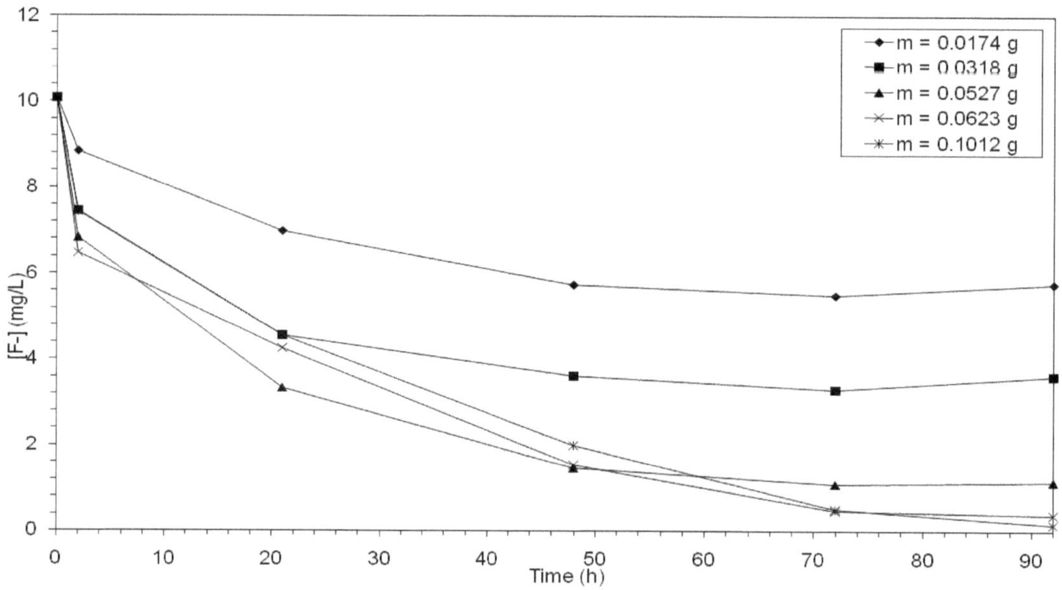

Figure 8. The evaluation of fluoride concentration during defluoridation: Effect of contact time and sorbent dose on precipitated DCPD.

DCPD sorbent dose, the steady state is not reached after 72 h.

Figure 9 shows the evaluation of the pH of the aqueous solution. The evaluation of pH fits the evaluation of fluoride ion concentration (Figure 9). We observed that the pH solution decreases rapidly and continuously with an increase in the dose of the sorbent (Lerch et al., 1966) while studying the hydrolytic conversion of DCPD to apatite. Indeed in aqueous solution, the following reactions are proposed:

*Quick and partial dispersion of phosphate in aqueous medium:

$$CaHPO_4.2H_2O \Leftarrow\Rightarrow Ca^{2+} + HPO_4^{2-} + 2H_2O$$

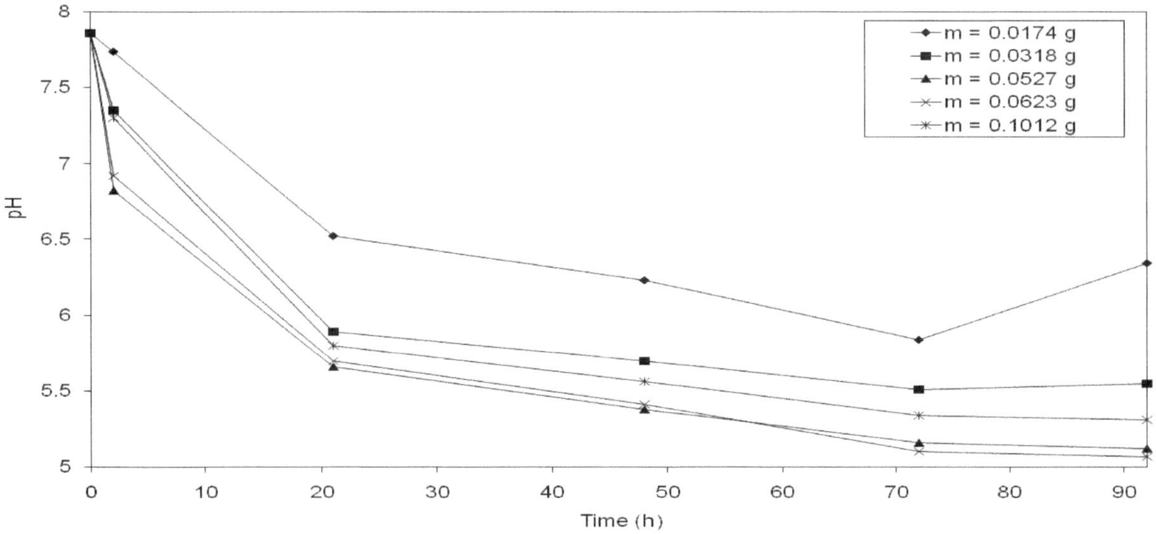

Figure 9. The evaluation of pH solution during the defluoridation: Effect of contact time and sorbent dose on precipitated DCPD.

*Quick hydrolytic conversion of HPO_4^{2-} ions:

$HPO_4^{2-} + H_2O \Leftarrow\Rightarrow H_2PO_4^{-} + OH^{-}$

*Low and concurential dismutation reaction:

$2HPO_4^{2-} \Leftarrow\Rightarrow PO_4^{3-} + H_2PO_4^{-}$

$2HPO_4^{2-} + 2H_2O \Leftarrow\Rightarrow 2H_2PO_4^{-} + 2OH^{-}$

*Precipitation of PO_4^{3-} ions:

$PO_4^{3-} + Ca^{2+} + HPO_4^{2-} + OH^{-} \Leftarrow\Rightarrow$ Apatitic calcium phosphate.

The decrease of pH is due to the presence of $H_2PO_4^{-}$ ions. Indeed this phosphate is acidic; its aqueous dissolution gives H_3PO_4 and a residual $CaHPO_4.2H_2O$ (Brown et al., 1959), reaches steady state at pH 5.5. In our experiments, DCPD reaches steady state after 72 h at pH 5.51 for 0.0318 g of sorbent; at pH 5.34 for 0.1012 g of sorbent; pH 5.16 for 0.0527 g of sorbent; pH 5.10 for 0.0623 g of sorbent. The pH solution plays an important role by controlling the adsorption at the water adsorbent interface and the dissolution-reprecipitation mechanism (Meenakshi et al., 1991). In the presence of fluoride ions and at these low pH, hydroxyapatite is faster when converted to fluorapatite because the solubility of the two salts differs increasingly with lower pH (Larsen et al., 1992). At high pH, the solubility of the two salts does not differ greatly (Featherstone et al., 1990), which explains why the uptake of fluoride in apatite is slow in this pH range. The dissolution of hydroxyapatite-reprecipitation of fluorapatite principle would be expected to operate most efficiently at low pH. As we have seen above, one disadvantage of such a defluoridation mechanism is that phosphate may be left in solution (the formation of H_3PO_4) and thus may favor bacterial growth.

L-Hap sorbent - effect of contact time and dose: In order to obtain the optimum L-Hap sorbent dose defluoridation and contact time, experiments were carried out with various dosages of L-Hap ranging from 0.0174 to 0.1012 g with 10 mg/l as initial fluoride concentration. We have investigated the sorption of fluoride ion on L-Hap as a function of contact time in the range of 0 to 72 h at room temperature and as a function of sorbent dose. The effect of fluoride removal capacity of L-Hap with contact time and sorbent dose is shown in Figure 10. It was observed that fluoride removal capacity increases with contact time and contrary to the fluoride ion concentration (Figure 11), the fluoride removal capacity decreases with increase in sorbent dose, where saturation is reached after 48 h for 0.0174; 0.0318; 0.0527 and 0.0623 g amount of L-Hap sorbent dose. The maximum of fluoride removal capacity is found to be 18.06 $mg.g^{-1}$ at a sorbent dose of 0.0174 g. For 0.1012 g amount of DCPD sorbent dose, the fluoride removal capacity stays increasing and reached 72 h for 8.00 $mg.g^{-1}$. Indeed at low sorbent dose, the fluoride uptake capacity is high because of the better utilization of the available active sites and at high sorbent dose, too many sites are available for limited quantity of fluoride and the lower equilibrium concentration of fluoride for sorption becomes negligible (Sanjay et al.,

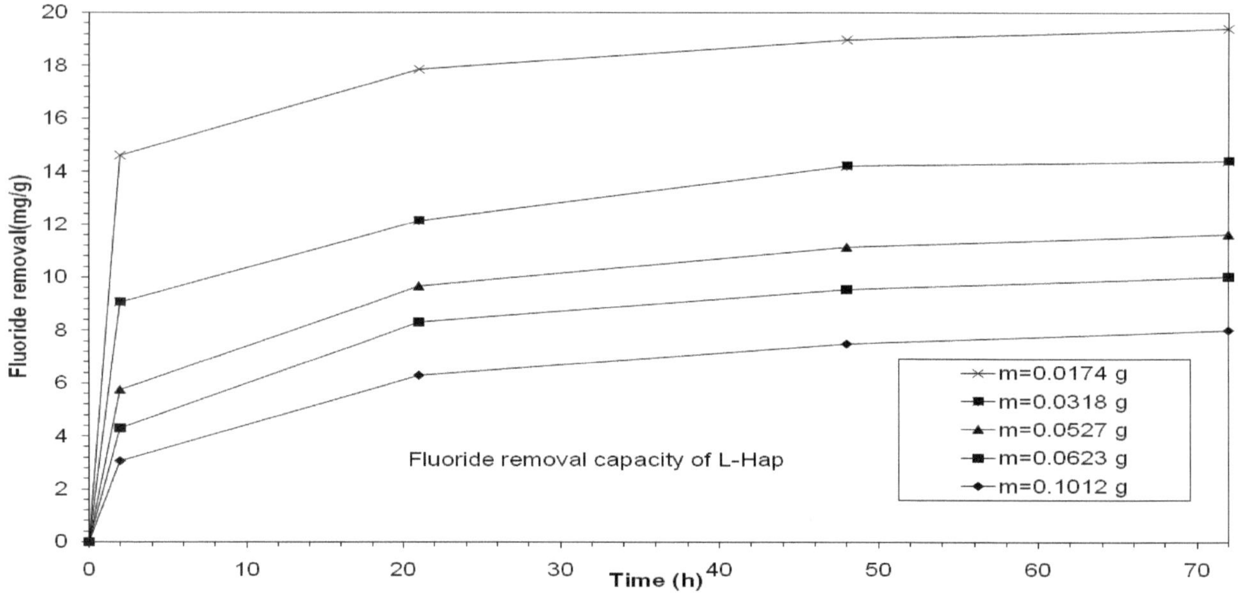

Figure 10. Effect of contact time and sorbent dose on the fluoride removal capacity of precipitated L-Hap.

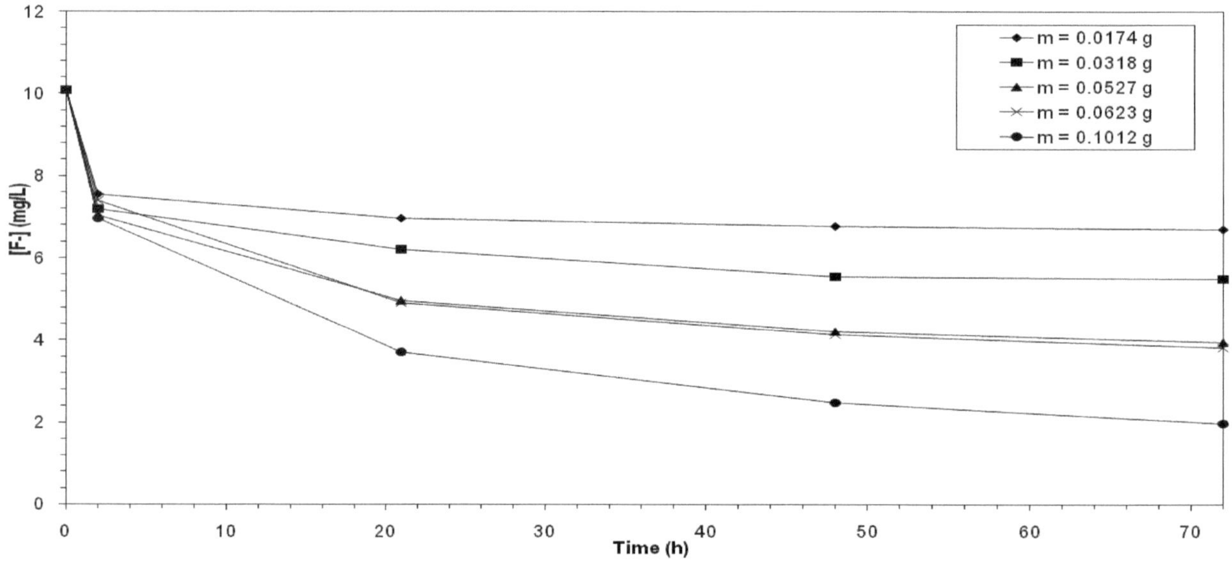

Figure 11. The evaluation of fluoride concentration during defluoridation: Effect of contact time and sorbent dose on precipitated L-HAp.

2009). Similar behavior has also been reported previously for other adsorbent (Kamble et al., 2007). The evaluation of fluoride ion concentration with contact time and sorbent dose is shown in Figure 11. We observed that the fluoride ion concentration decreases instantaneously and continuously within 2 hours, due to more active sites when increasing the dose of sorbent (Boualia et al., 1993). Within the first two hours, the sorption process is essentially controlled by ion exchange mechanism because of the rapidity of the phenomenon and because of the pH solution increases instantaneously (Figure 12). This phenomenon can be described by the following reaction:

$$Ca_{10}(PO_4)_6(OH)_2 + 2F^- \Rightarrow Ca_{10}(PO_4)_6F_2 + 2OH^-$$

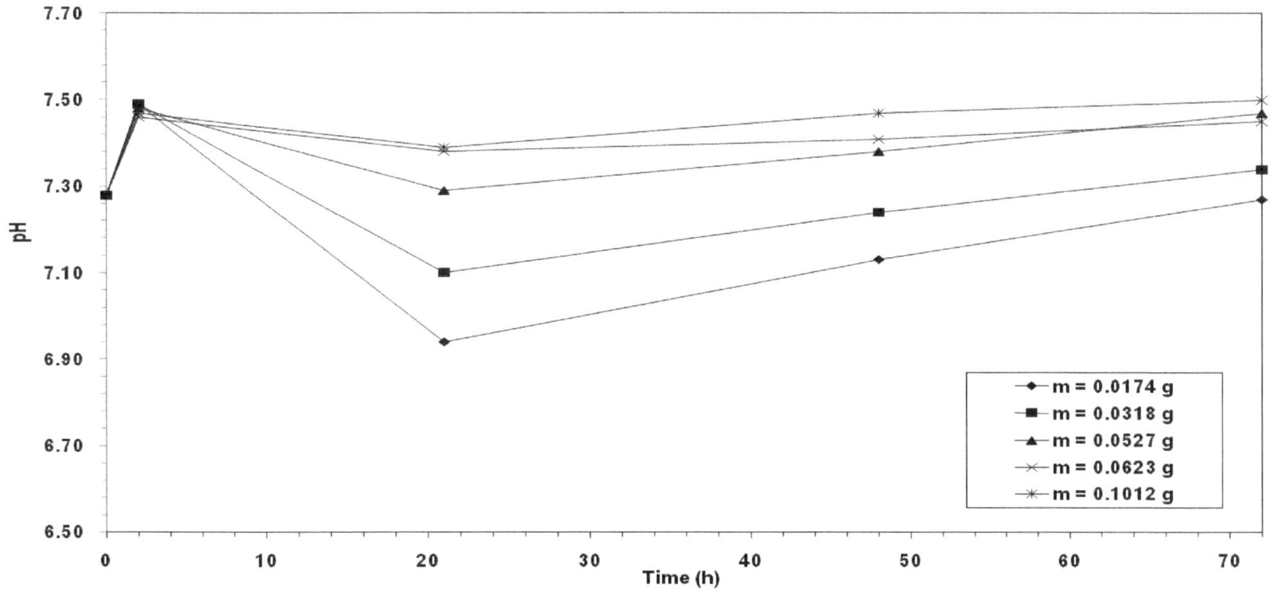

Figure 12. The evaluation of pH solution during the defluoridation: Effect of contact time and sorbent dose on precipitated L-HAp.

The uptake of fluoride can be controlled by adsorption or the dissolution-and-recrystallization mechanism. From 2 to 20 h, the fluoride ion concentration decreases with a slow rate with an increase in the dose of the sorbent, suggesting that the process is controlled by the dissolution-and-recrystallization mechanism because of the pH solution decreases in the same shape (Figure 12). This phenomenon can be described by the following reactions:

*Dissolution of Apatitic calcium phosphate:

Apatitic calcium phosphate $\Leftarrow\Rightarrow PO_4^{3-} + Ca^{2+} + HPO_4^{2-} + OH^-$

*Conversion of HPO_4^{2-} ions:

$2HPO_4^{2-} \Leftarrow\Rightarrow PO_4^{3-} + H_2PO_4^-$

*Formation of fluoroapatite:

$10Ca^{2+} + 6PO_4^{3-} + 2F^- + H_2PO_4^- \Rightarrow Ca_{10}(PO_4)_6F_2 + H_2PO_4^-$

The decrease of pH is due to the presence of $H_2PO_4^-$ which renders the solution acidic (Brown et al., 1959). From 20 to 72 h, the fluoride ion concentration decreases with a average rate with the growth in the dose of the sorbent, suggesting that the process is again controlled by ion exchange mechanism because of the pH solution increase in the same shape (Figure 11). This phenomenon can be described by the following reaction:

$Ca_{10}(PO_4)_6(OH)_2 + F^- \Rightarrow Ca_{10}(PO_4)_6FOH + OH^-$

These schemes have been proposed by Sairam et al. (2008) without precision during the process. The fluoride ion concentration reached at 72 h is [F⁻] = 6.70 mg/l for 0.0174 g amount of L-Hap sorbent dose, [F⁻] = 5.5 mg/l for 0.0318 g amount of L-Hap sorbent dose, [F⁻] = 3.83 mg/l for 0.0623 g amount of L-Hap sorbent dose, [F⁻] = 3.83 mg/l for 0.0527 g amount of L-Hap sorbent dose and [F⁻] = 1.98 mg/l for 0.1012 g amount of L-Hap sorbent dose. According to these results, the minimum fluoride ion concentration reached at 72 his [F⁻] = 1.98 mg/l for 0.1012 g amount of L-Hap sorbent dose. This value does not satisfy the level of fluoride within the tolerance limit (WHO guideline value = 0.8 mg/l).

At high pH, the solubility of the two salts does not differ greatly (Featherstone et al., 1990) (Figure 12), which explains why the uptake of fluoride in apatite is slow in this pH range.

Fluoride poisoning Koundoumawa field water

A second set of tests, with fluoride poisoning Koundoumawa field water during rainy season (Table 1) ([F⁻] = 2.3 mg/l) has been performed in the laboratory at the same conditions as the synthetic water. With the aim

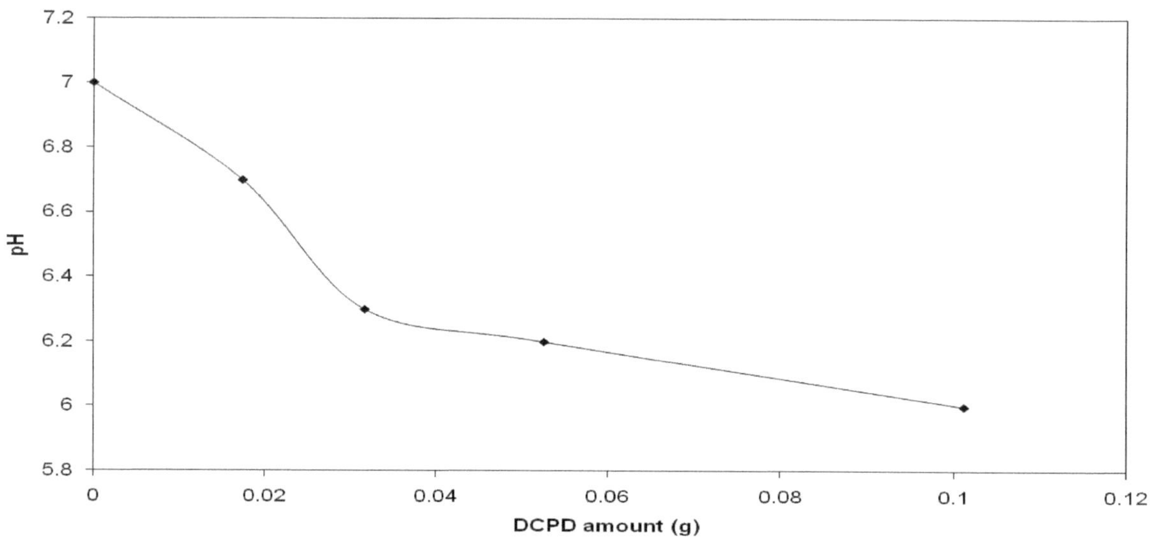

Figure 13. The evaluation of pH Koundoumawa field water during defluoridation: Effect of sorbent dose on precipitated DCPD.

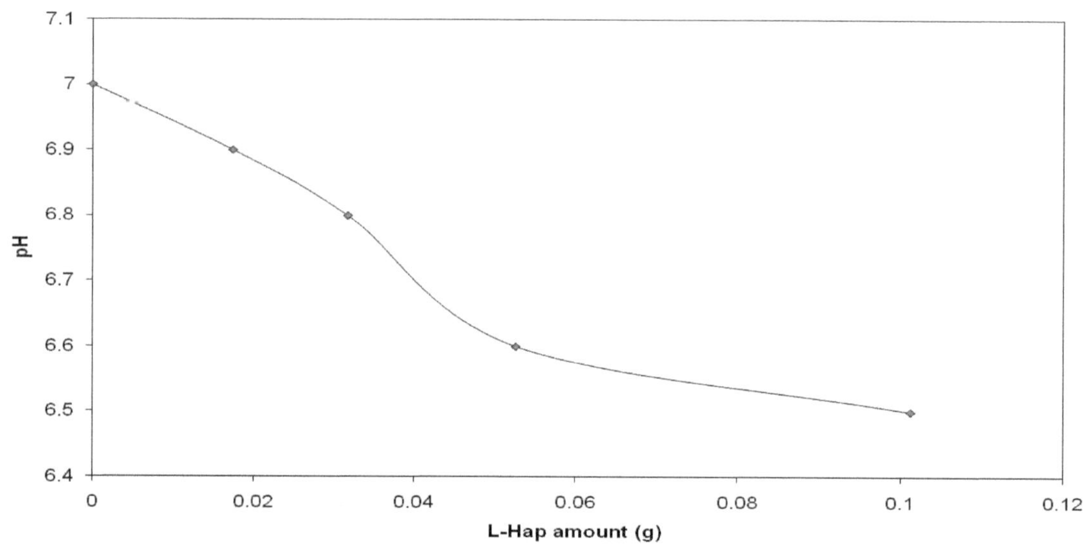

Figure 14. The evaluation of pH Koundoumawa field water during defluoridation: Effect of sorbent dose on precipitated L-HAp.

to obtain the maximum defluoridation, the contact time was fixed at 72 h. The results are presented in Figures 13, 14 and 15.

DCPD sorbent - Effect of amount: Figure 13 gives the evaluation of the pH of the aqueous solution. We observed that the pH solution decreases rapidly and continuously with increase in the dose of the DCPD sorbent. This phenomenon observed during synthetic solution defluoridation is in conformity with the observation (Lerch et al., 1966), suggesting that the process is controlled by the dissolution-and-recrystallization mechanism. The minimum pH obtained is 6.0 for 0.1012 g of sorbent.

L-Hap sorbent - Effect of amount: Figure 14 shows the evaluation of the pH of the aqueous solution. We observed that the pH solution decreases rapidly and

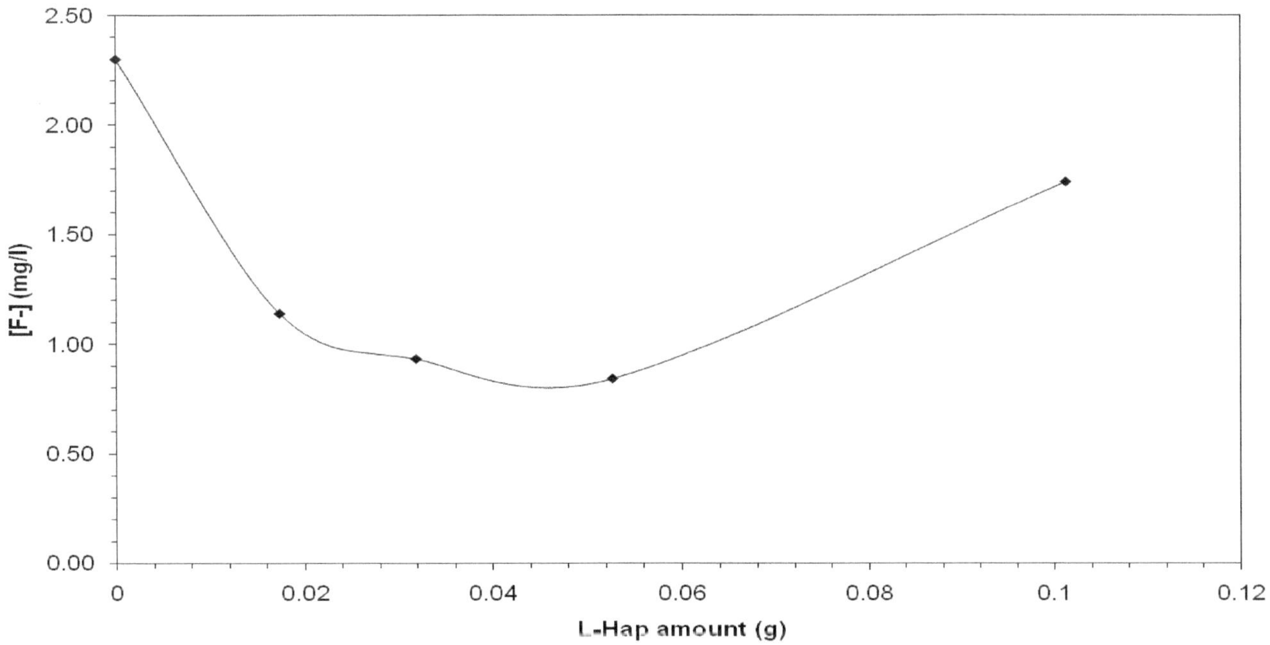

Figure 15. The evaluation of fluoride concentration of Koundoumawa field water during the defluoridation: Effect of sorbent dose of precipitated L-HAp.

continuously with an increasing dose of the L-Hap sorbent, contrary to synthetic solution. With the fluoride poisoning Koundoumawa field water, the defluoridation by L-Hap sorbent is mainly controlled by the dissolution-and-recrystallization mechanism, contrary to synthetic solution, where ion exchange is the dominant phenomenon. The minimum pH obtained is 6.5 for 0.1012 g of sorbent.

The evaluation of fluoride ion concentration with sorbent dose is shown in Figure 15. We observed that the fluoride ion concentration decreases with an increasing dose of sorbent. This phenomenon is still observed until the addition of 0.0527 g amount of L-Hap sorbent dose. The minimum fluoride ion concentration reached is 0,84 mg/l. This value satisfies the level of fluoride within the tolerance limit (WHO guideline value = 0.8 mg/l). After this dose, there is an increase of fluoride concentration, confirming the recrystallization-and-dissolution mechanism. For 0.0527 g amount of L-Hap sorbent with contact time of 72 h, the fluoride removal capacity of L-Hap is 11.63 mg/g for synthetic water and 2.80 for Koundoumawa field water. We can conclude that the presence of high concentration of bicarbonate ion in Koundoumawa field water is responsible of this low fluoride removal capacity of L-Hap during defluoridation of this water. Similar results for bicarbonate ion have also been reported (Dilip et al., 2010; Sairam et al., 2008; Sanjay et al., 2009).

Conclusion

From the present results, we can conclude as follows:

1. DCPD and L-Hap sorbents are appropriate for the defluoridation of synthetic solutions and Koundoumawa fluoride poisoning field waters.
2. The DCPD shows higher fluoride uptake capacity for defluoridation of synthetic water as compared to L-HAp. The uptake of fluoride in acidic pH is higher as compared to alkaline pH for synthetic solution.
3. The L-HAp shows higher fluoride uptake capacity for fluoride poisoning Koundoumawa field waters as compared to DCPD.
4. Within the first two hours, the sorption process is essentially controlled by ion exchange mechanism.
5. From 2 to 20 h, the sorption process is controlled by the dissolution-and-recrystallization mechanism.
6. From 20 to 72 h, the sorption process is again controlled by ion exchange mechanism
7. The fluoride removal capacity of DCPD and L-hap decreases with a raise in sorbent dose contrary to the fluoride ion concentration.
8. With the fluoride poisoning Koundoumawa field water, the defluoridation by L-Hap sorbent is mainly controlled by the dissolution-and-recrystallization mechanism, contrary to the synthetic solution, where ion exchange is the dominant phenomenon.

Conflict of interests

The authors did not declare any conflict of interest.

REFERENCES

Amit B, Eva K, Mika S (2011). Fluoride removal from water by adsorption - A review. Chem. Eng. J. 171(3):811-840.

Anirudhan TS, Maya RU (2007). Arsenic(V) removal from aqueous solutions using an anion exchanger derived from coconut coir pith and its recovery. Chemosphere, 66(1):60-66.

Boualia A, Mellah A, Aissaoui T, Menacer K, Silem A (1993). Adsorption of organic matter contained in industrial H_3PO_4 onto bentonite: batch-contact time and kinetic study, Appl. Clay Sci. 7:431-445.

Brown WE, Lehr JR (1959). Applictation of phase rule to the chemical behaviour of MCPM in soils. Soil Sci. Soc. Amer. Proc. 23:7-12.

Casciani FS, Condrate SrRA (1980). The infrared and Raman Spectra of Several Calcium Hydrogen Phosphates, Proc. 2nd Internatl., Congress on Phosphorus compounds, Boston, pp.175-190.

Chow LC, Markovic MI (1997). Calcium phosphates in biological and industrial systems. Boston: Kluwer Academic Publishers, pp. 67-84.

Diaz-Nava C, Olguin MT, Solache-Rios M (2002). Water defluoridation by Mexican heulandite-clinoptilolite, Sep. Sci. Tech. 37:3109-3128.

Dilip T, Sadhana R, Raju K, Siddharth M, Subrt J, Nitin L (2010). Magnesium incorporated bentonite clay for defluoridation of drinking water. J. Hazard. Mater. 180:122-130.

Fan X, Parker DJ, Smith MD (2003). Adsorption kinetics of fluoride on low cost materials, Water Res. 37:4929-4937.

Featherstone JDB, Glena R, Shariati M, Shields CP (1990). Dependence ofiln vitro demineralization of apatite and remineralization of dental enamel on fluoride concentration. J. Dent. Res. 69:620-625.

Hammari LEL, Laghzizil A, Barboux P, Lahlil K, Saoiabi A (2004). Retention of fluoride ions from aqueous solution using porous hydroxyapatite: structure and conduction properties, J. Hazard. Mater. 114:41-44.

JCPDS (Joint Committee on Powder Diffraction Standards) cards, (DCPD card number 09-0077).

JCPDS (Joint Committee on Powder Diffraction Standards) cards, (L-Hap card number 9-432).

Jiménez-Reyes M, Solache-Ríos M (2010). Sorption behavior of fluoride ions from aqueous solutions by hydroxyapatite, J. Hazard. Mater. 180(1-3):297-302.

Kalimuthu P, Natrayasamy V (2014). Synthesis of alginate bioencapsulated nano-hydroxyapatite composite for selective fluoride sorption. Carbohydrate Polymers, 112:662-667.

Kamble SP, Jagtap S, Labhsetwar NK, Thakare D, Godfrey S Devotta S, Rayalu SS (2007). Defluoridation of drinking water using chitin, chitosan and lanthanum-modified chitosan. Chem. Eng. J. 129:173-180.

Kevin JR, Kenneth TS (2014). Measurement of fluoride substitution in precipitated fluorhydroxyapatite nanoparticles. J. Fluorine Chem. 161:102-109.

Larsen MJ, Pearce EIF, Jensen SJ (1993). Defluoridation of Water at High pH with Use of Brushite, Calcium Hydroxide, and Bone Char. J. Dent. Res. 72:15-19.

LeGeros RZ, Lee D, Quirolgico G, Shirra WP, Reich L (1983). In vitro formation of Dicalcium Phosphate Dihydrate $CaHPO_4.2H_2O$, Scanning Electron Microscopy, pp. 407-418.

Lerch P, Lemp R, Kraheebuhl U, Bosset J (1966). Etude de l'hydrolyse de l'orthophosphate dicalcique dihydrat´e. Chimia, 20:430-432.

Low KS, Lee CK, Leo AC (1995). Removal of metals from electroplating wastes using banana pith, Bioresour. Technol. 51:227-231.

Lusvardi G, Malavasi G, Menabue L, Saladini M (2002). Removal of cadmium ion by means of synthetic hydroxyapatite,Waste Manag. 22:853-857.

Maiti GC, Freund F (1981). Influence of fluorine substitution on the proton conductivity of hydroxyapatite. J. Chem. Soc. Dalton Trans. 4:949-955.

Mameri N, Yeddou AR, Lounici H, Lounici D, Belhocine H, Grib BB (1998). Defluoridation of serpentrional Sahara water of North Africa by electrocoagulation process using bipolar aluminium electrodes, Water Res. p.32.

Manzola AS, Mgaidi A, Laouali MS, El Maaoui M (2014). On precipitated calcium and magnesium phosphates during hard waters softening by monosodium phosphate. Desalination and water treatment: 52(25-27):4734-4744.

Masamoto T, Tetsuji C (2004). Reaction between calcium phosphate and fluoride in phosphogypsum. J. Eur. Ceramic Soc. 26(4-5):767-770.

Masanobu K, Ryohei I, Koji I (2013). Preparation and evaluation of spherical Ca-deficient hydroxyapatite granules with controlled surface microstructure as drug carriers. Mat. Sci. Eng. C. 33(4):2446-2450.

Medellin-Castillo NA, Leyva-Ramos R, Padilla-Ortega E, Ocampo-Perez R, Flores-Cano JV, Berber-Mendoza MS (2014). Adsorption capacity of bone char for removing fluoride from water solution. Role of hydroxyapatite content, adsorption mechanism and competing anions. J. Indus. Eng. Chem. 20(6):4014-4021.

Meenakshi S, Anitha P, Karthikeyan G, Appa Rao BV (1991). The pH dependence of efficiency of activated alumina in defluoridation of water, Indian. J. Environ. Prot. 11:511-513.

Meenakshi S, Viswanathan N (2007). Identification of selective ion exchange resin for fluoride sorption, J. Colloid Interface Sci. 308:438-450.

Mourabet MM, El Rhilassi A, El Boujaady H, Bennani-Ziatni M, El Hamri R, Taitai A (2012). Removal of fluoride from aqueous solution by adsorption on hydroxyapatite (HAp) using response surface methodology, J. Saoudi Chem. Soc. doi:10.1016/j.jscs.2012.03.003.

Nash CL, Liu JC (2010). Removal of phosphate and fluoride from wastewater by a hybrid precipitation–microfiltration process. Separation and Purification Technology, 74(3):329-335.

Okazaki M, Aoba T, Doi Y, Takahashi J, Moriwaki J (1981). Solubility and crystallinity in relation to fluoride content of fluoridated hydroxyapatites. J. Dent. Res. 60:845-849.

Okazaki M (1992). Heterogeneous synthesis of fluoridated hydroxyapatites. Biomat.13:749-754.

Rapport Mission Internationale d'Enquête de la Fédération Internationale des Ligues des Droits de l'Homme (2002), N° 341 octobre.

Rao NV, Mohan R, Bhaskaran CS (1998). Studies on defluoridation of water, J. Fluorine Chem. 41:17-24.

Sairam CS, Natrayasamy V, Meenakshi S (2008). Defluoridation chemistry of synthetic hydroxyapatite at nano scale: Equilibrium and kinetic studies, J. Hazard. Mater. 155(1-2):206-215.

Sandrine B, Ange N, Didier B, Eric C, Patrick S (2007). Removal of aqueous lead ions by hydroxyapatite: equilibria and kinetic process, J. Hazard. Mater. 139:443-446.

Sanjay PK, Priyadarshini D, Sadhana SR, Nitin KL (2009). Defluoridation of drinking water using chemically modified bentonite clay. Desalination 249(9):687-693.

Sekar C, Kanchana P, Nithyaselvi R, Girija EK (2009). Effect of fluorides (KF and NaF) on the growth of dicalcium phosphate dihydrate (DCPD) crystal. Mater. Chem. Phys. 115(1):21-27.

Smiciklas I, Dimovic S, Plecas I, Mitric M (2006). Removal of Co^{2+} from aqueous solutions by hydroxyapatite, Water Res. 40:2267-2274.

Sumit G, Rajkamal B, Sampath KTS (2012). Effects of nanocrystalline calcium deficient hydroxyapatite incorporation in glass ionomer cements. J. Mechan. Behav. Biomed. Mater. 7:69-76.

Sundaram CS, Natrayasamy V, Meenakshi S (2008). Uptake of fluoride by nano-hydroxyapatite/chitosan, a bioinorganic composite. Bioresource Technology, 99(17): 8226-8230.

Taewook Y, Chulki K, Jaeyoung J, Il Won K (2012). Regulating fluoride uptake by calcium phosphate minerals with polymeric additives. Colloids and Surfaces A: Physicochem. Eng. Aspects 401:126-136.

Takahashi T, Tanase S, Yamamoto O (1978). Electrical conductivity of some hydroxyapatites. Electrochim Acta. 23:369-373.

Tse-Ying Liu, San-Yuan Chen, Dean-Mo Liu, Sz-Chian Liou (2005). On the study of BSA-loaded calcium-deficient hydroxyapatite nano-carriers for controlled drug delivery. J. Contr. Release 107(1):112-121.

UNICEF (1998). Fluoride in water: An overview." WES (Warter Environment & Sanitation) Section, Programme Division, UNICEF 3 UN Plaza, New York, NY, USA 10017 Phone: (212) 824-6000 Fax: (212) 824-6480

Vega ED, Pedregosa JC, Narda GE, Morando PJ (2003). Removal of oxovanadium (IV) from aqueous solutions by using commercial crystalline calcium hydroxyapatite, Water Res. 37:1776-1782.

World Health Organisation (WHO) (1996). Fluoride, Guidelines for Drinking Water Quality, vol. II, 2nd ed., Geneva, pp. 231-237.

Wen L, Lei Z, Longhua P, Christian R (2011). Fluoride removal performance of glass derived hydroxyapatite. Mater. Res. Bull. 46(2):205-209.

Xiaolin Y, Shengrui T, Maofa G, Junchao Z (2013). Removal of fluoride from drinking water by cellulose@hydroxyapatite nanocomposites. Carbohydr. Polymers 92(1):269-275.

Yulun N, Chun H, Chuipeng K (2012). Enhanced fluoride adsorption using Al (III) modified calcium hydroxyapatite. J. Hazard. Mater. pp. 233-234:194-199.

Yu W, Ningping C, Wei W, Jing C, Zhenggui W (2011). Enhanced adsorption of fluoride from aqueous solution onto nanosized hydroxyapatite by low-molecular-weight organic acids, Desalination, 276(1-3):161-168.

Changes in microbial and soil organic matter following amendment with olive mill wastewaters

Ali Mekki, Abdelhafidh Dhouib and Sami Sayadi

Laboratoire des bioprocédés, Pôle d'Excellence Régionale AUF (PER-LBP), Centre de Biotechnologie de Sfax, BP: « K », 3038 Sfax, Tunisie.

In this work, the occurrence of the untreated and treated olive mill wastewaters (OMW) amendment on the soil organic matter recycling and on its microbial biomass evolution were investigated. Compared to the control, soils amended with untreated and treated OMW showed high levels of organic and mineral matters. Soil amended with untreated OMW presents low levels of total and inorganic nitrogen (0.38 ± 0.03 and 0.08 ± 0.02 mg g^{-1} dry soil). Treated OMW had a little content of pollutants (COD = 4 g l^{-1}± 0.4; phenolic compounds = 0.6 g l^{-1}± 0.04) and organic matter brought by these residues was rapidly mineralized in the soil. The number of heterotrophic bacteria was increased (from 54 ± 5 .10^5 CFU g^{-1} dry soil in control soil to 123 ± 11. 10^5 CFU g^{-1} dry soil) in response of the OMW amendment, mainly following C/N ratio correction. The amendment of the soil with untreated OMW improved the soil carbon content (2.18 times higher) while the specific respiration remained very low. However, the amendment with treated OMW positively affects the soil specific respiration that increases from 6.1 in control soil to 9.75 in soil amended with treated OMW. This phenomenon was accompanied by an enhancement (from 12 ± 2 .10^4 CFU g^{-1} dry soil in control soil to 83 ± 5. 10^4 CFU g^{-1} dry soil in soil amended with OMW) of nitrifiers number , urease and ammonium oxidases activities.

Key words: Olive mill wastewaters, soil, organic matter, nitrifiers, mineralization.

INTRODUCTION

Olive oil industries are of fundamental economic importance for many Mediterranean countries that account for approximately 95% of the worldwide olive oil production (Dhouib et al., 2006a). Olive mill wastewaters (OMW) are the liquid by-product generated during olive oil production (Kapellakis et al., 2006). They are characterized by the following chemical properties: a very high content of organic matter (COD between 60 and 185 g l^{-1}; BOD_5 between 14 and 75 g l^{-1}), a low pH, and high polyphenols, potassium and phosphorus contents (Rinaldi et al., 2003). Extremely high organic load and the toxic nature of olive mill wastewaters (OMW) prevent their direct discharge into domestic wastewaters treatment systems. To solve the problems associated with these wastewaters, different disposal methods have been proposed (Sayadi and Ellouz, 1995; Paredes et al., 2000; Casa et al., 2003; Dhouib et al., 2006b). However, the most frequently used method nowadays is the direct application to agricultural

Table 1. Physico-chemical characteristics of untreated and treated olive mill wastewaters.

Characteristics	Untreated OMW	Treated OMW
pH (25°C)	5.1 ± 0.2	8.2 ± 0.2
Electrical conductivity (25°C) (dS m^{-1})	8.9 ± 0.1	14.1 ± 0.1
Chemical oxygen demand (g l^{-1})	72 ± 2.8	4 ± 0.4
Biochemical oxygen demand (g l^{-1})	13 ± 0.9	1.8 ± 0.14
COD/BOD_5	5.53 ± 0.72	2.2 ± 0.32
Salinity (g l^{-1})	6.75 ± 0.66	12.1 ± 1.1
Water content (g l^{-1})	948 ± 17.2	984 ± 19.5
Total solids (g l^{-1})	52 ± 2.98	15.9 ± 0.7
Ash (g l^{-1})	8 ± 0.43	10.15 ± 0.5
Volatile solids (g l^{-1})	44 ± 1.85	4.8 ± 0.22
Total organic carbon (g l^{-1})	25.52 ± 1.18	3.2 ± 0.12
Total nitrogen Kjeldahl (g l^{-1})	0.6 ± 0.06	0.21 ± 0.03
Carbon/Nitrogen	43 ± 1.8	15 ± 0.65
P (mg l^{-1})	36 ± 3.6	15 ± 1.1
Na (g l^{-1})	0.94 ± 0.09	0.86 ± 0.08
Cl (g l^{-1})	1.6 ± 0.15	1.3 ± 0.11
K (g l^{-1})	8.8 ± 0.8	5.34 ± 0.6
Ca (g l^{-1})	1.2 ± 0.11	3.2 ± 0.2
Fe (mg l^{-1})	32 ± 2.9	38.3 ± 3.3
Mg (mg l^{-1})	187 ± 18.8	281 ± 27.1
Ortho-diphenols (g l^{-1})	9.2 ± 1.8	0.6 ± 0.04
Toxicity by LUMIStox (% I_B)	99 ± 9	30 ± 2.9

soils as organic fertilizers (Fiestas and Borja, 1992; Nieto and Garrido, 1994; Mekki et al., 2006b; Mekki et al., 2007a). OMW physico-chemical composition and characteristics are well documented (Fiorentino et al., 2003; Allouche et al., 2004; Obied et al., 2005). Little is known, however, of the impact of OMW on the chemical, biochemical and biological properties of soil (Aggelis et al., 2003; Mekki et al., 2007b). Indeed, the effect of OMW addition on soil organic matter cycles is one of topics that need to be addressed from an agricultural point of view.

The present work was aimed at evaluating the occurrence of untreated and treated OMW amendment on the soil organic matter recycling and then on its microbial biomass evolution. The main goal is to know how much the OMW treatment application and its incubation in soil can remove its toxicity and their effects on the soil matter recycling.

MATERIALS AND METHODS

Untreated OMW origin

Untreated olive mill wastewater was taken from a three-phase discontinuous extraction factory located in Sfax, Tunisia.

Treatment of untreated OMW

The treated OMW was obtained with an integrated process based on aerobic fungal pre-treatment using *Phanerochaete chrysosporium* DSMZ 6909 followed by a decantation step then anaerobic digestion (Sayadi and Ellouz, 1995). The characteristics of the treated and untreated OMW are given in Table 1.

Physico-chemical analyses

The pH and the electrical conductivity were determined according to Sierra et al. (2001) standard method. Organic matter (OM) was determined by combustion of the samples in a furnace at 550°C for 4 h. Total organic carbon was determined by dry combustion (TOC Analyser multi N/C 1000). Total nitrogen was determined by Kjeldahl method (1883). Chemical oxygen demand (COD) was determined according to Knechtel method (1978). Five day biochemical oxygen demand (BOD_5) was determined by the manometric method with a respirometer (BSB-Controller Model 620 T WTW) and phenolic compounds (ortho-diphenols) were quantified by means of Folin-Ciocalteu colorimetric method using caffeic acid as standard (Box, 1983). The absorbance was determined at l = 765 nm.

Study sites and sampling

The soil of the study area was described in a previous study (Mekki et al., 2006 a). The field was divided in 5 plots. S_1 and S_2 were amended with 100 m^3 ha^{-1} of untreated OMW. The amendment is achieved without (S_1) or after (S_2) correction of the C/N/P ratio to 100/10/1. S_3 and S_4 were amended with 400 m^3 ha^{-1} of treated OMW. The amendment is achieved without (S_3) or after (S_4) correction of the C/N/P ratio to 100/10/1. The fifth plot, plot S_c, was not amended and served as a control.

Amendment was realized in February 2006. Soil samples (S_1, S_2,

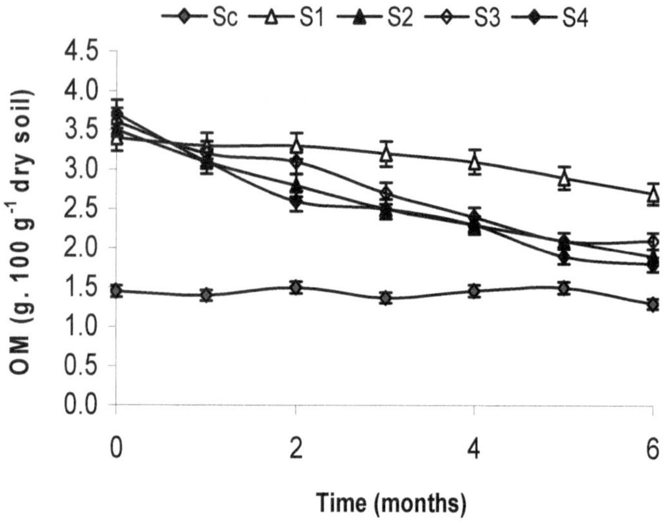

Figure 1. Soil organic matter evolution as a function of amended OMW and time.

S_3, S_4 and S_c) were collected monthly (during 6 months) from different parts of each plot from 0 to 20 cm deep, using a soil auger. All soil samples, taken from each plot were then mixed, air-dried, sieved with a mesh size of 450 µm and stored at 4°C prior to use. Water content was immediately determined before air-drying the sample.

Physico-chemical soils analyses

pH and electrical conductivity were determined according to Sierra et al. (2001) standard method. Organic matter was determined after furnacing samples at 550°C for 4 h. Total nitrogen was determined by the Kjeldahl method (1883). Inorganic nitrogen was determined by the Kandeler method (1995). Organic nitrogen is determined by difference between total and inorganic nitrogen.

Biological soil analyses

10 g of each soil sample was suspended in an erlenmeyer flask containing 90 ml of a sterile solution (0.2% of sodium polyphosphate $(NaPO_3)_n$ in distilled water, pH 7.0). The flask was shaken at 200 rpm for 2 h. Serial 10-fold dilutions of the samples in a 0.85% NaCl solution were plated in triplicate on PCA at 30°C for total bacterial counts. Nitrifiers and denitrifiers microorganisms were enumerated by the most probable number (MPN) procedure (Trolldenier, 1995). The soil biological activity was achieved by measuring respirometric evolution in the aerobic condition (Öhlinger, 1995). Urease and ammonium oxidases activities were determined according to Kandeler et al. (1999).

Statistical analysis

For physico-chemical analyses, three replications were used for each parameter. For biological soils analyses, each soil sample was analysed in duplicate and the dilution series were plated in triplicate for each medium. Data were analysed using the ANOVA procedure. Variance and standard deviation were determined using Genstat 5 (second edition for windows).

RESULTS AND DISCUSSION

Treated and untreated OMW characteristics

The physico-chemical characteristics of untreated and treated OMW are summarized in Table 1. The high pollutant load of untreated OMW and its acidity could be observed. However, treated OMW was a slightly alkaline effluent, rich in inorganic loads such as potassium, calcium, magnesium and iron. Its content of phenolic compounds was lower than 1 g l^{-1}, reflecting a significant reduction of its toxicity from 99%I_B before treatment to only 30%I_B after treatment application (Table 1).

OMW impact on the physico-chemical soil parameters

Untreated and treated OMW effects on some soil physico-chemical parameters such as pH, electrical conductivity (EC), organic matter (OM), soil nitrogen and soil C/N ratio have been studied. Results showed that several chemical and biochemical properties of the investigated soil changed in response to the application of OMW. The addition of treated or untreated OMW with or without C/N ratio correction didn't show a significant effect on the initial soil pH. Indeed in spite of the initial untreated OMW acidity, the follow-up of this parameter during 6 months showed a weak reduction of the soil pH (0.2 units). Then, the acidity of the untreated OMW was compensated by the soil carbonate alkalinity as was shown by Sierra et al. (2001), whereas the addition of treated OMW caused a weak increase. Indeed, the alkalinity of this waste was not buffered by the soil components.

In the same way, the count of OMW induced an increase of the soil electric conductivity. This increase is proportional to the added OMW quantity. The increase of the soil Salinity could result from the main ionic species, sodium chloride and sulphate, coming from the treated or untreated OMW (Zanjari and Nejmeddine, 2001).

The studied soil is initially poor in organic matter. OMW addition improves the soil organic and mineral matters amounts. The organic matter supplement brought by 400 m^3 ha^{-1} of treated OMW is comparable to the one brought by 100 m^3 ha^{-1} of untreated OMW (Figure 1). The follow-up of the biodegradation kinetics of this organic matter brought by OMW shows that for the same quantity of added organic matter, soil receiving the treated OMW (S_3) presents a potential of biodegradation 3 times higher than the one of soil receiving the untreated OMW (S_1). The correction of the C/N ratio accelerates this biodegradation for soil irrigated with untreated OMW (S_2) and show a low effect in the case of soil receiving the treated OMW (S_4) (Figure 1).

The increase of organic and mineral nutrient contents may have a beneficial effect on the soil fertility. Nevertheless, the difficulty of untreated OMW organic matter biodegradation can be explained by its richness in phenolic compounds which are toxic for the soil biological

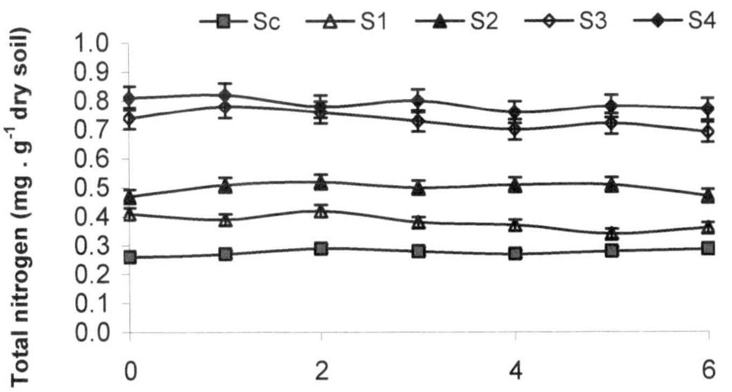

Figure 2. Soil total nitrogen evolution as a function of amended OMW and time.

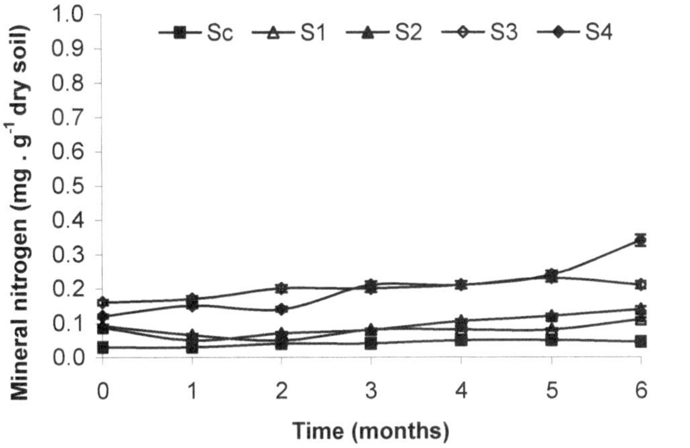

Figure 4. Soil inorganic nitrogen evolution as a function of amended OMW and time.

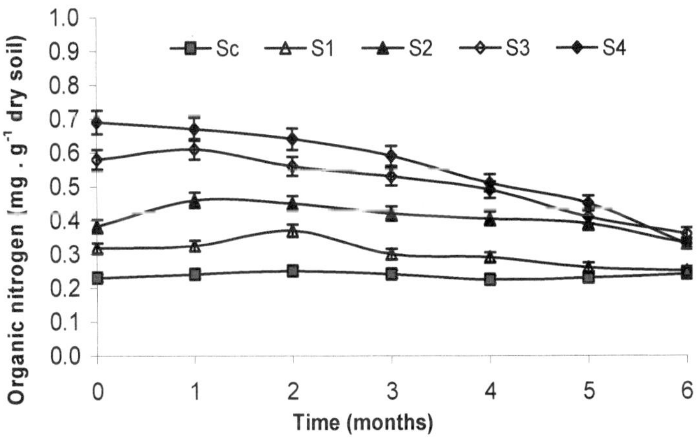

Figure 3. Soil organic nitrogen evolution as a function of amended OMW and time.

activity (Cox et al., 1998).

The OMW nitrogen content remained negligible compared to the other elements as the carbon and the potassium (Table 1). Control soil (S_C) is initially poor in nitrogen. Soil irrigated with the treated OMW presents more important total nitrogen content in that of soil receiving untreated OMW (Figure 2). The weak reduction of the total nitrogen or even its constancy during the time can be explained by the retraining of the different soil nitrogen shapes.

The organic nitrogen brought by the untreated OMW presents a slow mineralization. Indeed the affluence in suspended matters of untreated OMW enhances the immobilization of soil nitrogen. However, the organic nitrogen brought by the treated OMW turns quickly into inorganic nitrogen, what explains the important reduction of the content in organic nitrogen of soil irrigated with treated OMW (Figures 3 and 4).

The slow mineralization of the untreated OMW organic nitrogen is due to the composition of this material that facilitates its adsorption to soil particles since the untreated OMW is very rich in suspended matters. Negative effects of raw OMW on soil properties have also been recorded, including the immobilization of available nitrogen (Kissi et al., 2001).

OMW impacts on the biological soil parameters

OMW effects on the soil aerobic heterotrophic bacteria, nitrifiers and denitrifiers microorganisms and on the soil enzymatic and respirometric activities have been studied.

The aerobic heterotrophic bacteria counted on the studied soil are relatively weak (10^5 to 10^6 CFU $g^{-1}$of dry soil). Results showed an initial increase in total number of heterotrophic bacteria after the OMW amendment, mainly following C/N ratio correction. In line with this finding, Paredes et al. (1987) reported also an increase in the total viable counts in the soil polluted with OMW (Table 2).

The control soil was very poor in organic matter and particularly in nitrogen, so the number of nitrifiers microorganisms is very weak. The OMW addition increases in a meaningful manner their number. This increase is more remarkable in the case of soil receiving treated OMW (Figure 5). However, from the 4th month after OMW addition, a reduction of these microorganisms number in all tested samples was observed. This decrease could be explained by the reduction of the soil organic nitrogen content. As it is the case for the nitrifiers, the OMW addition improves the presence of denitrifiers whose number increases correlatively with the OMW quantity (data not shown).

Indeed, nitrifiers and denitrifiers microorganisms play a critical role in the natural nitrogen cycle (Oved et al., 2001; Mendum and Hirsch, 2002). This microflora could be affected by a variety of chemical conditions including aromatic compounds and salts. The number of nitrifiers shifted from the CFU g^{-1} number ranging from 11.2 to 12.4 x 10^4 in the control soil to CFU g^{-1} number ranging

Table 2. Aerobic heterotrophic bacteria count in the different plots and time (CFU $10^5 g^{-1}$).

Time (month)	March	April	May	Jun	Jul	August
Sc	54±5	49±5	52±5	49±5	50±5	51±5
S_1	75±7	78±7	72±7	70±6	63±6	64±6
S_2	88±8	103±9	98±8	93±9	96±9	94±8
S_3	117±10	114±10	108±10	106±1	105±10	98±9
S_4	123±11	126±11	117±11	109±10	111±10	107±1

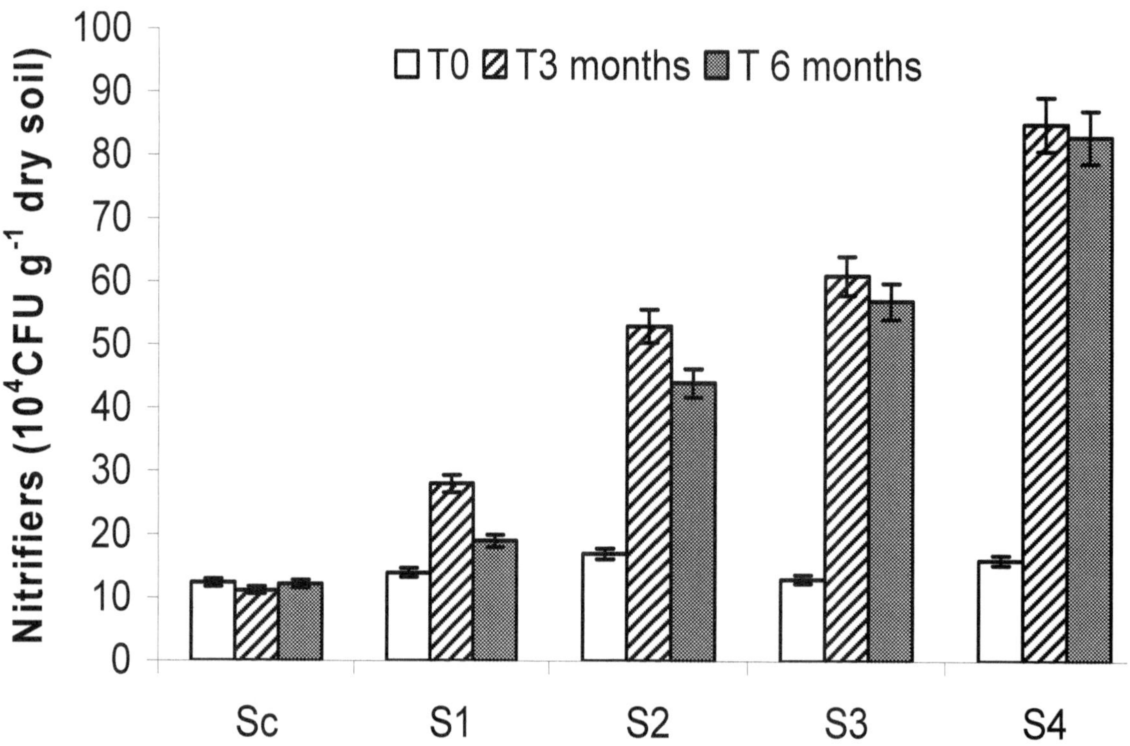

Figure 5. Nitrifiers numbers evolution as a function of amended OMW and time.

from 13 to 61 10^4 in the soil amended with treated OMW. Additionally, the decomposition of OMW by soil microbes could have induced oxygen depletion in the surface soil, thereby inhibiting aerobic microbial activity (Kowalchuk et al., 2000; Schloter et al., 2003; Gianfreda et al., 2006). Besides, the OMW addition improves the presence of denitrifiers whose number increases correlatively with the OMW quantity. Indeed, these microorganisms appear more resistant to the poisonous compounds that the nitrifying bacteria.

The need to measure the activities of a large number of enzymes has been emphasized to provide information on soil microbial activity (Sukul, 2006). Urease and ammonium oxidases constitute the two major enzymes of nitrogen metabolism. Urease plays a key role in the nitrogen cycle; it contributes to the transformation of the organic nitrogen in ammoniacal assimilated nitrogen ($N\text{-}NH_4$). Ammonium oxydases assures the transformation of the product of the ammonification in assimilated nitrogen by plants ($N\text{-}NO_3$) (Tscherko et al., 2003). Soils urease and ammonium oxidases activities are stimulated distinctly in soils irrigated with treated OMW, whereas the untreated OMW addition inhibits these two enzymatic activities (Table 3).

In this context, Deni and Penninckx (1999) reported that addition of hydrocarbon to an uncontaminated soil stimulated immobilization of nitrogen and reduced nitrification and soil urease activity.

The soil respirometric activity constitutes an important parameter to understand its biologic activity (Javorekova et al., 2001). A respirometric test was achieved on soils during 49 days. The specific respiration rate expressed as the ratio of $C\text{-}CO_2/Ctot$ for the different soil samples is shown in Figure 6. The amendment of the soil with untreated OMW increased the carbon content while the specific

Table 3. Urease and ammonium oxidases activities detected in soils 6 months after OMW application.

Enzymatic activities	Urease (µg NH_4 g^{-1} 2 h^{-1})	Ammonium oxidases (µg NO_2 g^{-1} 24 h^{-1})
Sc	30±2.8	0.22±0.02
S_1	18±2	0.17±0.01
S_2	29±2.5	0.21±0.02
S_3	84±7	0.69±0.06
S_4	98±8.2	0.89±0.07

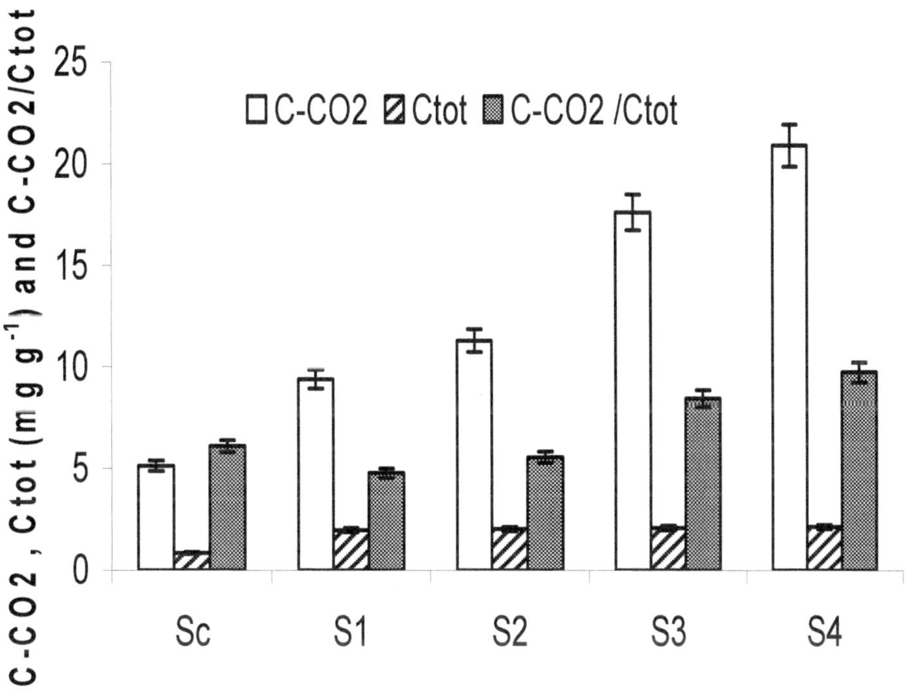

Figure 6. Specific respiration $C\text{-}CO_2/C_{tot}$, cumulative $C\text{-}CO_2$, and total carbon C_{tot} of the soil samples studied.

respiration remained very low. Indeed, Rinaldi et al. (2003) reported that the affluence of the untreated OMW organic matter in toxic phenolic compounds makes difficult its biodegradation in the nature. However, the amendment of the soil with treated OMW positively affects its specific respiration. The initial correction of the C/N ratio enhances the specific respiration rate (Figure 6).

Conclusion

Olive mill wastewaters constitute a serious environmental problem. Our results seem to confirm that the impact of OMW on soil properties was the result of opposite effects, depending on the relative amounts of beneficial and toxic organic and inorganic compounds present.

Soil organic matter recycling is assured by its autochthonous microflora. A contaminated soil constitutes an environment limiting for the microbial normal development, and therefore an unbalance of the soil matter cycles. OMW treatment before their application on soil is therefore necessary to limit negative impact on the soil biological activity.

Treated olive mill wastewaters with white-rot fungi followed by anaerobic digestion contains again relatively high amounts of dissolved and suspended organic matter in a large volume of water and are potential candidates for use as liquid organic amendment especially for soils and crops. The use of these waters for soil amendment improves its organic pool, bring assimilated nitrogen and stimulate the organic matter mineralization.

Conflict of Interests

The author(s) have not declared any conflict of interests.

ACKNOWLEDGMENTS

This research was funded by E.C program "Medusa water" contract ICA-CT-1999-00010 and contracts programme (MESRS, Tunisia).

REFERENCES

Aggelis G, Iconomou D, Christou M, Bokas D, Kotzailias S, Christou G, Tsagou V, Papanikolaou S (2003). Phenolic removal in a model olive oil mill wastewater using *Pleurotus ostreatus* in bioreactor cultures and biological evaluation of the process, Water Res. 37:3897-3904.

Allouche N, Fki I, Sayadi S (2004). Toward a high yield recovery of antioxidants and purified hydroxytyrosol from olive mill wastewaters, J. Agri. Food Chem. 52:267-273.

Box JD (1983). Investigation of the Folin-Ciocalteau phenol reagent for the determination of polyphenolic substances in natural waters, Water Res. 17:511-522.

Casa R, D'Annibale A, Pieruccetti F, Stazi SR (2003). Reduction of the phenolic components in olive-mill wastewater by enzymatic treatment and its impact on durum wheat (*Triticum durum Desf.*) germinability, Chemosphere. 50:959-966.

Cox L, Celis R, Hermosin MC, Beker A, Cornejo J (1998). Porosity and herbicide leaching in soils amended with olive-mill wastewater, Agr. Ecosyst. Environ. 65:151-161.

Deni J, Penninckx MJ (1999). Nitrification and autotrophic nitrifying bacteria in a hydrocarbon-polluted soil, Appl. Environ. Microbiol. 56 (9):4008-4013.

Dhouib A, Aloui F, Hamad N, Sayadi S (2006 a). Pilot-plant treatment of olive mill wastewaters by *Phanerochaete chrysosporium* coupled to anaerobic digestion and ultrafiltration, Process Biochem. 41:159-167.

Dhouib A, Ellouz M, Aloui F, Sayadi S (2006 b). Effect of bioaugmentation of activated sludge with white-rot fungi on olive mill wastewater detoxification, Lett. Appl. Microbiol. 42:405-411.

Fiestas JA, Borja R (1992). Use and treatment of olive mill waste-water: current situation and prospects in Spain, Grasas y Aceites. 43:101-106.

Fiorentino A, Gentili A, Isidori M, Monaco P, Nardelli A, Panella A, Fabio T (2003). Environmental effects caused by olive mill waste waters: Toxicity comparison of low-molecular–weight phenol components, J. Agri. Food chem. 51:1005-1009.

Gianfreda L, Piotrowska A, Iamarino G, Antonietta Rao M (2006). Short-term effects of olive mill waste water (OMW) on chemical and biochemical properties of a semiarid Mediterranean soil, Soil Biol. Biochem. 38:600-610.

Javorekova S, Stevlikova T, Labuda R, Ondrisik P (2001). Influence of xenobiotic on the biological soil activity. J. Central Eur. Agric. 2:3-4.

Kandeler E (1995). Total nitrogen. In: Schinner F., Ohlinger R., Kandeler E., Margesin R. (Eds.). Methods in Soil Biology, Springer. Berlin, pp. 406-408.

Kandeler E, Stemmer M, Klimanek EM (1999). Response of soil microbial biomass, urease and xylanase within particle size fractions to long-term soil management, Soil Biol. Biochem. 31, 261-273.

Kapellakis IE, Tsagarakis KP, Avramaki C, Angelakis AN (2006). Olive mill wastewater management in river basins: A case study in Greece, Agri.Water Manag. 82:354370.

Kissi M, Mountadar M, Assobhei O, Gargiulo E (2001). Roles of two white-rot basidiomycete fungi in decolourization and detoxification of olive mill waste water, Appl. Microbiol. Biotechnol. 57:221-226.

Kjeldahl J (1883). A new method for the determination of nitrogen in organic matter, Anal. Chem. 22:366.

Knechtel RJ (1978). A more economical method for the determination of chemical oxygen demand, Water Pollut. Control. (May/June) pp.25-29.

Kowalchuk GA, Stienstra AW, Heilig GH, Stephen JR, Woldendorp JW (2000). Molecular analysis of ammonia-oxidizing bacteria in soil of successional grasslands of the Drentsche A (The Netherlands), FEMS Microbiol Ecol. 31:207-215.

Nieto Martinez L, Garrido H (1994). El Alpechin: Un problema medioambiental en vias de soluci6n (1), Quibal. 41:755-765.

Mekki A, Dhouib A, Sayadi S (2006 a). Changes in microbial and soil properties following amendment with treated and untreated olive mill wastewater, Microbiol. Res. 161:93-101.

Mekki A, Dhouib A, Aloui F, Sayadi S (2006 b). Olive wastewater as an ecological fertiliser, Agron. Sustain. Dev. 26:61-67. DOI: 10.1051/agro: 2005061.

Mekki A, Dhouib A, Sayadi S (2007 a). Polyphenols dynamics and phytotoxicity in a soil amended by olive mill wastewaters, J. Environ. Manage. 84:134140.

Mekki A, Dhouib A, Fki F, Sayadi S (2007 b). Assessment of toxicity of the untreated and treated olive mill wastewaters and soil irrigated by using microbiotests, Ecotoxicol. Environ. Saf. doi:10.1016/j.ecoenv.2007.04.008.

Mendum TA, Hirsch PR (2002). Changes in the population structure of b-group autotrophic ammonia oxidizing bacteria in arable soils in response to agricultural practice, Soil Biol. Biochem. 34: 1479-1485.

Obied HK, Allen MS, Bedgood DR, Prenzler PD, Robards K, Stockmann R (2005). Bioactivity and analysis of biophenols recovered from olive mill waste, J. Agri. Food chem. 53:823-837.

Öhlinger R (1995). Soil respiration by titration. In: Schinner F., Öhlinger R., Kandeler E., Margesin R. (Eds.). Methods in soil biology. Springer, Berlin, pp. 95-98.

Oved T, Shaviv A, Goldrath T, Mandelbaun RT, Minz D (2001). Influence of effluent irrigation on community composition and function of ammonia-oxidizing bacteria in soil, Appl. Environ. Microbiol. 67: 3426-3433.

Paredes MJ, Moreno E, Ramos-Cormenzana A, Martinez J (1987). Characteristics of soil after pollution with waste waters from oil extraction plants, Chemosphere. 16:1557-1564.

Paredes C, Roig A, Bernal MP, Sanchez-Monedero MA, Cegarra J (2000). Evolution of organic matter and nitrogen during co-composting of olive mill wastewater with solid organic wastes, Biol. Fert. Soils. 32:222-227.

Rinaldi M, Rana G, Introna M (2003). Olive-mill wastewater spreading in southern Italy: effects on a durum wheat crop, Field Crops Res. 84: 319-326.

Sayadi S, Ellouz R (1995). Roles of lignin peroxidase and manganese peroxidase from *Phanerochaete chrysosporium* in the decolourization of olive mill wastewaters, Appl. Environ. Microbiol. 61: 1098-1103.

Schloter M, Dilly O, Munch JC (2003). Indicators for evaluating soil quality, Agr. Ecosyst. Environ. 98:255-262.

Sierra J, Marti E, Montserrat G, Cruanas R, Garau MA (2001). Characterization and evolution of a soil affected by olive oil mill wastewater disposal, Sci. Total Environ. 279:207-214.

Sukul P (2006). Enzymatic activities and microbial biomass in soil as influenced by metalaxyl residues, Soil Boil. Biochem. 38:320-326.

Trolldenier G (1995). Nitrifiers by MPN method. In: Schinner F., Öhlinger R., Kandeler E., Margesin R. (Eds.). Methods in soil biology. Springer, Berlin, pp. 32-36.

Tscherko D, Bölter M, Beyer L, Chen J, Elser J, Kandeler E, Kuhn D, Blume H-P (2003). Biomass and enzyme activity of two soils transect at King George Island, Maritime Antarctica; Arctic, Antarctic, and Alpine Research. 35:34-47.

Zanjari B, Nejmeddine A (2001). Impact of spreading olive mill waste waters on soil characteristics: Laboratory experiments, Agronomie. 21:749-755.

Effects of different NaCl Concentrations on germination and seedling growth of *Amaranthus hybridus* and *Celosia argentea*

Amukali, O. [1], **Obadoni, B. O.**[2] **and Mensah, J. K.**[2]

[1]Geography and Environmental Management Department, Niger Delta University, Wilberforce Island, Bayelsa State, Nigeria.
[2]Botany Department, Faculty of Natural Sciences, Ambrose Alli University, Ekpoma, Edo State, Nigeria.

Salinity refers to the salt content of any given system. By nature, arid soils are naturally saline. Soils could also acquire salinity due to some agricultural practices like irrigation. There is need therefore for search on halophytes that could adapt to such soils and be used to reclaim soils contaminated with salts. Thus, this present study focused on investigating *Amaranthus hybridus* and *Celosia argentea* with respect to their potentials and suitability for use in saline environments. This study was conducted in two phases; germination (laboratory-based) and seedling tests (field-based). Shoot length, root length, root/shoot ratio, total length, fresh weight, dry weight, dry matter content, leaf area index, leaf number as well as relative water content were parameters used for assessing results for the field-based test. Germination results showed that only the control (0 mM) showed 100 % germination in both species. Germination percentages decreased steadily with increasing NaCl concentrations in both species. Growth was steadily stimulated in both species at lower NaCl levels with best growth stimulation at 50 mM NaCl but was adversely affected by higher NaCl concentration levels (75, 100 and 150 mM). Both species showed almost the same phenomenon for fresh weight, dry weight, dry matter, leaf area index, number of leaves as well as relative water concentration. These show that both species would best be cultivated under moderate than low or high saline concentration. The findings in this study showed that it is best for seedlings of both species of plants to be raised in nurseries free of NaCl to attain 100% germination and after which, the seedlings could then be transplanted to moderately saline soils for maximum growth and development. This has strong implications for maximizing food productivity and ensuring food security.

Key words: Germination, early seedling growth, NaCl, salinity, *Amaranthus hybridus, Celosia argentea*.

INTRODUCTION

Salinity refers to the salt content of any given system. According to Herr (2005), salinity results from the build-up of the following minerals which are deposited by evaporating water: Na+, K+, Ca^{2}+, Mg^{2}+, Cl^{-}, HCO^{3-}, and

SO_2^{4-}, respectively. Addition of salts to water lowers its osmotic potential, resulting in decreased availability of water to the root cells (Sairam et al., 2002). High salt concentrations could affect various physiological processes in plants. Ratnakar and Rai (2013) stated that high salt concentration hampers vital processes such as seed germination, seedling growth and vigour, vegetative growth, flowering as well as fruit set. Thus, high salt concentration ultimately reduces crop yield and the quality of the produce (Sairam and Tyagi, 2004).

According to Galston et al. (1980), soil salinity is becoming a serious worldwide problem as more land is irrigated thoroughly and heavily fertilized. Salinity is one of the major abiotic stresses in arid and semi-arid regions that substantially reduce the yield of major crops by more than 50% (Bray, 2000). Salinity affects agricultural production in a large proportion worldwide (Byordi et al., 2010). Salinity affects about 7% of the world's land area, amounting to about 930 million ha (Munns, 2002), thus seriously limiting crop production, especially the sensitive ones (Zadeh and Naeini, 2007). The problems of salinization are increasing, either due to bad irrigation, drainage or agricultural practices (Al-Seedi and Gatteh, 2010).

Cuartero et al. (2006) opined that although salt stress affects all growth stages of a plant, but that seed germination and seedling growth stages are known to be more sensitive for most plant species. In addition, Ibid (2006) further stressed that germination and seedling stages are predictive of plant growth responses to salinity.

Amaranthus hybridus is a robust annual herb (Olorode, 1984), that is cultivated for its nutritional value (Oguntona, 1998) and used as food by man and other animals. *C. argentea*, on the other hand, is a short-lived annual herb, slow growing and more drought-resistant than *A. hybridus* (Ogunwenmo et al., 2010). Both *A. hybridus* and *C. argentea* belong to the family, Amaranthaceae (Dutta and Dutta, 2008).

Byordi (2010) studied the influence of salt stress on seed germination, growth and yield of Canola cultivars. It was found that significant differences existed between influence of salt stress on the cultivars during germination and vegetative growth respectively. Al-Seedi and Gatteh (2010) while studying the effects of salinity on seed germination, growth and organic compounds of Mung bean plant [*Vignia radiate (L.) Wilczek*] found that an increase in the salinity caused a corresponding decrease in germination rate, growth parameters and the carbohydrate content of the Mung bean plant. Ratnakar and Rai (2013) observed while studying effects of NaCl salinity on seed germination and early seedling growth of *Trigonella foenum-Graecum L. Var Peb* that increasing NaCl concentrations caused a gradual decrease in root length, shoot length, fresh weight and dry weight of the growing seedling. In addition, Katambe et al. (1998) stated while studying the effects of germination and seedling growth of two *Atriplex* species *(Chenopodiaceae)* that NaCl caused a greater increase in nuclear volume than iso-osmotic PEG solutions. In addition, Ogunwenmo et al. (2010) studied effects of brewery, textile and paint effluent on seed germination of *A. hybridus* and *C. argentea* and stated that industrial effluents affected seed germination at different rates.

However, information pertaining to the effects of salinity due to varying concentrations of NaCl on *A. hybridus* and *C. argentea* are hard to come by. It is important to study the effects of different NaCl concentrations on germination and seedling growth of both crop species as that would help in establishing the tendencies of both crop species to be used in saline environments, most especially as there appears to be some level of universality in the occurrence of NaCl in most saline environments.

MATERIALS AND METHODS

Salt preparation

Salt solution of 1 M NaCl was prepared by dissolving 58.8 g of NaCl crystals in a universal bottle upon which distilled water was added onto to make up 1 L. Corresponding rations of 25/1000, 50/1000, 75/1000, 100/1000 and 150/1000 gave 25, 50, 75,100 and 150 mM respectively. Pure distilled water was considered to be 0 mM, thus represented the control. The experiment was conducted in two phases; germination and growth tests, respectively.

Germination test

This was conducted in the Botany Laboratory of Ambrose Alli University, Ekpoma, Edo State, Nigeria. It is located on latitude 06° 42 ´´N and longitude 06° 08 ´´E. It lasted for ten (10) days, from 20th June, 2001 to 30th June, 2001. Seeds of both *A. hybridus* and *C. argentea* used for the study were obtained from the Federal Department of Agriculture, Ubiaja, Edo State, Nigeria. Approximate seed sizes of both species were randomly selected and soaked in distilled water for 2 h before being transferred into glass Petri dishes. Ten (10) seeds of each crop species were sown into each glass Petri and these were replicated ten (10) times each for 0, 25, 50, 75, 100 and 150 mM respectively. Thus, total number of seeds per concentration (from 0 to 150 mM) amounted to 100 seeds each. The seeds were placed between folds of moistened filter paper in the glass Petri dishes at room temperature of 27.5°C. The seeds of both species in the glass Petri dishes were moistened every 12 h with varying concentrations (0, 25, 50, 75, 100 and 150 mM) of NaCl salt and observations were recorded every 24 h for radical emergence as indicative of germination. Seeds were considered to have germinated when up to 1 mm radicle emergence from the seed was noticed.

Seedling growth test

Black polythene bags measuring 25 x 25 cm were all filled with sandy-loamy soil from the Experimental Garden of Botany department, Ambrose Alli University, Ekpoma, Edo State, Nigeria. Fifteen seeds were randomly chosen and sown into each potted bag at a depth of 1 cm and after germination; only ten (10) seedlings considered 'healthier' were allowed to grow. Masking

Table 1. Effect of varying concentrations of NaCl on seed germination of *Amaranthus hybridus* in percentages.

NaCl (mM)	Day 1	Day 2	Day 3	Day 4	Day 5	Day 6	Day 7	Day 8	Day 9	Day 10
0	81	87	94	100	100	100	100	100	100	100
25	54	68	76	79	80	80	80	81	81	82
50	49	63	70	71	72	72	73	74	74	75
75	31	48	51	55	58	61	62	62	62	63
100	27	36	50	53	54	54	54	55	55	55
150	21	39	42	44	45	46	46	47	47	47

tapes were used to label the bags appropriately. Salt solutions of NaCl corresponding to 25, 50, 75, 100 and 150 mM were used for watering the plants in the potted bags on twelve (12) hourly bases, throughout the period of experimentation. Distilled water was labeled 0mM and considered as the control. The experimental period lasted for eight (8) weeks. At the end of the eighth (8th) week, results were collected for parameters like shoot length, root length, root/shoot ratio, total length, fresh weight, dry weight, dry matter content, leaf area index, leaf number as well as relative water content, respectively.

Shoot length, root length and total length were measured with a ruler. Shoot to root ratio was determined mathematically through simple ratio. Numbers of seeds were determined manually. Before shoot and root lengths were determined, seedlings were uprooted carefully to avoid breakage from the medium. Particles stuck to the roots and shoots were carefully removed before measurements were taken. Each seedling was then measured on a weighing balance for fresh weight while for dry weight, the same seedlings were dried at 80 °C till constant weights were obtained and records taken. Leaf area index was determined following the methods of Gunkel and Mulligan (1953) in which leaf discs were punched from five (5) leaves using a cork-borer of diameter 0.82 cm. The discs of known area of the punched leaves were dried to constant weight at 80°C for 24 h hand the total leaf area per plant was then calculated using the relationship:

$$\text{Leaf Area Index} = \frac{\text{Leaf Dry Weight} \times \text{Disc Area}}{\text{Disc Dry Weight}}$$

Relative water content was determined following the method of Handley and Jennings (1977). In which case, four (4) plants from each treatment in both plant species were separated into leaf, stem and root tissues. The materials were dried at 80°C for 24 h and their corresponding dry weights measured. It is then calculated using this relationship:

$$0\ \text{Relative Water Content} = \frac{\text{Fresh weight} \times \text{Dry weight}}{\text{Dry weight}} \times 100$$

RESULTS AND DISCUSSION

From Table 1, it was observed that the control experiment attained 100% germination on the fourth day after sowing. Also, only the control was found to attain 100% germination at the end of the period of experimentation unlike other salinity levels that could not attain 100% germination. Generally, it was observed that germination percentage decreased steadily with increasing NaCl concentrations. This is in agreement with the works of several authors; Nasir (2002), Al-Seedi (2004), Herr (2005), Al-Seedi and Gatteh (2010) and Ratnakar and Rai (2013) to mention but a few. This could be attributed to specific ion effect (Hassen, 1999) or osmotic stress (Zekri, 1993) which later led to reduced water intake into seeds for enhancement of germination.

In addition, Begum et al. (2010) stated that germination of seeds depends on the utilization of reserved food materials of the seed and, Ratnakar and Rai (2013) opined that salinity interferes with the process of water absorption by the seeds which subsequently inhibits the hydrolysis of food reserves which ultimately delays and decreases seed germination.

Table 2 shows that germination percentages were highest in the first three days after planting than the other days. The control attained 100% germination within the first three days whereas the other experimental sets could not attain up to 90% even on the tenth day after sowing. Also, apart from 25 mM NaCl level which attained 50% germination on the third day, no other experimental set (50, 75, 100 and 150 mM NaCl) attained up to 50% germination rate.

According to Ratnakar and Rai (2013), this may be due to the fact that the increased amount of NaCl disturbed ionic balance of plant cells and also caused imbalances in plant nutrients which must have affected germination percentages. In addition, it could be deduced that *A. hybridus* tolerated varying concentrations of NaCl better than *C. argentea* during seed germination. Hence, *A.hybridus* is a better halophyte than *C. argentea*.

Length measurements

According to Ratnakar and Rai (2013), root and shoot lengths are the most important parameters for studying salt stress. This is obvious as roots are in direct contact with soil salinity and the effects are then translocated and manifested along the shoot, thus the need for evaluation of salt stress effects in terms of root/shoot ratio as well as total lengths. Shoot length, root length and total length behaved in a similar pattern in both *A. hybridus* and *C. argentea* (Table 3). Growth was steadily stimulated in

Table 2. Effect of varying concentrations of NaCl on seed germination of *Celosia argentea* in percentages.

NaCl (mM)	Day 1	Day 2	Day 3	Day 4	Day 5	Day 6	Day 7	Day 8	Day 9	Day 10
0	86	93	100	100	100	100	100	100	100	100
25	41	47	50	51	52	52	53	54	55	55
50	18	29	35	40	40	41	41	42	43	43
75	14	26	29	31	32	33	33	33	33	33
100	11	16	18	20	22	22	22	22	22	22
150	6	11	12	14	15	15	15	15	15	15

Table 3. Effect of varying concentrations of NaCl on growth parameters of *Amaranthus hybridus*.

NaCl (mM)	Shoot length	Root length	Root /shoot ratio	Total length	Fresh weight	Dry weight	Dry matter	Leaf area index	Rel. water index	Leaf No.
0	19.21±0.24	5.06±0.12	0.26	24.29±0.53	7.24±0.32	0.68±0.01	9.39	2.03±0.11	757.11	10
25	12.43±0.21	5.87±0.18	0.47	18.37±0.41	4.22±0.37	1.46±0.03	34.61	3.08±0.09	2284.21	13
50	13.46±0.19	6.74±0.24	0.50	20.25±0.43	7.03±0.43	0.63±0.02	8.96	2.15±0.07	2166.67	14
75	10.81±0.18	4.79±0.15	0.44	15.66±0.31	6.84±0.39	0.56±0.02	8.19	1.62±0.06	1125.81	11
100	9.43±0.17	4.18±0.13	0.44	13.69±0.29	3.01±0.14	0.51±0.01	16.94	1.28±0.05	1105.71	10
150	7.97±0.15	3.34±0.09	0.44	11.42±0.26	2.43±0.09	0.49±0.02	20.17	1.04±0.03	244.83	8

both species at lower NaCl salinity levels with best growth stimulation at 50 mM NaCl where they achieved highest values respectively. However, both species were adversely affected by higher NaCl salinity levels (75 mM, 100 and 150 mM) where increased NaCl salinities caused a corresponding decrease in root length, shoot length as well as total lengths owing to adverse salinity effects on cell differentiation and elongation. Hence, at lower NaCl salinity levels, both studied parameters are good halophytes.

Osmotic differences could explain this phenomenon where by lower NaCl salinity levels (25 and 50 mM) positively influences solutes to readily cross the cell membranes into the cytoplasm of the cells but at higher salinity levels, active metabolic pumps prevents accumulation of these ions (Katembe et al., 1998). Built up toxic ions emanating from continuous exposure to higher NaCl levels could lead to decreased availability of some essential nutrients (Werner and Finkelstein, 1995). In addition, Heidari et al. (2001) while studying the effects of NaCl concentrations on *Helianthus annus* suggested that reduction in plant growth is due to decreasing turgur pressure in the soils under saline environment. Furthermore, evidence from root/shoot ratio showed that shoot length was more stimulated at 25 and 50 mM that root length where as the effects remained the same at higher salinity levels (75, 100 and 150 mM) in *A. hybridus* (Table 3) as compared to *C. argentea* where root length was more negatively affected than shoot length with increasing NaCl levels (Table 4).

Weight measurements

This includes fresh weight, dry weight and dry matter contents. Fresh weight was most stimulated at 0 mM NaCl level in *A. hybridus* whereas 50 mM NaCl concentration influenced fresh weight more than the other NaCl levels in *C. argentea*. Both species showed decreasing fresh weights with increasing NaCl levels, from 50 mM to 150 mM NaCl concentrations. This implies that 50 mM NaCl concentration most positively stimulated water uptake into cytoplasm of cells in both species, apart from 0 mM NaCl in *A. hybridus*. Reduced fresh weights could be attributed to water deficits as established by Cha-Um and Kirdmanee (2009) while studying maize seedlings under NaCl influences and Ratnakar and Rai (2013) who observed a similar trend while investigating *Trigonalla foenum-graecium L. Var. Peb* under NaCl exposure.

Dry weight

Dry weight was highest at 25 mM NaCl concentration in *A. hybridus* as compared to *C. argentea* where highest value occurred at 50 mM NaCl level. Also, both species showed decreasing dry weight values with increasing salinities from 25 mM NaCl in *A. hybridus* and 50 mM NaCl concentrations in *C. argentea*. This is in line with the study of Turan et al. (2009) while investigating NaCl salinity effects in maize plants. Dadkhan and Grifiths

Table 4. Effect of varying concentrations of NaCl on growth parameters of *Celosia argentea*.

NaCl (mM)	Shoot length	Root length	Shoot /root ratio	Total length	Fresh weight	Dry weight	Dry matter	Leaf area index	Rel. H_2O index	Leaf no.
0	15.33±0.32	5.21±0.19	0.34	20.54±0.27	3.61±0.31	0.38±0.01	10.53	1.36±0.12	545.16	10
25	16.97±0.35	5.84±0.21	0.34	22.85±0.29	3.04±0.27	0.41±0.01	13.49	2.88±0.15	1172.73	15
50	18.28±0.28	5.97±0.18	0.33	24.26±0.24	8.64±0.38	0.76±0.02	8.80	2.47±0.14	1142.86	16
75	13.46±0.22	4.07±0.12	0.30	17.54±0.21	5.58±0.27	0.33±0.02	5.91	2.24±0.12	1066.67	11
100	12.22±0.18	3.23±0.11	0.26	15.45±0.17	5.41±0.22	0.31±0.01	5.73	2.19±0.10	1004.35	10
150	12.16±0.13	3.04±0.09	0.25	15.21±0.13	3.24±0.18	0.21±0.01	6.48	2.11±0.08	900.08	9

(2006) attributed such a decrease in dry weight to greater reduction in uptake and reduction in utilization of mineral nutrients by plants under salt stress while Jafari et al. (2009) opined that it could be due to reduced rate of photosynthesis.

Dry matter was highest at 25 mM NaCl level in both species studied. There was a gradual decrease to 75 mM NaCl concentration until an increase started occurring from 100 up to 150 mM NaCl concentration in both species. This could be attributed to accumulation of salt particles within the tissues of both species.

Leaf area index

Mean leaf area index was found to be highest at 25 mM NaCl level in both species studied. After which, there was a corresponding decrease in average leaf area index in both species with increasing NaCl salinities. Reduction in leaf area index could have serious implications for chlorophyll content, thus a reduction in photosynthetic ability of plants. This explains the direct relationship existing in this study between leaf area index and parameters as root length, shoot length as well as total lengths of the species under study. Ungar (1991) opined that higher salinities cause decreases in assimilation of CO_2 through the effect of stomatal opening and sufficiency of photosynthesis process. High salinity could affect stroma volumes of chloroplasts (Price and Hendry, 1991) and the protein bonds of green pigments, thus causing a decrease in chlorophyll content (Rivera and Heras, 1973) in plants.

Relative water concentration (RWC)

RWC decreased with increasing salinities in both *A. hybridus* and *C. argentea* from 25 mM through 150 mM NaCl concentrations. This phenomenon could best be explained by salt induced stress since increasing NaCl levels have tendencies of decreasing water absorption into the cells of plants thereby causing physiological desiccations. This is in agreement with the findings of Abbas and Latif (2005) who established a decrease in the growth of jute seedlings under NaCl stress, which they attributed to low water absorption into cells as the salinity increased.

Leaf number

Both *A. hybridus* and *C. argentea* are cultivated chiefly for their vegetative parts, thus, it becomes pertinent to evaluate their leaf number. Highest number of leaves in both species was noticed among seedlings growing in 50 mM NaCl concentration. In addition, seedlings with higher total shoot lengths seemed to have higher number of leaves. The numbers of leaves first increased from 0 to 50 mM NaCl concentrations, before a gradual decrease from 75 mM through 150 mM NaCl levels were observed. This shows that moderate NaCl levels (50 mM) positively influenced leaf formation and development as compared to lower (0 and 25 mM) and higher (75, 100 and 150 mM) NaCl concentrations in both *A. hybridus* and *C. argentea*. This shows that both species would be most productive and best cultivated under moderate than low or high NaCl concentrations.

Conclusion

Findings in this study have shown that both *A. hybridus* and *C. argentea* are good halophytes that could sustainably be cultivated in farmlands contaminated with NaCl-induced salinities either due to natural aridity or man-made soil salinization through buildup of salts as a result of extensive use of fertilizers and/or irrigation practices. Although, both species responded to varying NaCl concentrations in different ways, however, both tended to show almost a similar trend.

On the one hand, germination of seeds of both species was adversely affected with increasing NaCl concentrations. Conversely, seedling growth was positively stimulated at moderate levels and adversely affected at varying rates at low (25 mM) and higher (100 and 150 mM NaCl levels in *A. hybridus* and 75, 100 and 150 mM

NaCl levels in *C. argentea*), respectively. Thus, it is being recommended that seedlings of both species of *A. hybridus* and *C. argentea* should be raised in nurseries free of NaCl influences (0 mM), that is, salt-free soils and as soon as maximum germination has been attained, the seedlings could then be transferred to moderately NaCl saline soils for maximum growth and development. There is therefore need to enlighten farmers on these findings to enhance maximum productivity and help ensure food security.

REFERENCES

Abbas SM, Latif HH (2005). Germination and Proteinpatterns of some Genotypes of two species of Jute as affected by NaCl stress. Pak. J. Biol. Sci. 8(2):227-234. http://dx.doi.org/10.3923/pjbs.2005.227.234

Al-Seedi SNN (2004). The effect of Salinity on Germination, Growth and Emergence of Mung Vignia radiate (L) Wilczek in different Soil textures. J. Thi-Qar Univ. 1:12-18.

Al-Seedi, SNN, Gatteh HJ (2010). Effect of Salinity on Germination, Growth and Organic Compounds of Mung bean Plant Vignia radiate (L.) Wilczek. J. Thi-Qar Univ. 6(5): ٧٨-٨٧.

Begum F, Ahmad IM, Nessa A, Sultana W (2010). The effect of Salinity on Seed Quality of Wheat. J. Bangl. Agric. Univ. 8(1):19-22. http://dx.doi.org/10.3329/jbau.v8i1.6392

Bray EA, Bailey-Serres J, Weretilnyk E (2000). Responses to Abiotic stress, pp. 1158-1203. In: Buchanan B, Gruissem W, Jones R (eds.). Biochemistry and Molecukar Biology of Plants. American Society of Plant Physiology, Rockville.

Byordi A (2010). The Influence of Salt stress on Seed Germination, Growth and Yield of Canola cultivars. Not. Bot. Hort. Agrobot. Cluj 38(1):128-133.

Cha-Um S, Kirdmanee C (2009). Effect of Salt stress on Proline accumulation, Photosynthetic ability and Growth characters in two Maize cultivars. Pak. J. Bot. 41(1):87-98.

Cuartero JMC, Bolarin MJ, Asins A, Moreno V (2006). Increasing Salt tolerance in Tomato. J. Exp. Bot. 57(5):1045-1058. http://dx.doi.org/10.1093/jxb/erj102

Dadkhan AR, Grifiths H (2006) The effect of Salinity on Growth, Inorganic ions and Dry matter partitioning in Sugar beet cultivars. J. Agric. Sci. Techno. 8:199-210.

Dutta AC, Dutta TC (2008). Botany for Degree Students. Oxford University Press, New Delhi. Pp. 221-264.

Galston AW, Davis PJ, Satter RC (1980). The life of the Grown Plants. Third edition. Prentice-Hall Inc., Englewood. 464p.

Handley J, Jennings D (1977). The effects of ions on the Growth and Leaf succulence of Atriplex hortensis var cuperata. Ann. Bot. 41:1109-1112.

Hassen KAK (1999). Effect of Salinity on Germination, Growth and Ionic content of three varieties of Barley Hordeum vulgaris L. J. Basrah Res. 2:87-98.

Heidari A, Toorchi M, Bandehagh A, Shakiba MR (2001). Effect of NaCl stress on Growth, Water relations, Organic and Inorganic osmolytes accumulation in Sunflower (Helianthus annus L.) lines. Univ. J. Envir. Res. Technol. 1(3):351-362.

Herr S (2005). Effects of soil Salinity on Seed Germination. AP Environmental Science, section 1. pp. 1-4.

Jafari MHS, Kafi M, Astaraie A. (2009). Interactive effects of NaCl induced Salinity, Calcium and Potassium on Physiological traits of Sorghum (Sorghum bicolor L.). Pakis. J. Bot. 41(6):3053-3063.

Katembe WJ, Ungar IA, Mitchell JP (1998). Effect of Salinity on Germination and Seedling Growth of two Atriplex species (Chenopodiaceae). Annals Bot. 82:167-175. http://dx.doi.org/10.1006/anbo.1998.0663

Munns R (2002). Comparative Physiology of Salt and water stress. Plant Cell Envir. 25:239-250. http://dx.doi.org/10.1046/j.0016-8025.2001.00808.x

Nasir SN (2002). Effect of salinity, growth and mineral contents of three varieties of Wheat Triticum durum L. J. Babylon Univ. 7(2):1334-1340.

Oguntona T (1998). Green Leafy Vegetables. In: Nutritional Quality of Plant Foods, Osagie, A. U and Eka, O. U. (eds.) PHRU, Department of Biochemistry, University of Benin, Benin city, Nigeria. ISBN: 978-2120-02-2, pp: 120-133.

Ogunwenmo KO, Oyelana OA, Ibidunmoye O, Anyasor G, Ogunnowo, AA (2010). Effect of brewery, textile and paint effluents on seed germination of leafy vegetables- Amaranthus hybridus and Celosia argentea (Amaranthaceae). J. Biol. Sci. 10(2):151-156. http://dx.doi.org/10.3923/jbs.2010.151.156

Olorode O (1984). Taxonomy of West African Flowering Plants. Longman Group Ltd. London, ISBN: 0-582-64429-1, pp: 158.

Ratnakar A, Rai A (2013). Effect of NaCl Salinity on Seed Germination and early Seedling Growth of Trigonella foenum-Graecum L. Var. Peb., Octa J. Envir. Res. 1(4):304-309.

Rivera AM, Heras L (1973). Effect of different Salinity levels on Chlorophyll content, Mineral composition and Growth of Tetraploid rye Secale cereal. Ann Aula Dei. 12:100-108.

Sairam RK, Tyagi A (2004). Physiology and Molecular Biology of Salinity stress tolerance in Plants. Current sci. 86(3):407-421.

Sairam RK, Rao KV, Srivastava GC (2002). Differential response of Wheat Genotypes to long term Salinity stress in relation to Oxidative stress, Antioxidant activity and Osmolyte concentration. Plant Science. 163:1037-1046. http://dx.doi.org/10.1016/S0168-9452(02)00278-9

Turan MA, Elkarim AHA, Taban N, Taban S (2009). Effect of Salt stress on the Growth, Stomatal resistance, Proline and Chlorophyll concentrations on Maize Plant. Afri. J. Agric. Res. 4(9):893-897.

Ungar IA(1991). Ecophysiology of Vascular halophytes. Boca Raton: CRC Press. pp.209.

Price A, Hendry G (1991). Iron-catalysed oxygen radical formation and its possible contribution to drought damage in nine native grasses and three cereals. Plant Cell Environ. 14:477-484. Werner JE, Finkelstein RR (1995) Arabidopsis Mutants with reduced response to NaCl and osmotic stress. Physiologia Plantrarum 93: 659-666.

Zadeh HM, Naeini MB (2007). Effects of Salinity stress on the Morphology and Yield of two cultivars of Canola (Brassica napus L.). J. Agron. 6:409-414. http://dx.doi.org/10.3923/ja.2007.409.414

Zekri M (1993). Effects of Salinity and Calcium on Seedling emergence, Growth, Sodium and Chloride concentrations of Citrus rootstocks., Proc. Fla. State Hort. Soc. 106:18-21.

Heavy metal content in fish and water from River Niger at Agenebode, Edo State, Nigeria

Wangboje, O. M. and Ikhuabe, A. J.

Department of fisheries, university of Benin, PMB 1154, Benin City, Edo State, Nigeria.

The concentrations of Lead (Pb), Copper (Cu), Cadmium (Cd) and Zinc (Zn) in fish and water from River Niger at Agenebode, Nigeria were determined by atomic absorption spectrometric technique. The mean concentrations of heavy metals in water ranged from 0.00742 mg/L for Pb to 0.239 mg/L for Zn. The summary statistics for heavy metals in fish showed that the mean concentrations of metals ranged from 0.0291 mg/kg for Pb to 69.14 mg/kg for Zn, while the mean concentrations of metals in individual fish species ranged from below detection/quantification limit (BQL) for Pb in *Barbus occidentalis* and *Chrysichthys nigrodigitatus* to 128.87 mg/kg for Zn in *Hemichromis fasciatus*. The bioaccumulation quotient (BQ) values for metals in fish ranged from BQL for Pb in *B. occidentalis* and *C. nigrodigitatus* to 539.21 for Zn in *H. fasciatus,* while the condition factor for the fish species ranged from 0.32 in *Clarias gariepinus* to 2.07 in *Oreochromis niloticus*. The maximum acceptable risk (MAR) values for heavy metals in fish ranged from BQL for Pb in *B. occidentalis* and *C. nigrodigitatus* to 6.38 for Cu in *H. fasciatus*; while the Hazard Quotient (HQ) values for heavy metals in water ranged from 0.05 for Zn to 0.43 for Cd. In fish, the HQ values ranged from BQL for Pb in *B. occidentalis* and *C. nigrodigitatus* to 10.74 for Zn in *H. fasciatus*. The National Environmental Standard and Regulation Enforcement Agency (NESREA) were enjoined to ensure that the environmental safety of our inland water bodies is sustained in order to mitigate negative impacts on human health.

Key words: Heavy metals, fish, River Niger, maximum acceptable risk, hazard quotient.

INTRODUCTION

In today's environment, the potential for risks to humans and wildlife due to exposure to ever increasing levels of toxic chemical agents of both natural and synthetic origin is on the increase in scientific and regulatory interest (United States Environmental Protection Agency, US-EPA, 1998). Furthermore, environmental pollution is a worldwide problem with heavy metals belonging to the most important category of pollutants (Tabari et al., 2010). Heavy metals are classified as metallic elements which have relatively high atomic weight and are toxic at low concentrations (Butu and Igiusi, 2013).These metals are naturally found in the environment but can also be produced anthropogenically to levels high enough to cause disruptions amongst aquatic species such as, impaired metabolic functions, changes in distribution and abundance of populations (Elder, 1988).

All heavy metals are potentially harmful to most organisms at some levels of exposure and absorption (Adedeji and Okocha, 2011). The ingestion of heavy metals by fish via food and water may affect not only the productivity and reproductive capabilities of such fish, but ultimately affect the health of man that depends on these organisms as a major source of protein (Fonge et al., 2011). Many ecotoxicological studies in Nigeria have focused on aquatic bodies within or around major cities with little emphasis on aquatic media in rural areas, which may also be impacted by heavy metals. The River Niger is one of the major rivers in Nigeria that is rich in fisheries resources that are widely exploited as human food. The river at the Agenebode axis of Edo State, Nigeria was selected for the study as the area is densely populated and plays host to numerous cottage industries with expected varied levels of anthropogenic impacts on the river. The study specifically ascertained the concentrations of Cadmium (Cd), Copper (Cu), Lead (Pb) and Zinc (Zn) in water and fish from the River Niger at Agenebode in order to determine potentially hazardous levels of these heavy metals especially with regard to human health.

MATERIALS AND METHODS

The study area

The River Niger at Agenebode lies within Latitude 7.106° N and Longitude 6.696° E. Agenebode is the Administrative Headquarters of Etsako East Local Government Area of Edo State, Nigeria. According to the 2006 National Census, the population of the Local Government Area was put at 145, 996. The area has a size of 1,132.82 km^2 with a population density of 86 people per km^2 (Stephen and Lanihan, 2007). The people are actively engaged in commerce, fishing, farming and mining. The major agricultural products are groundnut, rice, cassava, yam, maize, fish, tomatoes and melon. The substratum of River Niger consists mainly of fine sand mixed with mud and pebbles. Decaying macrophytes and debris also form part of the substratum. Other settlements along the River Niger within the same Local Government Area include Emokweme, Ofugbo, Obadolo, Udochi and Udaba. Directly opposite Agenebode, on the other side of River Niger is the Idah Community in Kogi State. The climate of Agenebode is tropical with the dry season running from November to March, while the wet season runs from April to October. Annual temperature ranges from 23 to 35°C while annual humidity ranges from 67 to 94%. Two stations namely, Abattoir and Boat House stations were established for the study based on the level of human activities, nature of effluents and accessisibility. Activities around the Abattoir station include abattoir activities, cattle grazing, bathing of animals in the river, operation of pig pens and fishing; while at the boat house station fishing, farming, laundry, and boat maintenance work can be observed. Both dug-out canoes and boats fitted with petrol powered outboard engines can be found at the boat station. The sampling stations are located approximately 4 km apart.

Collection of water and fish samples

Water samples were collected at 30 cm depth below the water surface using pretreated polypropylene bottles of 500 ml capacity. Five (5) ml of concentrated nitric acid was applied *in-situ* to each bottled sample in order to keep the metals in solution. Water samples were transported to the laboratory and stored at -5°C in a freezer. The fish samples were netted using cast nets with the assistance of fishermen. Fish samples were washed in River water to remove adhering debris and were thereafter placed in labeled polythene bags and transported to the laboratory within 24 h in a Thermolineo® ice box in order to preserve their integrity.

Laboratory procedures

In the laboratory, fish samples were identified with taxonomic keys as published by Olaosebikan and Raji (1998) and Adesulu and Sydenham (2007). The identified fish species were *Clarias gariepinus* (Burchell, 1822), n=4, *Hemichromis fasciatus* (Peters, 1857), n=6, *Barbus occidentalis* (Boulenger, 1911), n=2, *Synodontis sorex* (Gunther, 1864), n=2, *Alestes baremose* (Joannis, 1835), n=8, *Chrysichthys nigrodigitatus* (Lacepede, 1803), n=4 and *Oreochromis niloticus* (Linnaeus, 1758), n=6. They were weighed whole in grams using an ATOM A-110C® electronic compact scale while their total lengths were recorded using a stainless steel ruler. Fish samples were oven dried at a temperature of 70°C for 48 h. Each sample was milled separately using a porcelain mortar and pestle and kept in foil paper prior to digestion. The following steps were applied in digesting water samples. The water was pretreated with 5 ml of diluted nitric acid. Two (2) ml of water sample was pipetted into a 250 ml conical flask after which 5 ml of 70% $HClO_4$ and 15 ml of 55% HNO_3 were added. The mixture was heated till the solution became clear. Five (5) ml of 20% HCl was added and content was poured into a 100 ml volumetric flask and made-up to mark with distilled water. The digest was stored in 100 ml plastic reagent bottles prior to atomic absorption spectrophotometer analysis (Galyean, 2010).

The following steps were applied in digesting fish samples. Fresh fish was dried to constant weight in a Gallenkamp® hotbox oven. Two (2) g of the dried fish sample was weighed into 250 ml conical flask, 5 ml of $HClO_4$ and 15 ml HNO_3 were added. The mixture was heated till the solution became clear. Five (5) ml of 20% HCl was added. The mixture was filtered into a 100 ml volumetric flask using No.42 Watman filter paper and made up to mark with distilled water. The digest was stored in a 100 ml plastic reagent bottle ready for Atomic absorption spectrophotometer (AAS) analysis (Galyean, 2010). Fish and water digests were analyzed for Pb, Cr, Cu and Cd by means of an Atomic Absorption Spectrophotometer (Unicam® 696 series) equipped with solar software using air acetylene flame. Concentrations of metals in water and fish were expressed in mg/L and mg/kg, respectively. Blanks, spiked samples, reference material analyses and duplicate analyses were performed for all analytes as part of the quality assurance procedures. All reagents used were of analytical grade (BDH). All glassware were washed with detergent and rinsed with distilled water before they were soaked in potassium permanganate solution overnight. They were then rinsed with distilled water before use to ensure the acquisition of high precision results.

Bioaccumulation quotient (BQ) for heavy metals in fish

The Bioaccumulation Quotient (BQ) for heavy metals in fish is

Table 1. Summary statistics for heavy metals (mg/l) in water.

Metals	Mean	Minimum	Maximum	SD (±)
Cd	0.0219	0.00	0.14	0.0376
Cu	0.182	0.08	0.30	0.0664
Pb	0.00742	0.00	0.017	0.0047
Zn	0.239	0.15	0.35	0.0571

Table 2. Mean heavy metal concentrations (mg/l) in water by month.

Months	Pb	Cu	Cd	Zn
March	0.00475 ± 0.001^{b}	0.1650 ± 0.002^{a}	0.0452 ± 0.01^{a}	0.2175 ± 0.015^{a}
April	0.00875 ± 0.02^{a}	0.1975 ± 0.03^{a}	0.00950 ± 0.01^{b}	0.2775 ± 0.05^{a}
May	0.0875 ± 0.03^{a}	0.1825 ± 0.001^{a}	0.0110 ± 0.03^{b}	0.225 ± 0.04^{a}

Means with the same letters are not significantly different (P>0.05)

Table 3. Mean heavy metal concentrations (mg/l) in water by station.

Station	Pb	Cu	Cd	Zn
Abattoir	0.06	0.18	0.01	70.02
Boat House	0.06	0.18	0.02	68.25

expressed below (Latif et al., 1982):

$$BQ = \frac{\text{Heavy metal concentration in fish}}{\text{Heavy metal concentration in water}}$$

Condition Factor for fish

This expresses the fatness, wellbeing and relative robustness of a fish. It is believed that a heavier fish of a given length is in better condition than a lighter fish of the same length (Baijot et al., 1997). The condition factor is expressed below:

$$CF = \frac{100\text{*Weight of fish (g)}}{\text{Length of fish (cm)}^3}$$

Maximum Acceptable Risk index (MAR) for heavy metals

The Maximum Acceptable Risk index (MAR), is a simplified representation of biomagnifications in food webs (Romijn et al., 1993). It is expressed by the following equation:

$$MAR = \frac{\text{Bioaccumulation Quotient (BQ) for chemical element in fish}}{\text{Dietary no observed effect concentration of chemical element in man}}$$

Where: MAR>1 = High MAR level and MAR<1= Low MAR level

Toxicity/Hazard Quotient (TQ) for heavy metals

The Toxicity/Hazard Quotient (TQ) for chemical elements is a comparison of the measured concentration of site-related elements in ecological matrices with specific health-based criteria (Newstead et al., 2002).

$$TQ = \frac{\text{Measured concentration of chemical element in ecological matrix}}{\text{Health based criteria}}$$

Where: TQ>1 = Toxicity/Hazard indicated.

Statistical method

Statistical software (GENSTAT® version 13.3 for Windows) was used for analyzing data. One-way Analysis of variance (ANOVA) was used to test for significant differences between mean values of heavy metals at 5% probability level. Duncan Multiple Range Test was used to separate significant means.

RESULTS

The mean concentrations of heavy metals in water ranged from 0.00742 mg/L for Pb to 0.239 mg/L for Zn (Table 1), while the mean concentrations of the heavy metals in water on a monthly basis ranged from 0.00475 mg/L for Pb in March to 0.2775 mg/L for Zn in April (Table 2). Station wise, the mean concentrations of heavy metals in water ranged from 0.01mg/L for Cd to 70.02 mg/L for Zn at the Abattoir station (Table 3). The summary statistics for heavy metals in fish showed that the mean concentrations of metals ranged from 0.0291 mg/kg for Pb to 69.14 mg/kg for Zn (Table 4); while on a monthly basis, the mean concentrations of metals ranged from 0.01850 mg/kg for Pb in May to 77.92 mg/kg for Zn in April (Table 5). Station wise, the mean concentrations

Table 4. Summary statistics for heavy metals (mg/kg) in fish.

Metals	Mean	Minimum	Maximum	SD(±)
Cd	0.0917	0.00	0.411	0.13
Cu	1.00	0.57	1.37	0.232
Pb	0.0291	0.00	0.09	0.0278
Zn	69.14	50.6	85.60	10.70

Table 5. Mean comparison of heavy metal concentrations (mg/kg) in fish by month.

Months	Pb	Cu	Cd	Zn
March	0.02600 ± 0.01^{b}	1.70 ± 0.22^{a}	0.05625 ± 0.02^{a}	57.76 ± 1.23^{b}
April	0.04275 ± 0.01^{a}	0.868 ± 0.05^{a}	0.17225 ± 0.02^{a}	77.92 ± 0.89^{a}
May	0.01850 ± 0.02^{b}	1.063 ± 0.34^{a}	0.04650 ± 0.01^{a}	71.78 ± 0.77^{a}

Means with the same letters are not significantly different (P>0.05).

Table 6. Mean heavy metal concentrations (mg/kg) in fish by station.

Station	Pb	Cu	Cd	Zn
Abattoir	0.03	0.96	0.13	70.03
Boat House	0.03	1.05	0.06	68.25

Table 7. Mean heavy metal concentrations (mg/kg) in specific fish species.

Fish species	Pb	Cu	Cd	Zn
H. fasciatus	0.04467	0.855	0.24847	128.87
O.niloticus	0.01500	0.967	0.04667	77.47
B. occidentalis	0.00*	0.957	0.08567	75.23
C. gariepinus	0.03667	0.967	0.07667	70.97
C. nigrodigitatus	0.00*	1.367	0.09167	65.70
S. sorex	0.04667	1.227	0.04667	62.37
A.baremose	0.05667	0.917	0.00*	52.23

*Below detection/quantification limit

Table 8. Bioaccumulation Quotient (BQ) values for heavy metals in fish species.

Fish species	Pb	Cu	Cd	Zn
H. fasciatus	6.02	4.69	11.34	539.21
O.niloticus	2.02	5.37	2.13	324.14
B. occidentalis	0.00*	5.25	3.91	314.77
C. gariepinus	4.94	5.31	3.50	298.94
C. nigrodigitatus	0.00*	7.51	4.18	274.89
S. sorex	6.29	6.74	2.13	260.96
A.baremose	7.64	5.03	0.00*	218.53

*Below detection/quantification limit

Table 9. Maximum Acceptable Risk (MAR) values for heavy metals in fish species .

Fish species	Pb	Cu	Cd	Zn
H. fasciatus	0.33	6.38	0.18	0.02
O.niloticus	0.99	5.59	0.94	0.04
B. occidentalis	0.00	5.71	0.51	3.50
C. gariepinus	0.41	5.64	0.57	0.04
C. nigrodigitatus	0.00	3.99	0.47	0.04
S. sorex	0.32	4.45	0.94	0.05
A.baremose	0.26	5.95	0.00	0.05

of heavy metals in fish ranged from 0.03 mg/kg for Pb at both stations to 70.03 mg/kg for Zn at the Abattoir station (Table 6).

The mean concentrations of metals in individual fish species ranged from below detection/quantification limit (BQL) for Pb in *B. occidentalis* and *C. nigrodigitatus* to 128.87 mg/kg for Zn in *H. fascistus* (Table 7); while the BQ values for metals in fish ranged from below detection/quantification limit for Pb in B. *occidentalis* and *C. nigrodigitatus* to 539.21 for Zn in *H. fasciatus* (Table 8). The Condition Factor for the fish species ranged from 0.32 in *C. gariepinus* to 2.07 in *O. niloticus* (Figure 1). The MAR values for heavy metals in fish ranged from below detection/quantification limit for Pb in *B. occidentalis* and *C. nigrodigitatus* to 6.38 for Cu in *H. fasciatus* (Table 9), while the HQ values for heavy metals in water ranged from 0.05 for Zn to 0.43 for Cd (Figure 2); while in the case of fish, the HQ values ranged from below detection/quantification limit for Pb in *B. occidentalis* and *C. nigrodigitatus* to 10.74 for Zn in *H. fasciatus* (Figure 3).

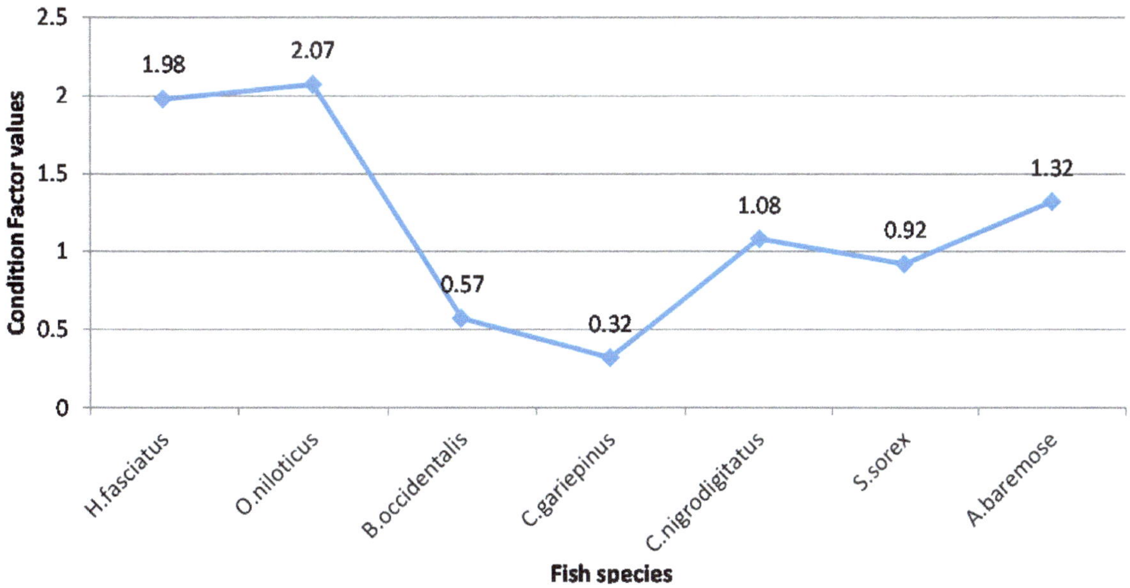

Figure 1. Condition Factor values for fish species.

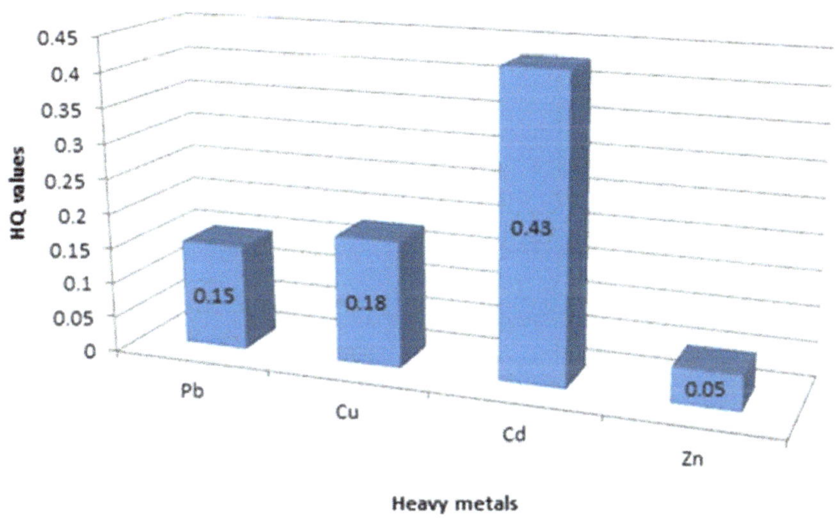

Figure 2. Hazard Quotient (HQ) values for heavy metals in water.

DISCUSSION

In this study, the mean concentrations of Zn in water were the highest across the sampled months, peaking in the month of April compared to the other metals. This finding may have been influenced by differences in precipitation patterns between the sampled months. The same metal had the highest concentrations at both stations, an indication that the water was more impacted with Zn and its associated compounds especially at the Abattoir station. The high Zn load recorded in water at the Abattoir station may be as a result of effluents and solid wastes emanating from the abattoirs which are completely void of proper waste disposal systems. Typical abattoir waste products include blood, stomach contents, bones, hoofs, hair, faecal material, urine and other unwanted parts of slaughtered cattle. Such waste products can impair water quality especially when their discharge into aquatic media continues unabated. Abattoir effluents are usually organically rich and may contribute to the eutrophication of the river.

The introduction of organic matter into natural aquatic media via eutrophication can provide additional binding sites for heavy metals. It has been recognized that under

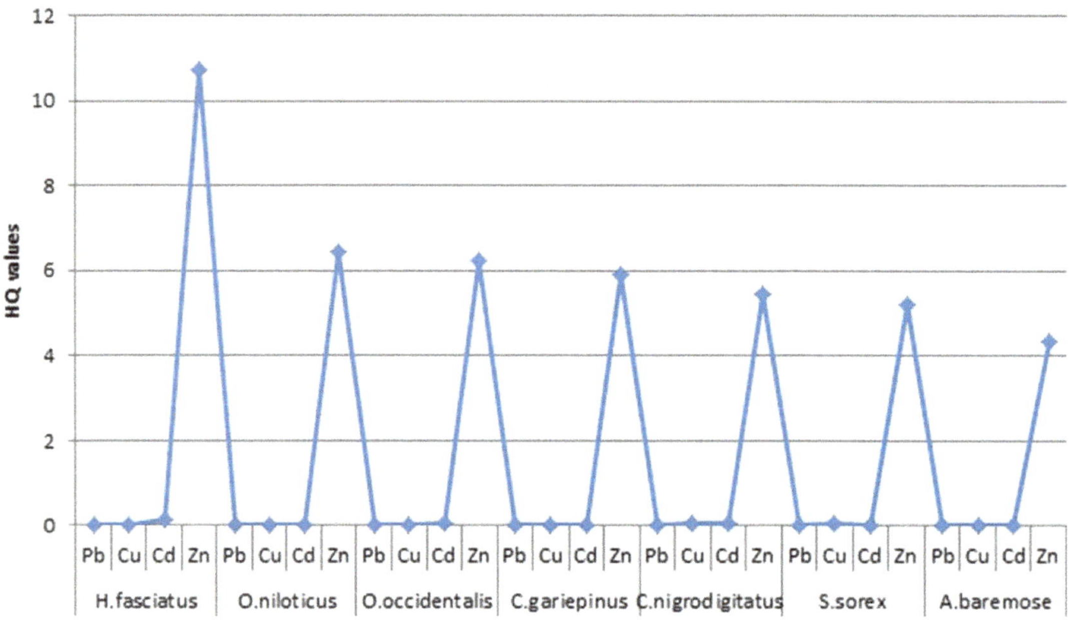

Figure 3. Hazard Quotient (HQ) values for heavy metals in fish species.

oxidizing conditions in natural waters, such organic matter can be degraded leading to the release of soluble metals into the aquatic medium (Tessier et al., 1979). Adekunle and Eniola (2008) observed that organic pollution of aquatic bodies can result from the discharge of raw effluents including sewage into such water bodies. Sangodoyin (1991) reported that effluents from domestic, agricultural and industrial sources can pollute receiving water bodies by altering their physical, chemical and biological properties.

The mean concentrations of heavy metals in water in this study compared well with the mean concentrations of metals recorded for water in some other studies on the African continent (Table 10). Not surprisingly, very high values of Zn were found in the captured fish species with fish at the Abattoir station generally recording higher values of Zn compared to the Boat House station. This finding may be attributed to a high bioavailability of the metal to fish in water especially at the Abattoir station. The highest mean value of Zn in fish was recorded for *H. fasciatus* an indication that this particular fish species was more impacted by Zn compared to the other fish species encountered in the study. The mean concentrations of Cd, Cu, Pb and Zn in *O. niloticus* from the Northern Delta Lakes in Egypt as reported by Saeed and Shaker (2008) were 0.014, 0.128, 0.019 and 0.450 mg/l respectively. These values were far exceeded by the mean concentrations of the aforementioned metals in the same fish species in this study, suggesting a higher impact by these metals on the Piscean community of River Niger.

The bioaccumulation pattern of heavy metals in fish varied with the highest BQ value recorded for Zn in *H. fasciatus* while Pb was not bioaccumulated by *B. occidentalis* and *C. nigrodigitatus.* The non-bioaccumulation of Pb by the aforesaid species may be attributed to the low levels of Pb in water compared to the levels of the other metals in the same medium. Bioaccumulation of metals can also be influenced by factors such as species size, species gender and seasons (Nussey et al., 2000). Bioaccumulation of metals in fish can occur especially when the rate of uptake far outweighs the rate of expulsion. Normally, fish have the ability to regulate the metal burden in their bodies but only up to a limit beyond which bioaccumulation comes into play. Heavy metals can be obtained by fish from the surrounding water, food and sediment. Nussey et al. (2000) reported that when heavy metals are absorbed by fish, they are transported by the blood to either a storage point such as bones or to the liver for transformation and storage.

The Condition Factor revealed that *C. gariepinus* was the least conditioned fish in the study while on the other hand *O. niloticus* was the best conditioned fish. Factors that may be responsible for this variation may include the availability of suitable fish food and water quality conditions as corroborated by Baijot et al. (1997). The application of the Condition Factor in this study was warranted in order to assess the general well-being of the investigated fish species especially in the light of anthropogenic contamination by heavy metals which may alter the chemistry of the aquatic medium in which fish thrives. High MAR values were recorded for Cu in all the fish species in this study with the highest MAR value occurring in *H. fasciatus.*

Table 10. Comparison of heavy metal concentrations (mg/l) in water with selected studies on the African continent.

Heavy metal	River Niger at Agenebode	Kpong Headpond Ghana (Biney, 199[illegible]).	Northern Delta Lakes, Egypt (Saeed and Shaker, 2008).	Ikpoba River Dam, Nigeria (Oronsaye et al., 2010).	Witbank Dam, South Africa (Nussey et al., 2000).	Doula Estuary, Cameroon (Fonge et al., 2011)
Cd	0.0219	< 0.010	0.019	0.035	ND	ND
Cu	0.182	< 0.02	0.186	0.025	ND	0.01-0.03
Pb	0.00742	< 0.02	0.064	0.060	0.08-0.25	0.01-0.03
Zn	0.239	< 0.02	0.177	0.490	ND	0.01-0.03

ND= Not Determined

The implication of this finding is that Cu has the greatest tendency to bio-magnify in man assuming these fish species were to be consumed especially *H. fasciatus*. In a study by Wangboje et al. (2014), high MAR values were recorded for Zn, Ni and Pb in the mangrove oyster (*Crossostrea gasar*) from Golubo Creek in the Niger Delta and the workers concluded that these metals conceivably had the greatest tendency to bio-magnify in man when the oysters are consumed. The HQ values calculated for heavy metals in water were all below unity, indicating that presently there are no toxicity/hazard risks associated with the metals in water as corroborated by Brooks et al. (2003). The HQ values for Zn in all fish species were above unity indicating a toxicity/hazard risk by the metal.

Similarly, Giri and Singh (2013) reported HQ values in shrimps that exceeded unity for Arsenic, Copper and Chromium. On the other hand, Bandowe et al. (2014) whom worked on Polycyclic Aromatic Hydrocarbons (PAHs) and trace metals in fish in Ghana, reported HQ values that were less than unity for Cd, Cu, Fe, Mn and Zn and observed that there was no danger to the fish consuming local population. In this study, the health based criteria employed in the evaluation of the risk indices were based on the World Health Organisation (WHO) maximum permissible limits for heavy metals in water and food (WHO, 2004). This is because when such stipulated limits are exceeded, unpleasant health impacts are expected to occur in man. Similarly, Gnandi et al. (2011) applied the WHO limit in calculating the Relative Health Factor (RHF), which is comparable to the HQ index applied in this study. In this study, with regard to human health risk, the mean concentration of Cd in water exceeded the WHO maximum permissible limit of 0.003 mg/l for Cd in drinking water. In the case of fish, the WHO maximum permissible limit of 75 mg/kg for Zn in fish food was exceeded by the mean concentrations of Zn in *H. fasciatus, O. niloticus* and *B. occidentalis.*

From the above findings, the heavy metals of ecological significance in this study are Zn, Cd, and Cu. Zinc is an essential element for both flora and fauna however, exposure to either deficient or high levels can have deleterious effects on an organism's health (Okafor and Opuene, 2007). Excessive intake of Zn by man can lead to vomiting, dehydration, abdominal pain, nausea, lethargy and dizziness (Agency for Toxic Substances and Disease Registry, 1994). Sources of this metal include batteries, refuse, pesticides, alloys, dyes and fossil fuels. Cadmium is a non-essential element that is of no use in the human body. It can be sourced from batteries, fossil fuel, fertilizers, plastics, alloys and paints. It has been reported that Cd can be assimilated from anoxic sediments with high organic matter content which generates the potential for bioaccumulation via dietary uptake (Muniz et al., 2004). According to the Erah et al. (2002), chronic exposure to Cd in humans results in kidney dysfunction, hypertension, anemia and liver damage. Copper is an essential element in the human body. The metal is present in pesticides, wood preservatives, alloys, fossil fuel, cooking utensils, piping and wiring. According to the Nigerian Industrial Standard for Drinking Water Quality (NIS, 2007), excess intake of copper can lead to gastrointestinal disorder. Other health problems associated with excess Cu uptake are anaemia, liver and kidney damage (Turnland, 1988).

Conclusion

The study successfully ascertained the concentrations of Cd, Cu, Pb and Zn in fish and water from River Niger at Agenebode, Edo State, Nigeria. The heavy metals of ecological significance in this study were essentially Zn, Cd, and Cu. The potential health hazards of these metals were also highlighted. Since heavy metals are non-biodegradable and toxic both to wildlife and man even at low concentrations, there is thus the need to ensure that fish and water from the river are safe for human consumption via a sustained monitoring programme that would take into consideration other chemical pollutants not covered in this study. The National Environmental Standard and Regulation Enforcement Agency (NESREA) are thus enjoined in this regard to ensure that the environmental safety of our inland water bodies is sustained in order to mitigate unwholesome health impacts on man.

Conflict of interest

The authors did not declare any conflict of interest.

ACKNOWLEDGEMENT

Special thanks to Martlet Environmental Laboratory, Benin City, for the use of their Atomic Absorption Spectrometric facility.

REFERENCES

Adedeji OB, Okocha RC (2011). Assessment level of heavy metal in prawns (*Macrobrachium Macrobrachion*) and water from Epe Lagoon. Adv. Env. Biol. 5(6):1342-1345.

Adekunle AS, Eniola ITK (2008). Impact of industrial effluents on quality of segment of Asa River within an industrial estate in Ilorin, Nigeria. New York Sci. J. 1(1):17-21.

Adesulu EA, Sydenham DHJ (2007). The freshwater fishes and fisheries of Nigeria. Macmillan Nigeria Ltd., Ibadan, Nigeria. 397 pp.

Agency for Toxic Substances and Disease Registry (1994). Toxicological profile for zinc. US Department of Health and Human Service (205)-88-0608.

Baijot E, Moreau J, Bouda S (1997). Hydrobiological aspects of fisheries in small reservoirs in the Sahel region. Technical Centre for Agricultural and Rural Co-operation (ACP-EU), Netherlands, 238 pp.

Bandowe BAM, Bigalke M, Boamah L, Nyarko E, Saalia FK, Wilcke W (2014): Polycyclic aromatic compounds (PAHs and oxygenated PAHs) and trace metals in fish species from Ghana (West Africa): Bioaccumulation and health risk assessment. Env. Int. 65:135-146.

Biney CA (1991). The distribution of trace metals in the Kpong Headland and Lower Volta River, Ghana. In: Perspectives in aquatic ecotoxicology. Edited by N.K. Shastree. Narendra Publishing House, Delhi, India.

Brooks B, Foran CM, Lapoint TW (2003). Aquatic ecotoxicolgy of fluorene. Tox. Let. 142(3):169-183.

Butu AW, Iguisi EO (2013). Concentration of heavy metals in sediment of river Kubanni, Zaria, Nigeria. Comp. J. Env. Earth Sci. 2(1): 10-17.

Elder JF (1988). Metal biogeochemistry in surface water systems. A review of principles and concepts. United States Geological Survey Circular 1013.

Erah PO, Akujieze CN, Oteze GE (2002). The quality of groundwater in Benin City: A baseline study on inorganic chemicals and microbial contaminants of health importance in boreholes and pen wells. Trop. J. of Pharm. Res. 1(2): 75-82.

Fonge BA, Tening AS, Egbe AE, Awo EM, Focho DA, Oben PM, Asongwe GA, Zoneziwoh RM (2011). Fish (*Arius heudelotii* Valenciennes, 1840) as bio-indicator of heavy metals in Doula Estuary of Cameroon. Afr. J. Biotechnol. 10(73): 16581-16588.

Galyean ML (2010). Laboratory procedures in animal nutrition and research. Department of Animal and Food Sciences. Texas Technical University, Lubbock. 193 pp.

Giri S, Singh AK (2013). Assessment of human health risk for heavy metals in fish and shrimp collected from Subarnarekha River, India. Int. J. Env. Health Res.Retrieved 5th December, 2013.

Gnandi K, Bandowe BAM, Dehenyn DD, Porrachia M, Kersten M, Wilcke W (2011): Polycyclic aromatic hydrocarbons and trace metal contamination of coastal sediment and biota from Togo. J. Env. Mon. 13:2033-2041.

Latif A, Khalaf AN, Khaliol BY (1982). Bioaccumulation of Cu, Cd, Pb and Zn in two cyprinid fishes of Iraq. J. Fish Bio. 13(2):45-64.

Muniz P, Danulat E, Yannicelli B, Garcia-Alonso J, Medina G, Bicego MC (2004). Assessment of contamination by metals and petroleum hydrocarbons in sediments of Montevideo Harbour (Uruguay). Environ. Int. 29:1019-1028.

Newstead JN, Cousins IK, Giesy JP (2002). Predicted distribution and ecological risk assessment of a segregated hydrofluoranthene in the Japanese environment. Environ. Sci. Technol. 36:4761-4769.

Nigerian Industrial Standard, NIS (2007). Nigerian Standard for Drinking Water Quality. Standard Organisation of Nigeria, NIS 554.

Nussey G, VanVuren JHJ, DuPreez HH(2000). Bioaccumulation of Cr, Mn, Ni and Pb in the tissues of the Moggel, *Labeo umbratus* (Cyprinidae) from Witbank Dam, Mpumalanga. Water SA. 26(2):269-284.

Okafor EC, Opuene K (2007). Preliminary assessment of trace metals and polycyclic aromatic hydrocarbons in sediment. Int. J. Environ. Sci. Technol. 4(2):233-240.

Olaosebikan BD, Raji A (1998). Field guide to Nigerian freshwater fishes. Federal College of Freshwater Fisheries Technology, New Bussa, Niger State, Nigeria. 106 pp.

Oronsaye JAO, Wangboje OM, Oguzie FA (2010). Trace metals in some benthic fishes of the Ikpoba river dam, Benin City, Nigeria, Afr. J. Biotechnol. 9 (51):8860-8864.

Romijn CAF, Luttik R, Van Demeent D, Canton JH (1993). Presentation of a general algorithm to include effect assessment on secondary poisoning in the derivation of environmental quality criteria (1) Aquatic food chains. Ecotox. Env. Saf. 26:61-85.

Saeed SM, Shaker IM (2008). Assessment of heavy metals pollution in water and sediment and their effect on *Oreochromis niloticus* in the Northern Delta lakes, Egypt. The 8th International Symposium on Tilapia in aquaculture, Egypt. pp. 475-490.

Sangodoyin AY (1991). Groundwater and surface water pollution by open refuse dump in Ibadan, Nigeria. J. Dis. Inno. 3(1):24-31.

Stephen O, Lenihan E (2007), Rural livelihood insecurity in Etsako East Area of Edo state, Nigeria. J. Sus. Dev. 5(1):2-7.

Tabari S, Saravi SSS, Bandy GA, Dehghan A, Shokrzadeh M (2010). Heavy metals (Zn, Pb, Cd and Cr) in fish, water and sediment sampled from Southern Caspian Sea, Iran. Tox. Ind. Health 26(10):649-656.

Tessier A, Campbell PGC, Bussan M (1979). Sequential extraction procedure for the speciation of particulate traces metals. Anal. Chem. 51(7):844-851.

Turnland JR (1988). Copper nutrition, bioavailability and influence of dietary factors. J. Am. Diet. Assoc. 1:303-308.

United States Environmental Protection Agency (1998). Guidelines for neurotoxicity risk assessment. EPA/630/R95/001F. USEPA Office of Research and Development, Washington DC 20460. 77 pp.

Wangboje OM, Oronsaye JAO, Okeke CE (2014). The use of the mangrove oyster (*Crossostrea gasar*) as a bio-indicator for chemical element contamination in the Niger Delta. Afr. J. Food Agric. Nutri. Dev. 14(3):8903-8919.

World Health Organisation (2004). Guidelines for drinking water quality, 3rd Edition, Vol. 1. Recommendations. World Health Organisation, Geneva.

Diversity, composition and abundance of macroinvertebrates associated with water hyacinth mats in Lake Victoria, Kenya

Patrick Oduor Orwa, Reuben Omondi, William Ojwang and Josephat Mwanchi

Kenya Marine and Fisheries Research Institute, P.O. Box 1881, Kisumu, Kenya.

This study is set to investigate the macroinvertebrate community structure within water hyacinth in the Kenyan waters of Lake Victoria. This is helpful in determining the relationship between water hyacinth and macroinvertebrates. A total of four replicates were taken from 18 sampling stations within the lake using a Ponar grab and a 500 µm scoop net. The samples were washed through a 300 µm sieve and sorted alive in the field. The organisms were then identified to genus level and further categorized into functional feeding guilds using available keys and literature. The vertical position of the organisms at a water hyacinth mat was also examined. The data was then analyzed for diversity and abundance. One way ANOVA was further done to test any significant variation in community attributes between stations within the lake. All the physic-chemical parameters measured varied significantly though pairwise comparison revealed that most stations were in same sub set. A total of 14 orders and 34 genera were identified during the study. Out of which, 13 genera which include *Chironomus*, *Lymneae*, *Hirudo*, *Tubifex*, *Platycnemis*, *Sympetrum*, *Trithemis*, *Melanoides*, *Biomphalaria*, *Ilybius*, *Elmis*, *Leptocerus*, and *Hexagenia* were found at every sampling station. Most of the taxa found were predators probably feeding on mosquito larvae which could explain their absence. There was no significant difference in the diversity between stations though they varied in abundance. It is thus concluded that water hyacinth influences the distribution of macroinvertebrates and can thus be used to predict their presence. The feeding guilds were however assigned using existing literature. It is therefore recommended that a further study be done to confirm the feeding habits and the presence of mosquito larvae.

Key words: Macroinvertebrates, hyacinth, Lake Victoria.

INTRODUCTION

Water Hyacinth (*Eichhornia crassipes*) can be described as a floating macrophyte with thick, glossy, round leaves, inflated leaf stems, and lavender flowers (Toft et al., 2003). It is a free-floating aquatic macrophyte that clusters in large mats that block shorelines (Rocha-Ramirez et al., 2007). It is sometimes found stuck in mud, appearing rooted and it is rarely found as a single plant (Villamagna, 2009). The foliage helps to provide shade and the roots provide filtration, a spawning area and refuge for the aquatic organisms like macroinvertebrates and fish (Schramm et al., 1987). The macrophyte absorbs excess nutrients from the water, helping to reduce algal blooms thus an excellent water clarifier (Rocha-Ramirez et al., 2007). Being a floating macrophyte it moves about with the filtrates thus expected to present a similar microhabitat condition wherever it goes. Studies done in other parts of the world have shown that specific macro-invertebrate taxa inhabit water hyacinth regardless of the region (de Marco et al., 2001; Toft et al., 2003; and Rocha-Ramirez et al., 2007). This study thus set out to confirm this and identify which taxa inhabit water hyacinth in the Kenyan part of Lake Victoria.

MATERIALS AND METHODS

Study Area

The Kenyan portion of Lake Victoria can conveniently be divided into Nyanza Gulf and open waters. The Nyanza Gulf receives water from several rivers and is eutrophic due to nutrients from these rivers while the expansive open waters have less nutrients. A total of 16 stations were sampled during the study between 18 and 26th December, 2013; 8 stations were in the Nyanza gulf while 8 were in the open waters of Lake Victoria (Figure 1). All the stations were geo-referenced using magellan global positioning system. The study was carried out in the early stages of water hyacinth re-infestation when most mats were found at the shallow littoral areas of the lake.

Sampling

Physico-chemical Parameters

Selected physic-chemical parameters were taken *in-situ* using appropriate meters. The parameters measured include water depth, Dissolved Oxygen (DO), turbidity, conductivity, pH, redox and temperature. At each station, these parameters were taken in triplicates and later averaged.

Macroinvertebrates

Sampling of macroinvertebrates was done in sixteen stations using a Ponar grab and a 500 µm mesh size scoop net. In each sampling station, three grabs and three scoops of macroinvertebrate samples were collected and washed at the site of collection sorted live and then preserved in absolute ethanol. In the laboratory they were identified to genus level. The vertical position of the macroinvertebrates was further examined to show which taxa were found at the benthos below the mat, at the roots and within the stem. This was done by taking samples at the benthos below the mat, examining the sediments trapped on the roots and opening the stems to see if there are any organisms within.

Data analysis

The data on physic-chemical parameters were summarized as mean ± SE and represented in a table. Variation between the stations was tested using one-way ANOVA. The identified benthic macroinvertebrates were then analyzed for taxon diversity (Shannon-Wiener, H'), abundance, evenness, and relative abundance for the lake at each sampling station. The data was represented in frequency tables and bar graphs. Log transformed one way analysis of variance was done to test if there is any variation between the stations sampled.

RESULTS AND DISCUSSIONS

Physico-chemical parameters

The values of the physic-chemical parameters are shown in Table 1. The DO values ranged between 1.04 ± 0.01mgL^{-1} and 6.85 ± 0.5 mgL^{-1} while turbidity ranged between 3.69 ± 0.21 and 312.5 ± 2.5. Conductivity hardly went below 100 µScm^{-1} except Bulwani and Got Kachola with 86.55 ± 0.45 and 96.0 ± 1.0, respectively while Ph ranged between weak acidic and weakly basic values. Temperature of these stations ranged between 23.4 ± 0.00 and 27.8 ± 0.00. Statistical tests revealed significant variations between the stations for all the parameters but on pairwise comparison, it was discovered that majority of the stations were at one sub set for all the parameters (Table 1). This result shows that there are similarities between the stations.

Macroinvertebrate composition

A total of 14 orders, 30 families and 34 genera were found during the sampling period (Table 2). Thirteen of these genera were found at all sampling stations as shown in Table 2.

Out of the orders sampled, Hemiptera, Pulmonata and Coleoptera had the highest number of genera with 5, 4 and 4, respectively. In terms of relative abundance, dipterans and Pulmonata were the most abundant while Hydracarina (water mites) were the least abundant (Figure 2). This is an indication that water hyacinth provides a favourable microhabitat for those macroinvertebrates and as a result they associate with the plant. Aquatic organisms such as macroinvertebrates live in selected areas within water bodies depending on the physical, chemical and biological characteristics (Mason 2002; Wang and Lyons, 2003). Due to this fact, they have been used to predict water quality status of different water systems (Masese et al., 2009; Aura et al., 2010). The taxa found represented the sensitive ones such as Trichopterans and the tolerant ones like dipterans and oligochaete worms.

The high diversity obtained can be attributed to favourable water quality conditions and availability of food. Water hyacinth, being a floating macrophyte traps a lot of matter as it floats which could eventually form food for organisms within. Rocha-Ramirez et al. (2007) argues that water hyacinth traps a lot of organic matter as it floats and moves with it from one point to another. Another reason for the high diversity could probably be its ability to protect the invertebrates from direct sunlight,

Figure 1. Map of Lake Victoria showing the sampling stations.

Table 1. Physico-chemical parameter values (Mean ± SEM) within water hyacinth mats in Lake Victoria (different superscript letters show stations that vary significantly).

Stations	Depth (m)	DO (mgl^{-1})	Turbidity	Cond (μScm^{-1})	pH	Temperature (°C)
Anyanga	1.2	6.85 ± 0.5^c	6.49 ± 0.68^a	108.5 ± 2.5^a	8.24 ± 0.25^c	27.55 ± 0.05^c
Asat	1.6	5.25 ± 0.67^c	54.6 ± 14.4^b	153 ± 0.5^c	7.45 ± 0.03^b	26.75 ± 0.05^c
Bulwani	2.3	1.31 ± 0.17^a	6.83 ± 0.32^a	86.55 ± 0.45^a	6.81 ± 0.11^a	23.4 ± 0.00^a
Got Kachola	0.5	1.04 ± 0.01^a	4.42 ± 0.01^a	96.0 ± 1.0^a	7.05 ± 0.07^a	24.25 ± 0.25^a
Goye	1.0	2.76 ± 0.32^b	3.69 ± 0.21^a	104.5 ± 0.00^a	6.82 ± 0.00^a	25.3 ± 0.00^b
Homa Bay	3.8	5.16 ± 0.18^c	27.0 ± 0.00^b	152.65 ± 1.95^c	7.81 ± 0.04^b	27.3 ± 1.50^c
Kendu Bay	1.5	4.53 ± 0.13^b	98.8 ± 19.2^b	157.9 ± 0.2^c	7.31 ± 0.16^b	28.3 ± 0.3^c
Kuja	1.0	2.06 ± 1.8^a	312.5 ± 2.5^d	116.15 ± 0.15^a	6.76 ± 0.47^a	23.9 ± 0.00^a
Lwanda Kotieno	1.4	6.72 ± 0.06^c	17.35 ± 4.45^b	124.5 ± 2.90^b	7.22 ± 0.23^b	27.65 ± 0.25^c
Mirunda	1.4	5.97 ± 1.35^c	247.5 ± 0.5^d	147.1 ± 1.20^b	7.68 ± 0.04^b	27.9 ± 0.8^c
Nyando	1.8	3.11 ± 0.67^b	56.5 ± 4.70^b	181.6 ± 3.10^d	6.98 ± 0.1^a	26.65 ± 0.25^c
Oluch	2.0	6.41 ± 0.56^c	182 ± 107^c	151.65 ± 0.85^b	7.89 ± 0.02^c	25.75 ± 0.45^b
Samunyi	1.3	1.56 ± 0.7^a	49.2 ± 23.5^b	156.6 ± 3.90^c	7.23 ± 0.18^b	24.05 ± 0.05^a
Sio	1.2	-	12.7 ± 1.10^b	113.6 ± 0.1^a	7.77 ± 0.19^b	27.55 ± 0.05^c
Sori	1.3	6.63 ± 0.03^c	6.99 ± 0.74^b	106.9 ± 0.1^a	7.33 ± 0.01^b	27.25 ± 0.05^c
Usenge	1.6	6.44 ± 0.00^c	3.02 ± 0.33^b	107.6 ± 0.00^a	7.64 ± 0.03^b	27.8 ± 0.00^c
F		10.27	11.58	264.38	6.66	12.84
p		<0.05	<0.05	<0.05	<0.05	<0.05

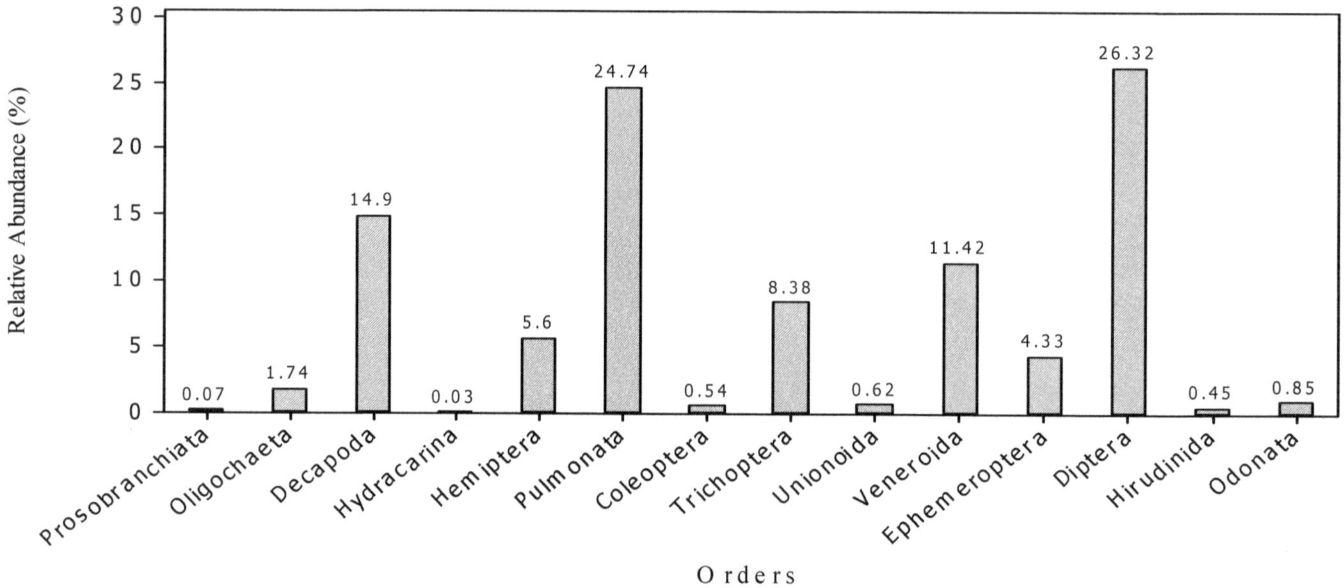

Figure 2. Relative abundance of Macroinvertebrate orders within water hyacinth mats in the Kenyan waters of Lake Victoria.

absorb excess nutrients and offer breeding grounds. Aquatic macrophytes like water hyacinth provide shade and the roots provide filtration, a spawning area for the aquatic organisms like macroinvertebrates and fish (Schramm et al., 1987). Water hyacinth could be very effective in nutrient absorption and as a result there were mayflies and caddis flies on the sediments beneath the mat.

For the case building invertebrates, the water hyacinth stem provided a soft material that is convenient to them. Some even burrowed through the soft stem and hatched their eggs inside. It can thus be summarized that a mat of water hyacinth provides food, shelter and breeding ground to macroinvertebrates as such a microhabitat on its own.

Table 2. Macroinvertebrate taxa sampled within water hyacinth mats during the study period.

Order	Family	Genera	Common name	Frequency
Coleoptera	Corculionidae	*Neochetina*	Introduced beetle	2
	Dytiscidae	*Ilybius*	Diving beetles	16
	Elmidae	*Elmis*	Riffle beetles	16
	Gyrinidae	*Gyrinus*	Whirligig beetle	8
Decapoda	Atyidae	*Caridina*	Freshwater shrimp	15
Diptera	Chironomidae	*Chironomus*	Midgefly larvae	16
	Culicidae	*Anopheles*	Mosquito larvae	2
	Ceratopogonidae	*Palpomyia*	Biting midges	7
Ephemeroptera	Ephemeriidae	*Hexagenia*	Giant mayfly	16
Hemiptera	Nepidae	*Ranatra*	Water stick	5
	Belostomatidae	*Sphaerodema*	Giant water bug	15
		Hydrocyrus	Giant water bug	5
	Mesovellidae	*Mesovelia*	Water Treaders	6
	Notonectidae	*Notonecta*	Backswimmers	5
	Corixidae	*Corixa*	Water boatman	3
Hirudinida	Hirudinidae	*Hirudo*	Medicinal Leech	16
Hydracarina			Water mites	8
Odonata	Aeshnidae	*Anax*	Blue emperor	6
	Gomphidae	*Ictinogomphus*	Tiger tail	1
	Libellulidae	*Sympetrum*	Skimmer dragonfly	16
		Trithemis	Dropwings	16
		Branchythemis		7
	Platycnemididae	*Platycnemis*	Featherleg damselfly	16
Oligochaeta	Naididae	*Tubifex*	Sludgeworm	16
	Lumbriculidae	*Lumbricus*	Aquatic earthworm	5
Prosobranchiata	Valvatidae	*Valvata*	Valve snail	3
Pulmonata	Lymnaeidae	*Lymnae*	Pond snail	16
	Physidae	*Physa*	Pouch snail	14
	Planorbidae	*Biomphalaria*	Ramshorn	16
	Thiaridae	*Melanoides*	Red-rimmed melania	16
Trichoptera	Rhyacophilidae	*Rhyacophila*	Caddis fly	7
	Leptoceridae	*Leptocerus*	Long horned	16
Unionoida	Unionidae	*Unodonta*	Pearly mussels	13
Veneroida	Sphaeridae	*Sphaerium*	Pill clams	15

Spatial variation

Macroinvertebrate abundance

Macroinvertebrate abundance was highest at Bulwani (1639 individuals) and lowest at Lwanda Kotieno (61). The beach stations (Asat and Anyanga) recorded higher abundances compared to those away from the shore. Despite this large variation in total abundance, box and whisker plots showed that the most of the whiskers intercept indicative of lack of significant differences between the stations (Figure 3).

Kruskal-Wallis test showed that the variation in abundance was not significant at 99% confidence level

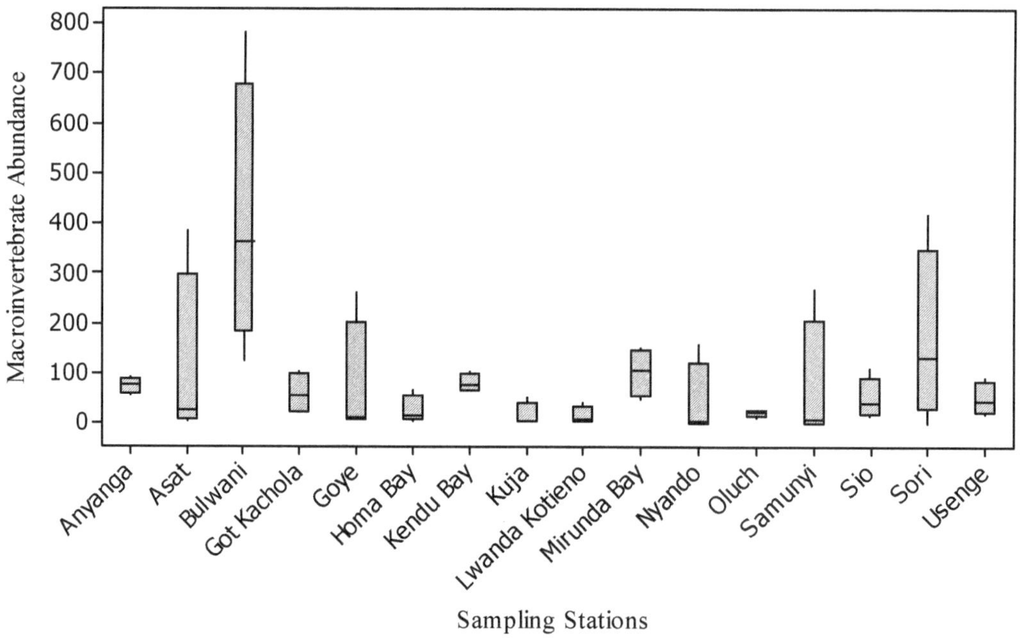

Figure 3. Macroinvertebrate abundance within the sampling stations represented in box plots showing dispersion within data sets.

Table 3. Macroinvertebrate community attributes within the study stations.

Stations	Diversity (H')	Evenness (e)	Dominance
Kendu Bay	0.799	0.37	0.65
Nyando	1.29	0.52	0.34
Lwanda Kotieno	1.36	0.55	0.36
Asat	0.67	0.24	0.74
Samunyi	0.81	0.45	0.63
Oluch	1.20	0.66	0.36
Homa Bay	1.27	0.71	0.36
Mirunda	1.19	0.47	0.46
Sori	0.60	0.18	0.73
Got Kachola	1.05	0.37	0.47
Kuja	1.50	0.90	0.24
Anyanga	0.64	0.27	0.73
Usenge	1.32	0.53	0.35
Goye	1.53	0.58	0.27
Bulwani	1.43	0.47	0.27
Sio	1.44	0.42	0.31

though significant at 95% ($H = 22.11$, $p = 0.033$). When the data was log transformed {Log(X+1)} then subjected to one way ANOVA, the variation was still not significant at 99% ($F = 2.02$, $p = 0.026$). Considering the abundance of the twelve common taxa, the variation was not significant at both 95% and 99% when subjected to one way ANOVA ($F = 1.98$, $p = 0.061$).

On pairwise comparison, DMRT revealed the stations responsible for the variation. Over 80% of the stations were at sub-set b. The stations that differed from the rest were Lwanda Kotieno, Kuja RM, and Oluch RM at sub-set a and Sori, Asat beach, and Mirunda Bay at sub-set c.

Macroinvertebrate diversity

Table 3 shows the values for Shannon-Wiener, evenness

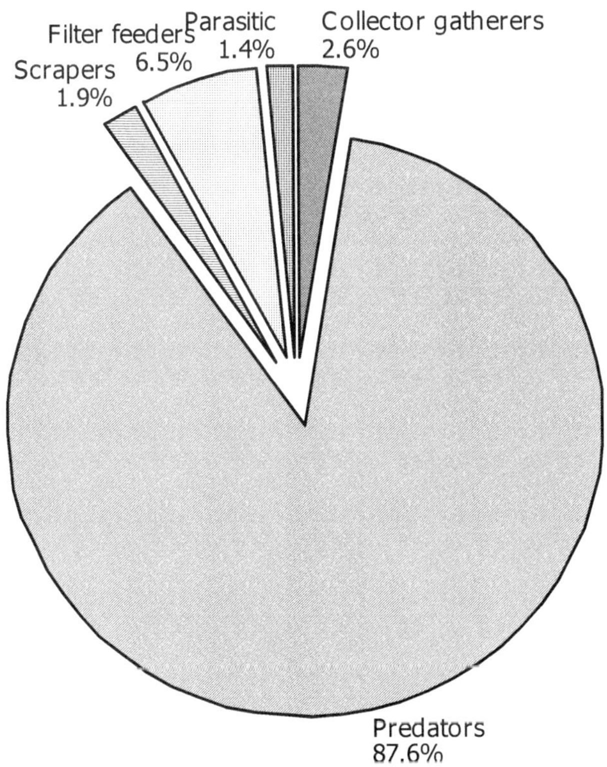

Figure 4. Proportion of Functional feeding of the sampled organisms derived from existing literature.

and dominance indices using the genera values for the stations sampled during the study period. Shannon-Wiener diversity index was between 0.6 and 1.53. Evenness index was highest at Kuja RM (0.9) and lowest at Sori (0.18). Apart from recording higher evenness index, Kuja RM also had the lowest dominance index which indicates that the organisms at the station coexisted well.

Vertical positioning of the macroinvertebrates

The common taxa were found at specific position in relation to the water hyacinth plant. On the benthic fauna below the plant mainly consisting of algae and sunken water hyacinth plant there were the *Hexagenia* spp, *Lymnae* spp and *Leptocerus* spp together with the cases. On the roots were *Tubifex* worms, *Chironomus* spp, *Hirudo* spp, *Melanoids* spp, *Lymnae* spp, *Biomphalaria* spp, *Platycnemis* spp, *Sympetrum* spp and *Trithemis* spp while on the stem there were the larvae of *Elmis* and *Ilybius* spp inside the stem of the plant.

Functional feeding groups

Over 80% of the macroinvertebrates sampled beneath the water hyacinth mat were predators mainly feeding on mosquito larvae (Figure 4). The other feeding guilds found according to literature were collector gatherers (2.6%), parasites (1.4%), filter feeder (6.5), and scrapers (1.9%).

Over 80% of the invertebrates were predators whose main preys are insect larvae. The Hemiptera taxa sampled (Notonectidae, Belostomatidae and Nepidae); Coleoptera (Dytiscidae and Gyrinidae); and Odonata (Gomphidae, Platycnemididae and Libellulidae) are believed to feed mostly on mosquito larvae. This could partly explain the absence of mosquito larvae in most stations despite common belief that mosquitos breed more in areas covered by water hyacinth mats. This corroborates the earlier findings of Ofulla et al (2010). The presence of filter feeders and gathering collectors such as the dipterans (Culicidae, Chironomidae); and some coleopterans (Elmidae) on the root of the plant indicates presence of algae and detritus that are trapped during their float. Research has shown that water hyacinth traps phytoplankton and detritus while floating in a water body or sedentary and lake shores (Rommens et al., 2003; Greenfield et al., 2007).

Conclusion

Based on the results and the existing literature, it is concluded that water hyacinth mat provides a microhabitat that favours the survival of several macroinvertebrate taxa. The plant absorbs excess nutrients to make water quality favourable and traps phytoplankton and detritus which in turn becomes food for aquatic life.

Recommendations

We hereby recommend the following based on the findings of this study:

1. Stomach contents of all predators within water hyacinth mats should be examined to check for presence of mosquito larvae.
2. The colonization of water hyacinth by the various macroinvertebrates should be studied to determine the trend and sequence of habitation.

Conflict of interest

The authors did not declare any conflict of interest

REFERENCES

Aura CM, Raburu PO, Herrmann J (2010). A Preliminary Macroinvertebrate IBI for biossesment of the Kipkaren and Sosiani Rivers, Nzoia River Basin, Kenya. *Lakes and Reservoirs.* 15(2):119-128.

de Marco P, Araujo MAR, Barcelos MK, dos Santos MBL (2001).

Aquatic invertebrates associated with the water-hyacinth (*Eichhornia crassipes*) in an eutrophic reservoir in tropical Brazil. Studies on Neotropical Fauna and Environment. 36:73-80.

Masese FO, Muchiri M, Raburu PO (2009). Macroinvertebrate assemblages as biological indicators of water quality in the Moiben River, Kenya. Afr. J. Aqua. Sci. 34:15-26.

Mason CF (2002). Biology of freshwater pollution. 4th Ed. Pearson Education Ltd. Essex. London. 38pp.

Ofulla AV, Karanja OD, Omondi R, Okurut T, Matano A, Jembe T, Abila R, Boera P, Gichuki J (2010). Relative abundance of mosquitos and snails associated with water hyacinth and hippo grass in the Nyanza gulf of Lake Victoria. *Lakes and Reservoirs: Research and Management* 15:255-271.

Rocha-Ramirez A, Ramirez-Rojas A, Chavez-Lopez R, Alcocer J (2007). Invertebrate assemblages associated with root masses of Eichhornia crassipes (Mart.) Solms-Laubach 1883 in the Alvarado Lagoon al System, Veracruz, Mexico. Aqua. Ecol. 41(2):319-333.

Rommens W, Maes J, Dekeza N, Inghelbrecht P, Nhiwatiwa T, Holsters E, Ollevier F, Marshall B, Brendonck L (2003). The impact of water hyacinth (*Eichhornia crassipes*) in a eutrophic subtropical impoundment (Lake Chivero, Zimbabwe). I. Water quality. *Archiv Fur Hydrobiologie* 158:373-388.

Schramm HL, Jirka KJ Jr, Hoyer MV (1987). Epiphytic macroinvertebrates on dominant macrophytes in two central Florida Lakes. J. Freshwater Ecol. 4 (2).

Toft JD, Simenstad CA, Cordell JR, Grimaldo LF (2003). The effects of introduced water hyacinth on habitat structure, invertebrate assemblages, and fish diets. *Estuaries* 26:746-758.

Villamagna, A. M. (2009). Ecological effects ofwater hyacinth (*Eichhornia crassipes*) on Lake Chapala, Mexico. D. Phil thesis, Virginia Polytechnic Institute and State University.

Wang L, Lyons J (2003). Fish and Benthic Macroinvertebrate Assemblage as Indicators of Stream Degradation in Urbanizing Watersheds. In *Biological Response Signatures: Indicator Patterns Using Aquatic Communities* (Simon, T.P., ed.), CRC Press, Boca Raton: pp. 113-120.

Influence of initial glycerol concentration upon bacterial cells adaptability and biodegradation kinetics on a submerged aerated fixed bed reactor using Biocell® (PE05) packing

B. Lekhlif[1,2] , A. Kherbeche[1,2,3], G. Hébrard[3,4,5], N. Dietrich[3,4,5] and J. ECHAABI[2]

[1]Research Team of Hydrogeology, Treatment and Purification of Water and Climate Change, Environmental Engineering Laboratory of EHTP, Km 7, Route d'ElJadida, B.P8108, Oasis, Casablanca, Morocco.
[2]Polymer Research Team, ENSEM, Hassan II University, Route d'ElJadida, B.P 8118, Oasis, Casablanca, Morocco.
[3]Université de Toulouse, INSA,UPS, INP, LISBP, 135 Avenue de Rangueil, F-31077 Toulouse, France.
[4]INRA, UMR792, Ingénierie des Systèmes Biologiques et des Procédés, F-31400 Toulouse, France.
[5]CNRS, UMR5504, F-31400 Toulouse, France.

The present paper reports an experimental work of the influence of initial substrate concentration on the adaptability of bacterial cells and the biodegradation kinetics of the substrate in a submerged aerobic fixed-film reactor, by studying some physicochemical parameters. The bioreactor used in this study is gotten from the biological aerated filter (BAF), but the used filter media is made of plastic of a large size than that used usually in BAF and with a high specific surface. For this purpose, various synthetic wastewaters were prepared based on a non-toxic substrate, in this case, glycerol, and some salts of nitrogen, phosphorus and some oligo-elements with different initial chemical oxygen demand (COD) of: 330 MgO$_2$/L (S1), 480 MgO$_2$/L (S2), 860 MgO$_2$/L (S3) and 1120 MgO$_2$/L (S4). Experiments have been carried out at laboratory scale in a cylindrical reactor, made of PVC (height of 1 m and diameter of 0.125 m). The pilot was filled by a Biocell® (PE05) packing fixed between two grids. The analysis of various physicochemical parameters, during biodegradation (COD, dissolved oxygen, pH, turbidity), showed that the performances of the submerged aerobic fixed-film reactor were influenced by the initial substrate concentration. In low concentrations, the adaptability of bacterial cells was easy and relatively quick. Biodegradation kinetics constants reached 0.407, 0.346, 0.341 and 0.232 d^{-1} respectively for synthetic wastewaters S1, S2, S3 and S4. It has been found that the physicochemical parameters could be used for monitoring the adaptability and the biodegradation process of the substrate. Turbidity was revealed as a good indicator for biofilm growth, mainly because of its decrease during the adaptation phase and its increase during the biodegradation phase. In the same time, the pH increased especially during biodegradation phase. In parallel, DO decreased gradually.

Key words: Submerged aerobic fixed-film reactor, adaptability, biodegradation kinetics, packing, synthetic wastewater, chemical oxygen demand (COD).

INTRODUCTION

Submerged aerobic fixed-film reactors are mainly used for carbonaceous, and ammonia removal in the aerobic treatment of urban wastewaters. They are also used in secondary or in tertiary treatment. Today, these

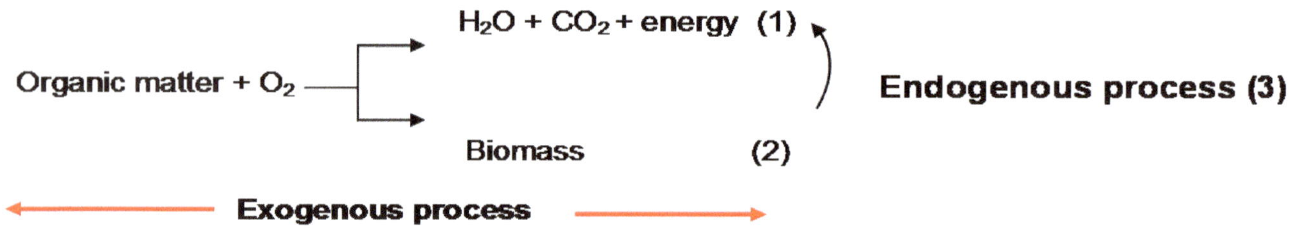

Figure 1. Removal of biological process of organic matter.

bioreactors have several applications in industry (Mendoza-Espinoza et al.,1999; Chaudhary et al., 2003). They combine compactness and high removal efficiencies in a large range of hydraulic and organic load. In secondary treatment, aerobic biological degradation of organic matters in submerged aerobic fixed-film reactor is made in the presence of bacteria according to the reactions in Figure 1.

This degradation involves three reactions in the same time: the first one (Reaction 1) corresponds to an oxidation of organic matter with H_2O and CO_2 as byproducts, it releases also energy, necessary to bacterial cells maintenance and growth; it uses external organic matter (Exogenous process). The second one (Reaction 2) is related to multiplication of bacterial cells using organic matters. The third one (Reaction 3) uses biomass when the dissolved substrate becomes rare (Endogenous process) and bacteria cells use their own reserves. The kinetics of the endogenous process reaction is relatively low as compared to the exogenous processes.

The packing filling the reactor allows the development of biofilm by its colonization over the available surface area, in which bacterial cells adhere to each other. These adherent cells are frequently embedded within a self-produced matrix of extracellular polymeric substance (EPS) allowing the formation of biofilm (Horan, 2003; Harvey et al., 2011).

It is well known that the performances of bioreactors depends on biofilm growth, which could be influenced by several factors, among others: packing characteristics (Mendoza-Espinoza, 1999; Cheremisinoff, 2002; Prenafeta-Boldú et al., 2008), aeration conditions (Kassab et al., 2010; Jin et al., 2012; Albuquerque et al., 2012), substrate concentration and its nature in wastewater (Amrouche et al., 2011; Chen et al., 2012).

The submerged aerobic fixed-film reactor performances could be assessed by the biodegradation yield (Y). It is expressed as follows:

$$Y\,(\%) = \frac{L_0 - L}{L_0} \qquad (1)$$

where: Y, Yield (%) removal of the substrate (in terms of BOD); L_0, Initial organic load of the substrate, expressed in BOD (MgO_2/L); L, final organic load of the substrate at time t, expressed in BOD (MgO_2/L).

The removal performances could also be assessed by calculating biodegradation kinetics constants of organic pollution. In batch reactor, the most significant period in the growth cycle is the exponential growth phase, when the population of biomass is perfectly adapted to the substrate. The first-order model, neglecting endogenous respiration, provides accurate simulations of biodegradation kinetics. It can be written as follows (Cheremisinoff, 2002; Mara, 2003):

$$\frac{dL}{dt} = -k_t \times L \qquad (2)$$

where:
t: time (d); kt: first-order BOD biodegradation kinetics constant expressed as (d^{-1}) depending on the temperature according to the following expression:

$$k_t = k_{t_0} \times \theta^{(T-T_0)} \qquad (3)$$

where: first-order BOD biodegradation kinetics constant at T0, expressed as (d^{-1}); : temperature coefficient, equal to 1.032, this value is in the range of those given by Lesouef et al. (1992) and Queinnec et al. (2006).
T: Temperature of the synthetic wastewater expressed in °C; T_0: Standard temperature at 20°C.

The integration of equation 2, taking into account equation 3, provide the expression of substrate degradation in time:

$$L = L_0 \times 10^{-[k_{t_0} \times \theta^{(T-T_0)}.t]} \qquad (4)$$

The linearization of this equation allows the determination of the constant using semi-logarithmic coordinates according to Equation 5.

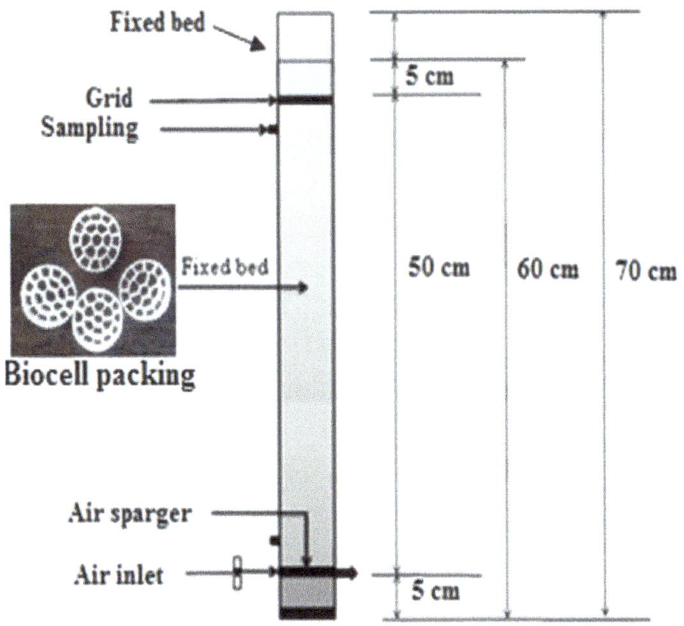

Figure 2. Experimental setup.

$$\log \frac{L}{L_0} = k_{t_0}\theta^{(T-T_0)} \times t \qquad (5)$$

This equation was used in this study. Glycerol substrate used is pure. So the COD will be considered almost equal to the BOD (Mara, 2003).Other physicochemical parameters such as conductivity, pH, dissolved oxygen and turbidity could be linked to the assessment of submerged aerobic fixed-film reactors performances (Mendoza-Espinoza et al., 1999; Akin et al., 2005; Albuquerque et al., 2012) and could give some explanations about the efficiency of this bioreactor against wastewater treatment. This work aimed to study the effect of initial concentration of the synthetic wastewater on the adaptability of bacterial cells on a submerged aerobic fixed-film reactor packing on surface area and the biodegradation kinetics; four tests were made with concentrations: 330 MgO_2/L (S1), 480 MgO_2/L (S2), 860 MgO_2/L (S3) and 1120 MgO_2/L (S4).

MATERIALS AND METHODS

Experimental setup

Experiments were carried out in a reactor operating in batch mode (Figure 2). It was a cylindrical column, with height of 1 m and diameter of 0.125 m, made of opaque PVC. It was filled with 5 L of the synthetic wastewater. The air was introduced at the bottom of the column through an air flow equal to 10 L/h, by a rectangular diffuser (8 x 1 cm).

The bioreactor was filled with Biocell® packing (Figure 2) with excellent physical characteristics (Table 1). It was supported by a grid placed just above the air sparger at the bottom of the column for maintaining the packing fixed. Another grid was used to fix this packing to avoid its flotation by the effect of crossing air bubbles.

Table 1. Physical properties of packing.

Name	Biocell® (PE05)
Color	White
Material	Virgin Polythene
Average particle diameter (mm)	25
Average particle height (mm)	12
Specific surface area (m^2/m^3)	~ 427
Density (kg/m^3)	95
Number of particles (/ m^3)	1.353×10^5
Bed porosity (%)	92.5

Synthetic wastewaters preparation and experiments conduct

Synthetic wastewaters were prepared in COD/N/P ratio of 100/4/1 (EPA, 1997; Mara, 2003) (Table 2), using distilled water. Pure glycerol was the organic substrate. Salts of sodium nitrate (nitrogensource) and potassium phosphate (phosphorus source) and some oligoelements ($FeSO_4$, $MgSO_4$, $MnSO_4$ and $CaSO_4$) were also added to synthetic wastewaters prepared. Doses of oligoelements were low (Mara, 2003; Jin, 2012).

To initiate the adaptation, inoculation by bacterial flora was made in the reactor. This bacterium was extracted from soil (Mara, 2003; Horan, 2003). Table 3 illustrates the physicochemical characteristics of synthetic wastewaters.

Submerged aerobic fixed-film reactor performances monitoring

Chemical oxygen demand (COD) and turbidity were monitored using a photometer type Palintest 7000. pH and dissolved oxygen were measured using a Hach sensor 40d-HQ -Multi parameters, which can also measure the temperatures of water and air. The conductivity was measured by Orion model 125. For measuring the COD concentrations, samples from the bioreactor were filtered through filters having a pore size of 0.45 μm. For the other parameters, measurements were realized directly in samples.

RESULTS AND DISCUSSION

COD and biodegradation kinetics evolution

COD results are presented in Figure 3. According to the figure, the same appearance was shown in COD curves. In fact, it shows three principal phases. The first slow phase correspond to the adaptation of bacteria cells to the substrate and also to the hydraulic batch mode conditions (adaptation phase). The time of this adaptation increased with the initial concentration as notified by some authors (Amrouche et al., 2011; Dey et al., 2010). This time was 1 day for S1 and S2, 2.5 days for S3 and 3.5 days for S4. After this time, the substrate concentration

Table 2. Synthetic wastewaters compositions.

Synthetic wastewaters	S1	S2	S3	S4
	(Mg/L)			
$C_3H_8O_3$	300	400	750	1000
$NaNO_3$	12	16	20	40
KH_2PO_4	3	4	5	10
$ZnSO_4.7H_2O$	0.3	0.4	0.8	1
$FeSO_4.7H_2O$	0.3	0.4	0.8	2
$MnCl_2$	0.3	0.4	0.8	3
$CaSO_4$	0.07	0.1	0.23	0.25
$MgSO_4.7H_2O$	0.03	0.04	0.1	0.1

Table 3. Physicochemical characteristics of synthetic wastewaters.

Parameters of synthetic wastewaters		S1	S2	S3	S4
COD	(MgO_2/L)	330	480	860	1120
Water T°	(°C)	25	25.5	15.1	14.3
Air T°	(°C)	25.3	26.2	14.5	14
Conductivity	(µS/cm)	1185	1320	1584	1955
Turbidity	(NTU)	10	28	28	30
pH	-	7.92	8.2	7.63	8.23
DO	(Mg/L)	-	10.32	10.28	9.35

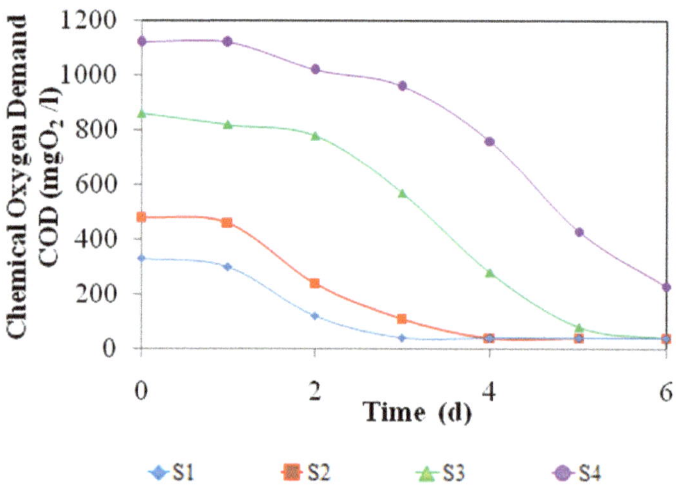

Figure 3. COD evolution versus time.

manifested a slow decrease. The second phase corresponded to the removal of the substrate. This removal depended on initial concentration. In low COD concentration (330 MgO_2/L), the time required was 3 days, moreover, this time was 4, 5 and more than 6 days, respectively for 480, 860 and 1120 MgO_2/L.

The third phase manifested a deceleration of the substrate concentration decrease. The COD tended to stabilize at times, an endogenous respiration has probably triggered after an almost total removal of the substrate (Reaction 3).

The biodegradation kinetics constants could be determined by linearization of the equation 5 and plotting log (L/L_0) against time as shown in Figure 4. Two different slopes of the glycerol biodegradation kinetics for each COD concentration removal have been noted. The first

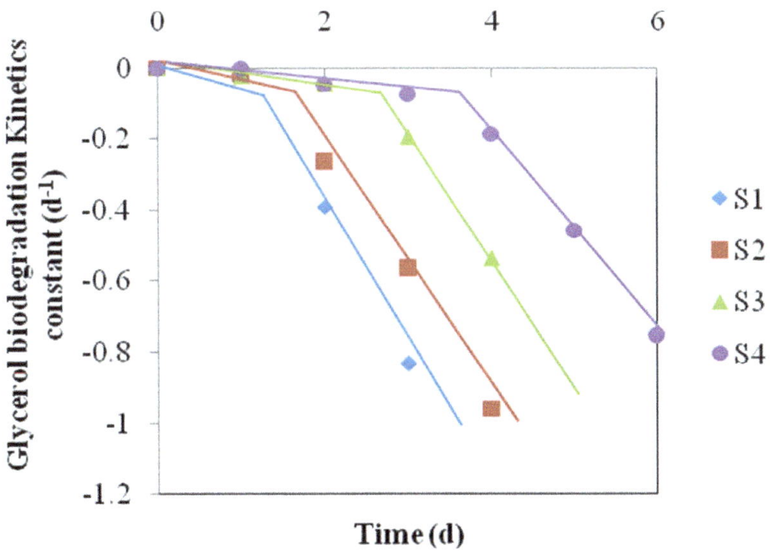

Figure 4. Semi-logarithmic glycerol biodegradation kinetics constants versus time.

Table 4. Biodegradation kinetics constants.

Synthetic wastewater (MgO_2/L)	S1(330)	S2 (480)	S3 (860)	S4 (1120)
Biodegradation kinetics k_t (d^{-1})	0.407	0.346	0.341	0.232

Table 5. Submerged aerobic fixed-film reactor purification yields (%) for S1, S2, S3 and S4.

Synthetic wastewater (MgO_2/L)	S1 (330)	S2 (480)	S3 (860)	S4 (1120)
Purification yield (%)	94.9%	91.7%	91.8%	79.5%

one was slight; it corresponded to the adaptation phase where a little part of substrate was consumed. The second slope was pronounced; it represented the biodegradation phase. The substrate removal was a function of the initial concentration of glycerol. Table 4 shows the values of biodegradation kinetics constants of degradation phase depending on the initial concentration. Biodegradation yields of these tests were illustrated in Table 5.

Both Tables 4 and 5 showed that the adaptability of bacterial cells (quantified by biodegradation kinetics constants and purifications yields) was better at a low initial concentration of glycerol. This adaptability behavior does not depend on the glycerol molecule (not toxic), unlike other tests where it is associated with the toxicity of the substrate when it becomes limited, as found by some authors (Duan, 2011; Marrot, 2006). This result is consistent with the study of Dey et al. (2010). Papadia et al. (2011), also highlighted the influence of the initial concentration of substrate on different biological treatment systems (trickling filter, activated sludge, bioflottation, flow jet aeration) working in continuous hydraulic regime. It has been shown that the increase of the organic load affected adversely the removal rate of the wastewater pollution. Silva et al. (2011) conducted tests to evaluate biodegradation kinetics of winery wastewaters in an aerobic batch, varying substrate concentrations (1, 3, 5 and 7 g/L). The evolution of COD removal efficiency presented a sharp drop of COD concentration at the beginning of the biodegradation curve, better for low concentrations, followed by a gentle drop.

Authors indicated that the adsorption, possibly occurring opposite intermediate adsorption of some compounds before biodegradation, is an important factor that may affect the biodegradation process and lead to exaggerated degradation slopes (Li et al., 2009). Matsuo et al. (2001) indicated that the initial concentration of the substrate influenced the properties of biofilm and its performance of purification. They indicated also that in the presence of heavily loaded waste water, a thick and less dense biofilm is formed; this caused a difficulty in the transfer of nutrient for the development of the bacteria inside the biofilm, unlike weakly charged wastewater.

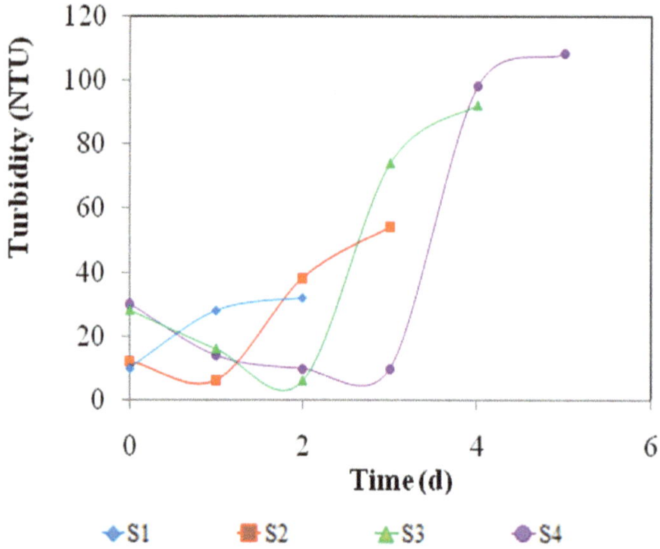

Figure 5. Turbidity evolution versus time.

Turbidity evolution

Turbidity results for the different synthetic solutions are presented in Figure 5. Results illustrated in Figure 4 shows that the turbidity followed the same trend for all initial concentrations tested, except for S1 (330 MgO_2/L).The turbidity decreased at the beginning for S2, S3 and S4, and then, it increased to reach a maximum. But for S1, turbidity increased probably because of the quick adaptation of bacteria as mentioned above, inducing an increasing in biomass formation (Reaction 2).

The different synthetic wastewaters have different initial turbidities, from 10 to 30 NTU. This was due to the quantity of matters, especially from soil, added to each synthetic wastewater, despite the filtration of synthetic wastewaters. On the adaptation phase, turbidity decreased probably by physical interception and the adsorption and flocculation of biofilm due to extracellular polymeric substances (EPS) secreted by bacteria (Zhang and Liu, 2005; Hongyuan et al., 2013).

The increase of turbidity coincided with the beginning of the degradation phase. These two processes occurred at the same time according to the Reactions 1 and 2. This increase may be due to some factors: augmentation of biomass quantity, a part of which can be detached by abrasion of biofilm due to the air flow or by gases (CH4, N2, CO_2, etc.) released by anaerobic/anoxic processes, generally produced when the thickness of biofilm became high. Turbidities reached a maximum, which depended on the initial concentration (Table 6).

Turbidity results confirm clearly those of COD, and allow the same conclusions concerning the behavior of batch bioreactor (Valentis, 1988; Kwok et al., 1998). So, the turbidity was a monitoring parameter which could characterize the stage of the biological process.

pH evolution

pH experimental results are presented in Figure 6. Figure 6 shows, at the beginning, an increase of pH for S1 and S2 tests, a slight decreased for S3 and a constant evolution for S4. After the adaptation phase, a pH increasing trend was observed for all tests. The pH behavior in test S1 could be probably explained by their rapid kinetics. It could be corroborated by COD and turbidity results of S1.

During biodegradation, the pH measured in synthetic wastewaters resulted from an equilibrium between bacterial cells catabolism process (CO_2 formation from Reaction 1) and from the denitrification process (increase of pH), as reported by some researchers (Fabregas, 2004; Akin et al., 2005). It could also be affected by the stripping of CO_2 (Cohen et al., 2004, Morales et al., 2013).

To explain pH behavior, the biofilm detachment caused a degradation of biomass protein. Akin et al. (2005) and Fabregas (2004) showed that this increase could be explained by the denitrification which occurred during anoxic phase in a SBR. The noted decrease of DO as shown above limited the oxygen transfer in the synthetic wastewater and promotes the conditions of the denitrification which occurred in the biological matter. The pH could be used to monitor the biological process. Its evolution depended on turbidity and the aeration of synthetic wastewater.

Dissolved oxygen evolution

DO results are illustrated in Figure 7. It has been clearly oxygen decreased continuously in S2 and S3 tests. Whereas, in S4 test, the observed initial DO was low

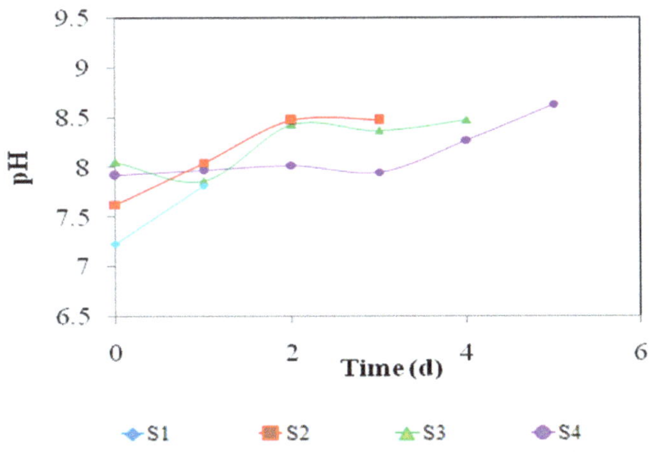

Figure 6. pH evolution versus time.

Table 6. Maximum turbidity versus initial concentration.

Synthetic wastewater (MgO_2/L)	S1 (330)	S2 (480)	S3 (860)	S4 (1120)
Maximum turbidity (NTU)	35	~ 50	~ 90	110

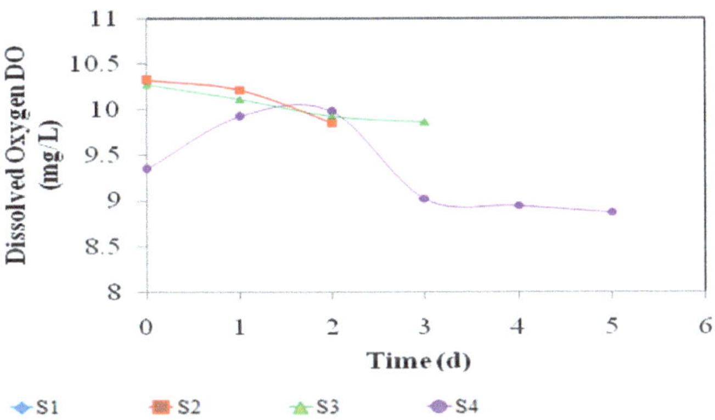

Figure 7. Dissolved oxygen evolution versus time.

Table 7. Dissolved oxygen concentration as function of initial concentration of substrate.

Test	Dissolved oxygen concentration
S2 (480 MgO_2/L)	10.32
S3 (860 MgO_2/L)	10.28
S4 (1120 MgO_2/L)	9.35

(Table 7). DO increase before a decrease. This behavior can be explained by the viscosity of S4 synthetic wastewater, due to its high concentration (1120 MgO_2/L), which prevents an efficient oxygen transfer (Stemmet et al., 2008; Kherbeche et al., 2013). But, during the adaptation phase a little part of substrate was consumed, so a DO increase was observed. After that, the DO decreased because of the substrate consumption by bacterial cells during biodegradation phase (Reaction 1).

The oxygen transfer could also be decreased because of the biological suspended matter, as reported by some authors (Rosenberger, 2003; Germain et al., 2007). It controlled both kLa20 and α-factor. Moreover, biological suspended matter accounted for the effects of the viscosity (Garcia-ochoa et al., 2000; Jin et al., 2001; Rosenberger, 2003). On the other hand, the α-factor was affected by the surfactants through EPS secreted by biological matter. Jimenez et al. (2014) showed a strong depression in bubble rise velocity and mass transfer in the presence of surfactants.

When transferring, oxygen must penetrate through the soluble microbial products (SMP) and then diffuse through the flock matrix (EPS) (Jimenez et al., 2014).

However, during biological process, the substrate could be transformed to some bio products which acted as surfactant agents, and affected the oxygen transfer (Garcia-ochoa et al., 2005; Painmanakul, 2005). So, the DO was correlated with the process biodegradation through its diminution during the process.

Conclusions

The main conclusions that could be drawn from this study are:

1. The various stages of the process of biological degradation occurring in a batch reactor are the adaptation, the degradation of the substrate and the endo genous

respiration.
2. The biological kinetics presents two slopes, the first one is slight and corresponds to the adaptation phase, the second one is accentuated, and it corresponds to the biodegradation phase.
3. When the substrate concentration is low, the duration of the adaptation phase is small and the elimination of the substrate is easy.
4. The tests gave excellent results in terms of the abatement rates: 94.2% (S1), 91.7% (S2), 91.8% (S3) and 79.5% (S4) which corroborated with calculated biological kinetics constants (0.407 d^{-1} (S1), 0.346 d^{-1} (S2), 0.341 d^{-1}(S3), and 0.232 d^{-1}(S4)). These results confirmed also the good adaptability at low concentrations of substrate.
5. The turbidity was revealed as a great indicator of bacteria metabolism. Its increase coincided with the beginning of the substrate degradation; several factors were acting in the same time: development of bacteria cells in which, a part was detached by abrasion due to the air flow or by the anaerobic/anoxic process occurring when the thickness of biofilm increased.
6. In the same time, the pH increased when the turbidity increased and dissolved oxygen decreased because of aerobic bacteria metabolism, striping of CO_2 and increasing in viscosity due to some metabolites produced during biological reaction.

Conflict of interests

The authors did not declare any conflict of interest.

REFERENCES

Akin B, Ugurlu A (2005). Monitoring and control of biological nutrient removal in a sequencing batch reactor. Process Biochem.40:2873-2878.

Albuquerque A, Makinia J, Pagilla K (2012). Impact of aeration conditions on the removal of low concentrations of nitrogen in a tertiary partially aerated biological filter. Ecol. Eng. 44:44-52.

Chaudhary DSVigneswaran S, Ngo H, Shim WG, Moon H (2003).Biofilter in Water and Wastewater Treatment. Korean J. Chem. Eng., 20:1054-1065.

Chen S, Hu W, Xiao Y, Deng Y, Jia J, Hu M (2012). Degradation of 3-phenoxybenzoic acid by a Bacillus sp. PLoS One.7: e50456.

Cheremisinoff NP (2002). Handbook of Water and Wastewater Treatment Technologies. Elsevier.

Cohen Y, Kirchmann H (2004). Increasing the pH of wastewater to high levels with different gases CO_2 stripping.Water Air Soil Pollut. 159:265-275.

Dey S, Mukherjee S (2010). Performance and kinetic evaluation of phenol biodegradation by mixed microbial culture in a batch reactor. Int. J. water Environ. Eng.. 2:40-49.

Duan Z (2011). Microbial degradation of phenol by activated sludge in a batch reactor. Environ. Prot. Eng. 37:53-63.

EPA (1997).Wastewater Treatment manuals - Primary, secondary and tertiary treatment. Environmental protection agency.

Fabregas TV (2004). SBR Technology for wastewater treatment: suitable operational conditions for a nutrient removal. Thesis.

Garcia-ochoa F, Castro E, Santos V (2000). Oxygen transfer and uptake rates during xanthan gum production. Enzym. Microb. Technol. 27:680-690.

Garcia-ochoa F, Gomez E (2005).Prediction of gas-liquid mass transfer in sparged stirred tank bioreactors. Biotechnol. Bioeng. 92:761-772.

Germain E, Nelles F, Drews A, Pearcec M, Kraumeb M, Reida E (2007). Biomass effects on oxygen transfer in membrane bioreactors. Water Res. 41:1038-1044.

Harvey G, Hasibul H, Dipesh D, Charles M, Hung YT (2011). Biofilm Fixed Film Systems. Water. 3:843-868

Hongyuan L, Wenchao G (2013). Research on biological aerate filter with volcanic filter for pretreatment of micro-polluted source water in lower temperature. Afr. J. Microbiol. Res. 7:4794-4800.

Horan N (2003). Handbook of Water and Wastewater Microbiology. Elsevier.

Jin B, Yu Q, Yan X, Van Leeuwen J (2001). Characterization and imporvement of oxygen transfer in pilot plant external air-lift bioreactor for mycelial biomass production. MicrobBiotechnol. 17:265-272.

Jin Y, Ding D, Feng C, Tong S, Suemura T, Zhang F (2012). Performance of sequencing batch biofilm reactors with different control systems in treating synthetic municipal wastewater.Bioresour. Technol. 104:12-18.

Kassab G, Halalsheh M, Klapwijk A, Fayyad M, Van Lier JB (2010). Sequential anaerobic-aerobic treatment for domestic wastewater - a review. Bioresour. Technol. 1013299. 101:3299-3310.

Kherbeche A, Milnes J, Jimenez M, Dietrich N, Hébrard G, Lekhlif B (2013). Multi-scale analysis of the influence of physicochemical parameters on the hydrodynamic and gas–liquid mass transfer in gas/liquid/solid reactors. Chem. Eng. Sci. 100: 515-528.

Kwok W, Picioreanu C, Ong S, Van Loosdrecht MC, Ng W, Heijnen J (1998). Influence of biomass production and detachment forces on biofilm structures in a biofilm airlift suspension reactor. Biotechnol. Bioeng. 58:400-407.

Lesouef A, Payraudeau M, Rogalla F, Kleiber B (1992). Optimizing nitrogen removal reactor configurations by on-site calibration of the IAWPRC activated sludge model. Water Sci. Technol. 25:105-123.

Li WW, Li XD, Zeng KM (2009). Aerobic biodegradation kinetics of tannic acid in activated sludge system. Biochem. Eng. J. 43:142-148.

Mara D (2003).Domestic Wastewater treatment in developing countries.

Marrot B, Barrios-Martinez A, Moulin P, Roche N (2006). Biodegradation of high phenol concentration by activated sludge in an immersed membrane bioreactor. Biochem. Eng. J. 30:174-183.

Matsuo T, Hanaki K, Takizawa S, Satoh H (2001). Advances in Water and Wastewater treatment technology - Molecular technology, nutrient removal, sludge reduction and environmental health.

Mendoza-Espinoza L, Stephenson T (1999). A review of biological aerated filters (BAF) for wastewater treatment. Environ Eng Sci. 16:201-216.

Morales N, Boehler MA, Buettner S, Liebi C, Siegrist H (2013). Recovery of N and P from urine by struvite precipitation followed by combined stripping with digester sludge liquid at full scale.Water. 5:1262-1278.

Painmanakul P (2005). Analyse locale du transfert de matière associé à la formation de bulles générées par différents types d'orifices dans différentes phases liquides Newtoniennes : étude expérimentale et modélisation. Thèse de Doctorat.

Papadia S, Rovero G, Fava F, Di Gioia D (2011). Comparison of different pilot scale bioreactors for the treatment of a real wastewater from the textile industry. Int. Biodeterior. Biodegradation. 65:396-403.

Prenafeta-Boldú FX, Illa J, van Groenestijn JW, Flotats X (2008). Influence of synthetic packing materials on the gas dispersion and biodegradation kinetics in fungal air biofilters. Appl. Microbiol. Biotechnol. 79:319-327.

Queinnec I, Ochoa J, VandeWouwer A, Paul E (2006). Development and calibration of a nitrification PDE model based on experimental data issued from biofilter treating drinking water. Biotechnol.Bioeng. 94:209-222.

Silva F, Pirra A, Sousa J, Arroja L, Capela I (2011). Biodegradation Kinetics of winery wastewater from port wine production. Chem. Biochem. Eng. 25:493-499.

Stemmet CP, Bartelds F, Van Der Schaaf J, Kuster BFM, Schouten JC (2008). Influence of liquid viscosity and surface tension on the gas–liquid mass transfer coefficient for solid foam packings in co-current two-phase flow. Chem. Eng. Res. Des. 86:1094-1106.

Valentis G (1988). Epuration par cultures fixées sur support géotextile. Ecole Nationale des ponts et chaussées. Thèse de Doctorat.
Zhang T, Liu X (2005). Effect of bidosicai activated carbon filter on treatment of micro-polluted source water. China Water and Wastewater. 21:39-40.

Diffusion coefficients of *Grewia* spp. biopolymers in water

Ndi. K. Sylvere, Kofa G. Patrice, Kameni Ngounou M. Bernard, Ngnié N. Patricia and Kayem G. Joseph

Industrial Filtration and Water Treatment (Chem. Eng.) Group, Department of Process Engineering, ENSAI, University of Ngaoundere, P.O. Box: 455, Ngaoundere, Cameroon.

Diffusion coefficients of biopolymers extracted from *Grewia* spp. barks in water were studied under various physicochemical conditions. Extractions were carried out on barks of 3 cm in length at different pH (5-9), temperature (25-50°C) and ionic strength (10^{-3}-10^{-2}M), using $CaCl_2$ and NaCl as background electrolytes. Factorial experimental plan design was used and diffusion coefficients were determined using second law of Fick equation, with linear and logarithmic approximation. Obtained results show that diffusion coefficients of *Grewia* spp. biopolymers are in the range of 9.78458E-06±1.6462E-08 cm^2/min, and are not significantly influenced by the pH, the ionic strength and the temperature. Maximum diffusion was obtained with $CaCl_2$ at high ionic strength. Both linear and logarithmic approximations were found to be suitable for diffusion coefficients and were dependent on the background electrolyte used.

Key words: Biopolymers, diffusion coefficient, *Grewia*.

INTRODUCTION

Different plants extracts have been used or tested as biopolymers for coagulation; *Moringa oleifera, Strychnos potatorum, Jatropha curcas sabdariffa, Curcuma angustifolia* (Pritchard et al., 2010), *Phaseolus vulgaris* and *Grewia*. *Grewia* spp. is a flowering plant from the family of *tiliacées* that grows in the shade of nearby rocks and reeds, native to warm tropical regions of Africa, Asia and Australia (Emeje et al., 2008). It is used as a thickener, stabilizer in food process and in some cases it may act as a coagulant /floculant (Ndi et al., 2013).

In water clarification processes, studies on efficient biopolymers for coagulation purposes are widespread due to the environmental impact of those coagulants (Chun-Yang, 2010). Biopolymers can be extracted from different parts of the plant including barks, leaves and roots but, when powders are used, fine particles are released, increasing the suspended material and the turbidity of the suspension, and may act as a support for bacterial growth (Kebreab et al., 2005). Therefore, purification of extracted biopolymers is often required. Unfortunately, these methods include a combination of lyophilisation, chromatography and dialysis, which are complex

Table 1. Factors range.

Factor	Values
pH	5- 9
Salt concentration (M)	0.001- 0.01
Temperature (°C)	25-50

Table 2. Complete factorial design matrix.

Run	pH	Ionic strength	Temperature
1	-1	-1	-1
2	1	-1	-1
3	-1	1	-1
4	1	1	-1
5	-1	-1	1
6	1	-1	1
7	-1	1	1
8	1	1	1

and expensive for implementation in rural areas of developing countries (Kebreab et al., 2005). Thus, looking for an optimum in the extraction of biopolymers with fewer amounts of fines released is of great importance for coagulation purposes. Using barks instead of powders as coagulant/flocculant may increase the settling velocity of flocs, reducing considerably the amount of fines released in the suspension, and further accelerate the clarification step.

During flocculation, the required amount of coagulant or flocculant for good clarification purpose is directly linked to the initial water quality. Thus, knowledge of biopolymer concentration in the suspension is of great importance for coagulation process. When barks are soaked in water, biopolymers are released from the bark to the suspension. This extraction process can be improved by agitation of the suspension, but in a house hold scale it is generally done without any agitation, as it is easily carried out, leading to natural diffusion of biopolymers from the bark to the suspension. As natural diffusion is a mass transfer process, induced by gradient concentration, there's need to evaluate the amount of biopolymers released within the contact period time of the bark in the liquid. This can be done knowing diffusion coefficients.

Diffusion coefficient is a key parameter of mass transfer process and describes how fast the amount of released biopolymers in the suspension is. Diffusion coefficient although generally assumed as constant depends on many parameters like ionic strength, pH, temperature and bark concentration (Giovana et al., 2011).

Diffusion coefficients are well determined by direct methods including technique of fluorescence recovery after photo bleaching (FRAP) which has been refined through the use of a laser scanning confocal microscope (LSCM) or by indirect methods like spectroscopy, chromatography and mass balance in the system (Matthew et al., 2000). When indirect methods are used, plotting the amount of biopolymer released as a function of the concentration as described by the second law of Fick equation gives the diffusion coefficient. Approximation of the Fick equation commonly used is linear and logarithmic ones which were found to be suitable for biopolymers diffusion in water system. The linear model corresponds to the earlier part of the slope of the diffusion process while logarithmic is mainly used when the required time for extraction is high (Poirot, 2007).

The aim of this study is therefore to evaluate diffusion coefficients of *Grewia* spp. biopolymers and to find suitable models describing it under different physicochemical conditions.

MATERIALS AND METHODS

Biological material

Grewia spp. was collected from Mokolo, a town located in the far Northern region of Cameroon. *Grewia* spp. barks of 3 cm length were used as biological material for biopolymer extraction. They were dried at 45°C using a forced convection dryer (CKA2000AUF) for 24 h, allowed to cool in a desiccator, and kept in sealed plastic bags prior to their use. The thickness (L) of *Grewia* spp. bark used was obtained by measuring the average size of 100 barks selected randomly.

Characterization

FTIR analyses were carried out before and after biopolymer extraction.

Polymer extraction

Polymer extraction was carried out in batch mode without any mechanical stirring of the solution, with distilled water of pH (5-9) adjusted using 1 M hydrochloric acid (Sherman Chemicals limited) or 1 M NaOH solutions (Prolabo), and ionic strength (0-0.01 M) adjusted using NaCl (Jeulin) and $CaCl_2$ (Jeulin) salts. Conductivities were monitored using conductimeter (HI 8733, HANNA Instrument) and biopolymer mass was determined after each extraction by drying in an oven (Memmert) at 105°C and weighing the resulting extract on a mass balance of 10^{-3} precision (Gibertini).

Extraction kinetics were done with a ratio of barks mass (5, 10 and 20 g) over extraction solvent volume of 250 mL.

Experimental design

Complete factorial design was used. Selected factors are presented in Table 1. The experimental matrix is described in Table 2, and the diffusion coefficient was defined as the response.

Diffusion coefficients determination

Diffusion coefficients were derived from the slopes of the following equations (Aghfir et al., 2008; Claudia et al., 2009). Taking into

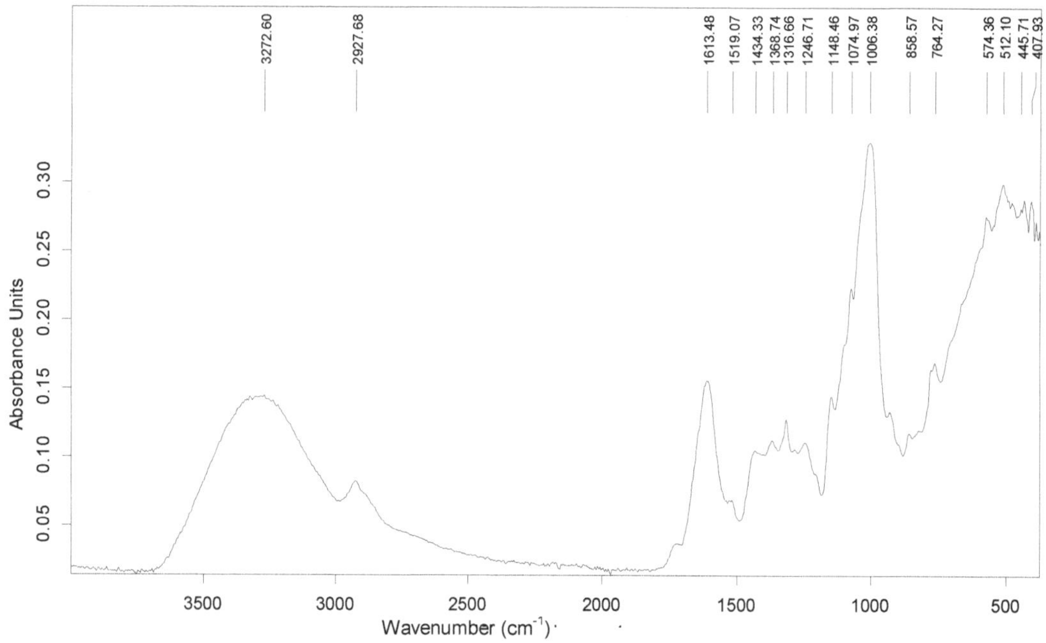

Figure 1. The FT-IR spectroscopic spectrum of the powders of *Grewia* unextracted.

consideration the hypothesis that:

1. The particles have the shape of a plate;
2. Uniform initial distribution of moisture;
3. Simplification of the movement of moisture by diffusion;
4. Negligible contracting;
5. Constant coefficient of diffusion.

The diffusion coefficient can be calculated by using the following equations for the two models:

$\frac{M_t}{M_\infty} = \frac{4}{L}\left(\frac{D_m.t}{\pi}\right)^{\frac{1}{2}}$ Equation1 (for the linear approximation),

$Ln\left(1 - \frac{M_t}{M_\infty}\right) = Ln\left(\frac{8}{\pi^2}\right) - \left(\frac{\pi^2 D_m.t}{4L^2}\right)$ Equation 2 (for the logarithmic approximation).

Where M_t is the mass of polymers at the moment t, M_∞ is the mass of polymers at infinite time, L is the thickness of the barks, *t* the time and D_m the multidirectional diffusion coefficient. The obtained quadratic equation was expressed as:

Diffusion coefficient = C_{ST} + C_A*A + C_B*B + C_C*C+ C_{AB} *A*B + C_{AC}*A*C - C_{BC} *B*C

Where, C_X denotes the coefficient factor related to the factor X studied.

RESULTS AND DISCUSSION

Characterization of *Grewia* spp.

The FT-IR spectroscopic spectra of the powders of *Grewia* unextracted and extracted are presented in the Figures 1 and 2, respectively. The main difference between these two spectra is at the level of the intensities of the peaks.

The spectra exhibit the typical bands and peak characteristic of polysaccharides. The broad band occurring at 3272.60 cm^{-1}, results from the presence of hydroxyl (-OH) group. The peak obtained at 2927.68 cm^{-1} results from stretching modes of the C-H bonds of methyl group ($-CH_3$). Natural gum usually contains fractions of sugar acid units which would usually impart a weakly anionic character to the gum macromolecule (Wang et al., 2003).

Absorption bands around 1618 and 1430 cm^{-1} are typical of carboxylate group of the galacturonic acid residues (Figueiro et al., 2004). The peak obtained at 1519.07 cm^{-1} is typical of carboxylic group. The peak obtained at 1074.97cm^{-1} result from the presence C-O-C group. The peaks obtained at 1720.23 and 1607.16 cm^{-1} results from the presence of C=O group. The wave numbers between 800 and 1200 cm^{-1} represents the finger print region for carbohydrates (Fillipove, 1992).

Physicochemical effects on released *Grewia* spp. biopolymers

Figure 3 presents the effect of bark mass/water solution ratio (5/250, 10/250 and 20/250) on biopolymer extraction over time on the ratio of biopolymer mass (EP) recovered over the initial mass of the bark introduced (BM). As shown, less than 4% of the bark mass is released in the

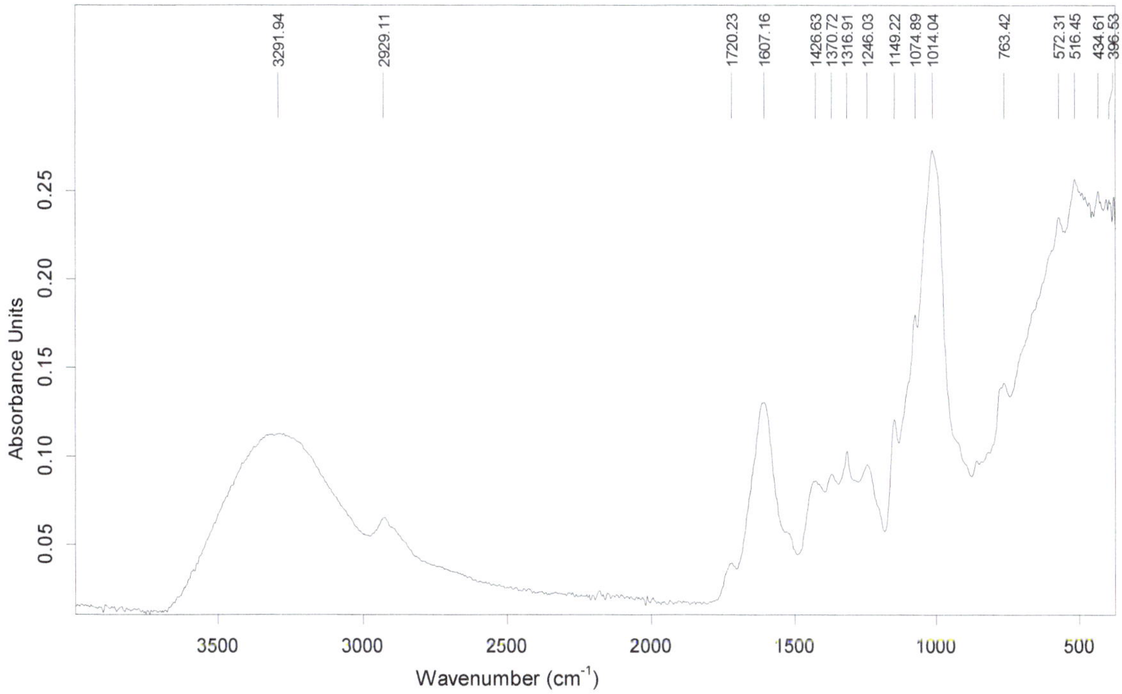

Figure 2. The FT-IR spectroscopic spectrum of the powders of *Grewia* extracted.

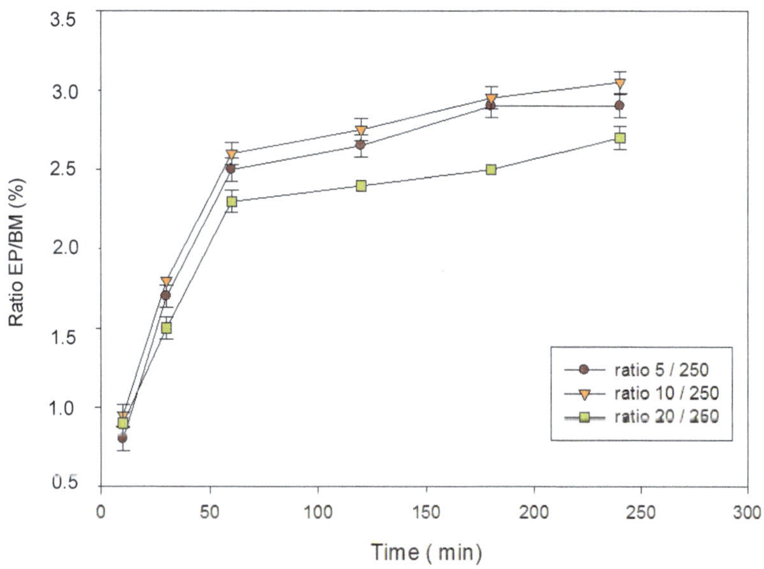

Figure 3. Kinetics of biopolymer extraction for different bark mass/water solution ratio.

suspension during batch operations whatever the ratio. From 0 to 60 min, rapid transfer of biopolymer from the bark to the suspension is observed. Further, the biopolymer is still recovered, but at a low rate. When 20 g of bark is introduced in 250 mL of suspension, the amount of polymer recovered slightly decreases. Polymer extraction in dilute solutions is coherent with other works related to polymer transfer. Reduction of the amount of polymer when bark ratio increases is due to the well-known resistance of transport due to the gradient of

Table 3. Factors coefficients and interactions for modeling of diffusion coefficient with NaCl for linear and logarithmic model, using complete factorial design plan.

Term	Linear model with NaCl		Logarithmic model with NaCl	
	coefficient	Probability	Coefficient	Probability
A:PH	2.93E-8	> 0.05	6.61E-7	> 0.05
B:FI	1.19E-6	> 0.05	1.26E-5	> 0.05
C:TEMP	3.99E-6	> 0.05	5.09E-5	> 0.05
AB	2.98E-8	> 0.05	-3.31E-6	> 0.05
AC	1.65E-7	> 0.05	3.31E-6	> 0.05
BC	4.44E-7	> 0.05	6.61E-7	> 0.05
Constant	1.02E-5		1.36E-4	

Table 4. Factors coefficients and interactions for modeling of diffusion coefficient with $CaCl_2$ for linear and logarithmic model, using complete factorial design plan.

Term	Linear model with $CaCl_2$		Logarithmic model with $CaCl_2$	
	Coefficient	Probability	Coefficient	Probability
A:PH	3.62E-7	> 0.05	3.31E-6	> 0.05
B:FI	2.27E-6	> 0.05	9.92E-6	> 0.05
C:TEMP	3.52E-6	> 0.05	4.69E-5	> 0.05
AB	1.65E-7	> 0.05	4.63E-6	> 0.05
AC	1.96E-7	> 0.05	-3.31E-6	> 0.05
BC	-8.78E-8	> 0.05	6.61E-7	> 0.05
Constant	1.07E-05		1.43E-4	

concentration in the solution which becomes low and then reduces the transfer of the polymer inside the suspension.

Modeling diffusion coefficients of *Grewia* spp.

Using complete factorial design followed with linear and logarithmic approximation, coefficients of factors involved in the extraction process and quadratic interactions are presented in Table 3 for diffusion with NaCl salt and in Table 4 for diffusion with $CaCl_2$ salt.

The obtained quadratic equation with NaCl was expressed as:

Linear model
Diffusion coefficient = 1.02E-5- 2.93E-8*A+1.19E-6*B+3.99E-6*C+ 2.98E-8*A*B + 1.65E-7*A*C + 4.44E-7*B*C

Logarithmic model
Diffusion coefficient = 1.36E-4+6.61E-7*A+1.26E-5*B+5.09E-5*C-3.31E-6*A*B+3.31E-6*A*C+1.98E-6*B*C

The obtained quadratic equation with $CaCl_2$ was expressed as:

Linear model
Coefficient diffusion = 1.07E-05+3.62E-7*A+2.27E-6*B+3.52E-6*C+1.65E-7*A*B+1.96E-7*A*C- 8.78E-8*B*C

Logarithmic model
Coefficient de diffusion = 1.43E-4+3.31E-6*A+9.92E-6*B+4.69E-5*C+4.63E-6*A*B-3.31E-6*A*C+ 6.61E-7*B*C

Tables 3 and 4 present the coefficients of parameters involved in the quadratic equation resulting from the factorial design plan. The observed probability of all the factors was found to be greater than 0.05, showing that at a probability of 95%, these factors did not have a great influence on the diffusion coefficient.

Table 5 presents the values of statistic parameter used to validate the model. Since the values obtained R^2, AADM, B_f is near the values expected, the mathematical model obtained is considered as valid.

But by taking into account the parameters of validation of the model, one notes that the model logarithmic model

Table 5. Validation of the models.

Parameter	Expected values	$CaCl_2$		NaCl	
		linear model	logarithm model	Linear model	logarithm model
R^2	1	0.989	0.999	0.995	0.988
AADM	0	0.030	0.005	0.000	0.040
B*f*	1	1.000	0.999	1.000	1.002
A*f1*	1	1.030	1.005	1.000	1.041

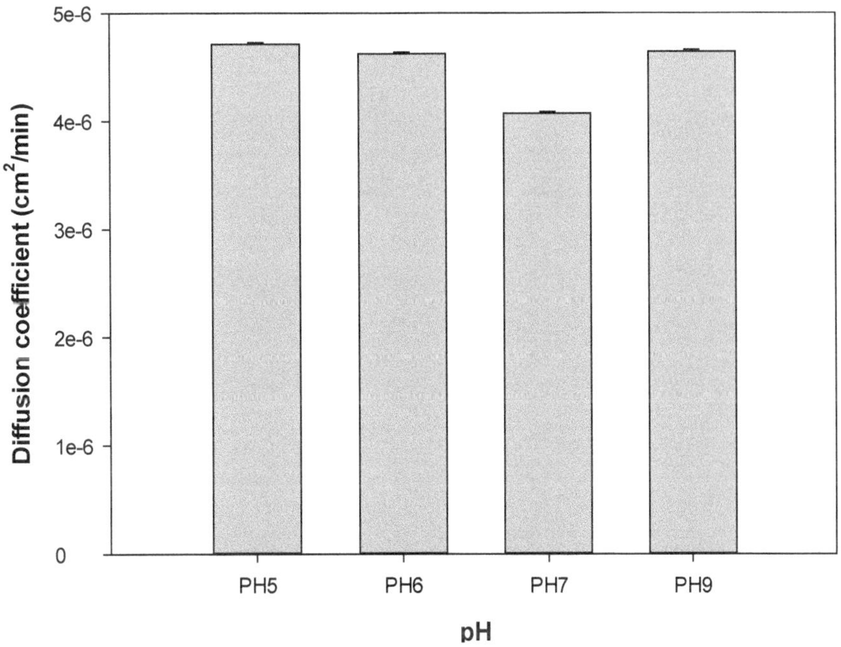

Figure 4. Biopolymers diffusion coefficients with the pH.

presents itself as an explanation of the interaction between the various factors of the extraction of polymers of *Grewia* in the presence of $CaCl_2$, whereas the linear model presents itself rather better than the extraction in the presence of NaCl.

Figure 4 presents the pH effect on extraction ratio on polymer removal and extraction time on polymer mass recovered over the mass of the bark as shown for a ratio bark/volume equal to 10/250. At the beginning of the diffusion of polymers into the solution (from 0 to 60 min), less influence of pH on polymer transfer is observed. Net pH influence is found after 120 min of extraction, and the lowest extraction ratio is obtained at pH 7, while the highest one is obtained at pH 6.

Plants polymers are mainly polysaccharides. They possess a high tendency to associate, caused by the abundance in hydroxyl or amino groups present in the macromolecules that easily undergo hydrogen bonding, and are very sensitive to pH variations. In acidic solutions, protonation of carboxylic and amino groups will increase electrostatic repulsion and solubilize the polymer, increasing its diffusivity (Nandhini and Abhilash, 2010; Lawrence et al., 2006). This behavior is also observed at high pH (9), but in this case, deprotonation is involved. At neutral pH 7, hydrogen bonds existing inside the polymers will increase the size of polymer which will result in a decrease of polymer diffusivity.

Figure 5 presents the effect of salts and salt concentration on polymer. Without any salt, the released amount of polymer is low (2.5 %) and increases with the type of salt and the concentration. As ionic strength increases, the solubility of the polymers increases as predicted by the salting in theory (Djamen, 2011). The solubilisation will induce a rapid diffusion of the polymer in the medium. In this particular case, traction is more pronounced with Ca than Na salts. These differences could be explained by the ionic conformation of the different salts used; Ca ions of greater ionic radius than

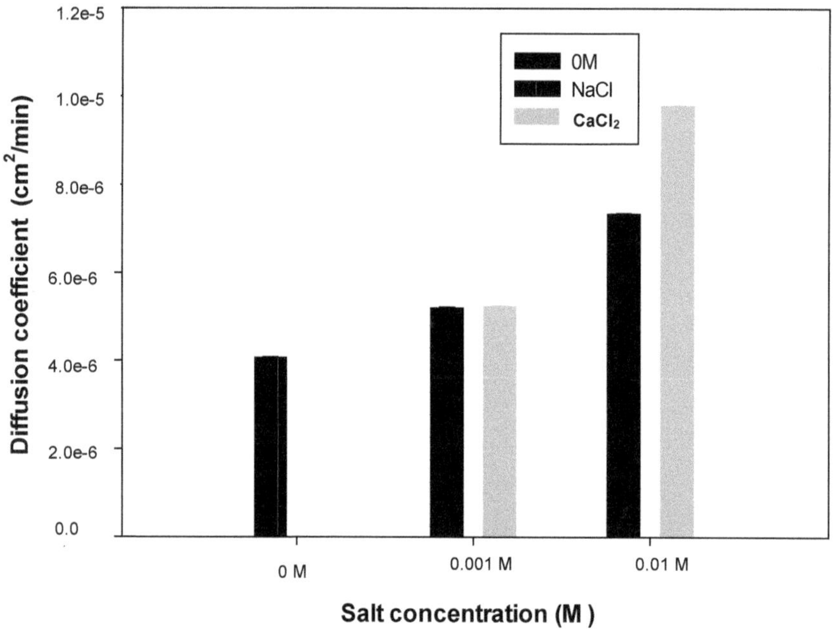

Figure 5. Effect of salts on diffusion coefficient.

Na ones release their solvation water more quickly, thus increasing the solubility of polymers and in fine the diffusion coefficient (Djamen, 2011).

From the above results it is obvious that physicochemical properties of the solution greatly affect the yield of polymer extraction.

Conclusion

The principal objective of this work was to determine the diffusion coefficient of polymers of *Grewia* spp. under various operating conditions. The results show that the polymers of *Grewia* spp. diffuse more quickly in the salt solutions and is effective at high temperatures than in distilled water; with diffusion coefficients of 9,78458e-06±1,646e-08, 7,3456e-06±1,746e-08 and 4,07001e-06±1,55077e-08 cm^2/min, respectively with $CaCl_2$, NaCl and distilled water. These results also show that the diffusion coefficients of *Grewia* spp. polymers in the presence of $CaCl_2$ are higher than those obtained in the presence of NaCl, and that it is not preferable to extract polymers from *Grewia* spp. with neutral pH. According to the type of salt, one can make use of either the linear model or the model logarithmic curve for the expression of the interaction between the parameters of extraction.

Conflict of interests

The authors have not declared any conflict of interests.

REFERENCES

Aghfir A, Akkad S, Rhazi M, Kane CSE (2008). Détermination du coefficient de diffusion et de l'énergie d'activation de la menthe lors d'un séchage conductif en régime continu. Revue des énergies renouvelables 11:3385-394.

Chun-Yang Y (2010). Emerging usage of plant-based coagulants for water and wastewater treatment. Process Biochem. 45:1437-1444. http://dx.doi.org/10.1016/j.procbio.2010.05.030

Claudia M, Rascon-Chu A, Jorge AE, Ana Luisa M (2009). Maize arabinoxylan gels as proteindelivery matrices. Molecules 14:1475-1482. http://dx.doi.org/10.3390/molecules14041475

Djamen MCN (2011). Etude structurale et rhéologique des systèmes mixtes caséinates/carraghénanes. Thèse de doctoratUniversité du Maine.

Emeje M, Isimi C, Kunle O (2008). Effect of Grewiagum on the mechanical properties of Paracetamol tablet formulations. Afr. J. Pharm. Pharmacol. 2(1):1-6.

Figueiro SD, Moreira RA, Sombra ASB (2004).On the physicochemical and dielectric properties of glutaraldehyde cross-linkelgalactomannan-collagen films. Carbohydr. Polym. 56:313-320. http://dx.doi.org/10.1016/j.carbpol.2004.01.011

Fillipove MP (1992). Practical infrared spectroscopy of pectin substance.Food Hydrocoll. 6:115-142. http://dx.doi.org/10.1016/S0268-005X(09)80060-X

Giovana D, Ferreira M, Tessarol C, Caciano PZ (2011). Evaluation of water, sucrose and NaCl effective diffusivities during osmotic dehydration of banana (Musa sapientum, shum.). LWT Food Sci. Technol. 44:82-91. http://dx.doi.org/10.1016/j.lwt.2010.06.011

Kebreab A, Ghebremichaela Gunaratnab K, Hongbin H, Harry B, Gunnel D (2005). A simple purification and activity assay of the coagulant protein from MoringaOleifera seed. ISRN KTH/LWR/PHD 1013-SE.

Lawrence AA, Mohammednur Desta S, Shiferaw M (2006). Effect of ionic strength and/or pH on Extractability and physico-functional characterization of broad bean (Vicia faba L.) protein concentrate. Food Hydrocoll. 20:1124-1134. http://dx.doi.org/10.1016/j.foodhyd.2005.12.010

Matthew D, Burke JO, Mohan S, Saad A (2000). Diffusion of Macromolecules in Polymer Solutions and Gels: A LaserScanning Confocal Microscopy Study.Macromolécules 33:7500-7507. http://dx.doi.org/10.1021/ma000786l
Nandhini V, Abhilash M (2010). Study of hydration kinetics and rheological behaviour of guar gum. Int. J. Pharma Sci. Res. 1(1):28-39.
Ndi KS, Ngnie NP, Kofa GP, Amba EV, Kayem GJ (2013). Optimization of polymer extraction from Grewia spp. Plant. SFGP2013 - XIVe congrès de la Société Française de Génie des Procédés.
Poirot R (2007). Méthodologie pour le passage en continu d'extraction de soluté à partir de matière végétale. National Polytechnique de Toulouse école doctorale: mécanique, énergétique, génie civil, procédés spécialité: génie des procédés et environnement.
Pritchard A, Craven T, Mkandawire B, Edmondson A (2010). A study of the parameters affecting the effectiveness of Moringa Oleifera in drinking water purification. Phys. Chem. Earth 35: 791–797. http://dx.doi.org/10.1016/j.pce.2010.07.020
Wang O, Ellis PR, Ross–Murphy SB (2003). Dissolution kinetics of guar gum powders: effect of concentration and molecular weight. Caborhydr. Polym. 53:75-83. http://dx.doi.org/10.1016/S0144-8617(03)00009-2

Thermodynamic parameters for the adsorption of volatile n-alkane hydrocarbons on water hyacinth (*Eichhornia crassipes*) root biomass: Effect of organic solvent and mineral acid treatment

Netai Mukaratirwa-Muchanyereyi[1,2], Jameson Kugara[1] and Mark Fungayi Zaranyika[1]

[1]Chemistry Department, University of Zimbabwe, P. O. Box MP 167, Mount Pleasant, Harare, Zimbabwe.
[2]Bindura University of Science Education, Bindura, Zimbabwe.

The use of lignocellulosic fibres as biosorbents has both environmental and economic benefits. They are biodegradable, safe to use and most importantly renewable and less costly than synthetic fibres. The objectives of the present work were to study the performance of lignocellulosic fibres from water hyacinth (*Eichhornia crassipes*), a pervasive fresh water aquatic weed widely distributed in the tropics, as a biosorbent for n-alkane hydrocarbons, n-hexane to n-nonane. The free energy (ΔG_a), enthalpy (ΔH_a) and entropy (ΔS_a) for the adsorption of n-alkanes hexane to nonane on ground dried water hyacinth (*E. crassipes*) root biomass were studied between 40 and 70°C column temperature using inverse gas chromatography, before and after treatment of the root biomass with mineral acid and organic solvent. The free energy of adsorption (ΔG_a), enthalpy of adsorption (ΔH_a) and entropy of adsorption (ΔS_a) values of -20.42 to -35.992 kJ.mol^{-1}, -21.18 to -33.704 kJK^{-1}mol^{-1} and +0.0023 to +0.0041 kJ.mol^{-1} respectively, were obtained for the adsorption of the n-alkane series on untreated water hyacinth root biomass. Mineral acid and organic solvent treatment leads to greater adsorption bond strength, with ΔG_a and ΔH_a values of -36.08 to -38.52 kJ.mol^{-1} and -34.01 to -36.33 kJK^{-1}mol^{-1}, respectively, after mineral acid treatment, and -40.98 to -46.7 kJ.mol^{-1} and -38.35 to -43.7 kJK^{-1}mol^{-1}, respectively, after organic solvent treatment. The results show that mineral acid and organic solvent treated water hyacinth root biomass has potential as biosorbent for the adsorption of volatile n-alkane hydrocarbons. Factors affecting the adsorbent-adsorbate interaction are discussed.

Key words: Water hyacinth, inverse gas chromatography, adsorption free energy, adsorption enthalpy, hydrocarbons.

INTRODUCTION

The use of lignocellulosic fibres as biosorbents has both environmental and economic benefits. They are biodegradable, safe to use and most importantly renewable and therefore much less costly than synthetic fibres.

It is for these reasons that lignocellulosic fibres are being targeted for use in various fields (Cordeiro et al., 2011a). Lignocellulosic fibres that have been in commercial use for a long time include flax (*Linum usitatissimim*), hemp (*Cannabis sativa*), kenaf (*Hibiscus cannabinus*), agave (*Agave Americana*), pine apple (*Ananas cmosus*) and sisal (*Agave sisalana*) (Cordeiro et al., 2011a). Non-conventional, low-cost lignocellulosic fibres that have been studied include agricultural wastes, fallen tree leaves, water hyacinth, algal biomass, among others (Rafatullah et al., 2010).

The performance of lignocellulosic adsorbents depends on the nature of the adsorbent and adsorbate. Lignocellulosic fibres can be considered as naturally occurring composites, consisting mainly of helically wound cellulose microfibrils bound together by lignin and hemicellulose. Hemicellulose molecules are hydrogen bonded to cellulose, and act as cementing matrix between cellulose microfibrils. It is highly hydrophilic, partly soluble in water, soluble in alkalis and easily hydrolysed in acids (John and Thomas, 2008; Spinace et al., 2009).

Lignin is a complex three dimensional hydrocarbon co-polymer of aliphatic and aromatic constituents of very high molecular weight. Lignin is completely amorphous, hydrophobic in nature, and is not hydrolysed by acids. It is not soluble in most acids, but soluble in alkalis (Spinace et al., 2009). The chemical composition of lignocellulosic fibres also depends on the conditions under which the plant grew, as well as the origin of the fibre, that is, whether leaf, bast, fruit or root. Crystallinity and surface composition also affect the adsorbents properties of lignocellulosic fibres. These properties can be affected by any chemical treatment or modification that the fibres are subjected to.

Lignocellulosic fibres contain strongly polarized hydroxyl groups, hence natural untreated lignocellulosic fibres are hydrophilic (Cordeiro et al., 2011a). Alkaline treatment leads to the removal of lignin and hemicelluloses which are amorphous, resulting in the increase in crystallinity of the fibres.

On the other hand, dilute acid treatment removes hemicelluloses components of the fibres, leading to the exposure of cellulose (Irfan et al., 2011). Subjecting lignocellulosic fibres to extraction with organic solvents has been reported to lead to an increase in the dispersive component of surface energy as a result of the removal of low molecular components at the fibre surface, such as fatty acids, and exposure of cellulose chains (Belgacem et al., 1995, 1996; Gamelas, 2013).

The properties of adsorbents are often studied using inverse gas chromatography (IGC) (Gamelas, 2013; Dorris and Gray, 1980; Schultz et al., 1987; Askin and Yazici, 2005; Reidel and Mutuana, 2006; Cordeiro et al., 2011a, b; Steele et al., 2008; Bilgic and Askin, 2003; Tshabalala and Wang, 2004; Widegren and Bruno, 2011; Kunaver et al., 2004; Kandor and Dallos, 2014). IGC is in fact gas-solid chromatography in which the interactions between the adsorbent used as solid stationary phase, and the adsorbate used as eluent, are the subject of investigation. This is done by injecting volatile molecules of the eluent and noting the retention times. The thermodynamic properties of the interaction between the solid stationary phase and the eluent are then inferred from the retention volume. The thermodynamic parameters measured are the adsorption free energy, adsorption enthalpy, adsorption entropy and the dispersive component of surface energy of the adsorbent (Reidl and Matuana, 2006). The enthalpy of adsorption is a thermodynamic measure of the strength of the adsorption interaction. IGC experiments are usually performed using a series of n-alkanes (hexane to decane) as non-polar probes; and a series of polar probes consisting of acidic probes (dichloromethane and trichloromethane), basic probes (diethyl ether and tetrahydrofuran) and amphoteric probes (acetone and ethyl acetate) (Shi and Qi, 2012). The choice of IGC probe thus represents the properties of the adsorbate as regards polarity and acid-base properties.

In this paper we report the results of the work we carried out to evaluate the thermodynamic parameters for the adsorption of volatile n-alkane by water hyacinth root biomass. Water hyacinth *(Eichhornia crassipes)* is a floating aquatic weed belonging to the pickerelweed family, *Pontederiaceae,* found throughout the tropics. It is listed as one of the most productive plants on earth, and is considered the worst aquatic weed, that can tolerate a considerable variation in nutrients, temperature and pH (Rajamohan, 2009). Water hyacinth is cheap and easily available, and therefore qualifies as a non-conventional low-cost adsorbent (Tahir and Alam, 2014). It has flourishing roots composed mainly of cellulose and lignin (Zheng et al., 2009).

The adsorbent properties of water hyacinth biomass towards heavy metals have been studied by several researchers (Zheng et al., 2009; Hasan et al., 2010; Ibrahim et al., 2012; Mahamadi and Nharingo, 2007, 2010a, b), and was reviewed by Mahamadi (2011). The review covered sorption dynamics, effect of physical and chemical treatment, pH, temperature, initial metal concentration and biosorbent dose, and concluded that the plant showed excellent biosorbent properties towards several metals of environmental concern, and that pH, temperature, adsorbent dose significantly affected its capacities as metal biosorbent. The review also concluded that further investigations were required on the structural properties of the biosorbent and biosorbent modification. The biosorption of organic compounds by water hyacinth biomaterial has also been studied. Tarawou et al. (2007) and Rajamohan (2009) studied the adsorption of methyl red and congo red dyes by water hyacinth roots as a function of pH, contact time and dye concentration, and showed that equilibrium was reached in about 3 h and fitted the Redlich-Peterson and Freundlich isotherms. The optimum pH was found to be

6.0. Khan et al. (2012) and El-Khalary (2007) studied the adsorption of methylene blue onto water hyacinth biomaterial.

Adsorption equilibrium data fitted the Langmuir isotherm, consistent with monolayer adsorption mechanism. Water hyacinth has also been studied for the removal of phenols from polluted water by Wolverton and McKown (1976), who concluded that 2.75 g of water hyacinth biomass could absorb 100 mg of phenol per 72 h from distilled water and river water.

The aim of the present study was to evaluate the thermodynamic parameters (free energy, enthalpy and entropy) for the adsorption of n-alkane hydrocarbons on untreated, acid treated and solvent extracted water hyacinth root biomass. Hydrocarbons are important water and air pollutants in the petroleum production and distribution industries.

Drinking water sources are often polluted by hydrocarbons in run-off from petroleum fuel distribution centres (storage depots and service stations) and precipitation of emissions from vehicles and air planes, as well as emissions from boats. The thermodynamic parameters for the adsorption of hydrocarbons by water hyacinth root biomass have not been studied previously.

Theoretical

Adsorption partition coefficient, K_R

In gas solid chromatography, the distribution of a solute between the solid stationary phase, S, and the mobile phase, M, at constant temperature and pressure, corresponds to an equilibrium when the solute free energy is at a minimum (Laub and Pecsock, 1978). Under these conditions, its chemical potential, μ in the mobile phase is equal to that in the stationary phase, that is,

$$\mu_S = \mu_M \qquad (1)$$

The chemical potential is related to solute activity, *a*, by Equation 2:

$$\mu_i = \mu_i^o + RT \ln a_i \qquad (2)$$

Therefore assuming $a_i \approx C_i$, where, *C* = concentration, Equation 1 becomes,

$$\mu_S^o + RT \ln C_S = \mu_M^o + RT \ln C_M \qquad (3)$$

Therefore,

$$\frac{\mu_S^o - \mu_M^o}{RT} = \ln C_M - \ln C_S = \ln\left(\frac{C_M}{C_S}\right)$$

And

$$\frac{C_M}{C_S} = \exp\left(\frac{\Delta\mu^o}{RT}\right) = K_R \qquad (4)$$

Where, $\Delta\mu^o = \mu_S^o - \mu_M^o$

Adsorption thermodynamic parameters

From thermodynamics, the adsorption free energy, ΔG_a, is given by:

$$\Delta G_a = -RT \ln K_R \qquad (5)$$

It can be shown that in gas solid chromatography (GSC), K_R is also given by (Laub and Pecsock, 1978):

$$K_R = \frac{V_N}{A_S} \qquad (6)$$

Where, V_N = net retention volume of mobile phase, and A_S = surface area per gram of adsorbent. Hence Equation 5 can be expressed as:

$$\Delta G_a = -RT \ln\left(\frac{V_N}{A_S}\right) = -RT(\ln V_N - \ln A_S) \qquad (7)$$

Therefore

$$\ln V_N = -\frac{\Delta G_a}{RT} + C \qquad (8)$$

where C = $\ln A_S$ = a constant. Hence a plot of ln V_N versus $1/T$ should yield a linear plot of
slope = $-\Delta G_a/R$. The adsorption enthalpy and entropy are obtained from Equation 9:

$$\Delta G = \Delta H - T\Delta S \qquad (9)$$

MATERIALS AND METHODS

Equipment

A Varian 3400CX gas chromatograph equipped with a flame ionization detector (Varian Chromatographic systems, Walnut Creek, California, USA) was used. The IGC experiments were carried out using stainless steel columns 90 cm x 2.1 mm ID (Sigma-Aldrich, Germany) packed with 150 - 212 μm ground untreated, acid and solvent treated dried water hyacinth root biomass. High purity nitrogen from a nitrogen gas generator (M4NT-

Table 1. BET analysis results.

Parameter	Natural	Nitric acid treated	Solvent extracted
Surface area (m^2/g)	3.8189	4.1978	5.0007
Pore volume (cm^3/g)	0.022364	0.024752	0.022228
Pore size (nm)	22.85328	22.94710	17.68746

Table 2. Mean retention times (mins) (n = 3) for volatile n-alkanes on untreated water hyacinth root biomass as IGC stationary phase.

Temperature	n-hexane	n-heptane	n-octane	n-nonane	n-decane
40	0.70±0.00	1.03±0.10	1.87±0.07	4.40±0.32	8.94±0.46
50	0.66±0.04	0.84±0.01	1.41±0.06	2.92±0.03	5.80±0.34
60	0.55±0.02	0.67±0.03	0.95±0.03	1.75±0.13	4.01±0.20
70	0.54±0.00	0.61±0.02	0.87±0.02	1.31±0.01	2.49±0.17

1, Toyo Electric Ltd, Japan) was used as carrier gas. Sieves, 150 and 212 µm (BS410/1986, Endecotts Ltd, London, England) were used to sieve the water hyacinth root biomass after grinding. A Hamilton GC syringe, 1.0 µL, (Sigma Aldrich, Germany) was used for the IGC injections. Brunauer-Emmett-Teller (BET) surface parameters measurements were carried out using a TriStar Model 3000 instrument (Micrometrics Instrument Corporation, Georgia, USA).

Water hyacinth root biomass

Water hyacinth plants were collected from Waerera River, Bindura, Zimbabwe, at Universal Transverse Mercator (UTM) coordinates of (318500;8078400). At the laboratory, the plants were washed with tap water several times, and then washed with distilled water. The roots were separated from the tops, cut into pieces and air dried for several days. The root samples were then ground using a mortar and pestle (Rajamohan, 2009), and then sieved first through the 212 µm sieve, and then the 150 µm sieve, retaining the 150 to 212 µm fraction.

IGC probes

N-hexane, n-heptane, n-octane, n-nonane and n-decane (Analytical grade, Sigma-Aldrich, Germany) were used as non-polar probes (Cordeiro et al., 2011a). Methane (99.99%, Fedgas, Johannesburg, South Africa) was used as the non-interacting probe (Cordeiro et al., 2011a).

Procedures

Solvent extraction

Samples of dried water hyacinth root biomass, 150 to 212 µm particle size, were successively extracted for 24 h in a Soxhlet apparatus, with toluene/ethanol (2:1,v/v) followed by ethanol. Each sample was filtered under suction, rinsed with boiling water, and then transferred to an Erlenmeyer flask containing boiling water. The flask was placed in a hot water bath and left to boil for approximately 1 h. The extracted sample was filtered under suction and air dried for 48 h. The dry sample was stored in a sealed glass jar.

Acid treatment

The water hyacinth root biomass powder was acid-washed by soaking in 0.1 M HNO_3 for 24 h, followed by washing with deionised water, and drying at 65°C for 24 h (El-Khalary, 2007).

Brunauer-Emmett-Teller (BET) surface area analysis

The surface area, pore volume and pore size of the water hyacinth biomass before and after treatment was measured by N_2 adsorption/desorption measurements using the TriStar instrument model 3000 at 373K. Before measurements, all samples were dried at 90°C for 48 h, then 0.1740 to 0.1857g of the sample was put in the tubes of the instrument and degassed for 24 h using a vacuum. The results obtained are shown in Table 1.

Inverse gas chromatography

IGC chromatograms were obtained under the following GC conditions: carrier gas flow rate: 6.4 ml/min; detector and injector temperature: 220°C; column temperature: isothermal at 40, 50, 60 and 70°C. The column was conditioned for 2 h prior to injecting the probes. 0.2 µl of the probe was injected each time. At least three injections were made for each probe at each column temperature. Figure 1 shows typical chromatograms obtained for n-heptane on untreated, acid treated and organic solvent treated water hyacinth root biomass. Table 2 shows typical mean retention times obtained on untreated water hyacinth root biomass. Figure 2 shows typical regression curves obtained when *lnVn* values for n-hexane and n-heptane are plotted as a function of *1/T* for the untreated water hyacinth root biomass. The ΔG_a values calculated for the untreated, acid and solvent treated water hyacinth root biomass from the slopes of the plots of *lnVn* as a function of *1/T* according to Equation 8 are shown in Table 3.

Equation 9 shows that the values of ΔH_a and ΔS_a for the adsorption of the n-alkanes can be obtained from the slope and intercept respectively of the plot of $\Delta G_a/T$ as a function of 1/T. Linear plots (R^2 = 0.9983 to 0.9995) were obtained in all cases.

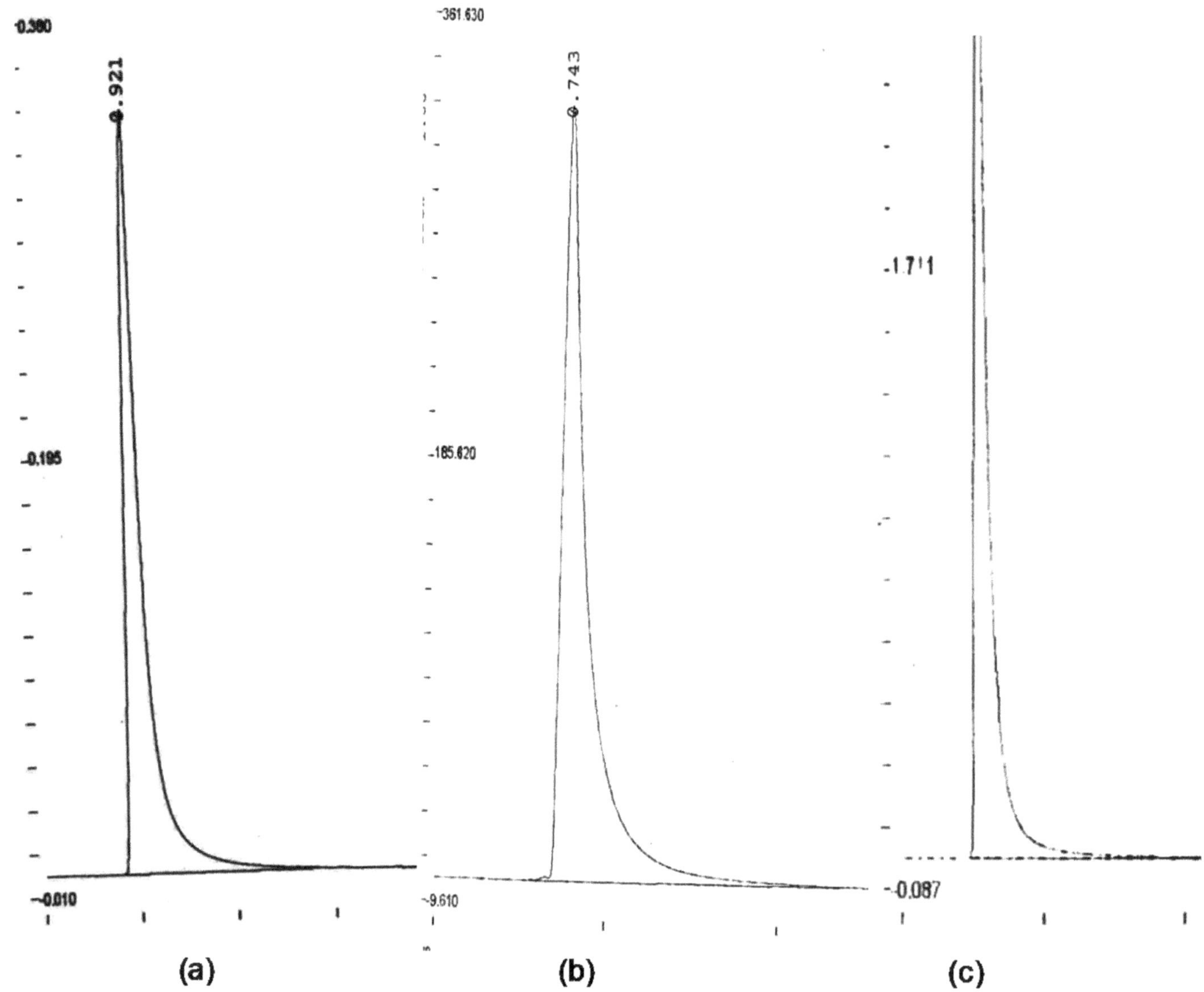

Figure 1. Chromatogram of n-heptane at a flow rate of 6.4 ml/min: (a) Untreated at 40°C, (b) solvent treated at 70°C and (c) nitric acid treated at 70°C.

The ΔH_a and ΔS_a values obtained are shown in Table 3.

RESULTS AND DISCUSSION

Adsorption thermodynamic properties

Untreated water hyacinth root biomass

From Figure 1, it is apparent that sharp peaks were obtained for n-alkane hydrocarbons, although they show some tailing, due to surface adsorption. The retention data obtained (Table 2), shows good reproducibility, with relative standard deviation in the range 0.0 to 10%. The plot of lnV_n versus $1/T$ for all the alkane probes used in the IGC experiments showed good linearity (R^2 = 0.9032 - 0.999), as predicted by Equation 8 (Figure 2). Adsorption free energy values, ΔG_a, obtained for the non-polar probes range from -20.42 kJ.mol^{-1} for hexane to -44.44 kJ.mol^{-1} for nonane (Table 3). When $\Delta G_a/T$ was plotted as a function of $1/T$ for the n-alkanes hexane to nonane, linear plots were obtained (R^2 = 0.9983 -0.9995) (Figure 3), from which the value of ΔH_a are obtained as -21.18, -30.74, -33.62 and -45.84 kJK^{-1}mol^{-1}, respectively for n-hexane, n-heptane, n-octane and n-nonane. The corresponding ΔS_a values obtained are +0.0023, +0.0036, +0.004 and +0.0041 kJ/mol respectively. Similar values, -17 to -25 kJmol^{-1} for ΔG_a, -36 to -52 kJmol^{-1} for ΔH_a and -0.0657 to -0.0896 kJmol^{-1} for ΔS_a for n-heptane to n-decane, were reported by Dorris and Gray (1980) for thermomechanical pulp (TMP) fibres. The values

Table 3. ΔG_a ($kJ.mol^{-1}$), ΔH_a ($kJK^{-1}.mol^{-1}$) and ΔS_a ($kJ.mol^{-1}$) values calculated for the adsorption of n-alkanes hexane to decane on untreated, acid treated and solvent treated water hyacinth root biomass.

Biomass treatment	Alkane	ΔG_a	ΔH_a	ΔS_a
	n-Hexane	-20.422	-21.182	+0.0023
	n-Heptane	-29.546	-30.739	+0.0036
Untreated	n-Octane	-32.287	-33.621	+0.004
	n-Nonane	-44.439	-45.837	+0.0041
	n-Decane	-35.992	-33.704	+0.0071
	n-Hexane	-38.521	-36.332	+0.0069
Acid treated	n-Heptane	-38.793	-36.332	+0.0077
	n-Octane	-38.186	-35.935	+0.007
	n-Nonane	-36.077	-34.014	+0.0041
	n-Hexane	-40.980	-38.349	+0.0082
Solvent treated	n-Heptane	-39.303	-37.04	+0.0071
	n-Octane	-43.650	-40.988	+0.0083
	n-Nonane	-46.704	-43.704	+0.0094

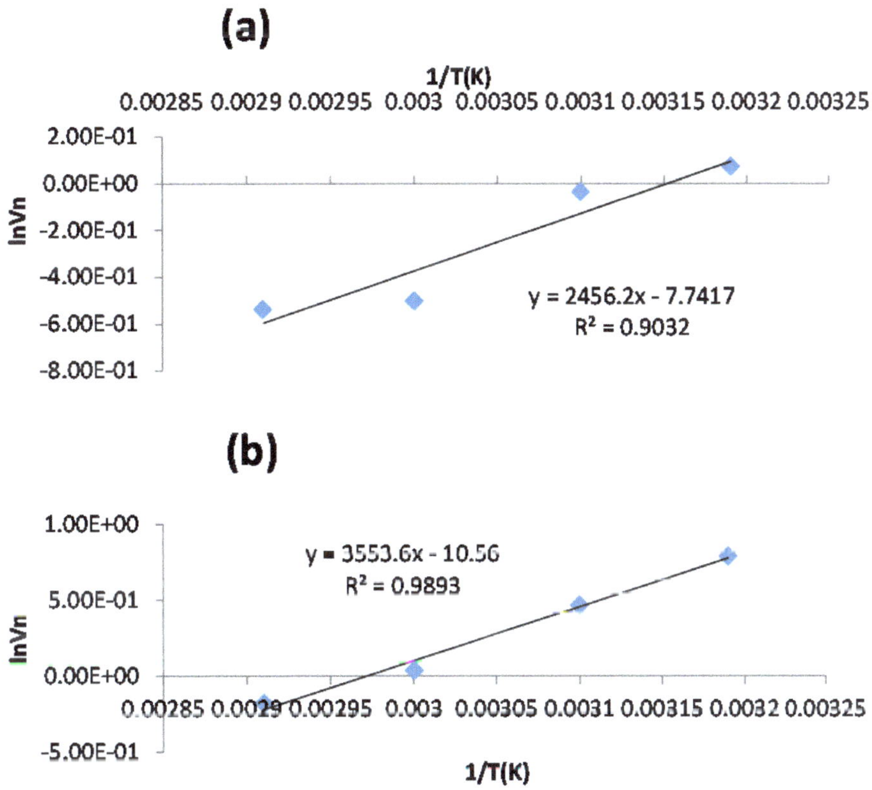

Figure 2. Plot of *lnV*$_n$ versus *1/T* for (a) n-hexane, (b) n-heptane for untreated water hyacinth root biomass.

obtained for ΔG_a. ΔH_a and ΔS_a show that adsorption occurs spontaneously and is exothermic, and involves very litle entropy change. The little positive entropy change exhibited may be attributed to the slight change

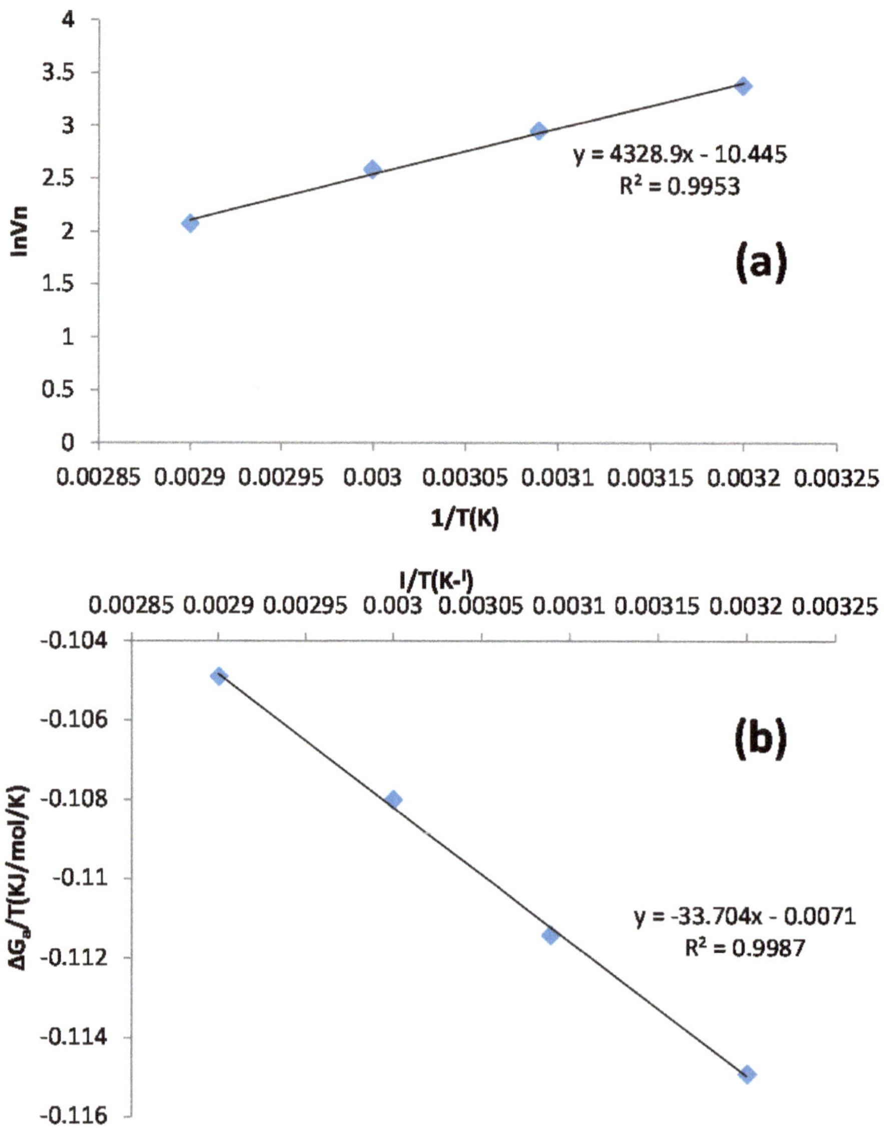

Figure 3. Plot of (a) *lnVn* versus *1/T* and (b) *ΔGa/T* versus *1/T* for n-decane on untreated water hyacinth root biomass.

in the volume of the stationary phase upon adsorption of the adsorbate.

Treated water hyacinth root biomass

Table 3 shows that higher $-\Delta G_a$ and $-\Delta H_a$ values, 36.077 to 38.783 $kJ.mol^{-1}$ and 34.014 to 36.332 $kJ.K^{-1}mol^{-1}$ respectively, were obtained for the acid treated water hyacinth root biomass samples, as compared to 20.442 to 35.992 $kJ.mol^{-1}$ and 21.182 to 33.704 $kJ.K^{-1}mol^{-1}$ respectively, for the untreated sample. Thus treatment of water hyacinth root biomass with dilute mineral acid leads to stronger adsorption bonds as compared to untreated water hyacinth root biomass. The increase in $-\Delta G_a$ and $-\Delta H_a$ upon dilute mineral acid treatment of water hyacinth root biomass is attributed to the hydrolysis of hemi-celluloses. Hemicellulose molecules are normally H-bonded to cellulose. Hydrolysis must free O-H groups in the cellulose microfibrils for H-bonding with the n-alkane probes.

Tables 3 also shows that $-\Delta G_a$ and $-\Delta H_a$ values at 40.98 to 46.704 $kJ.mol^{-1}$ and 38.349 to 43.704 $kJ.K^{-1}mol^{-1}$, respectively, for the solvent extracted water hyacinth root biomass, are higher than those for the acid treated water hyacinth root biomass, showing that organic sol-vent extraction leads to higher adsorption bond strength than obtained for both untreated and mineral acid treated water hyacinth root biomass. We attribute this increase in the adsorption bond to the removal of the protective waxy

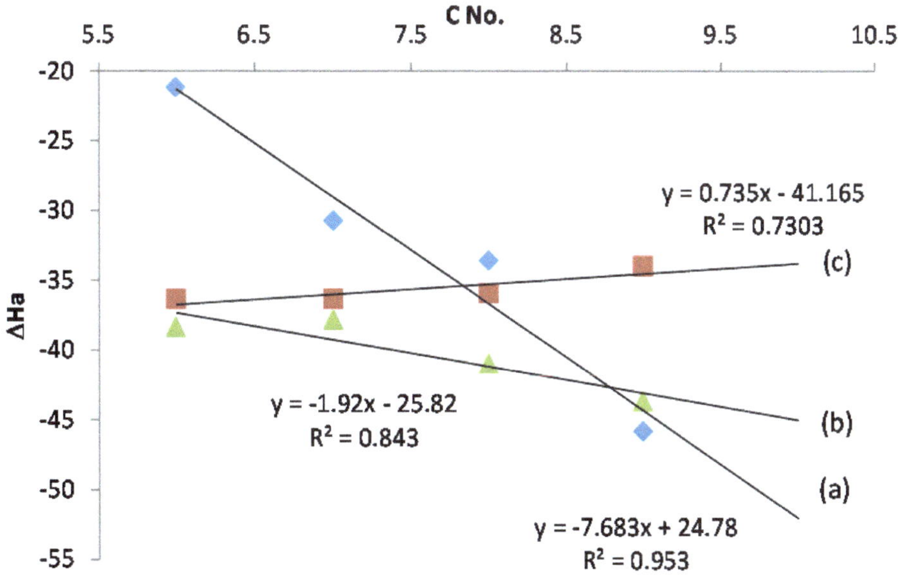

Figure 4. Plot of ΔH_a as a function of carbon number (C No.) for the adsorption of n-alkanes hexane to nonane on (a) untreated, (b) acid treated, and (c) solvent treated water hyacinth root biomass.

layer that normally covers the fibres, thereby deactivating the surface (Cordeiro et al., 2011a). Cordeiro et al. (2011a) have suggested that if extraction removes this waxy layer, a significant change in adsorption properties can be expected.

Figure 4 shows regression plots of ΔH_a as a function of n-alkane carbon number 6 to 9, for (a) untreated, (b) acid treated and (c) organic solvent treated water hyacinth root biomass. The regression curve for untreated root biomass is fairly linear with a slope of -7.683 and an R^2 value of 0.953. Linearity of ΔH_a with *C*-number has been reported previously for cellulose by Dorris and Gray (1980), who interpreted the slope of the ΔH_a versus *C*-number plot, of -5.3 kJmol^{-1}, as giving the enthalpy per *C*-centre or CH_2 group. Dorris and Gray (1980) concluded that the constant increment per CH_2 group in the enthalpy of adsorption was consistent with a flat orientation of the isolated n-alkane molecule on the surface of the cellulose fibres. Such an orientation permits maximum number of contact points with the adsorbent surface. Thus from Figure 4, the enthalpy for adsorption per CH_2 group is -7.683 kJmol^{-1} for untreated water hyacinth root biomass. The value of ΔH_a obtained by Doris and Gray (1980) suggests that contact occurs via the CH group orbitals and involves -5.3 kJmol^{-1} enthalpy, while the value of ΔH_a obtained in present study suggests that contact via CH_2 group orbitals involves -7.7 kJmol^{-1} enthalpy. Figure 5 shows that adsorption contact occurs via CH_2 group orbitals, which is consistent with a saw-tooth conformation or orientation of the n-alkanes on the surface of the water hyacinth root biomass adsorbent, whereas for the systems studied by Doris and Gray (1980), the alkanes assume the flat orientation in which contact occurs predominately via the CH group orbitals. It would appear that whether the n-alkane assumes the flat orientation or the saw-tooth orientation, depends on the physical properties of the adsorbent, such as particle size. Smaller particles result in a tighter column packing that would favour the flat orientation, while larger particles result in a looser packing that might favour the saw-tooth orientation.

Figure 5 shows very good agreement between the experimental ΔH_a value and that predicted from the $\Delta H_{a(CH2)}$ value for n-heptane. For n-hexane, n-octane and n-nonane, the experimental value deviates from that predicted on the basis of the $\Delta H_{a(CH2)}$ value by 8.7% respectively for n-hexane and n-octane, and 16.2% for n-nonane. We attribute the deviation, in the case of n-octane and n-nonane, to the a small fraction of molecules adsorbed in the flat orientation. The deviation of the experimental ΔH_a value from the predicted value for n-hexane can only be attributed to experimental error.

In Figure 4 the slopes of the regression curves for acid and solvent treated root biomass are -1.92 and 0.735, respectively. Since the experimental -ΔH_a values for the acid and solvent treated water hyacinth root biomass are much higher than those for the untreated biomass, regression slopes more negative than -7.683 are predicted if the n-alkane adsorbate assumes the flat or saw-tooth adsorption orientation.

The deviation of the experimental regression slopes from the predicted slopes can only be interpreted to mean that the n-alkane adsorbates do not assume the flat or saw-tooth adsorption orientation on the acid and

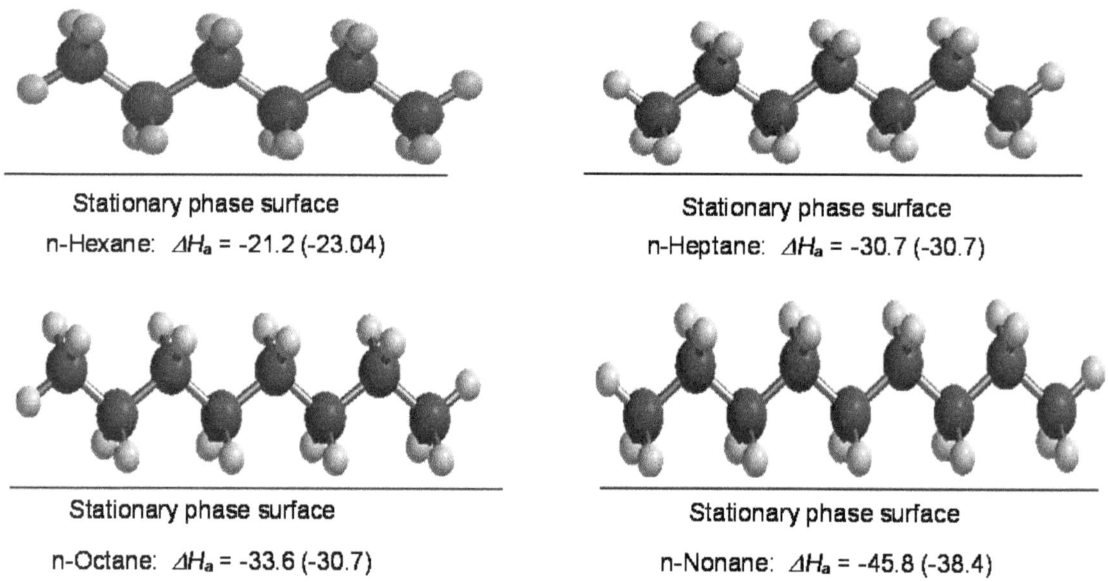

Figure 5. Adsorption of n-hexane, n-heptane, n-octane, and n-nonane by ground dried water hyacinth root biomass stationary phase during inverse gas chromatography: Possible conformation on solid phase stationary (ΔH_a = experimental value (predicted value); structures generated using the Spartan '08 V1.2.0 software, Wavefunction Inc., CA). SPS = Stationary phase surface.

Table 4. Regression coefficient (R^2) for the correlation of ΔH_a with water hyacinth root biomass powder surface area, pore volume and pore size.

N-alkane	Surface area	Pore volume	Pore size
n-Hexane	0.6673	0.1269	0.3371
n-Heptane	0.6037	0.1741	0.2762
n-Octane	0.9554	0.0014	0.7274
n-Nonane	0.1592	0.9747	0.4616

solvent treated root biomass.

Table 1 shows that nitric acid treatment increased the surface area and pore volume of the root biomass by 9.9 and 10.7% respectively, but had no effect on pore size. Organic solvent treatment on the other hand increased the surface area and pore size by 30 and 22.6%, respectively, but had no effect on pore volume. For n-hexane and n-heptane, Table 4 shows that the increase in ΔH_a correlates with the increase in surface area that result from acid and solvent treatment (R^2 values of 0.6673 and 0.6037, respectively). For n-octane, the increase in ΔH_a correlates with the increase in both surface area and pore size (R^2 values of 0.9554 and 0.7274, respectively), while for n-nonane, the increase in ΔH_a correlates with the increase in pore volume.

Enthalpy data for the adsorption of n-alkane hydrocarbons has been determined for a number of adsorbents. Widegren and Bruno (2011) determined the enthalpy for the adsorption of n-hexane, n-heptane and n-octane on concrete, synthetic clay Lonite-Rd and high purity kaolin, while Bilgic and Askin (2003) determined the enthalpy for the adsorption of n-hexane on molecular sieve 3A, molecular sieve 5A and alumina. ΔH_a values obtained are summarized in Table 5. Comparison of ΔH_a data in Table 5 to ΔH_a data obtained in the present study for the adsorption of n-alkanes by water hyacinth root biomass in Table 3 shows that the water hyacinth root biomass compares well with the molecular sieves 3A and 5A, and with alumina, especially after mineral acid and organic solvent treatment.

Conclusions

From the foregoing discussion, we conclude that the enthalpy of the adsorption of n-alkanes on water hyacinth root biomass is comparable to the enthalpy of the adsorption on molecular sieves 3A and 5A, as well as alumina. This makes water hyacinth root biomass a good adsorbent for n-alkane hydrocarbons. The strength of adsorption bond is also reflected in the very low values of the entropy of adsorption, +0.0023 to +0.0071 $kJ.mol^{-1}$, exhibited for the adsorption of n-alkanes by the material. Treatment of the material with dilute mineral acid or organic solvent improves the strength of the absorption bond significantly.

Conflict of interest

The authors did not declare any conflict of interest.

Table 5. Values of $-\Delta H_a$ ($kJK^{-1}mol^{-1}$) reported in the literature for the adsorption of hydrocarbons on concrete, synthetic clay, kaolin and molecular sieves 3A and 5A.

n-Alkane	$-\Delta H_a$ ($kJK^{-1}mol^{-1}$)					
	Concrete[a]	Clay[a]	Kaolin[a]	3A[b]	5A[b]	Alumina[b]
n-Hexane	34.3	56.0	59	27.5	34.7	41.0
n-Heptane	36.5	61.6	69			
n-Octane	42.0	65.2	81			

Source: [a]Widegren and Bruno, 2011; [b]Bilgic and Askin, 2003.

ACKNOWLEDGEMENT

This work was supported by a grant from the Research Board of the University of Zimbabwe.

REFERENCES

Askin A, Yacizi DT (2005). Surface characterization of sepiolite by inverse gas chromatography. Chromatographia. 61:625-631.
Belgacem MN, Blayo A, Gandini A (1996). Surface characterization of polysaccharides, lignins, printing ink pigments, and ink fillers by inverse gas chromatography. J. Colloid Interface Sci. 182: 431-436.
Belgacem MN, Czeremuszkin G, Sapieha S, Gandini A (1995). Surface characterization of cellulose fibres by XPS and inverse gas chromatography. Cellulose 2:145-157.
Bilgic C, Askin A (2003). Evaluation of the thermodynamic parameters for the adsorption of some hydrocarbons on alumina and molecular sieves 3A and 5A by inverse gas chromatography. J. Chromatogr. A. 1006:281-286.
Cordeiro N, Gouveia C, John JM (2011a). Investigation of surface properties of physico-chemically modified natural fibers using inverse gas chromatography. Ind. Crop Prod. 33:108-115.
Cordeiro N, Gouveia C, Moraes AGO, Amico SC (2011b). Natural fibers characterization by inverse gas chromatography. Carbohyd. Polym. 84:110-117.
Dorris GM, Gray DG (1980). Adsorption of the n-alkanes at zero surface coverage on cellulose paper and wood fibers. J. Colloid Interface Sci. 77: 353-362.
El-Khalary MI (2007). Kinetics and mechanism of adsorption of methylene blue from aqueous solution by nitric acid treated water hyacinth. J. Hazard. Mater. 147(1-2):28-36.
Gamelas JAF (2013).The surface properties of cellulose and lignocellulosic materials assessed by inverse gas chromatography: A review. Cellulose 20:2675-2693
Hasan SH, Ranjan D, Talat M (2010). Water Hyacinth biomass (WHB) for the biosorption of hexavalent chromium: Optimization of process parameters. Bioresources. 5:563-575.
Ibrahim HS, Ammar NS, Soylak M, Ibrahim M (2012). Removal of Cd(II) and Pb(II) from aqueous solution using dried water hyacinth as a biosorbent. Spectrochim. Acta., A96: 413-420.
Irfan M, Syed Q, Abbas S, Sher MG, Baig S, Nadeem M (2011). FTIR and SEM analysis of thermo-chemical fractionated sugarcane bagasse. Turk. J. Biochem., 36(4): 322 - 328.
John MJ, Thomas S (2008). Biofibres and Biocomposites, Carbohydr. Polym. 71:343-364.
Khan MR, Mozumder SI, Islam A, Prasad DMR, Alam MM (2012). Methylene Blue Adsorption onto Water Hyacinth: Batch and Column Study. Water Air Soil Pollut. 223:2943-2953.
Kondor A, Dallos A (2014). Adsorption isotherms of some alkyl aromatic hydrocarbons and surface energies on partially dealuminated yfaujasite zeolite by inverse gas chromatogr. J. Chromatogr. A. 1362:250-261.
Kunaver M, Zadink J, Planinsek O, Srcic S (2004). Inverse gas chromatography – a different approach to characterization of solids and liquids. Acta Chim. Slov. 51:373-394.
Laub RJ, Pecsok RL (1978). Physicochemical Applications of Gas Chromatography. John Wiley and Sons, New york.
Mahamadi C (2011). Water hyacinth as a biosorbent: A review. African J. Environ. Sci. Technol. 5(13):1137-1145
Mahamadi C, Nharingo T (2010a). Utilization of water hyacinth weed (Eichhornia crassipes) for the removal of Pb(II), Cd(II) and Zn(II) from aquatic environments: an adsorption isotherm study. Environ. Technol. 31(11):1221-1228.
Mahamadi C, Nharingo T (2010b). Competitive adsorption of Pb(II), Cd(II) and Zn (II) ions onto *Eichhornia Crassipes* in Binary and Ternary systems. Bioresour. Technol. 101:859-864.
Mahamadi C, Nharingo T (2007). Modelling the kinetics and equilibrium properties of cadmium biosorption by river green alga and water hyacinth weed. Toxicol. Environ. Chem. 89(2):297-305.
Rafatullah M, Sulaiman O, Hashim R, Ahmad A (2010). Adsorption of methylene blue on low-cost adsorbents: A review. J. Hazard. Mater. 177(1-3):70-80.
Rajamohan N (2009). Equilibrium studies on sorption of an anionic dye onto acid activated water hyacinth roots. Afr. J. Environ. Sci. Technol. 3(11): 399-404.
Reidel B, Mutuana LM (2006). Inverse gas chromatography of fibers and polymers. Encyclopedia of Surface and Colloid Science. 2006: 3018-3031.
Schultz J, Lavielle L, Martin C (1987). The role of the interface in carbon fibre-epoxy composites. J. Adhesion. 23:45-60.
Shi B, Qi D (2012). A method for improving the calculation accuracy of acid base constants by inverse gas chromatography. J. Chromatogr. A. 1231:73-76.
Spinace MAS, Lambert CS, Fermoselli KKG, De Paoli MA (2009). Characterization of lignocellulosic curaua fibres. Carbohydrate Polymers. 77: 47 - 53.
Steele DF, Moreton RC, Staniforth JN, Young PM, Tobyn MJ, Edge S (2008). Surface energy of microcrystalline cellulose determined by capillary intrusion and inverse gas chromatography. The Am. Assoc. Pharm. Sci. 10(3):494-503.
Tarawou T, Horsfall Jr. M, Vicente JL (2007). Adsorption of Methyl red by water hyacinth (*Eichhornia crassipes*). Biomass Chem. Biodivers. 4: 2236-2245.
Tshabalala MA, Karthikeyan KG, Wang D (2004). Cationized milled pine bark as an adsorbent for orthophosphate anions. J. Appl. Polymer Sci. 93: 1577-1583.
Tahir H, Alam U (2014). Lignocellulosic: Non-conventional low cost biosorbent for the elution of comassie brilliant blue (R-250). Int. J. Chem. 6(2): 56 - 72.
Widegren JA, Bruno TJ (2011). Enthalpy of adsorption for hydrocarbons on concrete by inverse gas chromatophy. J. Chromatogr. A. 1218: 4474-4477.
Wolverton BC, McKown MM (1976). Water Hyacinth for removal of phenols from polluted waters. Aquat. Bot. 2:191-201.
Zheng JC, Feng HM, Lam MH, Lam PK, Ding YW, Yua HQ (2009). Removal of Cu (II) in aqueous media by biosorption using water hyacinth roots as a biosorbent material. J. Hazard. Mater. 171:780 - 785.

27

Ecotoxicological effects of discharge of Nigerian petroleum refinery oily sludge on biological sentinels

Atuanyan Ernest and Tudararo-Aherobo Laurelta

Department of Microbiology, Faculty of Life Sciences, University Of Benin, Benin City, Edo State, Nigeria.

Ecotoxicological effects of the discharge of Nigerian petroleum refinery oily sludge on biological sentinels were examined. The ecotoxicological effects examined included acute toxicity tests on *Nitrobacter* sp., fresh water shrimp (*Desmoscaris trispinosa)* and brackish water shrimp (*Palaemoneles africanus*) from the aquatic environment. It also covered chronic toxicity tests on microbial nitrogen transformation activity in soil and the growth of the terrestrial fauna, earthworm (*Apporectoda longa*) in pristine soils spiked with predetermined concentrations of the sludge. Analysis of the Nigerian petroleum refinery oily sludge used in this study indicated that the sludge is slightly acidic with a high total petroleum hydrocarbon (TPH) content of 340,000 mg/kg made up mainly between 10-40 carbon unit compounds. The sludge reduced the growth of *Nitrobacter* sp. in aqueous medium and also caused chronic effect on microbial nitrogen transformation activity in soil because it exceeded the 25% inhibition limit for chemicals with the potential to cause chronic effects on soil microbial activities. Similarly, the sludge exhibited toxicity on fresh and brackish shrimp. The freshwater shrimp was however more affected with an LC_{50} of 1097.375 ± 0.62 mg/kg when compared with an LC_{50} of 1590. 37±0.92 mg/kg obtained for the brackish water shrimp. It also reduced the growth rate of the earthworm (*A. longa)* progressively as the sludge concentration increased.

Key words: Toxic effects, petroleum refinery oily sludge, biological sentinels.

INTRODUCTION

Petroleum sludge are oily and viscous residues, which are formed during production, transportation, refining of petroleum and storage. Petroleum refinery oily sludge is composed basically of oil, water and solids (Ururahy et al., 1998). The oil industry is responsible for the generation of high amounts of oily sludge as waste by-product. However, one of the problems faced by the oil industry is the safe disposal of the oily waste generated. It is estimated that approximately 1% of the total oil processed in a refinery is discarded as oily sludge (Ururahy et al., 1998). These oily wastes are expensive to store or destroy and previously contaminated areas have required expensive remediation processes to minimize contaminant dispersion.

Improper disposal of petroleum refinery oily sludge leads to environmental pollution, particularly soil contamination, and poses a serious threat to ground water. Many of the constituents are carcinogenic and immunotoxicants (Prospt et al., 1999). The polycyclic aromatic hydrocarbons (PAHs) have also been known to impair chemoreceptor functions in aquatic lives and hence lead to extinction of some species. They have also been known to bioaccumulate up the food chain, resulting in cancers and other genetic malfunctioning in man and other higher animals (Atlas and Bartha, 1992).

Several disposal options for petroleum refinery oily sludge disposal include thermal treatment (incineration) (Ururahy et al., 1998), landfills (Mishra et al., 2001) and biotreatment using the following methods: composting, land farming and biopile (Englert et al., 1993). Other conventional biological treatment methods include activated sludge and anaerobic digestion (Hazardous Waste Management, 1997). An interesting alternative to circumvent these problems is the use of bioreactors since optimum process conditions can be easily controlled, allowing higher quality final effluent in shorter times. However, they might have high costs (Oolman et al., 1996).

The Department of Petroleum Resources (DPR), an agency under the Federal Ministry of Environment (FMENV), which regulates activities in both the upstream and downstream petroleum sector, recommends that petroleum refinery oily sludge should be treated and disposed by method that shall not endanger human life and living organisms and cause significant pollution to ground and surface water (DPR, 2002, part iii sec. D 3.6.31, 12). Such approved methods include recycling (resource recovery), incineration, solidification, land farming (bioremediation) and land filling.

Toxicity tests are used to expose test organisms (fish, shrimps, microorganism, earthworms) to a medium– water, sediment, or soil and evaluate the effect of contamination on the survival, growth, reproduction, behavior and or other attributes of these organisms. These tests may help to determine whether the contaminant concentrations in a site's media are high enough to cause adverse effects in organisms. Acute toxicity tests are short term tests that measure the effects of exposure to relatively high concentrations of chemical. Chronic toxicity test, on the other hand, generally are longer– term tests that measure the effects of exposure to relatively lower, less toxic concentrations. Toxicologists have based their selection of test organisms (sentinels) on several factors; sensitivity to variety of substances, availability, representativeness of a variety of ecosystems, and ease of maintenance and culture under laboratory conditions. Sentinels are biological indicators that can help define the ecotoxicological effects of environmental contaminants (USEPA, 2004). In view of the problems encountered in the management of petroleum refinery sludge, this study examined the acute and chronic ecotoxicological effects of Nigerian Petroleum refinery oily sludge on biological sentinels.

MATERIALS AND METHODS

Sample collection

The petroleum refinery oily sludge used for this study was collected from petroleum refinery oily sludge holding tank of Warri Refinery and Petrochemical Company Ltd (WRPC), Warri, Ekpan, Delta State, Nigeria in 2 L glass bottle and preserved at 4°C until required for use. WRPC lies within the coordinates 5° 32' 15"N 5° 41'41"E.

The *Nitrobacter* sp. bacteria used was isolated from Aladja River, Aladja (5° 14'4"N, 60 15' 18"E) Delta State, in southern Nigeria. DSMZ heterotrophic nitrobacter medium was used for the isolation of bacteria. Isolates that were grayish, mucoid, flat, Gram negative, pear shaped and aerobic were selected according to the scheme of Colwell and Zambrushki (1972). Subcultures were made into slants of DSMZ– nitrobacter agar and stored at 4°C until required for use.

The shrimps were collected from the wild at Abua and Bomadi in Rivers and Delta states of Nigeria, respectively. Coordinates for Bomadi, Delta State are 5° 10' 0 N, 5° 56' 0"E. Abua in Rivers State is 10b miles from Port Hacourt with coordinates 5° 18' 30"N, 6° 25' 0"E Transportation of the shrimp to the laboratory was in air-inflated bags with the organisms' habitat water. Physicochemical parameters of the water were determined using standard methods (APHA, 1998).

The earthworm (*Aporrectoda longa)* commonly found in southern Nigeria was collected from a farm at Ubogo, Delta Southern Nigeria. The worms were collected according to the method described by Terhivuo et al. (1994) and Spiegel (2002). They were collected by digging and hand sorting from subsurface litters and taken to the laboratory for identification. They were then washed free of adhering soil particles and left on moist filter paper for voiding. Earthworms were selected based on their maturity (shown by the presence of clitellum) and liveliness (active response when anterior segment is prodded). The physicochemical parameters of the native soil were determined prior to the test.

Acute toxicity of petroleum refinery oily sludge pollution on the bacteria (*Nitrobacter* sp.), brackish water and freshwater shrimps

A range finding test was first conducted using three concentrations after which a definitive test was conducted with five geometric concentrations based on the result of the range finding test. The methods of Duffus (1980), Wang (1984) and APHA (1998) were adapted with some modifications. A fresh dilution and culture was made from the *Nitrobacter* sp. slant. A loopful of the bacteria was collected from the slant and dislodged in 20 ml peptone water and allowed to stand for a few hours at 30°C. The stock culture was prepared by inoculating 180 ml of sterilized peptone water with 20 ml of the activated culture. The test concentrations were prepared using the dilution water from the bacteria's habitat, sterilized at 121°C for 15 min.

Concentrations of sludge used were 5000, 2500, 1250, 625 and 312.5 mg/l. One hundred milliliters (100 ml) of each test concentration was then put into a 250 ml conical flask and sterilized at 121°C for 15 min. On cooling, 10 ml of the cell suspension was added to each flask containing the different sludge concentration and control (sterile dilution water). This was done in triplicate. The flasks were shaken thoroughly to mix and were incubated at 30± 2°C to determine the number of viable cell at 0 (start), 8, 16 and 24 h. 0.1 ml of each test concentration and the control was collected from the test solution and dispersed onto the surface of an already prepared DSMZ nitrobacter agar plate. The plates were then incubated at 30°C for 24 h. The viable cells were counted and recorded. The EC_{50} was determined using the probit method of analysis (Finney, 1978). The following parameters were determined on the test solutions; pH, conductivity, dissolved oxygen, TDS, ammonia, alkalinity and sulphide. The methods used were adapted from APHA (1998).

Acute toxicity of freshwater shrimp (*Desmoscaris trispinosa*) and brackish water shrimp (*Palaemonetes africanus*) exposed to petroleum refinery oily sludge was determined using the method recommended by OECD #218 (2004). The sediment used in the toxicity test was collected with the aid of hand held van Veen type grab. After the collection, the sediment was sieved through a 500 µm mesh using dilution water in order to remove any organisms, which may interfere with the test. The sediment was allowed to settle overnight and the supernatural water decanted. The sediment was then stored in the dark at 4°C until required for the experiment. Approximately 24 h prior to testing, the sediment samples were removed from the refrigerator storage and allowed to equilibrate to room temperature and weighed.

A preliminary range-finding test was conducted prior to the actual test to assist in determining the appropriate dilutions. The range finding test was conducted using a broad concentration range (100, 1000 and 1000 0mg/kg) and the test was terminated in 24 h. In the definitive test, the concentrations selected were based on the mortality values obtained from the range finding and were in appropriate logarithmic dilution series. The 10-day static sediment bioassay was conducted by placing the weighed sediment into triplicate sets of 5 L amber coloured glass tanks. Concentrations of 625, 1250, 2500, 5000 and 10000 mg/kg of sludge were prepared in three replicates and properly homogenized with the sediment and were spread evenly in each tank. The sediment was then overland with 2 L of water from the organism's habitats. The contents of the containers were left to settle for 2 to 3 ho prior to the addition of the test organisms. Shrimps were collected with a 500-µm mesh sieve and placed in dilution water to rinse off any debris. Ten (10) shrimps were gently transferred into each glass test-tank containing test chemical (sludge) and control. The overlaying water was gently aerated for the 10-day's exposure period. Observations for mortality in the test – vessels were made and records taken for the numbers of shrimp which were swimming, crawling on the surface, loss of appendages emergency of organisms from sediment, immobilized (lying on the sediment surface but obviously still alive) or dead. Dead shrimps were removed at each observation. After 10 days, the sediments were sieved and the number of dead shrimps recorded. Average mortality in the bioassay, that is, the total number of organism used on day 0 was used to estimate the average percentage mortality in the bioassay at day 10. The LC_{50} was also determined by the probit method (Finney, 1978). Controls with clean sediment (without sludge) were conducted along with the treated sludge (Environment Canada, 1992). The size of test shrimps used was 0.156±0.03 g and 2.65±0.36 cm in length for the freshwater and brackish water shrimp, respectively. The physicochemical parameters of the test solutions were determined 24 hourly for 96 h.

Chronic toxicity (sublethal) effects of petroleum refinery oily sludge pollution on nitrogen transformation activity in the soil

The OECD TG 216 (2000) test method was used for this test. The effect of the petroleum refinery oily sludge on the nitrifying bacteria, *Nitrobacter* sp. was determined. This test was used to detect long-term (chronic) adverse effects of petroleum refinery oily sludge to the process of nitrogen transformation activity in the soil. Pristine soil was taken from a depth of 0 to 20 cm from a garden in Orhuwhorun town in Delta state, Nigeria and was transported in an ice – chest at 4°C to guarantee the initial soil properties were not significantly altered. In the laboratory soil, samples were kept in the refrigerator at 4± 2°C when they could not be used immediately. The soil was dried, sieved and amended with 5 g/kg compost and treated with five concentrations (375, 6250, 12500, 25000 and 50,000 mg/kg) of petroleum refinery oily sludge or left untreated (control) after day 0, 7, 14 and 28, treated control samples were extracted and analyzed for ammonia, nitrate and *Nitrobacter* sp. counts, the rate of ammonia nitrate formation in treated soil was compared with rate in the controls and the percent deviation of the treated from control was calculated. Enumeration of nitrogen transformation bacteria (*Nitrobacter* sp.) was also done to correlate the microbial growth with the transformed nitrogen. Results from the test of multiple were analyzed using a regression model (ANOVA) and the EC_{50} was calculated. All analyses were done by ASTM method. The control contained only the soil. A geometric series of five concentrations was used. Three replicates for both treatments and control were used.

The test was carried out in the dark at 25± 20°C where 90 ml of sterile distilled water was added to each tank to achieve moisture content of between 40-60%, while 60 ml was added to the 50,000 mg/kg test – tanks to achieve the same percentage of moisture content. The content of the tank was mixed thoroughly and covered with perforated polythene to prevent excessive evaporation of water and volatile fractions. Moisture content of between 40-60% of the maximum water holding capacity was maintained during the test by watering at intervals with distilled water. The duration of the test was 28 days. Composite soil sampling was done on days 7 and 28 and the soils samples were analyzed for some physico-chemical parameters such as pH, total petroleum hydrocarbons, total organic carbon, polyaromatic hydrocarbon nitrate and ammonia. microbiological parameters included enumeration of total heterotrophic bacteria and *Nitrobacter* sp. count. Enumeration of *Nitrobacter* sp. was done with DSMS heterotrophic nitrobacter medium.

The quantity of ammonia and nitrate formed and *Nitrobacter* sp. counts obtained in each replication test soil were recorded, mean values of all replicates were determined and a dose response curve was prepared for the estimation of the effective concentration causing 50% reduction (EC_{50}value). The rate of nitrate formation in treated samples was compared with the rate in the controls, and percent deviation/inhibition of the treated from the control was calculated after 28 days using the formula below (Grunditz and Dalhammar, 2001):

$$\text{Inhibition } (\%) = \frac{C_{ref} - C_{sample}}{C_{ref}} \times 100$$

Where C_{ref} is concentration of nitrate formed in control, C_{sample}

Chronic toxicity effects of petroleum refinery oily sludge on the growth and survival of earthworm (*A. longa*)

Once organisms were obtained, they were identified and maintained in the laboratory using the procedures described in ASTM standard E2172 – 01 (ASTM, 2001). The selected worms were acclimatized for 1-7 days in the soil from the organisms' habitat. During this period, the worms were fed with cellulose. Cellulose was prepared in advance by shredding white, kaolin based paper, followed by converting it to pulp by mixing with distilled water, and subsequently drying at 30°C for 48 h. The weight of the worms was between 300 and 600 mg. Test conditions used were: temperature 20 ± 2°C, light-dark cycles; 16 and 8 h. The test earthworms used for this study were ecologically relevant to the Niger Delta of Nigeria. They possess the following characteristics, red – violet in colour, anterior black segment, a prolobous postonium, 8-14 cm in length, about 149 – 157 segments, a bar-shaped tubercular and alternately paired ternatogential tumescences.

Four concentrations (375, 750, 1500 and 3000 mg/kg) of the petroleum refinery oily sludge were prepared for the definitive test after a preliminary range-finding test was conducted for two days. Five hundred grams (500 g) of soil were mixed with various test concentrations of the sludge and 20 g of cellulose. Cellulose was added to the soil as food for the earthworms. These were manually homogenized and distilled water (80 ml) was added to achieve 45% moisture content in one litre (I L) amber-coloured glass jars. A blank (control) containing only cellulose, water and soil was also prepared. Test tanks were prepared in triplicates per concentration. Prior to use for the test, chosen worm were stored for 24 h on a damp filter paper to void contents of the stomach and intestinal tract. The ten (10) earthworms selected were placed on the surface of the control and test soil samples and were allowed to ingest and burrow into the test medium. The distribution of individual earthworms among the test chambers was randomized. The test medium and control were analyzed for pH, TPH content, total organic carbon $(TOC)_g$ metals (chromium, cadmium, nickel, iron and copper), at the start of the experiment and weekly for 28 day.

Death was the primary criterion used in this test guideline to evaluate the toxicity of the test substance. Earthworm in the test and control chambers were observed weekly for 28 days and the number alive were recorded and the dead removed. In addition to death, weight loss, behavioural symptoms and pathological symptoms were recorded. Each test and control chambers were for dead or affected earthworms and observations recorded on day 7, 14, 21 and 28 days after the beginning of the test. Missing earthworms were considered dead. Mortality was assessed by empting the test medium on a glass or other inert surface and the earthworms were sorted from test mixture and their reactions were tested by a gentle mechanical stimulus. Any adverse effects (eg weight loss, behavioural or pathological symptoms) were noted and reported. The 28 day test result would be unacceptable if more than 20% of control organisms died or the total mean weight of the earthworm in the control containers declined significantly during the test (by 30%). The sublethal effects and growth (fresh weight) data were used to determine the EC50 using the Probit software. ANOVA was used to test for significant differences between treatment means and the control. At the end of the test, worms were removed from each jar, washed, dried, counted and weighed. Observations such as motility, light sensitivity and physical qualities (discolouration), morphology (open wounds) were documented to provide some indication of toxic response. The worms were then depurated for 24 h to void contents of the intestinal tract and subsequently rewashed and reweighed. The worms were analyzed for TPH and metals concentrations.

RESULTS

The physicochemical and microbial qualities of Nigerian petroleum refinery oily sludge shown in Table 1 indicates the sludge was acidic with a pH value of 5.81 ± 0.28 and total petroleum hydrocarbon (TPH) of 340000 ± 50000 mg/kg. Polyaromatic hydrocarbon (PAH) content of the sludge was 0.075 ± 0.02 mg/kg while the value for ammonium was 21.65 ± 1.21 mg/kg. The heterotrophic bacteria and fungi counts were 5.86E + 05 and 4.72E + 05 cfu/g, respectively. Hydrocarbon degrading bacteria and fungi counts were 2.85E + 02 and 2.75E + 02 cfu/g, respectively.

As shown in Table 2, the counts of *Nitrobacter* sp. after 24 h exposure to five concentrations of petroleum refinery oily sludge used dropped from 4.92E + 07 to 2.52E + 06 cfu/ml from the lowest concentration (312.50 mg/l) to the highest concentration (5000 mg/L) of the petroleum refinery oily sludge, respectively. *Nitrobacter* sp. counts obtained in the experimental control were 2.68E + 09 cfu/ml.

As shown in Table 3, the chronic toxicity effects of petroleum refinery oily sludge on nitrogen transformation activities in soils showed that the percentage inhibition of nitrogen transformation in petroleum refinery oily sludge contaminated soils in relation to the control increased with increasing sludge concentration. The increase ranged from 18.7 to 79.38% from the lowest concentration of 3750 mg/kg to the highest of 5000 mg/kg, respectively.

Since in our analysis of metal contents of sludge, zinc recorded a high concentration of 100.62 mg/kg in relation to other metals analyzed, it could have contributed to the observed inhibitory effect of the sludge on the organisms used in this study as observed in similar studies by Wang and Reed (1984). They noted that a high concentration of metal cations inhibits microbial activities by causing damage or inactivating one or more critical enzymes resulting in formation of an inactive complex between the metal cations and an active enzyme. TPH contains toxic compounds such as PAHs and these have also been implicated in the inhibition of nitrification process (Suschka et al., 1996; Dokaniakis et al., 2005)

Chronic toxicity profile of nitrogen transforming bacteria exposed to petroleum refinery oily sludge for 28 days recorded an EC_{50} of 13761.059 mg/kg as show in Table 4. This indicates that at the obtained EC_{50}, there would be 50% inhibition of nitrogen transformation activity in the soil. The recommended limit for non-chronic effect on nitrogen transformation activity in soil is ≤ 25% (ISO/DIs 14248; 1995). This would be a sludge concentration of 502.41 mg/kg. The sludge concentration recorded as the EC_{50} and above would result in serious inhibition of nitrogen transformation activity in soils and subsequently result in soil infertility.

The mean acute toxicity profile of the fresh and brackish water shrimp exposed to varying concentrations of petroleum refinery oily sludge in the sediment shown in Table 5 indicate that there were no death or physiological changes in the negative control for the 10-days test duration. The control shrimps appeared active and healthy (responsive to stimuli) throughout the test period. The test organisms exposed to the various sludge concentration had higher mean percentage mortality by day 10 in the fresh water test (40, 57, 67, 80 and 100%) than in the brackish water test (33, 50 63, 73 and 93%) in 625, 1250, 2500, 5000 and 10,000 mg/kg respectively (Figure 1). These results indicated that mortality increased with increased sludge concentrations and exposure duration. The toxicity profile indicated the estimated mean LC_{50} at day 10 for the fresh water shrimp was 1097.375 ± 0.620 mg/kg while that for brackish water shrimp was 1590.376 ± 0.920 mg/kg. GESAMP rating indicates both LC_{50} to be hazardous (Table 5).

Effect of petroleum refinery oily sludge on organism in the terrestrial environment was also determined with earthworm (*A. longa)* bioassay shown in Table 6.

Table 1. Physicochemical and Microbiological properties of Nigerian Petroleum refinery oily sludge.

Parameter	Mean (± S.E).
pH	5.81± 0.28
Conductivity, us/cm^2	466.65 ± 25.25
Sulphate, mg/kg	4.83 ± 0.64
Nitrate, mg/kg	26.40 ±1.02
Phosphate, mg/kg	7.73 ± 0.88
Total Nitrogen, mg/kg	0.12 ± 0.5
Total Petroleum Hydrocarbon, mg/kg	340,000 ± 50,000
Polyaromatic Hydrocarbon, mg/kg	0.075 ± 0.02
Hydrocarbon Degrading fungi, *cfu/g*	2.85 x 10^2
Ammonium, mg/ kg	21.65 ± 1.21
Copper, mg/kg	5.53 ±0.20
Chromium, mg/kg	8.68 ± 0.03
Nickel, mg/kg	3.36 ± 0.02
Cadmium, mg/kg	0.32 ± 0.5
Zinc, mg/kg	100.65 ± 2.30
Barium, mg/kg	0.31 ±0.04
Heterotrophic Bacteria, cfu/g	5.86E +05
Heterotrophic Fungi, cfu/g	4.72E+ 05
Hydrocarbon Degrading Bacteria, cfu/g	2.75E + 05

Table 2. Growth of *Nitrobacter Sp* Exposed to Petroleum refinery oily sludge after 24hrs

Exposure time (h)	Control Counts (CFU/ml)	Microbial Counts CFU/ml Concentration of sludge (mg/l)				
		312.50	625	1250	2500	5000
0	2.88E +07	7.40E +05	5.10E + 05	3.60E + 05	2.40E + 05	2.20E + 05
	2.96E + 07	7.22E + 05	5.42E + 05	3.36E + 05	2.62E + 05	1.82E +05
8	6.78E + 08	5.36E + 06	4.88E + 06	2.88E + 06	6.88E + 05	3.96E + 05
	7.26E + 08	6.12E + 06	5.22E + 06	2.24E + 06	7.12E + 05	4.22E + 05
24	2.16E + 09	5.36E + 07	3.64E + 07	1.36E + 07	7.20E + 06	2.40E + 06
	2.68E + 09	4.92E +07	3.12E + 07	1.42E + 07	6.88E + 06	2.52E + 06

Table 3. Percentage (%) inhibition of Nitrate formation in Petroleum refinery oily sludge contaminated soil

Sludge concentration (mg/kg)	% inhibition
3750	18.66
5250	33.04
12500	51.70
25000	59.82
50000	79.38

Earthworms are associated with a healthy soil and their absence is an indication of poor soil health (Doube and Schmidt, 1997; Edwards and Shipitalo, 1998; Parmelee et al., 1998). This chronic effect studies of petroleum refinery oily sludge on earthworms showed that the sludge led to a reduction of growth progressively as the concentration of the sludge increased (Figure 2). The growth rate was inhibited from 86.71 (375 mg/kg) to 37.92 (3000 mg/kg). While 825.03 mg/kg was obtained as the EC_{50} at the end of 28 days (Table 6). This concentration reduced the growth rate of the test earthworms by 50%. In comparison with toxicity ratio of chemicals to earthworms (Davies, 2003), the EC_{50} value indicates the sludge as slightly toxic (Table 6). The values obtained as biocentration factor (BCF) for TPH concentration in the earthworm ranged from 1.22 to 5.17,

Table 4. Chronic Toxicity Profile of Nitrogen Transforming Bacteria Exposed to Petroleum Refinery Oily Sludge for 28 Days.

Days limit	EC_{50} (mg/kg)	Confidence	Probit equation	Slope
7	Could not be determined growth growth inhibition <50%	Could not be determined inhibition <50%	Could not be determined growth inhibition <50%	Could not be determined growth inhibition <50%
14	Could not be determined growth growth inhibition <50%	Could not be determined inhibition <50%	Could not be determined growth inhibition <50%	Could not be determined growthinhibition <50%
21	24627	19317-33858	-0.110 + 1.164X	7.82
28	13761	11796- 17418	-0.881+1.415x	5.005

Table 5. Acute toxicity profile of fresh and brackish water shrimp at day 10 exposure to Petroleum refinery oily sludgein comparison with GESAMP toxicity rating (1997).

Test shrimp	Duration	GESAMP (1997) Toxicity Rating for Damage to Living Resources			
		LC_{50} (mg/kg)	Rating (Days)	Toxicity status	96h LC_5
Fresh water shrimp	10	1097.375 ± 5 0.62	Very highly toxic	5	<0.10
Brackish water shrimp	10	1590.376 ± 0.92	Highly toxic	4	0.10-1.0
			Moderately toxic	3	1.0-10
			Slightly toxic	2	10-100
			Practically non-toxic	1	100-1000
			Non-hazardous	0	>1000

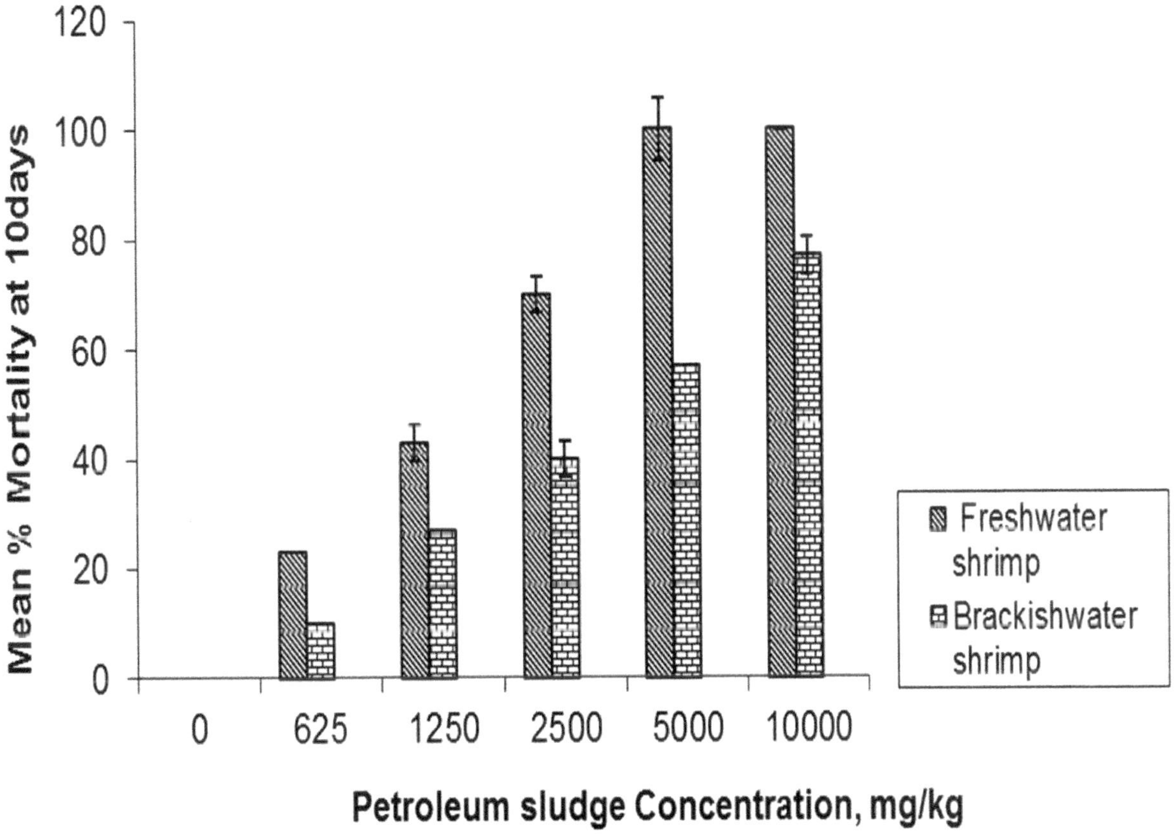

Figure 1. Mean percentage (%) mortality of shrimp exposed to petroleum refinery oily sludge after 10 days.

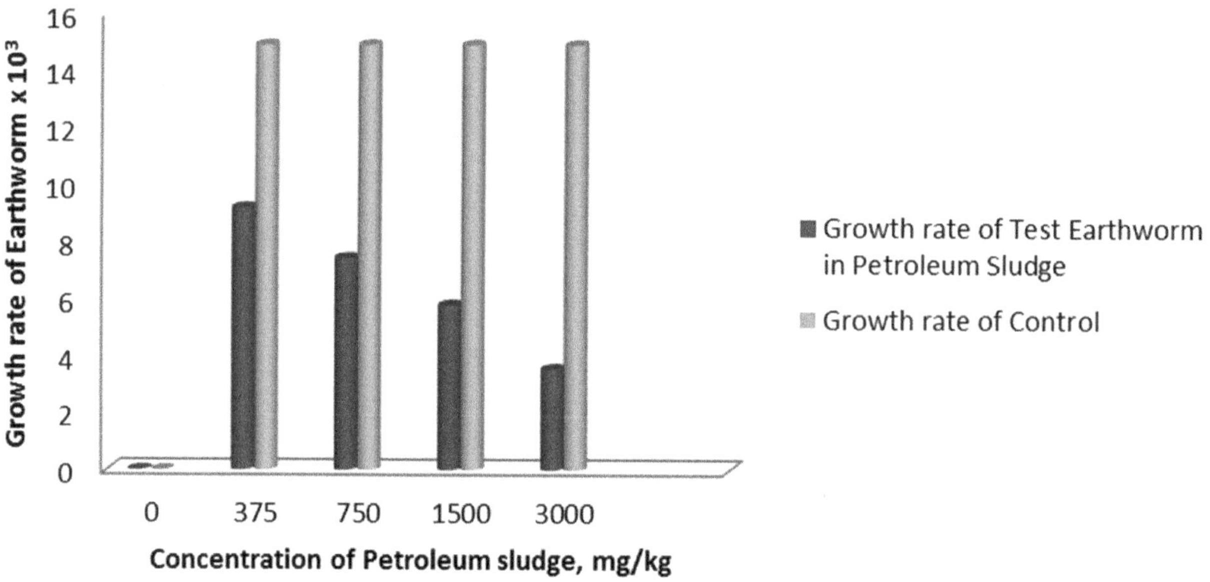

Figure 2. Growth rate of earthworm in petroleum refinery oily sludge.

Table 6. Chronic toxicity profile of Petroleum refinery oily sludge on the growth rate *of Apporectoda longa.*

Test sample	Time (days)	EC_{50}, mg/kg	Earthworm' Toxicity		
			Rating	Designation	EC50 (mg/kg)
Petroleum Sludge	7	Cannot be determined not up to 50% deaths	1	Super toxic	< 1.0
Petroleum Sludge	14	Cannot be determined not up to 50% deaths	2	Extremely toxic	1.0-10
Petroleum Sludge	21	1655	3	Very toxic	10-100
Petroleum Sludge	28	825.04	4	Slightly toxic	100-1000

Table 7. Bioconcentration factor of TPH in Earthworm exposed to Petroleum refinery oily sludge for 28 days

TPH concentration (mg/kg)	TPH concentration in Earthworm	BCF
375	1938	5.168
750	2052	2.738
1500	2660	1.773
3000	3651	1.2173

from the lowest to highest sludge concentration (Table 7). This indicates that the sludge would be bioaccumulated into the tissues of terrestrial organism as sludge concentration increases.

DISCUSSION

The discharge of untreated petroleum refinery oily sludge into the environment has been shown to have acute and chronic effects on the biotic and abiotic components of the aquatic and terrestrial environments (Wang and Reed, 1984; Ma and Ortolano, 2002; Tang et al., 2012). The acute effect of petroleum refinery oily sludge pollution on *Nitrobacter* sp. in the aquatic environment was conducted since the nitrification process is a function of enzyme activity and its measurement has been used as an indicator of pollution (Williamson and Johnson, 1981; Wang and Reed, 1984). The decline in the as an indicator of pollution (Williamson and Johnson, 1981; Wang and Reed, 1984). The decline in the *Nitrobacter* sp. counts as the concentration of petroleum refinery oily sludge increased could be due to the toxic effect of sludge resulting from sludge concentration as earlier reported by Okpokwasili and Odokuma (1997).

The genus *Nitrobacter* belongs to a variety of nitrate-oxidizing bacteria which are responsible for the second step of the nitrification process (oxidation of nitrite to nitrate). This bacterium was used for the chronic toxicity test of petroleum refinery oily sludge on nitrogen transformation activities in soil. This second step of nitrification is particularly sensitive. Inhibition of this step under uncontrolled conditions may lead to accumulation of nitrite nitrogen which is toxic (Dokaniakis et al., 2005).

As stipulated in the test guideline, OECD TG 216 (2000), since the difference between the lowest and highest percentage inhibition is greater than 25%, the sludge has the potential to inhibit nitrogen transformation. The observed increase in inhibition of transformed nitrogen as the concentration of petroleum refinery oily sludge increased could be due to the increase of some physicochemical properties of sludge such as total petroleum hydrocarbon (TPH) and metals (Wilde et al., 1983; Okpokwasili and Odukuma, 1997). Wang and Reed (1984) noted that a high concentration of metal cations inhibits microbial activities by causing damage or inactivating one or more critical enzymes resulting in formation of an inactive complex between the metal cations and an active enzyme. TPH contains toxic compounds such as PAHs and these have also been implicated in the inhibition of nitrification process (Suschka et al., 1996; Dokaniakis et al., 2005).

The obtained LC_{50} values for the acute toxicity test of the test sample on the shrimp indicated that the freshwater test organisms were more adversely affected by the sludge than the brackish water shrimps. Buikema et al. (1982) observed that the higher the LC_{50} value, the lower the toxicity or sensitivity of the test organism and vice versa. The difference in the response of the fresh and brackish water shrimps may be attributed to the osmoregulatory demand of the different environments.

The reduction of growth at higher concentration for the chronic toxicity test of petroleum refinery oily sludge on earthworms showed it reduced growth progressively as the sludge concentration increased and could eventually lead to death. The mechanism of toxicity of hydrocarbon to earthworms was observed to be based on the ability of hydrocarbons to bind to the Polar Regions in biogeneous membrane and to disorganize them (Krab et al., 2000). MacGeer et al. (2003) recorded an inverse relationship between BCF and exposure concentration of the test chemical on earthworms and attributed this to the lipo-philic nature of the sludge.

Conclusion

In conclusion, the findings from this research indicate that petroleum refinery oily sludge if not properly treated before disposal into the recipient environment could pose serious threat to the physiological and reproductive functions as well as survival of aquatic and terrestrial organisms. Also, xenobiotic compounds such as PAHs that are bioaccumulated in tissues of aquatic or terrestrial organisms could move up the food chain in higher organisms such as man and lead to conditions such as cancer, infertility among others.

Conflict of interests

The authors did not declare any conflict of interest.

ACKNOWLEDGEMENTS

The authors acknowledge Thermosteel Nigeria Ltd., Warri, Delta State Nigeria for making their laboratory available to carry out a major part of this research. Our gratitude also goes to the supervisor Professor E. I. Atuayan of Microbiology Department, University of Benin; Professor L. I. N. Ezemonye and Professor J. O. Olomukoro both of Animal and Environmental Biology Department of the University of Benin and Dr. D. F. Ogeleka of Chemistry Department, Federal University of Petroleum Resources, Effurun for their technical advice.

REFERENCES

American Public Health Association (APHA) (1998). Standard, Methods for the Examination of Water and Waste Water 20th Edition Published by American Public Health Association, Washington DC. 1220p.

American Society for Testing and Materials (ASTM). Standard E 2172-01 (2001). Standard guide for conducting laboratory soil toxicity with nematode *Caenorhabditiselegans*. Annual Book of ASTM Standard 11: 1606-1616.

Atlas RM, Bartha R (1992). Hydrocarbon biodegradation and oil spill bioremediation; in KC Marshall (Ed). Advances in Microbial Ecology. New York. Plenum Pres. Vol. 12, pp. 287-338.

Buikema AL, Nederlehner BR, Cairn J (1982). Biological Monitoring Part IV – Toxicity Testing. Water Resources 16:239-262.

Colwell RR, Zambruski MS (1972). Methods of Aquatic Microbiology University Park Press, Baltimore. 461p.

Davies NA, Hodson ME, Black S (2003). The influence of time on lead toxicity and bioaccumulation determined by the OECD earthworm toxicity test. Environ. Pollut. 121(1):55-61.

Department of Petroleum Resources (DPR) (2002). Environmental Guidelines and Standards for the Petroleum Industry in Nigeria (EGASPIN) 314p.

Dokaniakis SN, Kornaros M, Lyberatos C (2005). On the effect of xenobiotic bacterial nitrite oxidation. Proceedings of the 9th International Conference on Environmental Sciences and Technology Rhodes Island, Greece.

Doube BM, Schmidt O (1997). Can the abundance or activity of soil macrofauna be used to indicate the biological health of soils? In: Biological Indicators of Soil Health. Eds. C. Pankhurst, B Doube and V. Gupta. CAB International New York. pp. 265-295.

Duffus JH (1980). Environmental Toxicology (Resource and Environmental Science Series). Arnold Publishers Ltd. 26p.

Edwards WM, Shipitalo MJ (1998). Consequences of earthworms in agricultural soils; Aggregation and porosity. In Earthworm Ecology (Eds C.A. Edwards) St. Lucie Press Boca Raton, Florida. pp. 147-161.

Englert CJ, Kenzie EJ, Dragun J (1993) . Bioremediation of petroleum products in soil, Principles & Practice of Petroleum Contamination, in soils. API, 40:111-129.

Environment Canada (1992). Biological Test Method; Acute test for sediment toxicity using fresh and brackish or estuarine amplipods. Conservation and protection, Ottawa, Ontario Report EPSI /RM/26.

Finney DJ (1978). Probit Analysis, Cambridge University Press, London. 333p.

Grunditz C, Dalhammar G (2001). Development of nitrification inhibition assays using pure cultures of *Nitrosomonas and Nitrobacter.* Water Res. 35:433-440.

Hazardous Waste Management System (1997). Environmental Protection Agency 62:29-32.

Krab J, Mathes K, Breckling B, Lonker O, Schulz – Berend V (2000). Scale – up biologischer Reinigungsvertahren – projeckt – Abachlubericht, bmb .F; Bremen.

Ma X, Ortolano L (2002). Environmental regulation in China, Rowman and little field. Publishers, Inc . 209p.

MacGeer JC, Brix KV, Skeaff JM, Deforest DK, Brigham SI (2003). Inverse relationship between bioconcentration factor and exposure concentration for metals, implication for hazard assessment of metals in the aquatic environment. Environ. Toxicol. Chem. 22:1017-1037.

Mishra S, Jyot J, Kuhad RC, Lal B (2001). Evaluation of inoculums addition to stimulate in – situ bioremediation of oily – sludge contaminated soil. Appl. Environ. Microbiol. 67:1675-1681.

Okpokwasili GC, Odokuma LO (1997). Response of Nitrobacter to oilfield dispersants and domestic detergents. Trop. Freshw. Biol. 6:65-74.

Oolman T, Baker RR, Rentro NL, Marshall GE (1996). Refinery Uses Bioslurry Process to Treat RCRA Wasters. Hydrocarb. Process. 71(76):18-20.

Organization for economic co-operation and Development (OECD) (2002). Guidelines for testing of chemical TG 216. C. 21 Soil microorganisms; Test Nitrogen Transform.

Organization for economic cooperation and Development (OECI) (2004) Guidelines for the Testing of Chemicals No 218; Sediment – water chironomid toxicity test using spiked sediment (Adopted April 2004).

Prospt TL, Lochmiller RI, Quals CW, Mcbee K (1999). *In-situ* (mesocosm). Assessment of immunotoxicity risks to small mammals inhabiting petrochemical waste sites. Chemosphere 38:1049-1062.

Spiegel H (2002). Trace element accumulation is selected bioindicators exposed to emission along the industrial facilities of Danube, lowland. Turk. J. Chem. 26:815-823.

Suschka J, Mrowies B, Kuszmider G (1996). Respose of Nitrobacter toxicity of oil field dispersants and domestic detergents. Trop. Freshw. Biol. 6:65-74.

Tang J, Lu X, Sun Q, Zhu W (2012). Aging effect of Petroleum Hydrocarbon in Soil under different Attenuation Conditions. Agric. Ecosyst. Environ. 149:109-117.

Terhivuo J, Pankakoski E, Hyvarinen H, Koivisto K (1994). PB uptake by ecologically dissimilar earthworms (Lumbricidae) species near a lead smelter in Southern Finland. Environ. Pollut. 85:87-96.

United State Environmental Protection Agency (USEPA) (2004). United States Environmental Protection Agency, understanding oil spills and oil spill response. htt:// www.fundyforum.com/profile archives/ profile 3.html.

Ururahy AFP, Marins MDM, Vital RL, Therezinha I, Pereira N (1998). Effect of Aeration on Biodegradation of Petroleum Wastes. Rev. Microbiol. 29:4.

Wang W (1984) Time response of Nitrobacter to toxicity. Environ. Int. 10:21-24.

Wang W, Reed P (1984). "Nitrobacter bioassay for aquatic toxicity." Toxicity screening procedures using bacterial systems. 309-325.

Wilde EW, Soracce RJ, Mayack LA, Shealy RL, Broadwell TL, Steffen RF (1983). Comparison of chlorine and chlorine dioxide toxicity to fathead Minnows and Bluegill. Water Res. 17(10):1327-1331.

Williamson KJ, Johnson OG (1981). A bacterial bioassay for assessment of wastewater toxicity. Water Res. 15:383-390.

Trace metals in intertidal sediment of mangrove-sheltered creeks in Niger Delta, Nigeria: Variability before and after crude oil spillage

Zabbey, Nenibarini[1] and Babatunde, B. Benard[2]

[1]Department of Fisheries, Faculty of Agriculture, University of Port Harcourt, PMB 5323, East West Road, Choba, Rivers State, Nigeria.
[2]Department of Animal and Environmental Biology, College of Natural and Applied Sciences, Faculty of Biological Sciences, University of Port Harcourt, Choba, Rivers State, Nigeria.

Trace metals (Zn, Cu and Pb) fluxes were studied in five intertidal flats at Bodo Creek, Eastern Niger Delta, Nigeria in 2006, and re-evaluated in 2010 following two major oil spills that occurred in the creek. This study is the first to look at trace metal loads in the interstitial sediments of Bodo creek. Standard methods were employed in the sampling campaign and analysis. The concentration of all metals under study were higher than stipulated limit by WHO (2006) during post spill analysis. Concentration of Zn did not vary between stations and months pre-spill but did in post-spill, having values slightly higher than WHO (2006) stipulated limit. During the pre-spill sediment analysis, Pb concentrations ranged from no detection to 0.0012 mg/kg in 2006, increasing to 0.169 mg/kg post spill in 2010. Cu content ranged from 0.1312 - 0.1858 and 0.3374 - 0.4504 mg/kg pre- and post-spill, respectively. In addition to oil-induced metal loadings, the metals concentration dynamics show strong correlation with season. Metals had higher concentrations during the dry season due to reduced water volume in the creek and consequently less dilution. Implication for uptake, bioaccumulation and biotransformation by food species in the creek catchment above biosafety limits for user population is stressed and measures to avert public health consequences are advocated.

Key words: Trace metals, inter-tidal sediment, Niger Delta, oil pollution

INTRODUCTION

The Niger Delta is the hub of oil and gas production in Nigeria. It is also Africa's largest wetland and the continent's mangrove dominant region. Oil pollution substantially degrades the delta network of alluvial swamps and lands, creeks and rivers. Apparently, there is no consensus on the number of oil spills and volume of oil spilling into the delta environment, as the operating companies and government of Nigeria keep conflicting data (Amnesty International, 2009). However, a recent report by UNEP (2011) put the annual average number of oil spills and volume spilled into the delta environment at 273 and 115,000 barrels, respectively, making the delta one of the most oil spill vulnerable areas in the world (Steiner, 2008).

The exploration and exploitation of oil interfere with ecological and biodiversity integrity of ecosystems arising from flaring of associated gases, oil spills, use of drilling chemicals, etc. These processes can release trace metals into coastal waters. Pollution studies have revealed elevated levels of zinc (Zn), copper (Cu), lead (Pb), chromium (Cr), nikel (Ni), cadmium (Cd) and vanadium (V) in Niger Delta sediments and some species of fauna, suggesting inputs from petroleum exploration and exploitation (Kakulu and Osibanjo, 1992; Horsfall and Spiff, 2002; Howard et al., 2006). The direct impact of petroleum exploitation on the environment was illustrated in a study where trace metal levels of the Warri River were assessed before and after dredging a canal leading to an oil well (Ohimain et al., 2008). They noted that percentage increase was several folds for lead, zinc, copper, iron, chromium, manganese and cadmium. Trace metals are associated with crude oil in variable concentrations depending on the geologic background of the oil (Onojake and Okonkwo, 2011). For Nigeria's 'sweet' and dominant crude petroleum, Bonny Light, the associated metals occur in the order Ni>V>Cd>Cu>Pb (Osuji and Adesiyan, 2005). The concentration profile of trace metals (Zn, Cu and Pb) in soft-bottom sediments was assessed by the lead author as part of a 2-year (April 2006 to May 2008) baseline ecological survey of the 'Dor Nwezor' section of Bodo Creek. Information on other physico-chemistry, macrozoobenthos (tree, epifauna and infauna) of the study area (Dor Nwezor section of the creek) have been documented (Zabbey et al., 2010; Zabbey, 2012; Zabbey and Hart, 2011; Zabbey and Malaquias 2013). Shortly after the above study, 2 major crude oil spills hit Bodo creek from August 2008 to February 2009 from a Trans-Niger pipeline traversing the creek to Bonny export terminal (CEHRD, 2008; Amnesty International and CEHRD, 2011). Information on the spills and livelihood implication for the locals are reported by Pegg and Zabbey (2013). In this study, we compared the pre-spill and post spill concentrations of Zn, Cu and Pb in the intertidal flats to highlight changes in metal loads in time and space.

MATERIALS AND METHODS

Study area

Occupying approximately 9,230 ha, Bodo Creek is a network of brackish water creeks flanking Bodo community on the upper reaches of the Andoni-Bonny estuarine system in Rivers State, Nigeria (Pegg and Zabbey, 2013). Four major channels conduct saline waters in and out of Bodo Creek: Dor Nwezor, Kpador, Koola Tobsoi and Koola Seato. These major waterways are interconnected by myriad of feeder channels, some of which terminate blindly in mangrove swamps (Onwugbuta-Enyi et al., 2008). This study was conducted at a protected mangrove swamp (Sivibilagbara) and four open and unvegetated low inter-tidal flats along the Dor Nwezor channel of Bodo Creek; approximately between latitude 04° 36' 29.7''N to 04° 35' 26.3''N and longitude 7° 15' 30.2''E to 7° 16' 50.9''E (Figure 1). Detailed information on some physico-chemical parameters in the creek had been documented (Onwugbuta-Enyi et al., 2008; Zabbey, 2012; Zabbey and Malaquias, 2013). Salinity of the creek interstitial water fluctuates between mesohaline (5 psu) and polyhaline (28 psu), while surface water values range from 6.2 - 22.7 psu. Surface and interstitial water temperature varied from 26.7 to 30.1 and 25 - 34°C, respectively.

Sampling locations

Station 1

This was the most upstream station and was located in the Sivibilagbara protected mangrove swamp (approximately latitude 4° 36' 29.7''N and longitude 7° 15' 30.2''E (Figure 1). The vegetation of Sivibilagbara was homogenously red mangrove (*Rhizophora racemosa*) with knitted structural architecture of prop roots and thick intertwined crowns. The swamp dimension is approximately 105 x 42 m (4,410 m^2). The sediment type was peaty clay (dominated by silt and clay) (Zabbey and Hart, 2014).

Station 2

This was located approximately 1,280 m downstream from station 1 (Figure 1) on an open, unvegetated tidal flat locally called Si Eeva. Facing downstream, the station lay to the left of Dor Nwezor main channel (latitude 4° 36' 12.7'' N and longitude 7° 16' 08.1" E). The riparian vegetation of this station was mainly stunted red mangrove, and dwarf, aged and unproductive coconut trees at the edges of the supralittoral shores. The substratum was sandy mud.

Station 3

This was located approximately 956 m downstream from Station 2, on a right-flanked tidal platform. The station was sited opposite a sprawling fishing settlement called Kozo (lat. 4° 35' 55.3'' N and longitude 7° 16' 33.8" E). The marginal vegetation was dominated by red mangrove (*R. racemosa*), with few stands of the white mangrove (*Avicennia germinas*) and mangrove sedge (*Paspalum vaginatum*) at the high intertidal zone. The date palm (*Phoenix reclinata*) and mango are amongst the mosaic admixture of plants at the supralittorial zone. The bottom was muddy sand.

Station 4

This was located 994 m downstream from station 3, having expansive unvegetated intertidal flat (latitude 4° 35' 32.4" N and longitude 7°16' 56.6" E). The sediment was sandy mud. Predominantly black mangrove and few stands of red mangrove and Nypa palm characterise the marginal vegetation.

Station 5

It is located on an expansive tidal mudflat (sandy mud) parallel to station 4. The marginal vegetation is dominantly black mangrove. The distance between stations 5 and 4 is approximately 256 and they are separated by the main creek channel. Station 5 was geo-located at latitude 4° 35' 26.3" N and longitude 7° 16' 50.9" E).

Field and laboratory procedures

Sediment samples were collected every month at low tide from 5

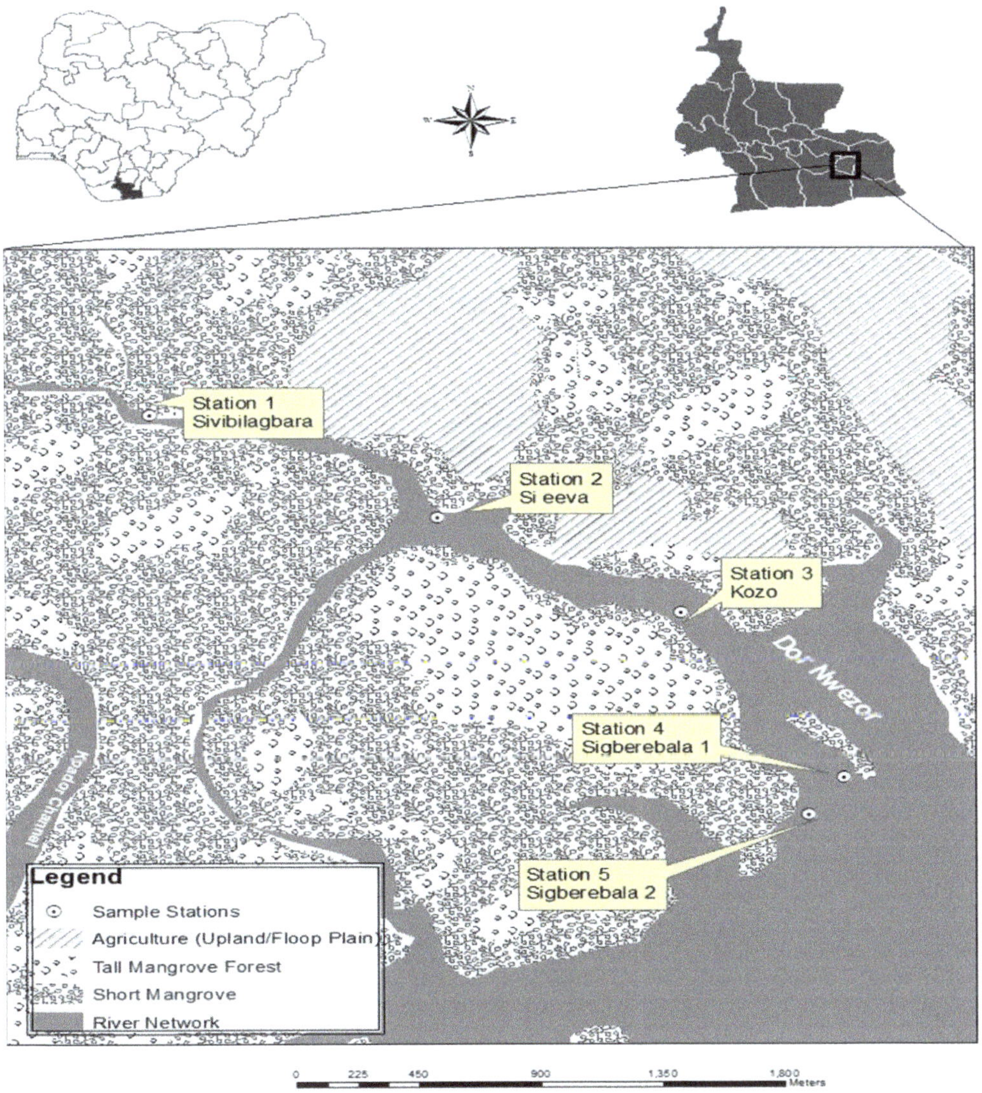

Figure 1. Map of Bodo creek in Rivers State, Nigeria.

intertidal flat locations along Dor Nwezor channels at Bodo Creek for 2 years (January – December 2006 and 2010), representing baseline and post spill respectively (Figure 1). Three replicate samples of the top 2 cm of sediment were collected with hand trowel and homogenised to form a composite sample per site during the sampling campaign. The sampling trowel was washed with the site water between replicates and with alcohol between stations (Chapman et al., 1996). The pooled sample of each location was covered in clean aluminium foil and transported in an ice-chest. In the laboratory, the sediment samples were analysed for Zn, Cu and Pb according to (APHA, 1995). Five grams of finely ground soil samples were weighed into 250 ml pyrex beakers and 10 ml of hydrogen peroxide (H_2O_2) added to it and allowed to warm for about 1 h during which most organic matter was destroyed. The content was transferred to savillex digestion bombs and concentrated hydrochloric (HCl) acid and nitric acid (HNO_3) were added in the ratio 5:10. The mixture was heated in a steam bath to a thick yellow liquid and the procedure repeated two times to get a clear solution indicating complete digestion of samples. Digested samples were cooled at room temperature, and filtered through a 0.45 µm membrane filter and diluted to 50 ml in volumetric flasks with double distilled water (Jin et al., 1999; Sastre et al., 2002). Following acid digestion, all samples were analyzed for the required metals (Zn, Pb and Cu) by flame atomic absorption spectrophotometer (Perkin Elmer Analyst AA 200 equipped with a high sensitivity nebulizer) calibrated by successive dilution of a 1000 mg/l multi - element instrument calibration standard solution.

All acid used were of analytical grade quality and control was assured by the use of procedural blanks and spikes. The spike recovery for each element was greater than 94%. All samples were run in triplicates and the relative standard deviation for the triplicate analysis was less than 10%. The instrument was calibrated with standard solutions prepared from Merck. The analytical blanks were run in the same way as the samples and concentrations were determined using standard solutions prepared in the same matrix. Doubled distilled water was used as solvent throughout the study. All glassware and other containers were thoroughly cleaned with 10% (w/v) nitric acid solution and finally rinsed with double distilled water several times and air dried prior to use. One-way analysis of variance (ANOVA) and Duncan's test (p=0.05) were used in order to assess whether trace metal concentrations varied significantly between sites and months.

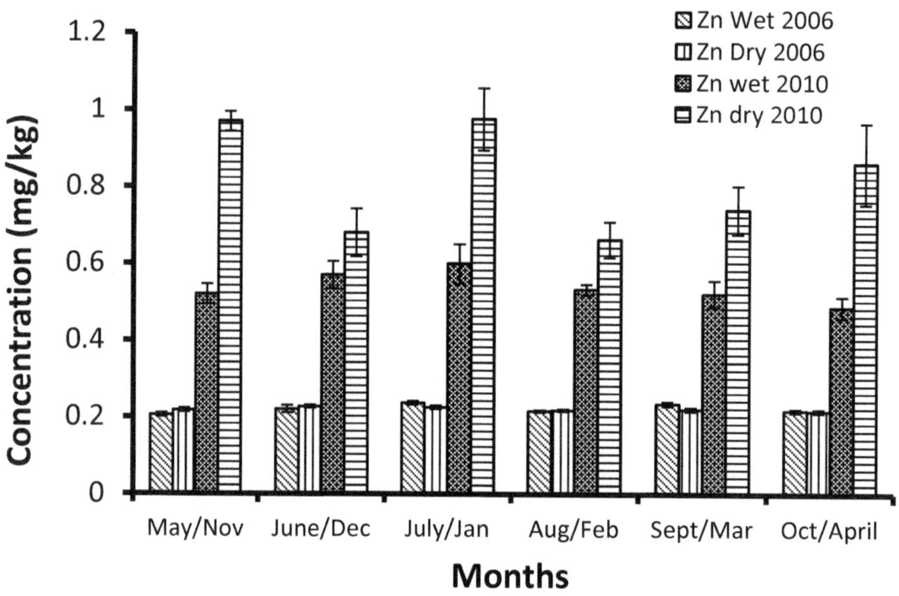

Figure 2. Mean monthly Zn concentrations in sediments of Bodo creek pre-spill and post spill (n = 5 for SE).

RESULTS AND DISCUSSION

Sediment characteristics at Station 1 had particles in the sand fraction (very coarse to fine sand) constituted 13% of the sediment, while clay represented 82.7%. The silt and clay proportions at Stations 2-5 ranged from 14.3 to 49.2%. Gravelly particles constituted 4 and 0.7% of sediments at Stations 1 and 5, respectively, whereas this class fraction was apparently lacking at Stations 2–4. The sand fraction ranged from 13.3% at Station 1 to 85.7% at Station 3. Particle size variation may influence trace metal loads at different stations studied. Silt and clay may retain and lock away metals preventing re-suspension while sand particles may readily release metals into solution during re-suspension. Generally, the levels of metal load in the present study may not be conclusive as low levels may simply mean more metal is locked away in silt and clay and not available during this survey. However, such contaminants may be re-suspended during dredging activities which are frequent at the study area.

The highest and lowest concentration values of Zn were recorded in the months of July and May (0.237 and 0.205 mg/kg, respectively) (Figure 2). The 2006 zinc monthly concentrations were not statistically different ($P = 0.105$; $P > 0.05$). There was also no statistical variation in the concentration of Zn at all stations between the weighted averages of the wet and dry seasons in 2006 ($P = 0.113$) (Figure 5). Zinc concentrations were higher in 2010 than 2006 at all stations and in all the months at statistically significant levels ($P = 0.0244$) (Figure 5). This variation may have been due to the oil spill incidents in Bodo creek in 2008 and 2009 (Pegg and Zabbey, 2013). Concentration of Zn at all stations in both years was also higher than stipulated limit (0.01 mg/kg) by WHO (2006). In 2010, Zn concentration values were highest and lowest in the months of January and October, respectively. Significant variation in contents were recorded between the wet and dry seasons in 2010 ($P = 0.0227$), with dry season values being higher at all stations (Figure 5). Seasonal variation may be attributed to less dilution of Zn in dry season leading to higher concentration and spatial variation may be due to varying impact from incidental inwelling sources at various stations. Lead concentrations did not vary significantly across the wet season months in 2006 ($P = 0.222$) ranging from below detection limit in August, September and October to 0.0012 mg/kg in May (Figure 3). During the wet season months in 2006, Pb was only detected at stations 2 and 3 in May, station 5 in June and stations 2, 3 and 4 in July. During the dry season months (November – April, 2006), Pd was detected only in November, March and April at all stations with the highest and lowest levels in March and April (0.0016 and 0.001 mg/kg), respectively. The concentration varied only numerically across the three months ($P = 0.104$; $P > 0.05$). Lead concentrations also did not vary significantly between the wet and dry seasons at all stations ($P = 1.00$). This result is similar to that reported for Pb (Onojake and Okonkwo, 2011) in surface and sub-surface soils in Bodo near the creek and lower than the limit of 0.01 mg/kg stipulated by WHO (2006). In 2010 however, Pb was detected at all stations and in all the months. Its concentration ranged between 0.074 mg/kg in October to 0.1206 mg/kg in July (wet season) and 0.066 to 0.169 mg/kg in March (dry season). The highest Pb concentration was recorded at station 2 in

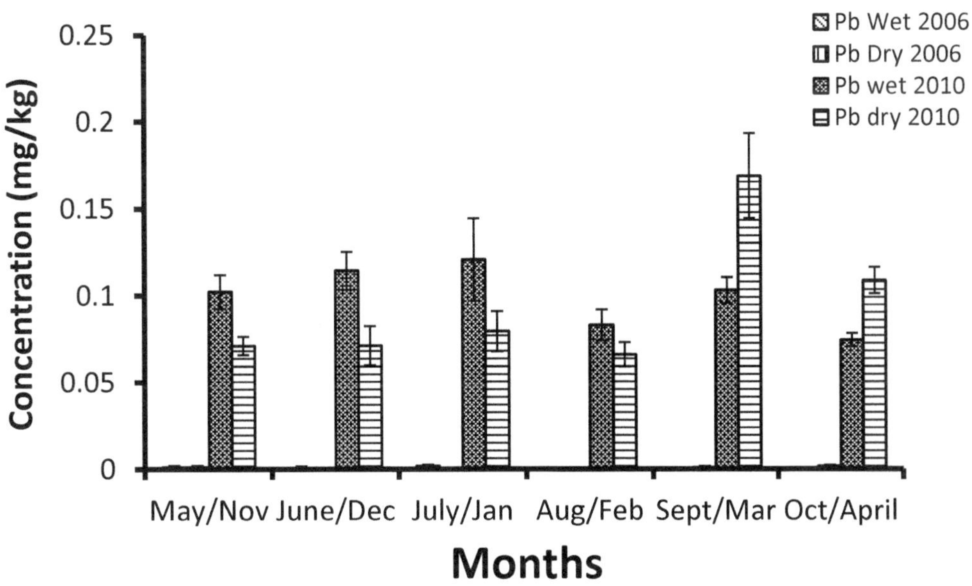

Figure 3. Mean monthly Pb concentrations in sediments of Bodo creek pre-spill and post spill sampling eras (n = 5 for SE)

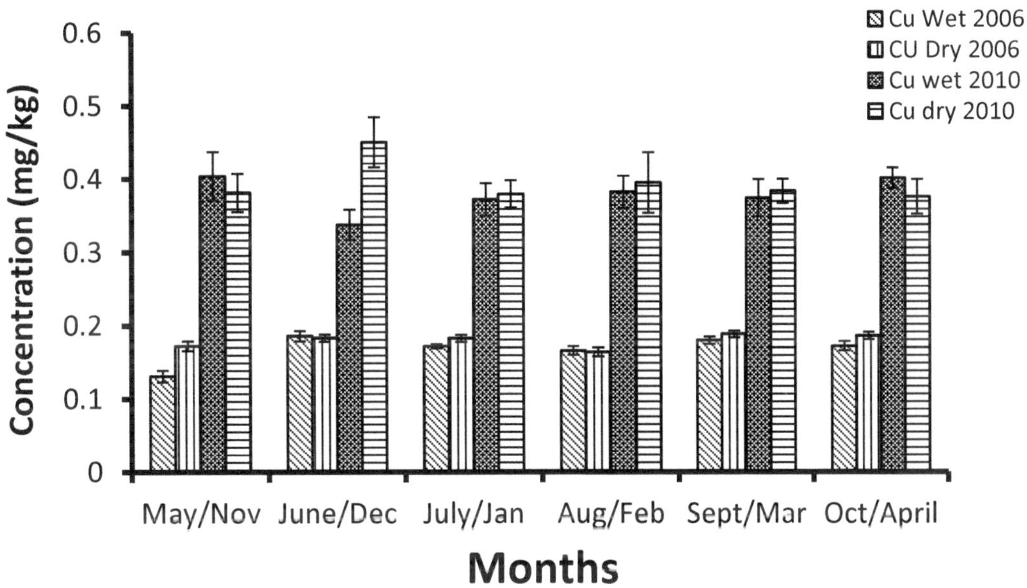

Figure 4. Mean monthly Cu concentrations in sediments of Bodo creek pre-spill and post spill sampling eras (n = 5 for SE).

wet season and station 1 in dry season in 2010 (Figure 6). Lead values varied significantly between wet and dry seasons in 2010 (P = 0.003). Lead values in 2010 were higher than those recorded in 2006 and higher than those reported by Onojake and Okonkwo (2011) and stipulated limit of 0.01 mg/kg by WHO (2006). This variation may have been due to the oil spill incidents in 2008 and 2009 (Pegg and Zabbey, 2013).

The concentration of Cu in all the stations investigated in 2006 ranged between 0.1312 and 0.1858 mg/kg with no statistical variation between the months (Figure 4). The highest and lowest values were recorded in March and May, respectively. There was no statistical variation in the concentration of Cu between the wet and dry seasons in 2006 (P =0.052). However, dry season recorded the highest value of Cu at station 2 while the

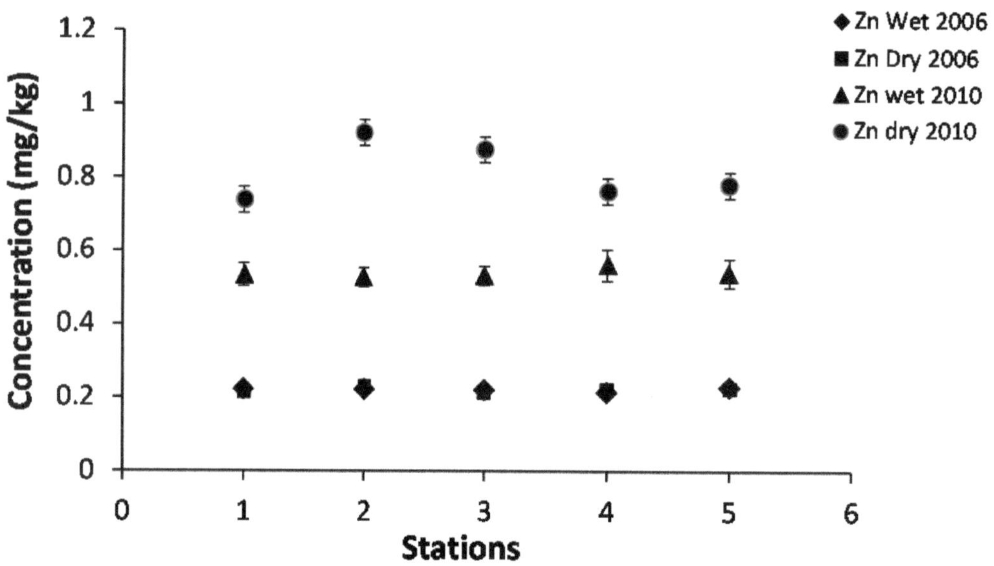

Figure 5. Mean Zn concentrations for wet and dry seasons in sediments of Bodo creek pre-spill and post-spill sampling eras (n = 5 for SE).

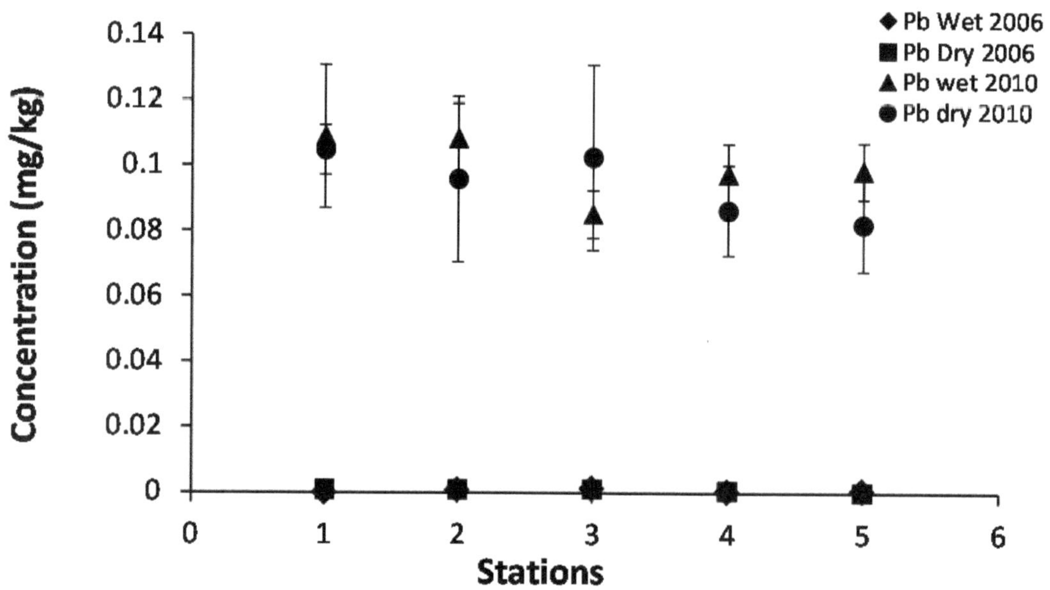

Figure 6. Mean Pb concentrations for wet and dry seasons in sediments of Bodo creek pre-spill and post-spill sampling eras (n = 5 for SE).

lowest value was recorded in wet season at station 3. In 2010, however, Cu concentrations ranged from 0.3374 mg/kg in June to 0.4504 mg/kg in December. Its concentration varied statistically between the wet and dry seasons (P = 0.001). The highest and lowest values were recorded at stations 5 and 1 in dry season and stations 3 and 2 in wet season, respectively; overall, dry season had higher values. Elevated concentration of metals in the dry season may be attributable to evaporation which leaves behind less water and thus more concentrated solution-dilution in the wet season was as a result of increased volume of exogenous water. Cu values in 2010 were higher than those in 2006 at all stations in all the months (Figure 7), increment was possibly due to the oil spill contamination. However, Cu values were below stipulated limit of 2 mg/kg by WHO (2006).

Seasonal variations in the concentrations of the metals investigated may be due to the effect of dilution during

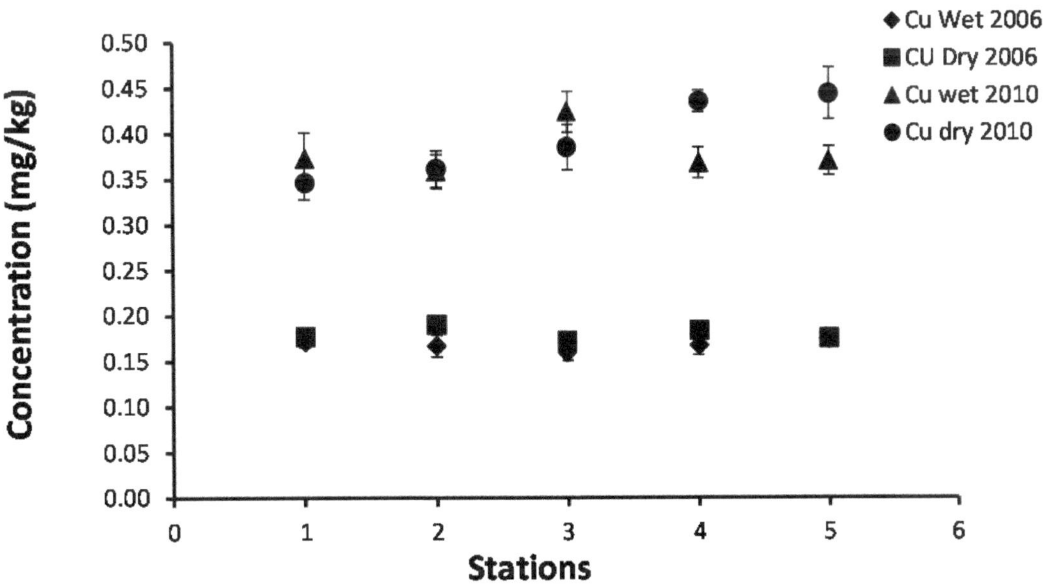

Figure 7. Mean of Cu concentrations for wet and dry seasons in sediments of Bodo creek pre-spill and post-spill sampling eras (n = 5 for SE).

the rains and evaporation during the dry season and this agrees with the report of Chindah et al. (2004). Spatial variation in trace metal load was not statistically significant (P>0.05), suggesting that common and anthropogenic sources contributed to the metal loading especially from the oil spills in Bodo creek in 2008 and 2009 (Pegg and Zabbey, 2013). Results of trace metals under study compares well with other reports on sediments in creeks and rivers in the Niger Delta that have been impacted by oil (Horsefall and Spiff, 2002; Obire et al., 2003; Chindah et al., 2004; Hart et al., 2005; Davies et al., 2006; Ubalua et al., 2007; Gideon and Chidiebere, 2008). However, some of the studies reported higher trace metal loads in similar oil contaminated environment in the Niger Delta, even much higher in organisms analysed for the same environment suggesting bioaccumulation of these metals (Chindah et al., 2004). In a similarly polluted river system in the Niger Delta, Horsefall and Spiff (2002) reported trace metal load in sediment and biota of the New Calabar river to be higher than those in the present study.

Conclusions

Results of trace metals compared between 2006 (pre-spill) and 2010 (post-spill) showed clearly that metal loads increased significantly after the oil spills in Bodo creek in 2008 and 2009. The present result shows that there are possibilities of high human exposure to trace metals in study area given that inhabitants of the study area consume biota from the studied creek. Further studies on trace metals in biota, particularly food species, and assessment of bioaccumulation factors are imperative to determine human exposure to trace metal loads detected in the present study following major oil spills. Some aquatic organisms have been reported to have capacity to concentrate trace metals such as Fe, Cu, Zn up to 10^5 times the concentration present in the water, making potential human health hazards associated with the consumption of contaminated sea food 20 to 40 times higher than the ingestion of contaminated water (WHO, 2011). Health effects of these trace metals (Pb, Zn and Cu) in humans have been demonstrated in acute toxicity, neurotoxicity and nephrotoxicity (Katz and Salem, 1993; ATSDR, 2000; Stift et al., 2000). Pb in particular is a confirmed carcinogen (Martin and Griswold, 2009; WHO, 2011).

Conflict of interests

The authors did not declare any conflict of interest.

REFERENCES

American Public Health Association (APHA) (1995). Standard methods for the examination of water and wastewater 19th Ed.

Amnesty International & Centre for Environment, Human Rights and Development (CEHRD) (2011). The true tragedy: delays and failures in tackling oil spills in the Niger Delta. Amnesty International Secretariat, London.

Amnesty International (2009). Nigeria: Petroleum, Pollution and Poverty in the Niger Delta. Amnesty International Secretariat, London.

ATSDR (2000) Toxicological profile for chromium. Atlanta, Georgia, United States Department of Health and Human Services, Public Health Service, Agency for Toxic Substances and Disease Registry.

Centre for Environment, Human Rights and Development (CEHRD) (2008). Persistent oil spillage at Bodo Creek; unprecedented impacts

on ecosystem stability, biodiversity and food security of Ogoni communities. www.cehrd.org
Chapman PM, Paine MD, Arthur AD, Taylor LA (1996). A triad study of sediment quality associated with a major, relatively untreated marine sewage discharge. Mar. Pollut. Bull. 31:47-64. http://dx.doi.org/10.1016/0025-326X(95)00108-Y
Chindah AC, Braide AS, Sibeudu OC (2004) Distribution of hydrocarbons and tracemetals in sediment and a crustacean (Penaeus notialis) from the Bonny River/New Calabar River Estuary, Niger Delta. Afr. J. Environ. Assess. Manag. 9:1-17.
Davies OA, Allison ME, Uyi HS (2006). Bioaccumulation of tracemetals in water, sediment and periwinkle (Tympanotonus fuscatus var radula) from the Elechi Creek, Niger Delta. Afr. J. Biotechnol. 5:968-973.
Gideon OA,Chidiebere GE (2008). The current pollution status of the new Calabar River in the Niger Delta region of Southern Nigeria: A survey of antibiogram profiles of its bacterial isolates. Afr. J. Environ. Sci. Technol. 2:134-141.
Hart AD, Oboh CA, Barimalaa IS, Sokari TG (2005), Concentrations Of Trace Metals (Lead, Iron, Copper And Zinc) In Crops Harvested In Some Oil Prospecting Locations In Rivers State, Nigeria. Afr. J. Food Agric. Nutr. Dev. 5:1-21.
Horsfall M, Spiff AI (2002) Distribution and partitioning of trace metals in sediments of the lower reaches of the New Calabar River, Port Harcourt, Nigeria. Environ. Manage. Assess. 78:309-326. http://dx.doi.org/10.1023/A:1019991020048
Howard IC, Horsfall M, Spiff IA, Teme SC (2006). Trace metals levels in surface waters and sediments in an oilfield in the Niger Delta, Nigeria. Glob. J. Pure Appl. Sci. 12: 79-83.
Jin Q, Liang F, Zhang H, Zhao L, Huan Y, Song D (1999). Application of microwave techniques in analytical chemistry, Pract. Trends Anal. Chem. 18(7):479-484. http://dx.doi.org/10.1016/S0165-9936(99)00110-7
Kakulu SE, Osibanjo O (1992). Pollution studies of Nigerian rivers: trace metal levels of surface water in the Niger Delta area. Int. J. Environ. Stud. 41:287-293. http://dx.doi.org/10.1080/00207239208710768
Katz SA, Salem H (1993). The toxicology of chromium with respect to its chemical speciation: A review. J. Appl. Toxicol. 13:217-224. http://dx.doi.org/10.1002/jat.2550130314
Martin S, Griswold W (2009). Human Health Effects of Trace Metals. Center for Hazardous Substance Research (CHSR).
Obire O, Tamuno DC, Wemedo SA (2003). Physicochemical quality of Elechi Creek in Port Harcourt, Nigeria. J. Appl. Sci. Environ. Manage 7:43-49.
Ohimain EI, Gbolagade J, Abah SO (2008). Variations in TraceMetal Concentrations Following the Dredging of an Oil Well Access Canal in the Niger Delta. Adv. Biol. Res. 2:97-103.
Onojake MC, Okonkwo V (2011). Trace Metals Associated with Oil Spillage: A case study. J. Chem. Pharm. Res. 3(6):742-751.
Onwugbuta-Enyi J., Zabbey N, Erondu ES, (2008). Water quality of Bodo creek in the lower Niger Delta Basin. Adv. Environ. Biol. 2:132-136.
Osuji LC, Adesiyan SO (2005). Extractable hydrocarbons, nickel and vanadium contents of Ogbodo-Isiokpo oil spill polluted soils in the Niger Delta, Nigeria. Environ. Mon. Assess. 110: 129 – 130. http://dx.doi.org/10.1007/s10661-005-6283-0
Pegg S., Zabbey N (2013). Oil and water: the Bodo spills and the destruction of traditional livelihood structures in the Niger Delta. Commun. Dev. J. 48:391-405. http://dx.doi.org/10.1093/cdj/bst021
Sastre J, Sahuquillo A., Vida, M, Rauret G (2002). Determination of Cd, Cu, Pb and Zn in environmental samples: microwave-assisted total digestion versus aqua regia and nitric acid extraction. Anal. Chim. Acta 462:59-72. http://dx.doi.org/10.1016/S0003-2670(02)00307-0
Steiner R (2008) Double standards? International standards to prevent and control pipeline oil spills, compared with shell practices in Nigeria. A report submitted to Friends of the Earth, The Netherlands.
Stift A., Friedl J., Langle F, Berlakovich G, Steininger R, Muhlbacher F (2000). Successful treatment of a patient suffering from severe acute potassium dichromate poisoning with liver transplantation. Transplantation 69:2454-2455. http://dx.doi.org/10.1097/00007890-200006150-00044
Ubalua AO, Chijioke UC Ezeronye OU (2007). Determination and assessment of tracemetal content in fish and shellfish in Aba River, Abia state, Nigeria. KMITL Sci. Technol. J. 7(1):31-45.
UNEP (United Nations Environment Programme). 2011.Environmental assessment of Ogoniland. Nairobi: UNEP.
WHO (2011). Adverse health effects of traceMetals in children. Children's Health and the Environment WHO Training Package for the Health Sector World Health Organization www.who.int/ceh
WHO (2006). Guidelines for drinking-water quality. Incorporating first addendum. Vol. 1, Recommendations. 3rd ed.
Zabbey N, Hart AI (2011). Preliminary checklist of macrozoobenthos of Bodo Creek in the Niger Delta, Nigeria. Niger. J. Fish. 8 (2):271-283.
Zabbey N, AI Hart (2014). Spatial variability of macrozoobenthic diversity on tidal flats of the Niger Delta, Nigeria: the role of substratum. Afr. J. Aquat. Sci. 39(1):67-76. http://dx.doi.org/10.2989/16085914.2013.869657
Zabbey N, Hart AI, Wolff WJ (2010). Population structure, biomass and production of the West African lucinid Keletistes rhizoecus (Bivalvia, Mollusca) in Sivibilagbara swamp at Bodo Creek, Niger Delta, Nigeria. Hydrobiologia 654:193-203. http://dx.doi.org/10.1007/s10750-010-0381-x
Zabbey N, Malaquias MA (2013). Epifauna diversity and ecology on intertidal flats in the tropical Niger Delta, with remarks on the gastropod species Haminoea orbignyana. J. Mar. Biol. Assoc. UK. 93:249-257. http://dx.doi.org/10.1017/S0025315411001615
Zabbey N (2012). Spatial and temporal variability in interstitial water quality of soft-bottom flats at Bodo Creek, Eastern lower Niger Delta, Nigeria. Trop. Freshw. Biol. 21(1):83-103 83

The effects of heavy metals concentration on some commercial fish in Ogun River, Opeji, Ogun State, Nigeria

Adeosun, F. I.[1], Akinyemi, A. A.[1], Idowu, A. A.[1], Taiwo, I. O.[1], Omoike, A.[2] and Ayorinde, B. J. O.[3]

[1]Department of Aquaculture and Fisheries Management, Federal University of Agriculture, PMB 2240, Abeokuta, Ogun State, Nigeria.
[2]Department of Biological Sciences, Bells University of Technology, PMB 1015, Ota, Ogun State, Nigeria.
[3]Tai Solarin University of Education, Ijagun, Ijebu-Ode, Ogun State, Nigeria.

A study was conducted on commercially important fish species of heavy metals in water bodies. The primary objectives were to find out the amount of heavy metals concentration in the fish species, sediments and water samples at the deep and shallow part of the river, to determine the toxicity and relationship between the collected samples and the rate at which it pollutes the water. Atomic absorption spectrophotometer (AAS) evaluates their wholesomeness for human consumption. Four heavy metals (lead, cadmium, copper and zinc) were analyzed and only copper and zinc were found to be present in the digested samples. The concentration of copper in the fish (flesh) was 0.09 mg/l and that of zinc was 0.25 mg/l. In the bone samples, copper was found to have 0.17 mg/l and zinc 0.22 mg/l. In the water samples analyzed, the concentration of copper is 0.035 mg/l while that of zinc is 0.047 mg/l. In the sediment sample, the amount of copper is 310.0 mg/kg while that of zinc is 2050.0 mg/kg. This however signifies that the fish species in the Ogun River as well as the water is suitable for consumption which was found to be less than the World Health Organization (WHO) maximum recommended acceptable limits in the food.

Key words: Heavy metals, fish species, water samples, toxicity.

INTRODUCTION

Metals are solids materials that are opaque, lustrous elements that are good conductors of heat and electricity. A study by Skeat (2005) showed that the heavy metals are members of a loosely defined subset of elements that exhibit metallic properties which include solubility in water, non- degradability and strongly attachment of polypeptides and protein. Over seventy five percent (75%) of everything on the planet Earth are directly or indirectly associated with metals. The recent activities of nature and that of human beings on most water bodies have led to a mass deposition of these minerals called metals.

Heavy metals occur naturally in the ecosystem with large variations in concentrations. Most metals in the streams and rivers come from industrial, municipal and urban run-offs which can be harmful to life (Tolcin, 2011). Increased urbanization of industrialization could be the cause for an increased level of trace metals, especially the heavy metals in our waterways (Njar and Al-Doush, 2012). The presence of toxic materials in ecosystems is presently related with increased concentrations of heavy metals ions, which enter water sources with sewage waters.

In recent years, world consumption of fish has increased simultaneously with the growing concern of their nutritional and therapeutic benefits. In addition to its important source of protein, fish typically have rich contents of essential minerals, vitamins and unsaturated fatty acids (Medeiros et al., 2012). The American Heart Association recommended eating fish at least twice per week in order to reach the daily intake of omega-3 fatty acids (Kris-Etherton et al., 2002).

However, fish are relatively situated at the top of the aquatic food chain; therefore, they can accumulate heavy metals from food, water and sediments (Yilmaz et al., 2007; Zhao et al., 2012). The content of toxic heavy metals in fish can counteract their beneficial effects; several adverse effects of heavy metals to human health have been known for long time (Castro-Gonzalez et al., 2008). This may include serious threats like renal failure, liver damage, cardiovascular diseases and even death (Al-Busaidi et al. 2011; Rahman et al., 2012). Therefore, many international monitoring programs have been established in order to assess the quality of fish for human consumption and to monitor the health of the aquatic ecosystem (Meche et al., 2010).

Heavy metals have various levels of toxicity as the surrounding environment affects it through certain factors. Even when it comes to fish, the toxicity of these heavy metals (lead, mercury, copper and zinc) plays certain roles that may even be hazardous. Sometimes, these heavy metals combine with soil organic matter and clay (WHO, 2007). As a result of drinking water and eating fish contaminated with heavy metals, diseases occur due to the bioaccumulation of these heavy metals in the body system of human being, thus leading to serious health problems and eventual death. Also, a lot of these metals are carcinogens and cannot be destroyed by heat.

Studies have been done to detect the presence of the heavy metal pollutants in water bodies in Nigeria (Olaifa et al., 2004; Omoregie et al., 2002; Oguzie, 2000). Ogun River with its tributaries is the main river that traverses the length and breadth of Opeji and Abeokuta metropolis in Ogun State where it serves as the source of domestic, agricultural and water consumption. It is the major source of freshwater fish for the inhabitants of Opeji, and is highly exploited by artisanal fishermen. It is also the final drain discharge for all waste-water from domestic and agricultural source within and around Opeji and environs.

The problem facing any living organisms will be related to measurable chemical or biological parameters such as a flow rate turbulence, inter and intra- specific competition, feeding behavior, disease, parasitism, commensalism and symbiosis. This work is evidence to ascertain the metallic concentration in most fish species at the River Ogun (Opeji). Some of these elements are actually necessary for humans in minutes or lower amounts (cobalt, copper, chromium, manganese, nickel) while others are carcinogenic or very toxic, affecting among others, the central nervous system (mercury, lead, cadmium, copper). In view of the strategic and important role of Ogun River to Opeji people and environs as a source of freshwater fisheries for the inhabitants, hence this study determine the amount of heavy metals (lead, cadmium, copper and zinc) concentration in the body parts of some commercial fish species, the water environment and the sediments (soil) samples of the River Ogun (Opeji village) and to evaluate the toxicity and relationship between the heavy metals concentration in the fish, water and sediments (soil).

MATERIALS AND METHODS

Description of the study area

Lower Ogun River is located in Abeokuta North Local Government of Ogun state. It lies between Longitude 3° 28'E to 3° 40'E and Latitude 7° 14'N to 7° 20'N of Abeokuta (Ogun State Bureau of Lands and Survey).

The study was carried out at lower Ogun River, Akomoje in Abeokuta, Ogun State. The river is located in Abeokuta North Local Government of Ogun State and lies between longitude 3°21'S and latitude 7°21'E North of Abeokuta with a size of 1000 hectares. Ogun River (Figure 1) as a perennial river in Nigeria has a coordinate of 3°28'E and 8°41'N from its source in Oyo State to 3°25'E and 6°35'N in Lagos State where it enters Lagos Lagoon. The dry season lasts from November to March while the wet season lasts from April to October. The annual rainfall ranges from 900 mm in the North of the River to 200 mm towards the South. Total annual potential evapotranspiration is 1600 and 190 mm. The Ogun River catchment is located in South West Nigeria, bordered geographically by latitude 6°26'N and 9°10'N and longitude 2°28'E and 4°4'E. The land is about 230 km^2. The relief is generally low, with the gradient in the North-south direction. The water source is from the Igaran hills at an elevation of about 540 m above the sea level and flows directly southward over a distance of 480 km before it discharge into the Lagos Lagoon. The major tributaries of the river are Ofiki and Opeki River.

Collection of samples

The study was carried out for 10 weeks between January and March 2011 by collecting water samples once in a month. The Fish samples and water parameters were determined according to APHA (1985) and Adeosun et al. (2011).

Fish samples were collected from Ogun River, Opeji village three times a week. The samples were collected for 10 weeks. Samples

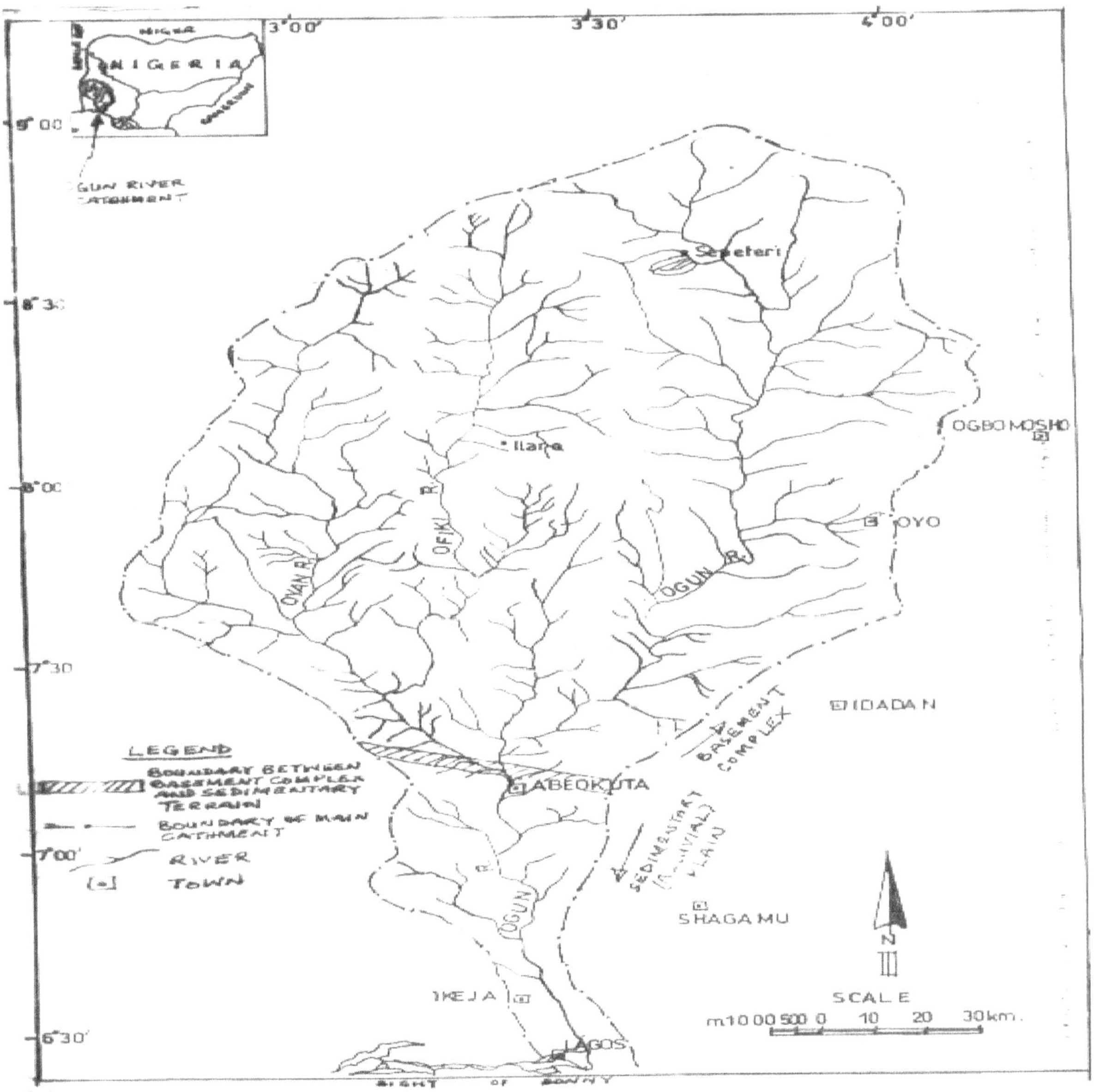

Figure 1. A map showing River Ogun.

of fish species were taken with the use of set gill nets and cast nets of three different mesh sizes of 1, 2 and 4 mm. Soil and water samples were collected from the deep and shallow parts of the water.

Ten (10) different fish species were harvested and brought to the laboratory and dissected with clean stainless steel instruments over a ten week period. One gram of each wet fresh tissue of fish, water sample and soil samples were weighed out and digested. The digests were allowed to cool, filtered, transferred to 100 ml volumetric flasks and made up to mark with 1% nitric acid (FAO, 1983). The digests were kept in plastic bottles and later, the heavy metal concentrations were determined using an atomic absorption spectrophotometer (AAS). This analysis were validated by diluting the salt solutions of the metals in various concentrations of 0.2, 0.4, 0.6, 0.8 and 1.0 ppm to enable the spectrophotometer to measure the metals from the samples of fish, water and soil. The following formula is used to calculate the concentration of heavy metals in the samples:

$$\mu/g = \frac{ppm \times v}{W \times 10^3}$$

Where ppm is mg/l (AAS reading), v is the volume of the digested sample and W is the weight of the sample used for digestion. The four different metals analyzed were lead (Pb), cadmium (Cd), copper (Cu) and zinc (Zn). The amounts of heavy metals concentration in the samples were then measured and recorded.

RESULTS

Table 1 shows that lead and cadmium are absent in the samples of fish while copper and zinc appear to be present in the flesh of the samples. From the result, it was observed that the average concentration of copper is

Table 1. Concentration of heavy metals in the flesh (tissue) of fish species.

Species	Pb (mg/l)	Cd (mg/l)	Cu (mg/l)	Zn (mg/l)
Hepsetus odoe	0.000	0.000	0.025	0.104
Chrysichthys nigrodigitatus	0.000	0.000	0.000	0.563
Tilapia mariae	0.000	0.000	0.020	0.145
Malapterurus electricus	0.000	0.000	0.069	0.340
Heterobranchus bidorsalis	0.000	0.000	0.054	0.203
Parachanna obscura	0.000	0.000	0.030	0.145
Tilapia zilli	0.000	0.000	0.045	0.202
Brycinus nurse	0.000	0.000	0.593	0.126
Heterotis niloticus	0.000	0.000	0.000	0.103
Oreochromis niloticus	0.000	0.000	0.015	0.520

Source: Field survey.

Table 2. Concentration of heavy metals in the bone of fish species.

Species	Pb (mg/l)	Cd (mg/l)	Cu (mg/l)	Zn (mg/l)
Hepsetus odoe	0.000	0.000	0.068	0.232
Chrysichthys nigrodigitatus	0.000	0.000	0.022	0.168
Tilapia mariae	0.000	0.000	0.058	0.271
Malapterurus electricus	0.000	0.000	0.058	0.100
Heterobranchus bidorsalis	0.000	0.000	0.031	0.249
Parachanna obscura	0.000	0.000	1.058	0.607
Tilapia zilli	0.000	0.000	0.054	0.213
Brycinus nurse	0.000	0.000	0.272	0.128
Heterotis niloticus	0.000	0.000	0.117	0.111
Oreochromis niloticus	0.000	0.000	0.000	0.129

Source: Field survey.

Table 3. Analysis of variance of heavy metals concentrations in the flesh and bone of the fish samples.

Source	Df	Sum of squares	Variance	F	F5%
Factors (flesh and bone)	1	0.336	0.336	7.23*	2.320
Treatment (metals)	3	2.960	0.987	21.44**	
Interaction (metals x factors)	3	2.624	0.875		
Residual (error)	57	1.973	0.035		
Total	64	7.893			

Lead (Pb) and cadmium (Cd) are absent because they were not observed.

found to be the highest in the flesh of *Brycinus nurse* amounting to 0.593 mg/l and is very minute or rather absent in the flesh of *Chrysichthys nigrodigitatus* while the average concentration of zinc was also found to be highest in the flesh of *Chrysichthys* spp. with an amount of 0.563 mg/l and lowest in the flesh of *Hepsetus odoe* containing about 0.104 mg/l.

In Table 2, lead and cadmium are absent as well while copper and zinc were observed to be present. The average concentration of copper is found to be highest in bone of *Parachanna obscura* with an amount of 1.058 mg/l and lowest or perhaps absent in the bone of *Oreochromis niloticus* while for zinc, it was observed to be highest in the bone of *P. obscura* having an amount of 0.607 mg/l and found to be lowest in the bone of *Malapterurus electricus* with an amount of 0.100 mg/l

Table 3 shows that at 50% significant level, it is clear that the interaction effect is significant. This effect was

Table 4. Distribution of average concentration of heavy metals in the flesh and bone of the fish species samples

	Flesh		Bone	
Treatment	Cu	Zn	Cu	Zn
Metals (mg/l)	0.09	0.25	0.17	0.22
95% CI	0.09±0.05	0.25 ± 0.05	0.17 ± 0.05	0.22 ± 0.05

95% CI means 95% confidence interval.

used to find out the average concentration of heavy metals in the fish samples (flesh and bone). From the analysis of the variance table, the independent effects of the factors and treatment are significant at the 5% level.

Table 4 shows that the average concentration of the metals in the flesh and bone of the ten fish species samples are 0.09, 0.25 and 0.17 and 0.22 mg/l, respectively. Also included are the confidence intervals of the average concentration of the metals in the flesh and bone of the fish samples.

DISCUSSION

The toxicity and accumulation of heavy metals in River Ogun (Opeji village) is very low and the villagers solely depend on it for their livelihood. The heavy metal concentration in the water body shows that it is within a safe limitation as it meets the recommended maximum acceptable limits of the World Health Organization (WHO, 2007).

The concentration of zinc was found to be the highest in the river followed by copper; this may be as a result of the fishing and agricultural activities and biological breakdown of rocks in the river (weathering). Based on the analysis ran on the concentration of heavy metals in the fish species, it should be a thing of note that the fishes at River Ogun (Opeji) are safe for consumption.

Due to the various activities such as dredging and clothes washing, the fish species in the River Ogun might go into extinction unless proper regulatory measures and action are taken by the state and federal government, the law makers of the country and IFSERAR (Institute of Food Security, Environmental Resources and Agricultural Research, FUNAAB) by passing out the information to the villagers, that their activities around the river can increase the concentration of heavy metals which can cause harmful effects on the fish species and the inhabitants when it exceeds the World Health Organization recommended maximum acceptable limits.

Conclusion

From the data presented, it can be concluded that the values of these heavy metals were found to be below the acceptable limit in fish food. However, these fish can be said to be wholesome for human consumption, there is need for caution as they have the potential to bio-concentrate some of these heavy metals in food chain over time.

Conflict of interest

The authors did not declare any conflict of interest.

REFERENCES

Adeosun FI, Omoniyi IT, Akegbejo-Samsons Y, Olujimi OO (2011). The fishes of the Ikere Gorge drainage system in Ikere, Oyo state, Nigeria: Taxonomy and distribution. Asiatic J. Biotechnol. Resour. 2 (04) 374-383.

Al-Busaidi M, P Yesudhason, S Al-Mughairi, WAK Al-Rahbi, KS Al-Harthy, NA Al-Mazrooei (2011). Toxic metals in commercial marine fish in Oman with reference to national and international standards. Chemosphere 85 (1):67-73. http://dx.doi.org/10.1016/j.chemosphere.2011.05.057

APHA (1985). Standard methods for the examination of water and waste water. 15th Edition. American Pub. Health. Assoc. Washington D.C. 1134 pp.

Kris-Etherton PM, Harris WS, Appel LJ (2002). Fish consumption, fish oil, omega-3 fatty acids, and cardiovascular disease. circulation, 106(21):2747-2757.

Meche A, MC Martins, BESN Lofrano, CJ Hardaway, M Merchant, L Verdade (2010). Determination of heavy metals by inductively coupled plasma-optical emission spectrometry in fish from the Piracicaba River in Southern Brazil. Microchem J. 94:171–174. http://dx.doi.org/10.1016/j.microc.2009.10.018

Medeiros RJ, LM dos Santos, AS Freire, RE Santelli, AMCB Braga, TM Krauss (2012). Determination of inorganic trace elements in edible marine fish from Rio de Janeiro State, Brazil Food Control, 23, pp. 535–541. http://dx.doi.org/10.1016/j.foodcont.2011.08.027

Njar DR, Al-Doush I (2012). Mercury content in skin lightening creams and potential hazards to the health of Saudi women. J. Toxicol. Environ. Health. 51(2):123-130.

Oguzie FA (2000). Distribution of heavy metals in water and sediment of lower Ikpoba River,Benin city. Nig. J. Appl. Sci. Environ. Manag. 4:55-60.

Olaifa FE, Olaifa AK, Adelaja AA, Owolabi AC (2004). Heavy metal contamination of clarias gariepinus from a lake and fish farm in Ibadan Nigeria. Afri. J. Biochem. Res. 7:145-140.

Omoregie E, Okonkwo MO, Eziashi AC, Zoakah AL (2002). Metal concentration in water colmn, benthic macroinvertebrate and tilapia from Delimi River. Nig. J. Aquat. Sci. 17:55-59.

Rahman MS, AH Molla, N Saha, A Rahman (2012). Study on heavy metals levels and its risk assessment in some edible fishes from Bangshi River, Savar, Dhaka, Bangladesh Food Chem. 134(4):1847-1854. http://dx.doi.org/10.1016/j.foodchem.2012.03.099

Skeat WW (2005). Concise Etymological Dictionary of the English Language. Cosimo, Inc... p. 622. ISBN 1-59605-092-6.

Tolcin AC (2011): "Mineral Commodity Summaries 2009: Zinc". United States Geological Survey. WHO contributors, (2007). "The impact of zinc supplementation on childhood mortality and severe morbidity". World Health Organization.

World Health Organisation, WHO (2007). "The impact of zinc supplementation on childhood mortality and severe morbidity". World Health Organization.

Yilmaz F, Özdemir N, Demirak A, Tuna AL (2007). Heavy metal levels in two fish species Leuciscus cephalus and Lepomis gibbosus. Food Chemistry. 100(2): 830-835. http://dx.doi.org/10.1016/j.foodchem.2005.09.020

Zhao S, C Feng, W Quan, X Chen, J Niu, Z Shen (2012). Role of living environments in the accumulation characteristics of heavy metals in fishes and crabs in the Yangtze River Estuary, China Mar Pollut Bull, 64:1163-1171. http://dx.doi.org/10.1016/j.marpolbul.2012.03.023

Study of the physico-chemical conditions and evaluation of the changes in eutrophication-related problems in El- Mex Bay

Hamdy A. Abo-Taleb[1] , M. El Raey[2], Mohamed M. Abou Zaid[1], Sawsan M. Aboul Ezz[3] and Nagwa E. Abdel Aziz[3]

[1]Faculty of Science, Al-Azhar University, Cairo- Egypt.
[2]Department of Environmental Studies, Institute of Graduate Studies and Research, Alexandria- Egypt.
[3]National Institute of Oceanography and Fisheries, Alexandria- Egypt.

El- Mex Bay is a relatively large coastal embayment west of Alexandria. The bay is an important fishery ground as well as recreation area. It includes both the Western Harbor and Dekhaila Harbor. It is one of heavily polluted areas on the Egyptian Mediterranean coast, receiving huge amount of agricultural, industrial, and sewage wastes from the adjacent Lake Mariut through El- Umoum Drain. According to different estimations in literatures the volume of the wastewaters varied between 7 × 10^6 and 8 × 10^6 $m^3 day^{-1}$, which is supposed to increase with the growing population density of Alexandria city. These conditions cause pronounced eutrophication and drastic changes. Eutrophication-related problems in El- Mex Bay of Alexandria were studied seasonally from autumn 2011 to autumn 2012. Geographical information system (GIS) and remote sensing techniques were used together with the ground-based surveys to assess the vulnerability of the most important physical and eutrophication parameters along El- Mex Bay coast. As a result of increasing population and industrial development, poorly untreated industrial waste, domestic sewage, shipping industry and agricultural runoff are being released to the bay. With the rapid increase in the industries and population, changes in water quality would have potential consequences for the large rapidly growing population of the Alexandria region. Recommendations for environmental recovery and restoration are proposed for preservation of El- Mex Bay and harbors in order to facilitate development of environmental and tourist activities.

Key words: El- Mex Bay, recreation area, El- Umoum drain, geographical information system (GIS, remote sensing, eutrophication, and restoration.

INTRODUCTION

The Egyptian Mediterranean coast receives huge volumes of wastewaters every year through the coastal lagoons and from other land-based effluents. These wastes are loaded by variable amounts and types of pollutants, in addition to great amount of nitrogenous and phosphorous compounds, which in turn cause high level of eutrophication along a significant part of the Mediterranean coast, particularly of both the Nile Delta region and Alexandria coast.

Eutrophication is an important problem to the Egyptian

Mediterranean coast, resulting in fundamental changes in the structure of the planktonic and benthic communities as well as fish mortality. Eutrophication was accompanied by the appearance of several harmful algal species at several hot spots along the coast. The level of eutrophic-cation demonstrated wide variation along the Egyptian coast relative to the variations in the volume and contents of discharged wastes. High nutrient levels and chlorophyll concentration are indication of Alexandria sea coast, like Mex Bay, Western Harbor, Abu Qir Bay, and others (Dorgham, 2011).

Eutrophication has become a persistent problem in El-Mex Bay of Alexandria and was recorded for the first time in 1985 (Dorgham, 2011). These problems came about as a result of the continuous enrichment of nutrients from different sources, including maritime activities, several land-based effluents consisting of mixed industrial, domestic and agricultural wastes as well as stored chemical fertilizers. Nutrient loads are directly dependent on human activities, which in turn depend on the growth of the world's human population. Consequently, human-induced eutrophication is in a way related to the increase in human population (De Jonge et al., 2002). In Alexandria City, the human population has just about doubled since the first record of eutrophication in El- Mex Bay in 1985 (Dorgham, 2011). This population increase has been associated with the intensive development of human activities, which directly or indirectly have led to the increase in nutrient enrichment in the bay and the consequent increase in the level of eutrophication during the past two decades.

Numerous studies have been conducted on the physical, chemical and biological characteristics of El-Mex Bay (Abuldahab et al., 1990; Soliman and Gharib, 1998; Gharib, 1998; El- Sherif, 2006; Dorgham et al., 2004; Hussein and Gharib, 2012; Hendy, 2013) and showed that, the continuous discharge of polluted water into the bay caused massive development of algal blooms and a gradual deterioration of water quality created (Zakaria et al., 2007; Mahmoud et al., 2009) identified four types of water in El-Mex Bay based on the salinity values: the mixed land drainage (L) of salinity < 10 ppt, mixed water (M) of salinity range 10 to 30 ppt, diluted sea water (D) of salinity range 30 to 38.5 ppt and Mediterranean Sea water (S) of salinity > 38.5. Youssef (2001) recorded the maximum average ratios of specific alkalinity at regions affected by the drainage water (El- Rayis, 1984; Abuldahab et al., 1986) and studied the levels of heavy metals in El-Mex Bay ecosystem including marine organisms of different trophic levels (Tayel and Shiradah, 1996; Halim et al., 1986) and also concentrations of heavy metals in fish tissues were studied. The majority of these studies were based on seasonal or bimonthly sampling. However, the rapid changes in water quality and biotic components require frequent follow-up at shorter time intervals. In the present work, the aim was to study the physico-chemical conditions and evaluate the changes in eutrophication-related problems in El- Mex Bay. Recommendations are presented to facilitate implementation of a coastal zone management program.

MATERIALS AND METHODS

Study area

El- Mex Bay is relatively large coastal embayment west of Alexandria, at longitude 29° 45′ and 29° 54′ E and latitude 31° 07′ and 31° 15′ N, from Agami headland (west) to the Western Harbor (east), with an average depth of about 10 m and surface area of about 19.4 km^2 (El- Sherif, 2006). The bay is an important fishery ground as well as recreation area. It includes both the Western and Dekhaila harbors. The bay is one of the heavily polluted areas on the Egyptian Mediterranean coast, receiving huge amount of agricultural, industrial, and sewage wastes from the adjacent Lake Mariut through El- Umoum drain. Accordingly to different estimations in literatures the volume of the wastewaters varied between 7 × 10^6 and 8 × 10^6 m^3 day^{-1} (Dorgham, 2011), which is supposed to increase with the growing population density of Alexandria city. In addition to El- Umoum Drain discharge, industrial wastes from several industries in the surrounding area, like chemical, chloro-alkali, tanneries, cement, and petroleum, are also discharged to the bay. These conditions cause pronounced eutrophication and drastic environmental changes.

Sampling design

Samples were collected seasonally during a year (from autumn 2011 to autumn 2012) at the selected stations. The stations were selected to cover all possible climatic and environmental characteristics of the different parts of the study area. Eight stations were chosen in the El- Mex Bay for the present study, the locations of the sampling stations are shown in Figure 1.

The samples for hydrographic measurements and nutrient analysis were collected. The water temperature was measured with an ordinary thermometer graduated to 0.1°C, the limit of visibility of water was measured by using a white enameled Secchi disc, the salinity measured by using Salinometer and the measurements were done by electrode (range 0-199.9 µm, 2-19.99 ms) model (Oyster, inspected 82738, Extech instruments, Germany). The pH values by using a pocket digital pH meter, and nutrients (phosphate, nitrate, nitrite, ammonia and silicate) were determined according to the methods mentioned in Grasshoff (1976). Chlorophyll *a* and dissolved oxygen content in the water was carried out according to Strickland and Parsons (1972).

The remotely sensed satellite imagery (LAND SAT 3) was found to be the most appropriate one for this study, as with its regional

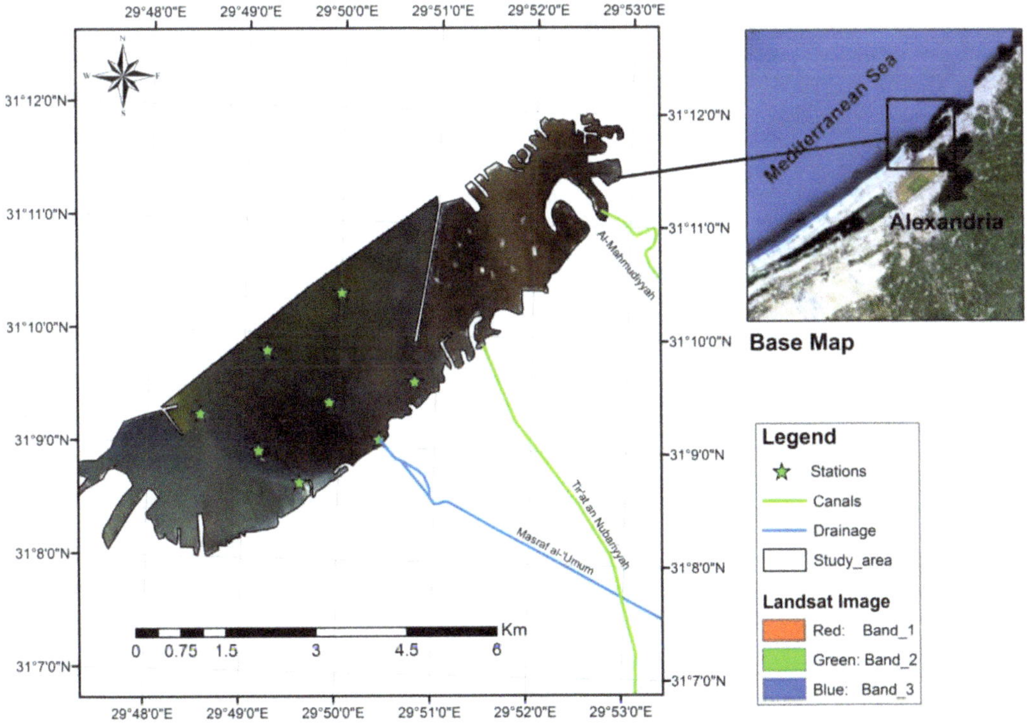

Figure 1. Base map showing the study sites.

coverage, all necessary map features were obviously clear and interpreted. A raster depending software was used for the purpose of digitizing the map features. ARC-GIS 10 and Envi software was used for this purpose for its high digitization capabilities, also in finalizing and visualizing data. The analysis and interpretation of physico-chemical parameters were done by ARC-GIS 10.

RESULTS

The hydrographic conditions and the eutrophication parameters varied widely during the study period. However, the water temperature did not deviate from the seasonal fluctuations normal on Egypt's Mediterranean coast (15 34°C) (El- Sherif, 2006). Although the bay is a shallow basin subject to vigorous mixing during much of the year, the largest value was reported in summer 2012 (32 °C) and the smallest one in winter 2012 (14.2°C) as shown in Figure 2.

Salinity is considered as a sensitive parameter for measuring the rate of dilution of seawater caused by land-based sources discharge and subsequently it reflects the degree of contamination in aquatic environment. Water salinity at El- Mex Bay showed wide variations from a minimum of 6.20 ‰ in front of El- Umoum Drain during winter 2012 to a maximum of 33.50 ‰ during autumn 2012 inside the sea (Figure 3), which directly reflects the changes in the volume and dispersion of the discharged wastewater through El-Umoum drain. This observation agrees with those found by several authors at the same area (Mahmoud et al., 2009). The seasonal values of salinity at El- Mex Bay water reflects the mixing of the bay water with fresh water. The salinity of the drain water, on the other hand, did not exceed 7‰. Generally the water salinity in the recent period decreased which insure the effects of land drainage.

The transparency of the bay water was relatively low for almost the whole year, with Secchi disc readings varying from 0.4 to 0.6 m. The water was more turbid (0.3- 0.4 m) in summer and autumn 2012, but relatively clear (1- 3 m) during winter and spring 2012 (Figure 4).

The pH values of water samples varied over rather a wider range (4.51- 8.01) with negligible differences between stations so, it was considered to range between slight alkaline and acidic (Figure 5).

Dissolved oxygen is considered as one of the most important and useful parameters in identification of different water masses and in assessing the degree of pollution especially with organic pollutants which affects fish and other marine life through oxygen reduction or depletion. The distribution pattern of DO at El- Mex Bay showed a wide variation and fluctuated between 2.52 to9.11 mg/L (Figure 6). The maximum value of DO recorded at station I during autumn 2012 can be attributed to lower water temperature (16.0°C) as well as agitation of water by strong winds (Nessim, 1994; Hendy, 2013). Drain water is characterized by relative lower

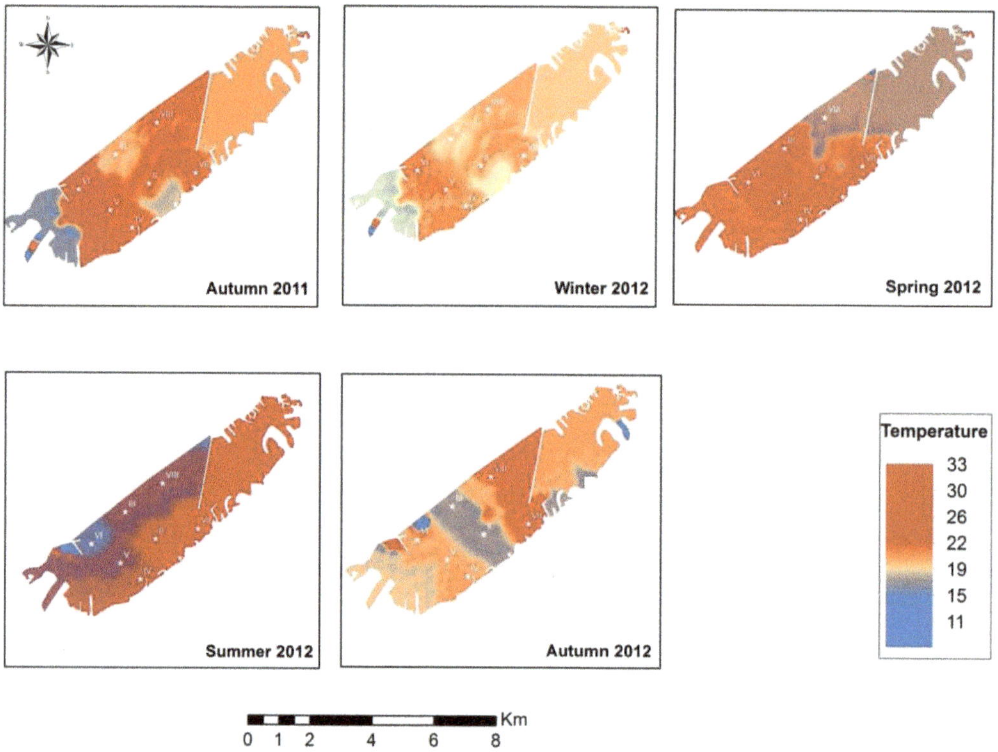

Figure 2. Variations of water temperature (°C) at different stations during study period.

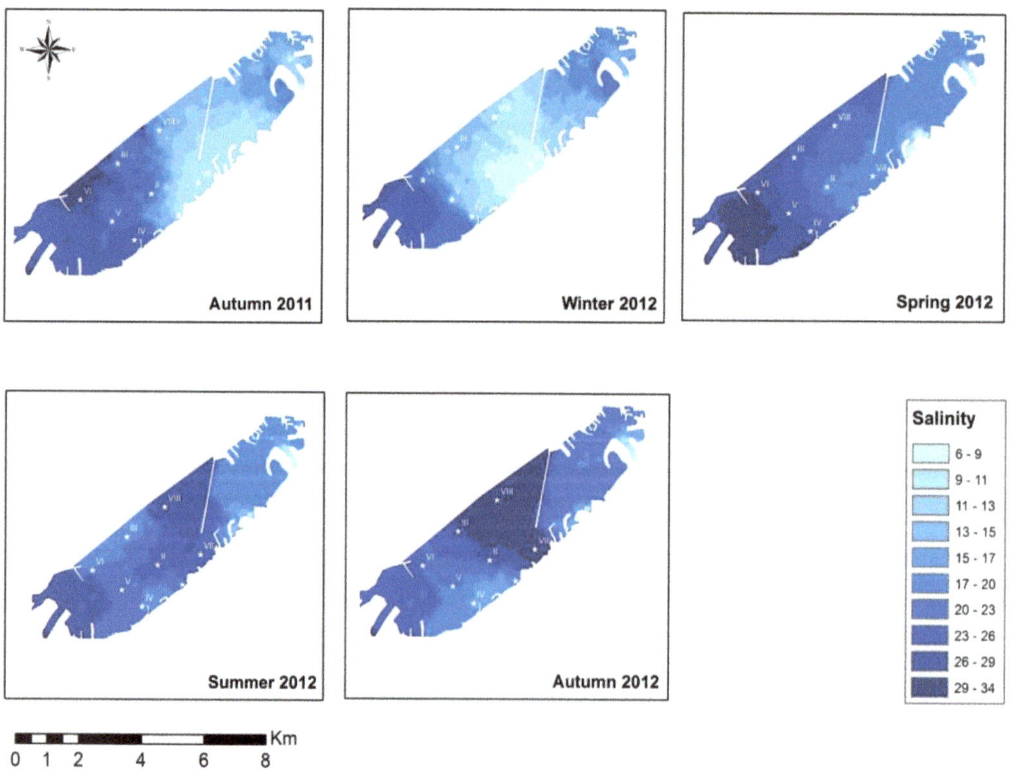

Figure 3. The seasonal distribution of salinity (‰) during study period at El- Mex Bay.

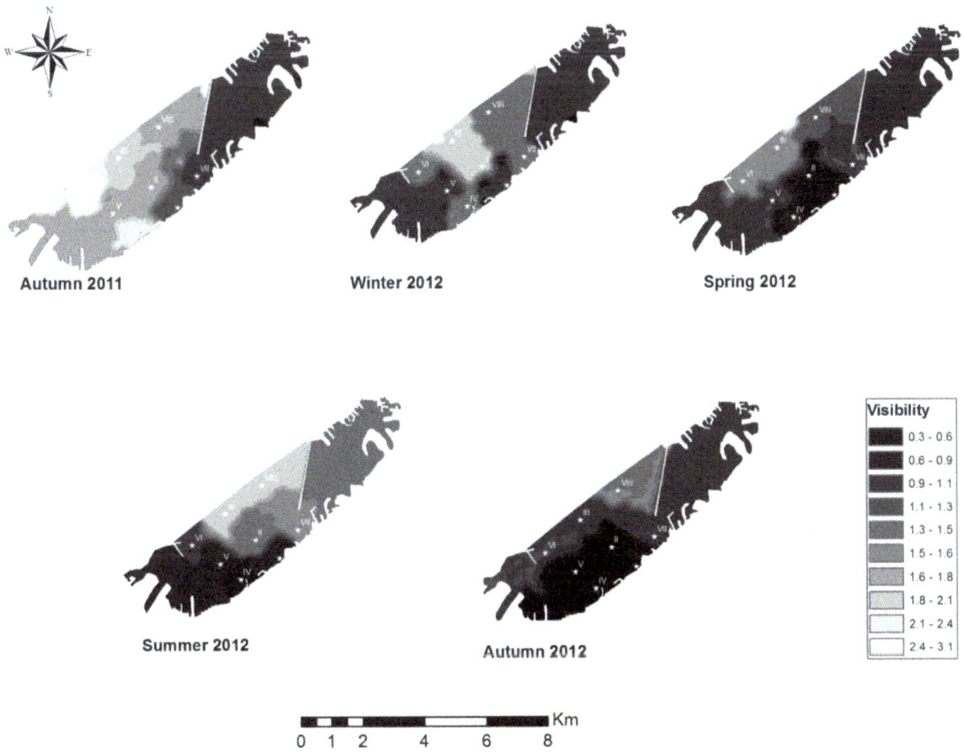

Figure 4. The seasonal distribution of transparency (m) during study period at El- Mex Bay.

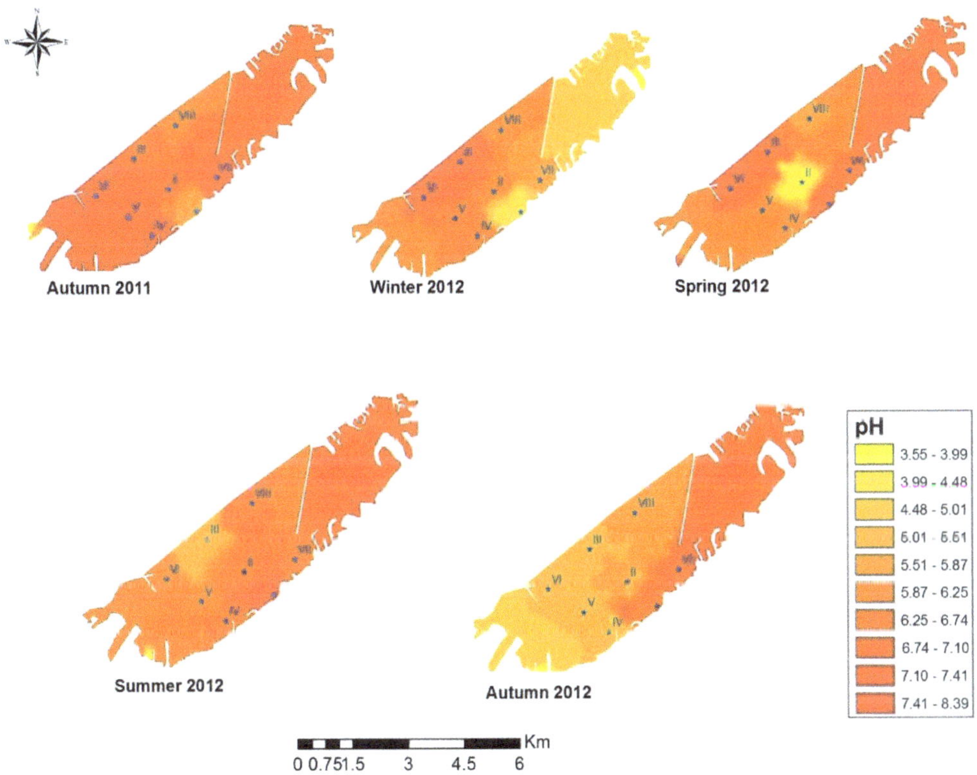

Figure 5. The seasonal distribution of pH during study period at El- Mex Bay.

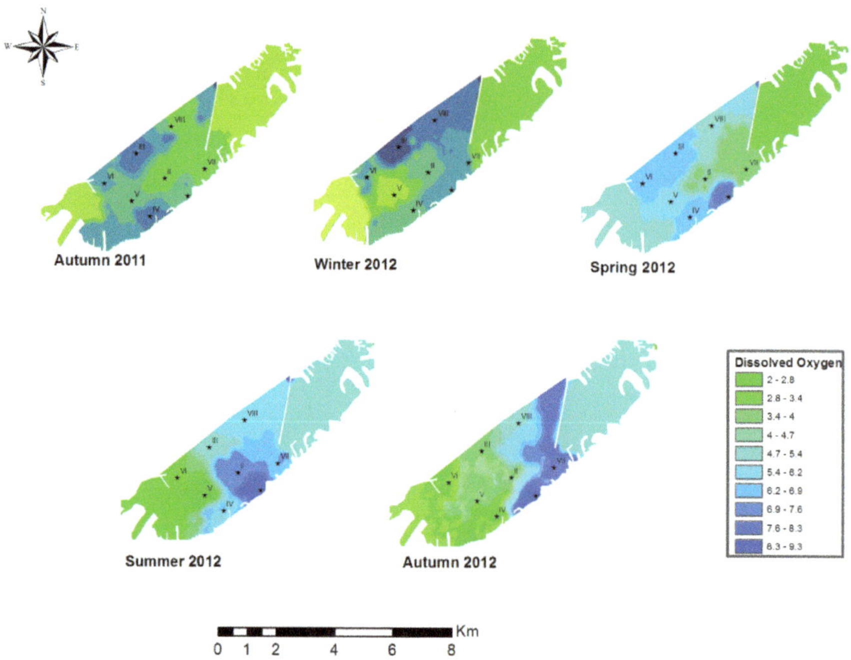

Figure 6. The seasonal distribution of DO (mg/l) during study period at El- Mex Bay.

oxygen content (5.8 mg/L), and complete depletion of oxygen was not observed among all samples but the values were found in under saturated conditions. The data showed sharp decrease in oxygen content of El-Mex Bay water during the study period that revealed the undesirable conditions.

The continuous nutrient enrichment resulting from the discharged wastes and other human activities raised the fertility in El- Mex Bay to a high level. However, the concentrations of nutrient salts displayed different ranges of variations relative to those occurring in the supplying sources.

Nitrite that appears in water results mainly from bio-chemical oxidation of ammonia (nitrification) or the reduction of nitrate (denitrification). During the period of study, station I during spring 2012 and autumn 2012 recoded higher contents of nitrite than those recorded in sea water (14.57 and 15.17 µg/l respectively), consequently, the discharged of El- Umoum drain water affected directly the in front area. The seasonal distribution of nitrite content was represented graphically in Figure 7. Due to the mixing of drainage water with seawater at El- Mex Bay, the values of NO_2 dropped to less than 0.23 µg/l during winter and spring 2012.

Nitrate is the most stable form of inorganic nitrogen in oxygenated water. It is the end product of nitrification process in natural water. Figure 8 illustrates that El-Umoum drain (station I) attained higher value of nitrate than most of the studied stations during survey with value of 20.09 µg/l, which means that El- Umoum drain possesses twice as much nitrate content as that of surrounding area. Several factors may affect the distribution of nitrate content in the investigated area, such as, the drainage water (the main factor), decom-position of organic matter, regeneration from suspended matter and bottom sediment as well as phytoplankton assimilation. The minimum values of NO_3 were observed during winter and spring 2012 (0.21 µg/l to 0.11 µg/l, respectively).

The environmental significance of phosphorus arises out of its role as a major nutrient for both plants and micro-organisms (Vanloon and Duffy, 2000). The mean concentration had a higher range and fluctuated between 0.84 (Okbah et al., 1999) and 3.34 µM (Said et al., 1991). Stations I and II (El-Umoum drain) attained higher values of reactive phosphate than all stations during the period of study with values 9.09 and 10.14 µg/l respectively (Figure 9).

The continuous discharging of polluted water into El-Mex Bay caused massive development of algal blooms and a gradual deterioration of water quality was created (Hussein and Gharib, 2012).

The land-runoff discharges from human settlements, certain industries and agricultural activities are largely the cause of man-made eutrophication in the Egyptian Mediterranean coastal waters of Alexandria. Controlled fertilization, mainly by nitrogen and phosphorous of infertile marine systems increases primary production, which can have consequence for fishery yield. Conversely, uncontrolled eutrophication of productive systems can lead to undesirable consequences (Hussein and

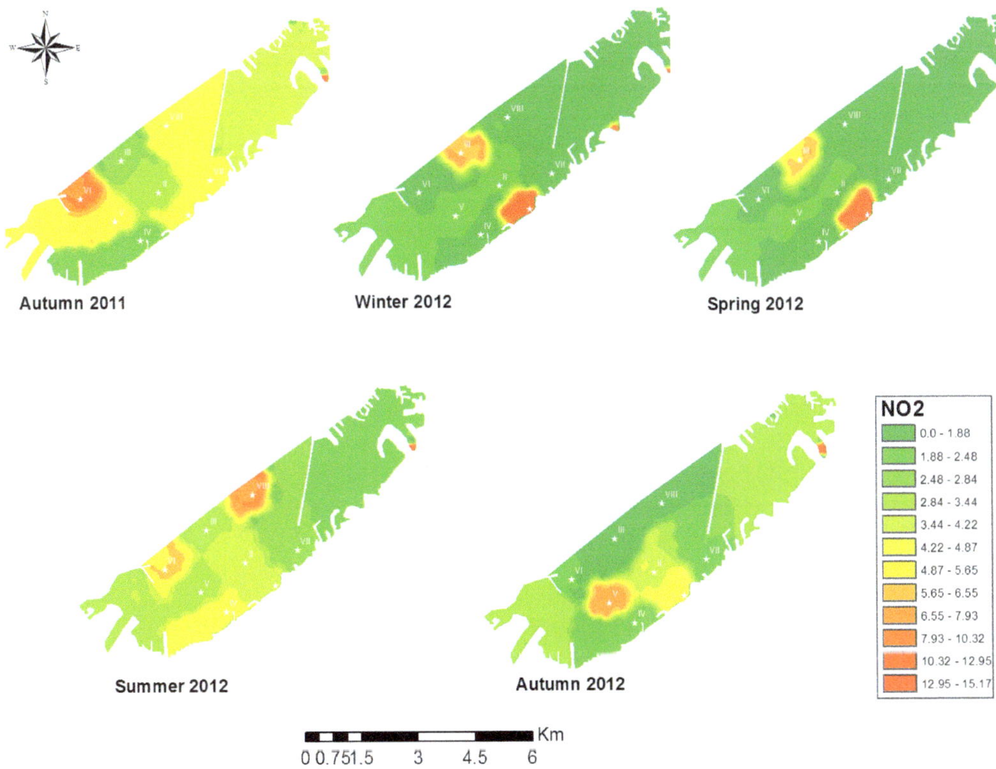

Figure 7. The seasonal distribution of NO_2 (µg/l) during study period at El- Mex Bay.

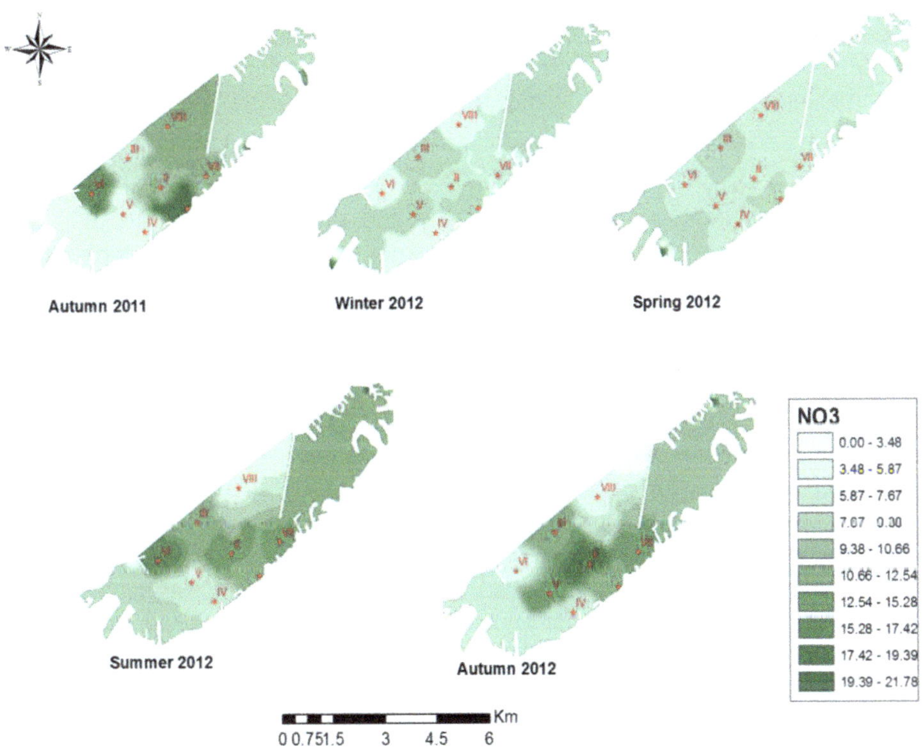

Figure 8. The seasonal distribution of NO_3 (µg/l) during study period at El- Mex Bay.

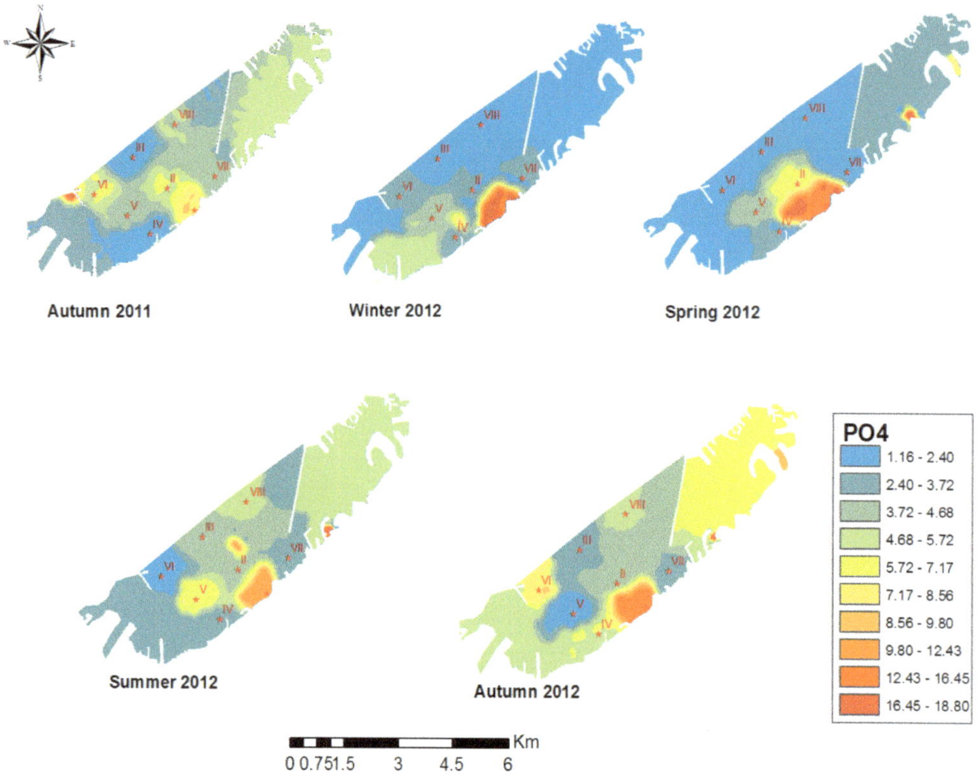

Figure 9. The seasonal distribution of PO_4 (µg/l) during study period at El- Mex Bay.

Gharib, 2012).

The maximum value of chlorophyll-a has recorded during 2 seasons summer and autumn 2012 with an annual average (34.38 and 29.92 µg/l, respectively). The high concentration of Chl-a content recorded in the water coincided with low salinity and high values of nutrient salts, which reflects such eutrophication condition caused by drainage effluents (Figure 10).

DISCUSSION

El-Mex Bay area has been and still being subjected to continuous major and drastic changes as a result of human activities; consequences of growing heavy industries (chloro-alkali plant, petrochemicals, pulp, metal plating, industrial dyes, and textiles) and uncontrolled disposal of resulting wastes, in addition coastal water of El- Mex Bay received huge amounts of untreated industrial wastes. The different human activities that bring large amounts of nutrient salts and harmful substances to the bay appear to have a pronounced impact on the physico-chemical characteristics.

The current regime there effectively controls the temporal and spatial variations of the ecological parameters. The predominant current directions were directed north-eastward under the influence of the north-westerly wind and sometimes directed south-eastward when the wind has southwesterly component in December 2012 and January 2012. The current and wind directions interpret that directions in 2012 and 2013 were not different from that in 1996 and 1997 whereas, the wind speed in winter 2012 was 27.36 knots and in January 2013 was 29.8 knots that exceed that which occurred in December 1996 which was 26 knots. Also, the current speed increased in winter 2013 and became 86.21 cm/s whereas winter in 1996 was 49.94 cm/s. El- Mex Bay and the two surrounding harbors (Western and El- Dekhela) are subjected to two main problems. Storm surges in association with spring tides which cause considerable trouble to coastal roads. Storm surges raise the sea level by 40 cm, causing overtopping of adjacent of beaches and structures. Storm waves in winter attack El- Mex Bay from the NW direction (Hendy, 2013). Moreover, the water exchange between the harbor and the sea exerts a considerable influence on the environmental and biological characteristics of the harbor, since one such cycle takes about 30 days to complete (El-Gindy, 1986; Hassan and Saad, 1996).

The spatial distribution of salinity was in fact a better reflection of the effect of El- Umoum Drain discharge into El- Mex Bay. In the long term, salinity decreased gradually from 39.6 ‰ in 1997 (Dorgham, 1997) to 33.50%‰ during

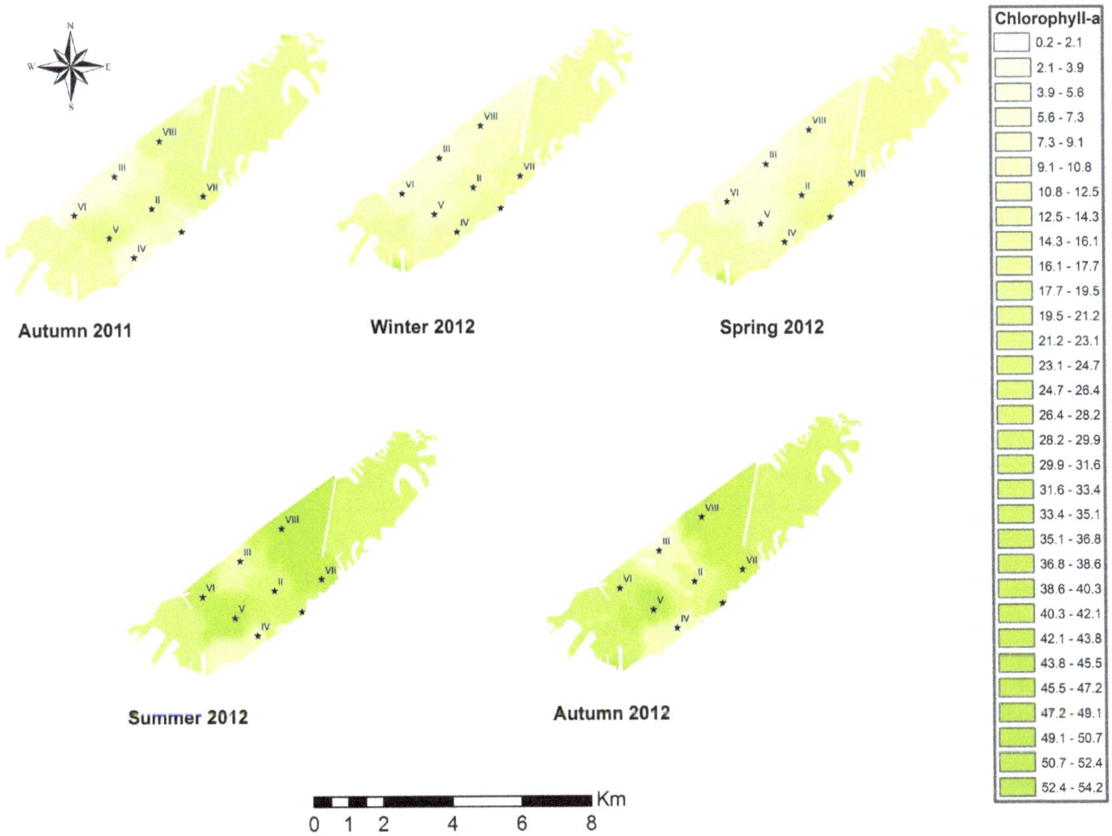

Figure 10. The seasonal distribution of chlorophyll- *a* (µg/l) during study period at El- Mex Bay.

the present study, indicating the chronic impact of the land-based effluents.

Although mixing processes caused by ship traffic and land-based runoff are the major cause of water turbidity in the bay, abnormally intensive phytoplankton blooms also substantially reduced water transparency; the two variables showed an inverse relationship all the year round.

The physico-chemical variables revealed that water temperature did not deviate from the normal seasonal fluctuations on the southeastern coast of the Mediterranean sea (15- 30°C). The pH values were always on the slight alkaline side and lower than that of the open sea. The decrease in the pH value coincided with the drop in oxygen content due to the effect of accumulating organic pollutants as well as the discharge of brackish water. El-Mex Bay demonstrated wide range of variations in its salinity on the spatial scale relative to the dispersal pattern of the discharged waste waters. The salinity of the near-shore waters sustained usually low values, increasing seaward to exceed 26.1 ‰ in the open part of the sea which reflects the effect of land drainage.

The water column of El- Mex Bay suffers from pronounced turbidity, particularly in front of the land runoff, whereas the Secchi disc readings were mostly < 1 m. Such turbidity is attributed to the strong mixing caused by discharged wastes, heavy traffic of fishing boats, and high count of plankton organisms. However, the open area of the bay shows comparatively high transparency (up to 5 m). These observations agreed with those of other studies (Mahmoud et al., 2009; Dorgham, 2011; Hendy, 2013).

In spite of the extremely high primary production in El-Mex Bay, dissolved oxygen was generally low along the water with a relatively small vertical gradient. The decrease in dissolved oxygen with depth is attributed to its consumption in oxidation of organic matter and the stagnation conditions prevailing in summer (Nessim and Tadros, 1986). The inter-annual records of dissolved oxygen testify to the continuing deterioration of water quality during the past two decades: its concentration fell from 1.6- 10.9 mg l^{-1} in 1997 (Labib, 1997) to <9.11 mg l^{-1} during the present study. It is therefore important to notice that the oxygen concentrations in the water of the study area was comparable to the threshold level of well oxygenation (<9.11 mg l^{-1}) proposed by Huet (1973) and near the bottom it resembled the hypoxia condition stated by Stachowitsch and Avcin (1988). Accordingly, El- Mex

Bay is categorized as an area with critical limits of dissolved oxygen (3.5-4.2 mg l^{-1}) necessary for healthy growth of biota in both cold and warm waters (Grundy, 1971; Arin 1974).

The seasonal distribution of nitrate in the water dropped in spring and summer owing to its intensive uptake by the abnormal phytoplankton blooms. Although the difference in nitrate between stations may be related to the effect of discharged wastes at the surface, it is probable that nitrification and mineralization of nitrate take place at different rates in the two layers. This concurs with the seasonally distribution of nitrate at all the stations, which reflects the ecological conditions prevailing at each one in different seasons. At stations I and II, nitrate attained its maximum concentration during summer and autumn 2012, when salinity dropped to its lowest value during the year; meanwhile, station I sustained the highest value for several seasons, possibly as a result of the dissolution of stored chemical fertilizers.

Nitrite levels in sea water can be regarded as a measure of the rate of nitrification and denitrification processes, which are also related to salinity variation. In the study area, the difference in nitrite between stations indicates that in the low salinity area, these processes occurred at the surface and the bottom at two different rates, while in the higher salinity area they took place at approximately similar rates. This suggests that nitrification occurred more rapidly in the former area, while denitrification was the faster process in the latter.

The value of the phosphate decreased gradually along the remaining stations due to detergents and decomposition of organic matter which is a general component of urban sewage may be the important source of reactive phosphate. The relative decrease of phosphate content in stations (III, V, VI, VII and VIII) may be attributed to several factors that lead to removal of phosphorus from the water, the consumption of phosphate by algae and aquatic plant phosphate adsorption on the clay mineral and suspended matter or precipitation by iron, calcium and aluminum.

The long-term observations of the nutritional conditions demonstrated wide variability in the spatial distribution in the bay, but the levels of all nutrient salts reflect high eutrophication. The markedly high nutrients reported during 1995 and 1996 (Soliman and Gharib, 1998; Gharib, 1998; Dorgham, 1997) reflect the large amount of nutrients reaching the bay through the discharged wastewaters, since the maximum values were reported in front of the land runoff. In contrast, the comparatively low concentrations during 2003- 2005 represented the amount of nutrients in area especially the locations far from entrance of El- Umoum Drain. It is clear that El- Mex Bay is characterized by great load of organic matter on the long-term scale. The phytoplankton demonstrated pronouncedly intensive growth in El- Mex Bay, maximizing the level of eutrophication condition, since the inter-annual records over the past three decades indicate pronouncedly high chlorophyll- a concentration in the bay (Dorgham, 2011).

Chlorophyll- a is considered as essential component responsible for photosynthesis process. It was primary photosynthesis pigment in all oxygen evolving photosynthetic plants. In the present study, we used chlorophyll-a concentration as an indicator of phytoplankton abundance and biomass in coastal waters. In the present study, the maximum chlorophyll- a concentrations was recorded at station (I) as 52.65 µg/l. The high concentration of Chl-a content recorded in the water coincided with low salinity, high temperature and high values of nutrient salts, which reflects such eutrophication condition caused by drainage effluents. These data agreed with that obtained by El-Sherif (2006) and Hendy (2013) where Chl-a ranged from 9.4 to 21.3 µg/l. Gharib (1998) observed that the phytoplankton abundance and the number of species increased consistently towards the outer region of El-Mex Bay, where the salinity was high.

Conclusion and recommendations

El-Mex Bay is located under stress condition due to the discharge of untreated domestic, industrial and agricultural effluents, beside the effect of ships movements from and to the harbors. Therefore, the condition at this bay is eutrophic and completely different from the open sea water. These are expected to continue and add to climatic influence of increasing temperature and rising sea level in the future.

As a consequence of rapid population growth, industrial development, untreated or poorly treated industrial waste, domestic sewage, industrial waste and agricultural runoff have moved to and through Mariut Lake south of the city and then released into The sea. This lake has also received a large loading of agricultural runoff through canals and drains. El- Mex Bay is subjected to severe environmental conditions.

Some recommendations includes:

1. the control of the discharge of drainage and sewage water into El- Umoum Drain, Mariut Lake and El- Mex Bay.
2. Water quality will continue deteriorating due to salt water intrusion; soil transport is still in progress. Water treatment plants and sewage systems are necessary for soil conservation, archeological sites protection and upgrading of health conditions.
3. Unplanned urbanization without consideration for the needed infrastructure is still progressing. This may lead to increased risk of extreme events and water pollution. A strong institutional capability in monitoring and assessment is necessary for law enforcement.
4. Low income, shortage of awareness and high unemployment rates are the main driving forces leading to low water quality due to careless behavior of coastal

area inhabitants. Planned industrial and tourism development, building a Marina and revival of fishing industries may be possible relief for upgrading conditions.

5. Pollution recovery and restoration by offshore extension of sewage pipelines.

6. Coastal management for future projects at Alexandria requires an environmental impact assessment study (EIA). This has to be enforced by monitoring and law enforcement and upgrading of awareness.

7. Establishment of management information system that stores all previous information and data which will help in future development, management and restoration of the Alexandria coastal zone.

Conflict of interest

The authors did not declare any conflict of interest.

REFERENCES

Abuldahab O, Halim Y, El-Rayis OA (1986). Mercury species in coastal marine organisms from different trophic levels west of Alexandria. FAO Fishery Report, 325 Suppl.:1-7.

Abuldahab O, Khalil AN, Halim Y (1990). Chromium fluxes through Mex Bay inshore waters. Marine Pollution Bulletin, 21:68-73. http://dx.doi.org/10.1016/0025-326X(90)90190-J

De Jonge VN, Elliott M, Orive E (2002). Causes, historical development, effects and future challenges of a common environmental problem: eutrophication, Hydrobiologia, 475/476, 1- 19. http://dx.doi.org/10.1023/A:1020366418295

Dorgham MM (1997). Phytoplankton dynamics and ecology in a polluted area on the Alexandria Mediterranean coast. Proceedings of the 3rd international conference on Mediterranean coastal environment, Qawra, Malta, 1:151-160.

Dorgham MM, Abdel-Aziz NE, El-Deeb KZ, Okbah MA (2004). Eutrophication problems in the Western Harbor of Alexandria, Egypt. OCEANOLOGIA, 46:25-44.

Dorgham MM (2011). Eutrophication problem in Egypt. Book: Eutrophication: causes, consequences and control, Ch. 8:171-194.

El- Rayis OA, Saad MAH (1984). Concentration of Cu, Cd, Zn, Fe and Mn in the River Nile water and its contribution to the Mediterranean Sea. VIIes Jurnees Etud. Pollutions, Lucerne, CIESM, pp.45-52.

El- Sherif Z (2006). Effects of Industrial, Touristic and Marine Transport on Physical, Chemical and Biological Characteristics of Water and Fish Populations, West of Alexandria.

El-Gindy AAH (1986). Estimation of water exchange and residence time of the water in the Western Harbor of Alexandria (Egypt), Rapp. Comm. Int. MerM'edit. (CIESM) No. 30(2).

Gharib SM (1998). Phytoplankton community structure in Mex bay, Alexandria, Egypt. Egypt. J. Aquat. Biol. Fish., 2, 81-104.

Grasshoff G (1976). Method of seawater analysis. Verlag Chemie Weinheim, New York, 317p.

Grundy RD (1971). Strategies for control of man-made eutrophication, Environ. Sci. Technol. 5 (12).1184-1190. http://dx.doi.org/10.1021/es60059a011

Halim Y, Abuldahab O, El-Rayis OA (1986). Chemical forms of mercury in flesh, gills and liver from fish species of different habitats from two localities, west of Alexandria. FAO Fishery Report, 325 Suppl.: 99-103.

Hassan MH, Saad NN (1996). Some studies on the effect of Alexandria Western Harbor on the coastal waters, pp. 464–472, Proc. 6th Int. Conf. 'Environmental protection is a must', NIOF/USPD, Alexandria.

Hendy DM (2013). Assessing the coastal vulnerability to climate change in El- Max Bay, Egypt. M.Sc. IGSR, 122 pp.

Huet M, (1973). Text book of fish culture. Breeding and cultivation of fish, Fish. News Books Ltd, Surrey, 436 pp.

Hussein NR, Gharib SM (2012). Studies on spatio-temporal dynamics of phytoplankton in El-Umum drain in west of Alexandria, Egypt. J. Environ. Biol. 33, 101-105.

Labib W (1997). Eutrophication in Mex Bay, Alexandria (Egypt). Environmental studies and statistical approach. Bull. Nat. Inst. of Oceanogr. And Fish., A.R.E., 23:49-68.

Mahmoud TH, Masoud MS, Shaltout NA, Hussein NR (2009). The biological pump of carbon dioxide in El-Mex Bay, Alexandria, Egypt. Egyp. J. Aqua. Res. 36(1):11-20.

Nessim RB, Tadros A B (1986). Distribution of nutrient salts in the water and pore water of the Western Harbor of Alexandria, Egypt, Bull. Natl. Inst. Oceanogr. Fish. (Egypt), 12:165-174.

Nessim RB (1994). Environmental Characteristics of Mex Bay, First Arab Conference on Marine Environment Protection, 5-7 Feb., Sheraton, Alexandria, 221-243.

Okbah MA, Tayel FTR (1999) Water Quality in The Coastal Area of Alexandria, Bulletin of National Institute of Oceanography and Fisheries, A.R.E., 25:89-102.

Said MA, El-Deek MS, Mahmoud TH, Shridah MMA (1991). Physico-chemical characteristics of different water types of El- Mex Bay, Alexandria, Egypt. Bull. Nat. Inst. Oceangr. and Fish., ARE, 17(1): 103-116.

Soliman AM, Gharib SM (1998). Water characteristics, phytoplankton and zooplankton population of El-Mex bay region. Bull. Fac. Sci. Alex. Univ., 38:45-66.

Stachowitsch M., Avcin A., 1988, Eutrophication induced modifications of benthic communities, UNESCO Rep. Mar. Sci., 49:67–80.

Strickland JDH, Parsons TR (1972). A practical handbook of sea water analysis, second edition. Fish. Res. Bd. Can. Bull., (176):310.

Tayel F, Shiradah M (1996). Fe, Cu, Mn, Pb and Cd in some fish species from western harbor of Alexandria, Egypt. Bulletin of the National Institute of Oceanography and fisheries, 22:85-96.

Vanloon R, Duffy A (2000). Zinc biochemistry and physiology and their arrangements. In: New trends in bioinorganic chemistry (ed R.J.P. Williams and J.R.R.F.D. Silva) Acad. Press, London: 11-57.

Youssef DH (2001). Total alkalinity, carbonate and speciation of bicarbonate in front of Alexandria, Egypt. Bulletin of the National Institute of Oceanography and Fisheries, 27:45-63.

Zakaria HY, Radwan A A, and Said MA (2007). Influence of salinity variations on zooplankton community in El- Mex Bay, Alexandria, Egypt, Egyptian Journal Of Aquatic Research, VOL. 33(3):52-67.

Permissions

All chapters in this book were first published in AJEST, by Academic Journals; hereby published with permission under the Creative Commons Attribution License or equivalent. Every chapter published in this book has been scrutinized by our experts. Their significance has been extensively debated. The topics covered herein carry significant findings which will fuel the growth of the discipline. They may even be implemented as practical applications or may be referred to as a beginning point for another development.

The contributors of this book come from diverse backgrounds, making this book a truly international effort. This book will bring forth new frontiers with its revolutionizing research information and detailed analysis of the nascent developments around the world.

We would like to thank all the contributing authors for lending their expertise to make the book truly unique. They have played a crucial role in the development of this book. Without their invaluable contributions this book wouldn't have been possible. They have made vital efforts to compile up to date information on the varied aspects of this subject to make this book a valuable addition to the collection of many professionals and students.

This book was conceptualized with the vision of imparting up-to-date information and advanced data in this field. To ensure the same, a matchless editorial board was set up. Every individual on the board went through rigorous rounds of assessment to prove their worth. After which they invested a large part of their time researching and compiling the most relevant data for our readers.

The editorial board has been involved in producing this book since its inception. They have spent rigorous hours researching and exploring the diverse topics which have resulted in the successful publishing of this book. They have passed on their knowledge of decades through this book. To expedite this challenging task, the publisher supported the team at every step. A small team of assistant editors was also appointed to further simplify the editing procedure and attain best results for the readers.

Apart from the editorial board, the designing team has also invested a significant amount of their time in understanding the subject and creating the most relevant covers. They scrutinized every image to scout for the most suitable representation of the subject and create an appropriate cover for the book.

The publishing team has been an ardent support to the editorial, designing and production team. Their endless efforts to recruit the best for this project, has resulted in the accomplishment of this book. They are a veteran in the field of academics and their pool of knowledge is as vast as their experience in printing. Their expertise and guidance has proved useful at every step. Their uncompromising quality standards have made this book an exceptional effort. Their encouragement from time to time has been an inspiration for everyone.

The publisher and the editorial board hope that this book will prove to be a valuable piece of knowledge for researchers, students, practitioners and scholars across the globe.

List of Contributors

Titilayo O. Aladesanmi
National Centre for Technology Management (NACETEM), Federal Ministry of Science and Technology. WZObafemi Awolowo University Campus, Ile-Ife, Nigeria

Oluwatosin G. Oladipo
National Centre for Technology Management (NACETEM), Federal Ministry of Science and Technology. Obafemi Awolowo University Campus, Ile-Ife, Nigeria

Godwin A. Ali
National Centre for Technology Management (NACETEM), Federal Ministry of Science and Technology. Obafemi Awolowo University Campus, Ile-Ife, Nigeria

Ramin Ezzati
Department of Plant Biology, Faculty of Biological Sciences, Kharazmi University, Tehran, Islamic Republic of Iran

Mukesh Kumar Napit
Department of Zoology Swami Vivekanand Government College Berasia, Bhopal (M.P.) India

Sahar Masud
Division of Livestock Production and Management, F.V. Sc & AH, SKUAST-Jammu, India

I. J. Singh
Department of Fishery Biology, C.F.Sc., G.B.P.U.A. & T. Pantnagar, Uttarakhand, India

Barthélemy K. Koffi
Laboratoire d'Hydrobiologie, UFR-Biosciences, Université Félix Houphouët-Boigny, 22 BP 582 Abidjan 22, Côte d'Ivoire

Benié R. D. Aboua
Laboratoire d'Hydrobiologie, UFR-Biosciences, Université Félix Houphouët-Boigny, 22 BP 582 Abidjan 22, Côte d'Ivoire

Tidiani Koné
Laboratoire d'Hydrobiologie, UFR-Biosciences, Université Félix Houphouët-Boigny, 22 BP 582 Abidjan 22, Côte d'Ivoire

Mamadou Bamba
Laboratoire d'Hydrobiologie, UFR-Biosciences, Université Félix Houphouët-Boigny, 22 BP 582 Abidjan 22, Côte d'Ivoire

Shazia Habib
Department of Environmental Science, University of Kashmir, Srinagar, India

A.R. Yousuf
National Green Tribunal, Government of India, India

Nike F. Aladetohun
Federal College of Fisheries and Marine Technology, Ahmadu Bello Way, Victoria Island, Lagos, Lagos State, Nigeria
Laboratory of Parasitology and Ecology of Parasites, Department of Zoology and Genetics, Faculty of Sciences and Technology, University of Abomey Calavi, Republic of Benin, West Africa

Nestor G. Sakiti
Laboratory of Parasitology and Ecology of Parasites, Department of Zoology and Genetics, Faculty of Sciences and Technology, University of Abomey Calavi, Republic of Benin, West Africa

Emmanuel E. Babatunde
Department of Marine Sciences, University of Lagos, Akoka, Lagos, Lagos State Nigeria

S. A. Mastan
Post-Graduate Department of Biotechnology, Post-Graduate Courses and Research Centre, DNR College, Bhimavaram - 534 202 A. P., India

Edet Okon Akpanyung
Department of Biochemistry, Faculty of Basic Medical Science, University of Uyo. P. M.B. 1017, Uyo, Akwa Ibom State, Nigeria

Udoudo Moses Ekanemesang
Department of Biochemistry, Faculty of Basic Medical Science, University of Uyo. P. M.B. 1017, Uyo, Akwa Ibom State, Nigeria

Ekam Ime Akpakpan
Department of Biochemistry, Faculty of Basic Medical Science, University of Uyo. P. M.B. 1017, Uyo, Akwa Ibom State, Nigeria

Ngozi Obiageli Anadoze
Department of Biochemistry, Faculty of Basic Medical Science, University of Uyo. P. M.B. 1017, Uyo, Akwa Ibom State, Nigeria

Shelar Piyusha Suresh
Department of Fisheries Environment, Fisheries College and Research Institute, Tuticorin, Tamil nadu- 628 008, India

S. Gauri
Department of Aquaculture, College of Fisheries, Ratnagiri-415612, Maharashtra, India

S. Desai
Department of Fisheries Engineering, College of Fisheries, Ratnagiri-415612, Maharashtra, India

M. Kavitha
Department of Fisheries Environment, Fisheries College and Research Institute, Tuticorin, Tamil nadu- 628 008, India

P. Padmavathy
Department of Fisheries Environment, Fisheries College and Research Institute, Tuticorin, Tamil nadu- 628 008, India

Steve O. Ngodhe
Department of Fisheries and Aquatic Sciences, University of Eldoret, P.O. BOX 1125, Eldoret, Kenya

Phillip O. Raburu
Department of Fisheries and Aquatic Sciences, University of Eldoret, P.O. BOX 1125, Eldoret, Kenya

Boaz K. Arara
Department of Fisheries and Aquatic Sciences, University of Eldoret, P.O. BOX 1125, Eldoret, Kenya

Patrick O. Orwa
Department of Fisheries and Aquatic Sciences, University of Eldoret, P.O. BOX 1125, Eldoret, Kenya

Alfred A. Otieno
Department of Fisheries and Aquatic Sciences, University of Eldoret, P.O. BOX 1125, Eldoret, Kenya

Osuala Fidelia Ijeoma
Department of zoology, Faculty of Science, University of Lagos, P.M.B. 101017, Lagos, Nigeria

Bawa Allha Kafilat Adebola
Department of zoology, Faculty of Science, University of Lagos, P.M.B. 101017, Lagos, Nigeria

Burnett J. W. Msika
Mines Department, P.O Box 251, Lilongwe, Malawi

John D. Kalenga Saka
Department of Chemistry, University of Malawi, Chancellor College, P.O Box 280, Zomba, Malawi

Zuze Dulanya
Department of Geography and Earth Sciences, University of Malawi, Chancellor College, P.O Box 280, Zomba, Malawi

A. A. Ayeloja
Department of Fisheries Technology, Federal College of Animal Health and Production Technology Moor Plantation, PMB 5029 Ibadan, Oyo State, Nigeria

F. O. A. George
Department of Aquaculture and Fisheries Management, Federal University of Agriculture, Abeokuta (FUNAAB) PO Box 2240 Abeokuta, Ogun State, Nigeria

A. Y. Shorinmade
Department of Science Laboratory Technology, Federal College of Animal Health and Production Technology Moor Plantation, PMB 5029 Ibadan, Oyo State, Nigeria

W. A. Jimoh
Department of Fisheries Technology, Federal College of Animal Health and Production Technology Moor Plantation, PMB 5029 Ibadan, Oyo State, Nigeria

Q. O. Afolabi
Department of Science Laboratory Technology, Federal College of Animal Health and Production Technology Moor Plantation, PMB 5029 Ibadan, Oyo State, Nigeria

K. D. Olawepo
Department of Fisheries Technology, Federal College of Animal Health and Production Technology Moor Plantation, PMB 5029 Ibadan, Oyo State, Nigeria

Grace Asio Ssanyu
Kayambogo University, P.O. Box 1 Kayambogo Uganda

Julius Kipkemboi
Egerton University, P.O. Box 536-20115, Nakuru Kenya

Jude M. Mathokoo
Global Research Akadamik and Monitoring Services (GRAMSs) P.O. Box 14824-20100, Nakuru Kenya

Sylvester B. Wandera
National Fisheries Resources and research institute, P.O Box 343 Jinja, Uganda

John Balirwa
National Fisheries Resources and research institute, P.O Box 343 Jinja, Uganda

Nike F. Aladetohun
Federal College of Fisheries and Marine Technology, Ahmadu Bello Way, Victoria Island, Lagos State, Nigeria
Laboratory of Parasitological and Ecology of parasites, Department of Zoology and Genetics, Faculty of Sciences and Technology, University of Abomey Calavi, Republic of Benin

Nestor G. Sakiti
Laboratory of Parasitological and Ecology of parasites, Department of Zoology and Genetics, Faculty of Sciences and Technology, University of Abomey Calavi, Republic of Benin

Emmanuel E. Babatunde
Department of Marine Sciences, University of Lagos, Akoka, Lagos State, Nigeria

H. M. Mzimela
University of Zululand, Private Bag, X1001, KwaDlangezwa 3886, South Africa

V. Wepener
Department of Zoology, University of Johannesburg, P.O. Box 523, Auckland Park 2006, South Africa

D. P. Cyrus
University of Zululand, Private Bag, X1001, KwaDlangezwa 3886, South Africa

Ssanyu Grace Asiyo
Kyambogo University, P. O Box 1 Kyambogo, Uganda

Julius Kipkemboi
Egerton University, P. O Box 536-20115, Nakuru, Kenya

Jude M. Mathooko
Global Research Akademik and Mentoring Services (GRAMSs) P.O Box 14824-20100, Nakuru, Kenya

John Balirwa
National Fisheries Resource and Research Institute, P. O Box 343 Jinja, Uganda

A. S. Manzola
Laboratoire de Chimie Analytique et Minérale, Faculté des Sciences et Techniques, Université Abdou Moumouni BP 10662, Niamey, Niger
Laboratoire des Traitements des Eaux Naturelles, CERTE, Technopôle de Borj Cedria, BP 273, 8020, Soliman,Tunisia

M. S. Laouali
Laboratoire de Chimie Analytique et Minérale, Faculté des Sciences et Techniques, Université Abdou Moumouni BP 10662, Niamey, Niger

M. Ben Amor
Laboratoire des Traitements des Eaux Naturelles, CERTE, Technopôle de Borj Cedria, BP 273, 8020, Soliman,Tunisia

Ali Mekki
Laboratoire des bioprocédés, Pôle d'Excellence Régionale AUF (PER-LBP), Centre de Biotechnologie de Sfax, BP: « K », 3038 Sfax, Tunisie

Abdelhafidh Dhouib
Laboratoire des bioprocédés, Pôle d'Excellence Régionale AUF (PER-LBP), Centre de Biotechnologie de Sfax, BP: « K », 3038 Sfax, Tunisie

Sami Sayadi
Laboratoire des bioprocédés, Pôle d'Excellence Régionale AUF (PER-LBP), Centre de Biotechnologie de Sfax, BP: « K », 3038 Sfax, Tunisie

O. Amukali
Geography and Environmental Management Department, Niger Delta University, Wilberforce Island, Bayelsa State, Nigeria

B. O. Obadoni
Botany Department, Faculty of Natural Sciences, Ambrose Alli University, Ekpoma, Edo State, Nigeria

J. K. Mensah
Botany Department, Faculty of Natural Sciences, Ambrose Alli University, Ekpoma, Edo State, Nigeria

O. M. Wangboje
Department of fisheries, university of Benin, PMB 1154, Benin City, Edo State, Nigeria

A. J. Ikhuabe
Department of fisheries, university of Benin, PMB 1154, Benin City, Edo State, Nigeria

Patrick Oduor Orwa
Kenya Marine and Fisheries Research Institute, P.O. Box 1881, Kisumu, Kenya

Reuben Omondi
Kenya Marine and Fisheries Research Institute, P.O. Box 1881, Kisumu, Kenya

William Ojwang
Kenya Marine and Fisheries Research Institute, P.O. Box 1881, Kisumu, Kenya

Josephat Mwanchi
Kenya Marine and Fisheries Research Institute, P.O. Box 1881, Kisumu, Kenya

B. Lekhlif
Research Team of Hydrogeology, Treatment and Purification of Water and Climate Change, Environmental Engineering Laboratory of EHTP, Km 7, Route d'ElJadida, B.P8108, Oasis, Casablanca, Morocco
Polymer Research Team, ENSEM, Hassan II University, Route d'ElJadida, B.P 8118, Oasis, Casablanca, Morocco

A. Kherbeche
Research Team of Hydrogeology, Treatment and Purification of Water and Climate Change, Environmental Engineering Laboratory of EHTP, Km 7, Route d'ElJadida, B.P8108, Oasis, Casablanca, Morocco
Polymer Research Team, ENSEM, Hassan II University, Route d'ElJadida, B.P 8118, Oasis, Casablanca, Morocco
Université de Toulouse, INSA,UPS, INP, LISBP, 135 Avenue de Rangueil, F-31077 Toulouse, France

G. Hébrard
Université de Toulouse, INSA,UPS, INP, LISBP, 135 Avenue de Rangueil, F-31077 Toulouse, France
INRA, UMR792, Ingénierie des Systèmes Biologiques et des Procédés, F-31400 Toulouse, France
CNRS, UMR5504, F-31400 Toulouse, France

N. Dietrich
Université de Toulouse, INSA,UPS, INP, LISBP, 135 Avenue de Rangueil, F-31077 Toulouse, France
INRA, UMR792, Ingénierie des Systèmes Biologiques et des Procédés, F-31400 Toulouse, France
CNRS, UMR5504, F-31400 Toulouse, France

J. ECHAABI
Polymer Research Team, ENSEM, Hassan II University, Route d'ElJadida, B.P 8118, Oasis, Casablanca, Morocco

Ndi. K. Sylvere
Industrial Filtration and Water Treatment (Chem. Eng.) Group, Department of Process Engineering, ENSAI, University of Ngaoundere, P.O. Box: 455, Ngaoundere, Cameroon

Kofa G. Patrice
Industrial Filtration and Water Treatment (Chem. Eng.) Group, Department of Process Engineering, ENSAI, University of Ngaoundere, P.O. Box: 455, Ngaoundere, Cameroon

Kameni Ngounou
Industrial Filtration and Water Treatment (Chem. Eng.) Group, Department of Process Engineering, ENSAI, University of Ngaoundere, P.O. Box: 455, Ngaoundere, Cameroon

M. Bernard
Industrial Filtration and Water Treatment (Chem. Eng.) Group, Department of Process Engineering, ENSAI, University of Ngaoundere, P.O. Box: 455, Ngaoundere, Cameroon

Ngnié N. Patricia
Industrial Filtration and Water Treatment (Chem. Eng.) Group, Department of Process Engineering, ENSAI, University of Ngaoundere, P.O. Box: 455, Ngaoundere, Cameroon

Kayem G. Joseph
Industrial Filtration and Water Treatment (Chem. Eng.) Group, Department of Process Engineering, ENSAI, University of Ngaoundere, P.O. Box: 455, Ngaoundere, Cameroon

Netai Mukaratirwa-Muchanyereyi
Chemistry Department, University of Zimbabwe, P. O. Box MP 167, Mount Pleasant, Harare, Zimbabwe
Bindura University of Science Education, Bindura, Zimbabwe

Jameson Kugara
Chemistry Department, University of Zimbabwe, P. O. Box MP 167, Mount Pleasant, Harare, Zimbabwe

Mark Fungayi Zaranyika
Chemistry Department, University of Zimbabwe, P. O. Box MP 167, Mount Pleasant, Harare, Zimbabwe

Atuanyan Ernest
Department of Microbiology, Faculty of Life Sciences, University Of Benin, Benin City, Edo State, Nigeria

Tudararo-Aherobo Laurelta
Department of Microbiology, Faculty of Life Sciences, University Of Benin, Benin City, Edo State, Nigeria

Nenibarini Zabbey
Department of Fisheries, Faculty of Agriculture, University of Port Harcourt, PMB 5323, East-West Road, Choba, Rivers State, Nigeria

B. Benard Babatunde
Department of Animal and Environmental Biology, College of Natural and Applied Sciences, Faculty of Biological
Sciences, University of Port Harcourt, Choba, Rivers State, Nigeria

F. I. Adeosun
Department of Aquaculture and Fisheries Management, Federal University of Agriculture, PMB 2240, Abeokuta, Ogun State, Nigeria

A. A. Akinyemi
Department of Aquaculture and Fisheries Management, Federal University of Agriculture, PMB 2240, Abeokuta, Ogun State, Nigeria

A. A. Idowu
Department of Aquaculture and Fisheries Management, Federal University of Agriculture, PMB 2240, Abeokuta, Ogun State, Nigeria

I. O. Taiwo
Department of Aquaculture and Fisheries Management, Federal University of Agriculture, PMB 2240, Abeokuta, Ogun State, Nigeria

A. Omoike
Department of Biological Sciences, Bells University of Technology, PMB 1015, Ota, Ogun State, Nigeria

B. J. O. Ayorinde
Tai Solarin University of Education, Ijagun, Ijebu-Ode, Ogun State, Nigeria

Hamdy A. Abo-Taleb
Faculty of Science, Al-Azhar University, Cairo- Egypt

M. El Raey
Department of Environmental Studies, Institute of Graduate Studies and Research, Alexandria- Egypt

Mohamed M. Abou Zaid
Faculty of Science, Al-Azhar University, Cairo- Egypt

Sawsan M. Aboul Ezz
National Institute of Oceanography and Fisheries, Alexandria- Egypt

Nagwa E. Abdel Aziz
National Institute of Oceanography and Fisheries, Alexandria- Egypt